Sketching Theoretical Biology

Sketching Theoretical Biology

Toward a Theoretical Biology

C.H. Waddington
editor

Volume 2

Routledge
Taylor & Francis Group

LONDON AND NEW YORK

First published 1969 by Transaction Publishers

Published 2017 by Routledge
2 Park Square, Milton Park, Abingdon, Oxon OX14 4RN
711 Third Avenue, New York, NY 10017, USA

Routledge is an imprint of the Taylor & Francis Group, an informa business

Library of Congress Catalog Number: 2009043982

Library of Congress Cataloging-in-Publication Data

Toward a theoretical biology. 2, Sketches.
    Sketching theoretical biology : toward a theoretical biology, volume 2 /
    C.H. Waddington, editor.
        p. cm.
    Includes index.
    ISBN 978-0-202-36319-6
        1. Biology--Congresses. I. Waddington, C. H. (Conrad Hal), 1905-1975. II. Title.

QH301.T68 2009
570--dc22

                                                                    2009043982

ISBN 13: 978-0-202-36319-6 (pbk)

Preface

In 1965 the Executive Committee of the International Union of Biological Sciences asked me to arrange a series of small Symposia at which biologists, mathematicians, theoretical physicists and others would explore the possibility of formulating, at least in broad outline, the structure of a discipline of General Theoretical Biology. It was felt that the time is approaching when such a subject might merit a recognized place within the field of biology, as Theoretical Physics has done for so many years among the physical sciences. The first Symposium was held at the Villa Serbelloni, Lake Como, at the kind invitation of the Rockefeller Foundation, from 28 August to 3 September 1966. From this meeting, there resulted a book of essays, published under the title *Towards a Theoretical Biology, I: Prolegomena* (Edinburgh University Press, and Aldine Publishing Company, Chicago, 1968). The present volume has issued from the second Symposium, which was held from 3 to 12 August 1967.

During our first meeting we were asking ourselves the entirely preliminary question – is there, or ought there to be, such a subject as General Theoretical Biology? Tentatively, we concluded that there might be. It would have to be quite different in character from Theoretical Physics. It would not seek for universal and eternal laws. Instead, it would accept that living systems are particular examples of some kind of 'organized complexity', and its task would be to ask 'what kind of complexity?', and 'what are the principles of its organization?' In this, the second Symposium, I think we began to find meaningful ways of formulating these questions, and to sketch out some general lines along which there is hope of answering them.

The purpose of this volume is not to record the very latest thoughts about Theoretical Biology, but rather to bring together a number of elements which, we thought, would be useful in the eventual construction of a coherent and comprehensive Theory of Biology. Some participants felt that the major contributions they have available at present had already been adequately expressed. I have therefore included a few already-published articles. (I have, with the authors' concurrence, considerably shortened the paper by *Gmitro* and *Scriven*, in which the basic ideas had been developed more thoroughly than is necessary for our purpose here.)

Further, this book emerges from conversations at a Symposium, which have a more informal character than the orthodox Scientific Paper, carefully tailored to the conventional format of scientific publication; and I offer the reader such non- (but I hope not sub-) standard items as Michael Arbib's Lecture Notes, my autobiographical Note, and the poems of Mary Reynolds.

Once again, we were fortunate enough to be the guests of the Rockefeller Foundation, and to enjoy the beautiful surroundings of the Villa Serbelloni and the hospitality of Dr and Mrs John Marshall, who understand so well how to create the best possible conditions for informal but intense discussions, so that a great deal of communication can be achieved within a comparatively few days. The I U B S, and in particular myself and all the participants at the meeting, would like to express our great gratitude for this most generous support. I am also very grateful to one of our fellow-guests, Mrs Mary Reynolds, for permission to print two poems she wrote during our stay at the Villa.

C. H. WADDINGTON
University of Edinburgh

Contents

Sketch of the Second Serbelloni Symposium

C. H. Waddington
University of Edinburgh

What we are trying to do at these Symposia is to bring into being a not yet existent academic discipline of THEORETICAL BIOLOGY. This should exist in the same sort of way that Theoretical Physics does today. That is to say, not, of course, as a closed body of doctrine with all the questions neatly answered. A few decades ago some biologists wrote as though they thought that physics was in this condition, and had in fact revealed the fundamental laws of physics and the basic constituents of matter. It is doubtful if physicists themselves ever thought this, and they certainly do not at the present time. They have, of course, many profound theorems such as quantum mechanics, but as we heard from David Bohm, it is by no means generally accepted as the final word of physics. Again the physicists know many 'fundamental particles', but now that they are looking inside the nucleus they are turning out new fundamental particles at the rate of several a year. It is very clear that they do not yet know *the* fundamental constituents of matter and it is becoming more and more probable that there are no such things, but that the material universe is open-ended to investigation in both directions towards the very small (sub-nuclear) and towards the very large (cosmological). The recognition of this largely removes the point of trying to distinguish between vitalist and mechanist theories of biology, where vitalism is defined as the notion that the objectively observable behaviour of living systems demands the postulation of entities not contemplated by the laws of physics. Since we do not know in full what entities are demanded by the laws of physics, the distinction is more or less inapplicable. The question to ask about biological theory is not whether it is vitalist or mechanist, but whether it is a useful scientific theory or not. Many useful biological theories cannot yet be expounded in terms of conventional physics, e.g. conditioned reflex learning, and genetics could not be until about 1960 at earliest.

The question for our Symposium therefore is not to produce a coherent body of theory, but rather to decide what are the type of questions which in biology might correspond to the kind of questions discussed by theoretical physicists in their field. Then, of course, we can best elucidate the nature of the questions by attempting some provisional answers of them.

In practice we have attempted to discover these fundamental questions by

1

approaching biology from the outside. We have done this from two directions: firstly from general metaphysics and secondly from considering the origin of life from non-living systems.

The metaphysical approach was little in evidence in the first Symposium, but much more so in the second, where it was strongly represented by the quantum physicist David Bohm. He urged us to begin by considering the nature of the basic entities that science is about. His thought followed what I should consider a very Whiteheadian line. The basic entities are events or experiences involving something going on and a percipient who discerns particular specifiable characteristics in what is proceeding. Bohm was very concerned with how we divide this unitary experience up into two sections, one which we consider objective, and the other subjective. He made the point that we are pushed by our language, which is heavily functional, to call things we can talk about easily objective and things we can't talk about easily subjective. In his view many 'subjective' concepts (which tend to be thought of in pairs of opposites such as harmony and conflict, beauty and ugliness, etc.) should be thought of as much more objective than they usually are, since they are also inherent in the crude basic experiences and it is only convention which leads us to take them out of the 'real world' and put them into ourselves. For instance, at one point I asked him whether 'noise' can be defined without reference to a biological function. He answered 'Not yet, but I think it should be.'

Coming from these very general questions nearer to biology we find that one of our main means of discerning specific characteristics in ongoing processes is a visual sense, i.e. the seeing apparatus, the eye. Here biology immediately brings us up against two very interesting and challenging points. In the first place, as Richard Gregory pointed out, the eye does not directly provide us with nearly as much information as we succeed in some peculiar way in extracting from it. For instance, the image on the retina is only two-dimensional, yet we can use it as though it were three-dimensional. Putting his point in a crude way for brevity, you have to set up a computer in the eye-brain system which could carry a working model of the three-dimensional world with only occasional checking, and the checking involves making the best bet you can of what the two-dimensional retinal image indicates. It may be that the necessity for this computer-like system to interpret visual signals is the basis for the whole development of intelligence and conceptual thought. Moreover, here we meet for the first time a type of ambiguity or indeterminacy not unlike the physicist's indeterminacy of velocity and position of a particle—an uncertainty which we shall keep on coming across in other contexts in many aspects of fundamental biology.

A second point about our perceptive system, which I have been making for many years, is that natural selection must have tailored our perceptive system, which is a product of evolution, so that it is appropriate to perceive the kinds of things that it pays off to perceive—or rather the kind of things it paid our mammalian ancestors to perceive. Relation between this point and David Bohm's questions about how we divide the primitive experience into subjective and objective sections was not pursued, but clearly needs thinking about. Moreover, here we have another of those circular or even knotted intellectual situations which are so characteristic of biology. We perceive only what it is good for us (evolutionarily) to perceive, but when we look at the process of evolution we realize that we could equally well say that our perceptive apparatus has progressed in evolution along the lines on which it was already best at functioning. We know what it is good for us to know, and it is good for us to know what we know.

The other approach to biology from outside, namely the problem of the origin of life and the criteria for recognizing life, also led us rather directly to the true subtleties of the evolutionary process. At the first Symposium, Cairns-Smith and Pattee had both given us examples of non-biological systems which exhibit the hereditary transmission of mutable information, Cairns-Smith in terms of crystal dislocations and Pattee of polymerization processes. Pattee in particular raised some very general theoretical points in this connection. He argued that a hereditary process is equivalent to a classification process, i.e. a choosing between sharply defined alternatives on inadequate evidence. He suggested that an enzyme choosing between possible substrate molecules is the simplest model of this. He went on to argue that any such classification process must depend on something over and above the basic quantum mechanical laws, and in particular demands the action of a non-holonomic constraint, which in the crudest terms may be thought of as a structure whose material integrity has a decay time very much longer than that of the process with which one is concerned. He also raised the question whether any quantum mechanical chemical system can possibly have the reliability required for biological hereditary transmission (the shade of Wigner rose at this point). But it is not clear to me whether the reliability need after all be so high. Cells we now know have quite a lot of repair enzymes for dealing with the results of unreliability; and in any case a very large number of cells die, both amongst micro-organisms and during the development of higher organisms.

We reached general agreement, however, that the mere hereditary transmission

3

of mutable genetic information is not really good enough to qualify as life. We need something that can really undergo evolution by the process of natural selection, and that involves something which interacts with the environment sufficiently for natural selection to be effective. Life requires not only transmission of a memory store, subject to occasional mistakes, which will presumably be encoded in some rather inert substance such as D N A, but also the production of something much more actively connected with the surrounding world which will take the form of more reactive substances, such as proteins. Life which is to engage in evolution requires in fact not only a genotype but a phenotype. The process by which the genotype becomes worked out into the phenotype can be referred to as 'epigenesis' ('metabolism', which thirty or more years ago was considered by most leading biologists as the discriminating characteristic of biology, 'local reversals of the Second Law' is no more than a short time-scale section of epigenetics).

▶ Following this line of thought there are three fundamental problems for theoretica l biology:

i. What is involved in replication, i.e. the hereditary transmission of specificity (we will go back to this old-fashioned term instead of using the current jargon 'information' because though hereditary specificity is merely information while it is being transmitted, as soon as it becomes active in determining the character of development it becomes instructions or 'algorithms').

ii. We need a theory of epigenesis which is sufficiently powerful and flexible to encompass the development of higher organisms as well as, say, the production of a single enzyme in a bacterium.

iii. How exactly does evolution work? And again we need an inclusive theory such that the simplest systems can be regarded as degenerate cases.

(i) The logical structure of the process of replication is by no means as simple as might be thought at first sight, or as one would think after looking at such model systems as a Xerox copying machine or Penrose's ingenious toys. At the first Symposium the 'Party Game Model' presented by Michie and Longuet-Higgins brought out some of the implications. At the second Symposium Michael Arbib expounded a computer programme for building a machine that would replicate itself—a synthesis of the approaches of Turing and von Neumann.

(ii) We actually spent more time on the metabolism-epigenetics problem. If we are to set up an inclusive theory in this field in which simple cases are degenerate examples, this theory must deal with situations in which systems containing large numbers (say 10^5) genes develop along time-trajectories which are to

4

some extent self-stabilizing, and which lead to a relatively small number of sharply separated quasi-stable (or adult) states.

There were several approaches.

(a) Elsasser has been rather obscurely worried about this problem for a long time. His original formulation in terms of an 'increase in information during development' was not very acceptable to most of the rest of us, who felt that information was not an applicable concept in this field, where we are really dealing with algorithms. However, Elsasser rephrased his basic uneasiness in terms of an 'immense number indeterminacy'. If 10^5 elements interact with one another, the total number of equations required to express the properties of the system are measured in such immense numbers that they are certainly in practice uncomputable, and, Elsasser argues, in principle uncomputable for the reason of what he calls the 'immense number phenomenon'. Thus the system he claims has a basic indeterminacy in it, of a kind which presents the same intellectual challenge as the usual quantum physicist's indeterminacy, although it is different in nature.

(b) The second approach is to develop a theoretical system in which one can describe these processes in terms of topological structures in a multi-dimensional function space. At the second Symposium we did not carry this much beyond the stage we had got at the first Symposium, where René Thom and Waddington had discussed such ideas as chreods, catastrophes, and the like.

(c) Another way of dealing with regularities emerging in multi-dimensional interacting systems is Gibbsian statistical mechanics. One of the highlights of the second Symposium was a convergence on to this type of mathematics from several sides. Ed Kerner, studying as a theoretical physicist looking at the problems of general ecology and interrelations between species, had made one of the crucial steps in applying Gibbs' ensemble theory to processes governed by non-linear feedback equations. Brian Goodwin had applied methods derived from Kerner to the problems of metabolism and epigenetics in cells and had emphasized the importance of oscillatory phenomena that are almost bound to arise. He defined indeed a series of concepts parallel to those of temperature, free energy, etc., but characterizing the oscillatory phenomena of cell behaviour and therefore qualified by the adjective 'talandic'. Jack Cowan, studying interactions between neurones and precisely defined anatomical regions of the human visual cortex, had found himself calling on essentially similar types of mathematics. Art Iberall had also come by a rather different route to a realization of the basic importance of the limit cycles of non-linear oscillators in the organization of many metabolic processes in animals. It seemed to emerge, since nearly all biological processes are controlled by feedback loops of one kind or another and occur as factors in enormously complex aggregates of interacting processes, that the properties of non-linear oscillators and the mathematics of the Gibbs ensemble form one of the most important methods by which theory can approach some of the most fundamental aspects of biology.

Another aspect of epigenetics to which we devoted some time was morphogenesis—the development of specific geometrical forms. Waddington outlined a scheme for classifying the types of processes by which biological forms can be brought into being. This has already been published in his *New Patterns in Genetics and Development* and will not be repeated in these notes. Scriven provided an extremely interesting discussion of the formation of regular morphological structures resulting from the breakdown of uniformity in originally homogeneous systems in which chemical or physical processes are actively proceeding. Examples are the arising of hexagonal cells subdividing large areas from which convection currents are being generated, the break-up of jets of liquid into separate drops, and so on. This was a generalization of Turing's fundamental work on the production of form within originally uniform systems. It would seem also to be related to the discussion of 'catastrophes' by René Thom, but unfortunately Thom was not present this year to take part in the discussion.

(iii) Another major topic for discussion was the theory of evolution and the adequacy of the present biological orthodoxy on this subject. Of course, nobody doubts that the introduction of Mendelism into evolutionary thinking resulted in very great progress, for instance in the realization that enormous stores of genetic variability are hidden in apparently uniform-looking populations, in understanding the genetic structure of special groups such as social insects, polyploid plants, and the like. Maynard Smith gave a very good account of some of these major advances and he expressed himself as in general satisfied with what he referred to as the theory of neo-Darwinism. Waddington, however, expressed the view that the term neo-Darwinism should really be reserved for the strictly theoretical work of mathematicians who have dealt with evolution in recent times. He argued that the most widely accepted mathematical formulations, those of Haldane and Fisher, are seriously defective in several ways—the criticisms apply to a much less extent to the much more subtle and profound work of Sewall Wright, though even he does not escape them entirely. According

6

to Waddington, the basic error in the current formulation is to attach coefficients of selective value to genotypes, whereas in fact it is the phenotype which is acted on in natural selection. This point again brings us from another angle face to face with what looks like an 'indeterminacy' situation. As we have seen above, Elsasser produced arguments based on the immense number phenomenon, that the relation between the genotype and the phenotype may be, in principle, indeterminate. This may or may not be justified, but in any case we have to realize that natural selection operates without attempting to resolve the relation: the phenotype provides the criterion which is used to decide at what rate the genotype will be transmitted to later generations.

▶ Waddington also pointed to three other questions which seem intellectually challenging to theorists. Two of these are concerned with the concept of random mutation, and the third with the concept of evolutionary value or evolutionary progress.

1. The effect of a gene mutation on the phenotype is determined by the interaction of the mutant gene with all the other genes and with the environment during epigenesis. Thus, if the epigenetic system has certain stabilities and instabilities built into it—as is obviously the case—the effect of random changes in genes will not be random by the time they are worked out into phenotypes.

2. If we are looking for an inclusive theory of evolution it must be able to deal with the evolution of higher organisms. In these the characters on which natural selection acts will often be such characteristics as the ability to run or fly fast, to eat certain kinds of food, etc. Two further orders of complexity are introduced. In the first place, a character such as the ability to run fast is affected by a very large number of genes. Waddington suggested that the contribution of an individual gene mutation to such an ability is not much more direct than the contribution of a particular random shaped pebble with the engineering qualities of a concrete member of which this pebble forms part of the aggregate. But secondly, running fast may be only one of the possible ways in which members of the species can score highly in the natural selection stakes. It might be equally effective to develop a greater ability to fight off the predator, rather than to escape from him by flight. The situation is therefore radically different from one in which there is a known fixed endpoint to be reached, and we ask whether the process of random search is an adequate method of reaching it. This is the paradigm situation envisaged by many mathematicians who approach the theory of evolution. According to this argument, it is basically mistaken and inapplicable. A somewhat better model (though still very imperfect) would be to consider

7

the natural selection within one generation as equivalent to one trick in a card game such as bridge. The gene pool of the population corresponds to a hand of cards which have indeed been produced by random processes of mutation. The question at each particular round of the game is whether the species has got a suitable card to play—has it got some small trumps when it needs them, so that it does not have to commit its high cards, and on other occasions does it have the necessary high card to get by ? The process of natural selection must in fact be envisaged in terms of games playing strategy.

3. The current theory of natural selection is in terms of the leaving of offspring in the next generation. It has little to say about the long-term prospects of a species which may find itself having to meet the challenge of a new Ice Age, new predator, new disease, and so on. Waddington argued that here again an adequate formulation would need to be in terms of an extended and improved games theory. Can we find a way of defining in the first place what is a good bridge hand, or a good set-up on the chess board ? The 'evolutionary potential' of a species cannot be defined in terms of the number of offspring it will leave in the next generation, but must be considered as something comparable to a hand of cards which will prove useful whatever cards the opponents may hold.

The problem of 'randomness' was approached by Pattee from another angle. 'Simple physical systems are perfectly reliable (electrons do not make mistakes). At some level of organization we speak of error. What is this level ?' It seems, in fact, to be some sort of intermediate level. The movement of individual molecules in a gas is perfectly determinate if we look at it in detail. It becomes 'random Brownian' movement or 'noise' when we consider the movements within a small volume of gas, but there is again nothing random about the overall gas pressure exerted on the walls of the container. Part of Waddington's argument above is that a great deal of evolution takes place at the level corresponding to the gas laws, so that the randomness of the mutation (cf. Brownian movements of the molecules) is not basically relevant. Again, the molecular biologists studying changes of nucleotide sequences in DNA can also not remain content with regarding them merely as random but would search for their deterministic mechanisms (cf. following in detail the collisions and flight path of a molecule in a gas).

At this point Bohm brought us back to fundamental physics and metaphysics, with the argument that all realms of existence contain some things which are fortuitous or contingent, and others which are necessary or organized. This he claims applies to all levels, including physics. Fundamental physical entities

8

such as quanta arise as organized entities from a lower level of fortuitous elements. This arising of the necessary or ordered from the fortuitous is what he regards as creation. His paradigm is a child starting to draw a random line, which may express something of his internal state ; then he sees he has drawn a square and he takes that into himself and begins to express it again by putting into his drawing definite corners and straightening the lines and so on. On a more down-to-earth biological level, consider a genetically variable population, members of which may try all sorts of ways of earning a living in the face of the environment surrounding them. When one way turns out to pay off well in terms of natural selection (e.g. escaping predators by flight rather than fighting them), then there will be natural selection for that particular solution of the problem. Then the population will find itself being drawn into a fast-running type of organization.

A Personal Overview

A. S. Iberall

General Technical Services Inc., Pennsylvania

It is possible to argue that the physicists succeeded in the aim of the conference: to consider and determine whether a theoretical biology might be brought into existence in the same sense that a theoretical physics has. However, despite their self-satisfaction, it is far from certain that the representatives of the other disciplines, particularly the biologists, would accept that conclusion. This personal summary, threading the proceedings in overview, may bind the diversity of physical positions that were expressed.

Bohm sounded a keynote of the troubled metaphysical foundations of modern physics. Partly pure professionalese, his views irritated the non-physicists, and, at the beginning, a substantial number of the physicists. Yet by the end of the conference his insistence that the physicists consider the faulty foundations of quantum mechanics, the foundations of thinking about space-time and of order itself, finally began to stroke sufficient chords among them into resonance. This arose because of the striking parallels between the quantization of elementary physical systems—in nuclear, atomic, molecular physics; and in the biological systems—in the molecular biology of the formation of life; in its evolution—in the internal cellular processes, in the cell, the internal organ systems, the total organism, and in the species in the ecosystem. It really wasn't until positions about the quantization at these other levels had emerged in the conference that any meaningful dialectic began with Bohm's point of view.

Consider, for example, one such exchange. Gradually a (vertical) hierarchy of organizational levels began to appear at the conference; and a (horizontal) heterarchy of diverse elements. One might see at most a finite matrix of this sort; or, by extension, an indefinite denumerable growth which might proliferate emanatively in time. Yet Bohm took exception to this. His point (in this interpretation) was essentially that ordering was non-denumerable. It is the abstraction provided by people that establishes autonomous hierarchical order. However, this ordering, from a 'prime ministering' or 'executing' function (using social system elements as analogue) down to the lowest 'functionaries', represents the orders of the hierarchy, while there is a flux of state identifications upward, which represents information. Such ordering information, however, represents timeless processes. It is the horizontal heterarchial order that gives

10

the time order (i.e. time is given by the processes of different variety).

The law of the hierarchy is to govern the hierarchy, and this must flow from the timeless order (e.g. consider the ordering of the king and the duke, etc.).

Now the concept of function cannot be described or defined in terms of function. It must have an aim which is not function. It is for this reason that a functionless (and thus timeless) order must exist. However, function and its opposite, malfunction, lead to harmony and conflict, as objective realities. 'Survival' involves harmonization (e.g. the continued existence of cars is harmonized by the traffic laws). Conflict is dissipative. Every view of function tends to maintain harmony and order.

Ultimately, it is the law of constancy of numbers, of a one-ness other than function, that establishes the relations of order, and it is the time space that determines the hierarchical space.

While no certain idea emerged whether Bohm had any significant positive idea for establishing an abstract structure for quantum mechanics (and no reader could judge this further from these remarks), nevertheless the attack he mounted on the problem of quantization is one of a number of paths that may have some meaning. (The reader may judge for himself from Bohm in July 1966 *Rev. Mod. Phys.*) His attack—it appears in this overview—is that time and space are not to be equated, but, instead, that 'timely' properties emerge from hierarchical ordering itself. Beyond this, Bohm's own remarks on his notion of order will have to speak for themselves.

It is certainly true that Bastin and Lieber were also concerned with such global questions, in seeking a line to view the physical-material universe, and thence, biological processes.

A similar message was received (1967) in a New York Academy of Science meeting ('Interdisciplinary Perspectives of Time', R. Fischer, ed. ; *Ann. N.Y. Acad. Sci., 138,* 367, 1967), particularly as speaker after speaker made his distinctions between physical time in its many faces (astronomical, radioactive, geological, etc.), and biological time (e.g. physiological, psychological).

This appears to me then to afford the transition to my contribution, which began from a very specialized problem. I had started, as a classical physicist (of a 'systems' persuasion), to try to clarify a few topics in the physics of regulation and control of the macroscopic biological system. Experimentally, instead of finding the small amplitude dynamics of linear regulation and control theory (for example, the variations in regulated metabolic power), I found characteristics much more similar to so-called bang-bang control theory. In a number of related studies I found the system loaded from one end to another with large-amplitude oscillations. 'Limit cycles !' I said. 'The system must be described by spectroscopy.' Thus, purely on empirical grounds, I was led to the same path that emerged in physics, which started from the chemical 'field' spectroscopy of individual atomistic species, and continued to the physical 'temporal' spectroscopy of atoms and molecules. My contribution, thus, is to offer the same path for a theoretical biology that developed as a theoretical physics, namely, that the gross 'material' system is made up of temporally quantized atoms in the case of simple solids, or of atomistic biochemical chains in the case of the biological system, whose everbeating cooperative effort represents the foundation for the characteristics of the overall system.

It is quite pointed of Bohm and others to have asked the question as to what is the foundation for quantization, since this has meaning at many different levels. It is important also to recognize that the ordering relations which may give the meaning at different levels may result in differences in detail and connection at various levels, yet possess great similarities. In the case of the gross biological system, I have chosen to insist that the spectroscopy is due to non-linear limit cycles at the biochemical level, largely.

It was most fortunate (for me) that Warren McCulloch had put me in touch with Goodwin a few years ago (and through whose good offices I have been able to attend this conference). It was clear that Goodwin had taken on for himself the same task at the 'atomistic' biological level, namely at the level of the biological cell, that I had taken for the system as a whole. His advantage was that whereas my chains (I prefer chains to loops or networks to emphasize its chained causal nature without stressing the fixity of the lumped element-to-element connection of networks) were much more complex thermal-mechanical-hydraulic-electrical-chemical systems sweeping cooperatively throughout the system—say as represented in particular by any endocrine system response—his chains were shorter enzyme-linked chains nearly available by a more direct spectroscopic separation. It is to his credit that his quick appreciation of the non-linear exposition of its dynamics became the most direct path to sink it into biological conscious-ness. This is shown by the acceptance of his book, *Temporal Processes in Cells,* and by the speed with which it has inspired other investigators. In these efforts, Goodwin and my colleagues have therefore been the 'practical' translators of from whence the foundations for a theoretical biology should come.

It was pleasing to Goodwin and me (in the important sense of its fittingness) to find in the remarks of Scriven that the modern chemical engineer was capable of following up the non-linear mechanics of the biological system with his unit processes, transport processes, and reaction systems. It is not surprising, for it was in the works and examples of such people as Onsager, Kirkwood, Eyring, etc., that people like myself learned enough about non-equilibrium mechanics to begin to tackle real systems.

Then Pattee, immersed at the lowest biological level question, the origin of life, better than anyone squared the circle. It turns out that here, too, at the lowest level of biochemical synthesis, the question is the non-linear dynamics of the chain. It is not sufficient to obtain a replication mechanism. A tactical polymerization by which aggraded forms may be created, by condensation reactions, is simply not sufficient. These will limit, reach a static, steady state. There must be a degradation step; and, further, it must be a step which produces a reliable gain from the energetic (free energy) changes. This requires specific catalysis ('No hereditary process can take place without uncatalyzed reactions').

From my point of view, Pattee's remarks are directed at the steps of a non-linear chain that can produce cyclic certainty from a DC potential bath of chemicals. It is not the DNA, RNA geometric specificity that provides the steps. They provide a geometric milieu over which the decisive steps can take place. Specific catalytic reactions are the escapement-like processes that make the non-linear clock run. (See, for example, Pattee *et al*, *Natural Automata and Useful Simulations*, Sparton, 1966.) The timing phases are possibly the many unit process steps involved in protein synthesis. Thus, Pattee's story is not the completion of the mystery of non-linear limits cycles—the spectroscopy of life formation and maintenance of the molecular biological level—but only its sketchy formal beginning. Room is still left for Nobel prizes at this level. (Arbib's contribution was to emphasize that there is a mathematical apparatus for doing the kind of thing that the complex biological system must do in self-replication. That is, he attempts an elegantly simple proof of the von Neumann result of whether a machine can be built that can reproduce machines as complex as itself.) We can only hope that as investigators like Pattee can elucidate more of the experimental problems of dynamic synthesis, that Arbib's algebraic semi-groups will be able to keep up, and perhaps sometime even surpass the experimental problems in predictive and prognostic value.

There is little doubt that the mathematical-computational complexity will be great. ('Isomorphic' mathematical theories will be quite important.)

While Kornacker directed his attention pointedly at the nervous system elements, he was really casting light on the one intermediate element which is needed to clarify the non-linear mechanisms of both the gross biological system and the microscopic cellular biological system; that is, the membrane. At present, his structure is still formal.

Guided by the statistical mechanics of thermal fluctuations and the formation of thermodynamic averages, he applies these ideas to cellular transport, specifically to electrical excitability of nerve membrane. The basic need, he shows, is to come up with a negative resistance within the membrane model. He finally proposes that a diode rectifier operating in a particular direction against a concentration gradient (sodium is his proposed gradient source) is the necessary and sufficient condition for a negative resistance.

While the details of his own particular problem are exceedingly clever, it is the broader implication that is really more interesting. His modelling contains some fertile seeds for the necessary instability to get the biological membrane to act as an active source for transport—whether at Goodwin's level of the cellular non-linear performance, or my level of the system response of such a system, say, as red cell-capillary interaction in the vascular beds.

Beyond that, Cowan suggests that the next few steps in understanding the electrical activity in the nervous system are coming along quite well. In the path in which Hebb has been so influential, a rough view exists of the relation of cortex and the reticular formation-thalamic link. Cowan has been developing a modelling of the cortico-thalamic nets, proposed as an improvement on the crude McCulloch-Pitts descriptions of processes in neural arcs. Thus, the high-frequency, tens of millisecond, responses in the brain are gradually coming into perspective.

For the gross nature of behaviour, McCulloch and I have recently sketched out a description of what a non-linear patterned model of man has to look like (copies were made available at the symposium, '1967 Behavioral Model of Man—His Chains Revealed', *NASA Contractors Report*, CR-858, July 1967)*. The model, physiological-psychological, is based on a central concept, homeo-kinesis, which echoes Waddington's homeorhesis. Waddington's thoughts long pre-date ours (apologies are in order for not having stumbled on his concept earlier). However, vanity and religion make me prefer or feel more comfortable with our concept (yet), for the ground we covered with it.

What is clearly missing is the 'theoretical' intermediate structure that covers the behavioural spectrum from 0·1 second to a few days. That it may be dominated

*The paper is now available in *Currents in Modern Biology, 1* (1968), 337.

14

by endocrine system and neuro-endocrine system responses is quite probable. However, we did not succeed at the conference in covering this ground. Although not represented at the conference, there are 'systems' scientists in endocrinology who have begun the laborious task of outlining the dynamic networks that govern their response. (Illustrative is the modelling of Yates and Urquhart. For example, Yates' most recent model of the adreno-cortical system was presented at an October 1967 conference on 'Hormonal Control Systems in Health and Disease', San Diego.)

However, the biologist claims there are two global questions which these statements of operating mechanisms do not answer:

1. What is the adaptive nature of the individual in his milieu?

2. What is the operative and adaptive nature of the species in their milieu?

The first question—which is where Waddington's and Maynard Smith's interests lie, and the 'real' reason for calling this conference—can be examined by default. Let us first pursue the second.

Fortunately, Kerner furnishes the beginnings of an answer. (I was very fortunate to attend a N.Y. Acad. Sci. conference—see *Ann. N.Y. Acad. Sci., 96*, 975, 1962— in which he presented a primary paper on the subject.) Following Volterra dynamics for interacting species, Kerner embeds this in the statistical 'mechanics' of Gibbs' ensemble theory. By this step, he reminds us that—if we can find the interparticulate forces—we can treat all kinds of ensembles, including biological, by this technique. I must confess that his paper inspired me, and in fact I think that I *have* furnished an answer for the 'effective' characterization of the inter-particle forces in biological spectroscopy. Just as in theoretical physics—if the energetics presented by the spectroscopy of the underlying atomistic particles can be identified, then in summation, thermodynamic functions can be set up (e.g. the free energy—see for example Landau, Lifshitz, *Statistical Physics,* 1958) and the equations of change can be derived. This technique avoids the description of the detailed 'force' laws (actually it embeds them in a quantized structure which is quite close). Thus, I believe a structure now exists to describe both the individual quantization and the population changes. Kerner has continued with his problem of the application to competition among species.

Thus, we approach the question Waddington wants answered: 'You physical fellows have had your fun, and I've paid for it, now where is the answer to my question? What is the theoretical basis for evolution?' (This is a case in which I hope I am only putting my words in his mouth!)

Maynard Smith said that the fitness of an individual species is governed by or related to the number of its children. We will go along with this. The fitness of a species depends upon its ability to achieve a non-decaying number existence in the ecology. It is not a question of brain size or abilities. It is a question whether, for changing ecologies, the genetic content, epigenetically enfolding, has enough survival value to produce a yield slightly greater than one-for-one. This is not a deterministic question, but an emanative evolutionary one. Thus one cannot state the course of future evolution. (Today's species may die; yesterday's certainly have died. This is well illustrated in Gaylord's *Life in the Past*, which shows the growing, peaking, and decline in many phyla.)

However, this again is a quantization problem, i.e. the formation of a stable cycle within a quasi-static changing milieu. The theoretical foundation here is whether the epigenetically produced branchings—of phenotype—are exceedingly rich for changing environmental conditions. This must be at the basis of a science of the theory of evolution. We physicists could start to examine such questions, but the hint is better directed to the geneticist. He knows so many more details than we; we would only fumble. However, he has to absorb from us the static problem, Arbib's theory of automata, and the dynamic problem—Bohm's, Pattee's, Goodwin's, our spatial and temporal quantization by dynamics.

As Pattee indicated on the lowest level, the genotype is the instruction for a working cycle, the phenotype is the enzyme that makes it carry over. In our terms, to illustrate non-linear quantization—using the theory of the (very good) clock as example (very good will mean highly reliable in hereditary terms)—the timing phase (more generally now the spatial phase with a rate governing reaction) is given by a linear isochronous element—pendulum, mass-spring, atomistic element. While this will create a cycling phase, it either will not sustain, or will die down; thus an escapement which injects 'impulsive energy' (the enzyme), carries it over. A variety of phenotypic escapements might do. This governs the direction of evolution.

The contribution of 'practical' Richard Gregory, at this point, is very impressive. We mathematical physical types, prepared to lay out the mathematical and electrical characteristics of the nervous system, and of the emergent patterns of behaviour, may sometimes be casual about one link—the sensory link—and its implied informational data processing. For the species to stay out of trouble—normally to have children and not get eaten—remote 'at-a-distance' processing is desirable. The remote sensory detector (why it arose in the genotype is Waddington's and Maynard Smith's business), the eye, proliferated the whole nervous system, says Gregory. The clue to its origin, likely, is the pervasive dark-light variation, that most common signal. Did this start the higher development ? Agreeing with Gregory, we are inclined to say, 'Yes'.

Thus, the fundamental pieces fall into line—obviously not in any well developed sense, but in outline. At this point now it can be (and is being) followed by competent workers clarifying the pieces. At what level description will subsequently break down is not for us to say. However, this era puts the foundation of the living system at every level into quantized limit cycles and their adaptive, emanative evolution.

Some incidental tidbits that concerned us also arose at the conference. In conducting a personal dialectic with Bohm, a thought struck me on a feature common to his presentation and mine, on the nature of creativity. Creativity in humans, to me, meant the establishment of novel emergent cyclic patterns in behaviour. To Bohm, the next instant of time itself was the only 'creative' emanative process. These, I suddenly saw, were the same. The 'creation' of time is the only continuing monotonous thing that is 'created'. In this sense, the 'creation' of man (as a linear chain) is the only act of creation of man to man. Kornacker and I worked this over to the conclusion:

The creation of time order is to the creativity of dynamic process as the hereditary (i.e. reproductive) sequence is to the creativity of evolution.

We presented this as one unifying theme.

Kornacker posed one other problem—of the definition of obscenity—which I attempted to dismiss, until he forced a meaningful dialectic. In my terms, a person whose 'body' image (of both internal and external cyclic events, i.e. his entire 'superego') is well composed and placid, will regard any presentation put before him which is too complex in structure and 'jittery' with regard to his 'ego' image (by definition, one's own ego image is stable, all else is moving— when this is upset, the individual is in trouble), as 'obscene'. To a person whose image is multi-dimensional and darting in time structure, almost no image can be obscene. This is the essence of the concept.

By this means we succeeded further in seeing ourselves, seeing others, and seeing the temporal-spatial problem of ideas, and creativity, and discovery in science, and the central problem of the conference.

Finally, Kornacker proposed, in apt summary, a (hopefully non-obscene) structural picture to unify the conference. I leave that summary to him (see p. 321).

17

Some Remarks on the Notion of Order

David Bohm

University of London

This conference is concerned with the question of whether the development of biology has now reached the point where a coherent over-all theory of the subject can begin to be formulated. In my view, such a theory would very probably have to involve the notion of *order,* in a way that is more fundamental than that in which order now enters into the theories of physics. Indeed, as I shall try to explain, our physical theories are at present in a state of flux, that may lead to radical changes in them, such that current fundamental ideas, based on measure and metric, may also have to be replaced by new ideas, based on order. So order may well be a fundamental notion that underlies both physics and biology, and permits them to be related in a deep and essential way, by making available a common language and concepted structure for the formulation of both.

In regard to this point, I would like to go even further, by emphasizing that order is something that is more fundamental and more universal than most of what has previously been generally regarded as basic in our thinking. This is because order is common not only to physics and biology, but also to all that we can know and all that we can perceive. Thus, there is the order of events in time, the order of cause and effect, and the manifold topological orders that constitute the essence of what is meant by space (e.g. order of inside and outside, right and left, up and down, open curves and cycles, etc.). Without this vast totality of topological orders, there would be no meaning to *measuring* intervals of time and space, nor even to the idea of continuity or discontinuity of these intervals. And then there are also the directly perceived orders of warm and cold, hard and soft, and shades of colour, as well as the tremendous possibilities for orders in the notes of the scale which are the basic content of music. There is the order of words (both temporal and syntactical) that makes communication possible and the order of feelings that is an inseparable part of the meaning of communications (e.g. pleasure and pain, interest and boredom, etc.). Indeed, wherever one looks, whether outwardly at nature, or inwardly at the thoughts and feelings that are the expressions of the operation of the mind, one finds that the essence of things is always in one kind of order or another. Thus, order may well be the basic factor which unites mind and matter, living and non-living things, etc.

David Bohm

Moreover, the notion of order is evidently more fundamental than other notions, such as, for example, that of relationships and classes, which is now generally regarded as basic in mathematics. To illustrate this by an extreme example, one can point out that to establish a new order of society would evidently be more fundamental than to establish new relationships and classes in society. And more generally, wherever two things are related, they are related by being comprehended within a totality of common or similar orders. So some sort of order, either tacit or explicit, is always a kind of ground or foundation that is logically prior to the notion of relationships.

If order is more fundamental than almost any other notion that we can think of, how then can we hope to define it ? That is to say, how are we to arrive at the essence of order, which must, as we have seen, in some ways transcend the whole field of what can be put into words ?

Of course, we cannot possibly obtain a complete verbal definition of order. Rather, we must begin with the fact that everyone already has a vast totality of tacit and implicit knowledge about order. What we can do with words is to 'point to' certain essential features of this tacit knowledge, and thus to bring out explicitly what is already implicit in the whole structure of our thinking and perception.

With all this in mind, I now propose that a good point of departure into this subject is to consider the notion that order is basically a set of *similar differences*. To illustrate what is meant, consider a geometric curve, which is in some way an *ordered* set of points. To describe this order, let the curve be approximated by linear chords of equal length. Then, intuitively, we can see in a general way that to obtain a regular curve rather than an arbitrary set of points the differences in the chords must be similar.

The simplest curve is a straight line. Here the successive chords differ only in position, being similar (and indeed the same) in direction. The whole curve is determined by the first chord. So we can call it a curve of *first order*.

The next curve is a circle. Here the chords differ both in position and in angle, but successive differences in angle are similar (and indeed equal). So a circle is determined by the first two chords and can be called a curve of *second order*.

The next curve is a spiral. Here the planes determined by successive pairs of chords are different, so that the curve turns into a third dimension. However, the differences of angle between the planes is similar (and indeed the same). Thus, a spiral is a curve of *third order* determined by its first three chords.

Evidently we can in this way define curves of higher and higher order, eventually

19

reaching curves of *infinite order*. Among these would be curves so complex and 'tangled up' that everyone would be inclined to call them 'random'.

The question of the meaning of the term 'randomness' has never been answered very clearly. Very often the quality of randomness has been equated with what is called 'disorder'. But if one thinks for a moment he will see that disorder, in the sense of the total absence of any kind of order whatsoever, is both a logical and a factual impossibility. Thus, if an object moves on what is called a random curve, it always moves in some kind of order. At least, after the motion has taken place, one can describe the order of the curve in question and distinguish it from any other curve which follows a different order. (Such a description might for example be made going through the analysis of similar differences, in the way described above.) Of course, the future order of a random curve would not be predictable (which is just another way of saying that a random curve is of infinite order). But it is evidently wrong to identify the totality of all possible order with nothing more than predictability (for example, the subsequent order in a musical composition is not predictable from what came before any given part, and yet it has a real order, considered as a totality).

It would seem then that it is a source of confusion to equate randomness with disorder, or even to say that disorder can exist in any context whatsoever. No matter what happens, it always has to happen in some kind of order, and what we have to do is to describe and analyse the order rather than to avoid the question by calling it disorder.

We have said that a random curve is one of infinite order. Evidently this is a necessary but not a sufficient condition for randomness. The attempt to get a full definition of the sufficient conditions for randomness leads to some very subtle questions, into which we cannot enter here. It may be pointed out, however, that one requirement on a random order is that it must eventually contain every possible kind of sub-order or partial order. In other words, a random order is, in some sense, *open*. In addition, it must satisfy certain statistical conditions on the partial orders (which allow the concept of probability to be applied). A great deal of work remains to be done in clarifying these questions. But a necessary condition to begin this work is that we cease to use the word 'disorder', which blocks the way to thinking about the *order of randomness*, because it formally and logically denies the existence of any and every kind of order that could possibly be conceived of.

We are now ready to consider the question of the *difference of orders*. First of all, it is evident that two geometrical curves have different similarities of their differences, and that this is their essential difference. To avoid confusion of terminology, let us say that there are 'differences$_2$ in the similarities of the differences$_1$', where the term 'differences$_1$' describes differences *in* a given curve, while 'differences$_2$' denotes differences *between* the two curves.

Two such different curves can then be related. The most elementary relationship is a one-one correspondence of their differences. Such a relationship is a set of similarities$_2$ in the differences$_1$ of the two curves. There can also be similarities$_2$ of the similarities$_1$ in the curves in question, as well as differences$_2$ in the similarities$_1$, etc. In this way we can explicate the full subtlety and complexity of the relationship of different curves.

Here it is important to emphasize that logically speaking one can properly relate only things that are different. Indeed, the Latin root of 'different' means 'to carry apart', while the root of 'related' or 'referent' is 'to carry back together'. That is to say, when things are 'carried apart' they will 'refer' to each other, and this is the essence of their relationship. So what is logically prior to relationship is difference and similarity, leading to order.

This is true perceptually as well as conceptually. Thus, in the corner of the eye, we can perceive that things have changed and are different before we know what it is that is different. As we turn the central part of the retina toward the thing that has changed, our perceptions sharpen up to reveal just what it is that has altered. Very probably the brain is thus enabled, step by step, to register the differences and the similarities, giving rise to a perception of order. This is then related to other orders that are stored in memory. In this way we can see what is happening and recognize it by referring it to the order of what is already known.

In discussing the subject of relationship, it is necessary to understand that two kinds of differences are always involved. First there are the constitutive differences, which determine the essence of the order of whatever we are talking about (in the case of the geometrical curve these are the differences in the chords). Then there are the distinctive differences, which determine and define how one order can be distinguished from another, and yet refer to the other through their mutual relationships.

Of course, the constitutive differences can always be related (or referred) to distinctive differences in another order. For example, the chords that constitute a curve can be related to another set of curves constituting a coordinate system or a reference frame (e.g. by giving the coordinates of each part of the various chords). So now we can focus on the distinctive differences between the chords

Some remarks on the notion of order

by calling attention to the differences between their coordinates, slopes, etc. But note first of all that to do this we have had to introduce the notion of constitutive differences in the coordinate curves themselves. In other words, underlying each analysis of distinctive differences there must be a set of constitutive differences *somewhere*, to which the distinctive differences can be referred. Thus, we never totally remove the need for constitutive differences, but at best only transfer the constitutive differences to another part of our conceptual structure. But secondly, granting this, we do not in general even reduce a given set of constitutive differences to distinctive differences by relating them to another referential order. Thus, with regard to the geometrical curve, mathematicians have seen the need to define and work out the intrinsic properties of the curve, i.e. those properties that are independent of the coordinate frame to which it is referred (also called the invariant features of the curve). So when one relates constitutive differences to distinctive differences, this is in general merely a descriptive process rather than an explanation of one in terms of the other.

What is needed in this field is to see clearly the difference between constitutive differences and distinctive differences, and then to see how these two kinds of differences are related. Indeed (as has already been pointed out), what is constitutive difference at one level corresponds to a related set of distinctive differences at another level, and so on in principle without limit. For this reason, orders tend to develop into indefinitely extending hierarchies.

To a certain extent, such hierarchies of order are introduced by us as a result of our analysis (e.g. geometrical curves are referred to coordinate frames which can in turn be referred to other frames of finer mesh, etc.). But, more generally, we find that there are certain natural hierarchies of order which reflect not mainly our particular procedures of analysis, but rather the existence of a real structure.

Structure can best be described as a constitutive order of constitutive orders (constitutive order being the result of a set of similar constitutive differences). To illustrate what this means, let us consider the structure of a house. One begins with the bricks, which are similar in size and shape but different in position and orientation. The similarity of these differences of the bricks leads to the order of the wall. The wall in turn becomes an element of a higher order, in such a way that the similar differences in the walls make the rooms. Likewise, the similar differences of the rooms make the house, those of the houses the streets, those of the streets the city, etc.

It is clear that the principle of structure is universal. Thus, the elementary particles are ordered to make the atoms, the atoms are ordered to make molecules the molecules make micro-objects, and so on to the planets, stars, galaxies, galaxies of galaxies. For living matter the molecules are ordered to make the components of cells. These are ordered to make the cells, these the organs, these the organisms, these the societies of organisms. And something similar goes on in perception and thinking. Indeed, even our most abstract concepts form structures in this way (e.g. a set of ordered classes makes a class of higher order, and in turn can be the beginning of a class of yet higher order, and so on without limit).

One of the most characteristic features of structure is that partial constitutive orders can be abstracted from it in such a way that distinctive differences and relationships show themselves naturally. For example, one can abstract a pair of parallel walls, whose distinctive differences and relationships are indicated by another pair of parallel walls that connect the end points of the first pair. In all structures one finds a very rich set of such cross-references of every kind. These cross-references are both inferences from the structure in question and indications that what we are dealing with is in fact a unified totality of structure rather than an arbitrary and fortuitous array of elements. So when we discover a set of data with very rich cross-references in it we try to find a structure, and if we succeed in doing this, we test our assumed structure by observing further cross-references in the data, often of new and hitherto unsuspected kinds.

To carry out this kind of inquiry adequately we need a language that describes order and structure properly. In my view we do not at present have such a language. Evidently the common language is inadequate, because its terms referring to order are extremely vague and confused. Indeed, it is to remedy this situation that I have called attention to the need to consider similar differences, different similarities, the hierarchy of orders, the constitutive and distinctive orders, etc.

One might then ask whether we could not describe orders and structures properly with the aid of mathematical language. However, I do not think existing forms of mathematics are really adequate for this purpose either. To be sure, something along this line is being done in *topology* and in information theory. But in both subjects, what is absent is an adequate notion of order. Indeed, the general mathematical notion of order is now formulated basically in terms of certain relationships. Thus, in what is called a lattice one defines an ordering relation, symbolized by $>$, which has, in essence, the same qualities as the notion of 'greater than'. This relation is assumed to be asymmetric, transitive, and reflexive, and further assumptions are made to allow for the many 'strands' of a partially ordered system.

23

Some remarks on the notion of order

The main difficulty with this notion is that it does not readily allow for the basically hierarchical possibilities of order, which lead to structure as an order of orders. In essence, it is order at one level only. As such, it is only a special case of the general descriptions of order questions, as similar differences and different similarities of the differences. Moreover, it does not permit a clear expression of the difference between constitutive differences and distinctive differences. Indeed, the relation > really refers only to distinctive differences. For it tacitly supposes that each element is considered separately as fully constituted, so that the order of elements is *external* to the elements themselves, and refers only to the way in which the elements are distinct from each other and yet related On the other hand, when for example we considered the differences in the chords that make up a curve, we were regarding the order as basic to what the curve is, and not as some purely external descriptive property that was being applied to show how a set of distinct but related points happened to be aggregated in such a way as to be distributed along a given curve.

What is needed is to develop a new mathematics of order and structure. This requires an extensive study, in which one slowly and carefully 'feels one's way' into the subject. It cannot properly be done solely by applying existing mathematics, because the latter does not have the right general structure. To this end it is necessary to formulate a new set of mathematical axioms which treat order and structure as bare concepts. After all, our present axioms do not explicitly define the basic elements in our mathematical thinking (for example, in geometry points and lines are taken as purely abstract words, defined only tacitly by the ways in which they are used). So there is no reason why we cannot introduce new axioms, in which the notions of order and structure, defined only tacitly and not explicitly, are taken as the fundamental points of departure for our thinking.

Indeed, I have been doing some preliminary work on this question. The main difficulty seems to be to develop a new structure of mathematical symbolism that takes into account the hierarchical potentialities of order and that does not tacitly commit one to the view that the world is composed of separate 'elements' whose orders and relationships are external to what these 'elements' are. In addition, it is necessary that the symbolism explicitly differentiates between constitutive differences and distinctive differences so that it will permit the expression of how these two kinds of differences are related in a vast set of cross-references of one aspect of structure to another.

Thus far, both in general terms and with reference to mathematics, we have been considering order and structure largely as static. But in reality, to do this is to abstract from a process of movement and development in which each order and each structure is always becoming different. What is essential to process is not merely that there is a change of order and structure, but that the differences are similar, so that the changes are themselves ordered. In other words, process is an order of change. And, needless to say, even the orders of change can themselves be ordered to form a larger hierarchy of process, which is an order of orders of change.

Consider, for example, the laws of motion in physics. To simplify this discussion we will consider changes taking place in equal but short intervals of time rather than trying to use the infinitesimal calculus. Now, in free space an object moves at uniform speed in a straight line. This means that successive differences in position define segments that are similar (and indeed equal) in magnitude and direction. Thus, the law of motion is just an assertion of similar differences, implying a certain linear order of change. In the presence of forces it is necessary only to go on to the second differences, i.e. to differences in successive segments, whose similarities define the acceleration of the body. Thus, Newton's laws of motion may be stated by saying that similar differences in the applied force always lead to similar differences in the acceleration.

Evidently the laws of physics are expressed in terms of a very simple kind of order, i.e. the order of mechanical motion of a body. In biology we may express the growth of an organism (the phenotype) in terms of a very rich and complex set of similar differences and different similarities in the changes of its various aspects and features. And as we go on to consider the growth and development of intelligent responses of the higher animals and man, we find yet higher orders of similar differences and different similarities.

The need to proceed in this way illustrates a very characteristic feature of process, i.e. that *the breaks or changes in the order of a given process can themselves be the basis of a higher order of process.* (This is the temporal counterpart of how the walls, which are the ordered set of bricks, are themselves the basis of the rooms, which are formed by the ordered set of walls.) Such a possibility can best be seen in music. Thus, there may be a short set of notes in a given order. This order changes, then changes again and again. But all the changes of order form a yet higher order, which constitutes a part of the *development* of the over-all theme. Each order of development itself changes in an ordered way to form a still higher order of development. And to the possibilities of going on with this process there is in principle no limit.

25

Some remarks on the notion of order

It seems clear that in biological and psychological processes something similar is involved. Thus, we see how on one level the DNA and the RNA function is an ordered way to build and maintain the cell. On the next level, the cells function so as to maintain the organs, the organisms, the society of organisms, etc. Here we are emphasizing not merely the order of orders of static structure, but also the order of orders of dynamic function which is needed for maintaining coherent growth and life itself. A similar emphasis is needed in psychological processes, which reveal an even greater richness and variety of orders of orders of function.

The study of order leads to particularly interesting questions when we consider the process of evolution. Of course, to a certain extent, evolution can be considered to be a set of changes within a given order of process, from one possibility to another. But I would like to suggest that there is another kind of evolution, which is the coming into being of a new and higher order of process. Thus, in music there can be variation on a given theme. But then there can be a basic change of order of the whole theme. And then there can be something yet more— an ordered series of such changes in this theme. This latter order is not only new relative to what was there before, but it is also evidently of a higher order. Likewise, we can think of the evolutionary process by considering not merely a set of variations on a particular kind of structure of organism, but also the coming into being of new orders, along with an ordering of the changes of order in the whole process.

For example, in a certain sense, intelligence enables man to order his physical actions in new ways. Over a short period of time, man's physical actions are, on the whole, not very different from those of the higher animals. But in the higher animals these actions vary in a way that tends to have no particular *intrinsic order* (i.e. an order not imposed mainly by the environment). However, in man, intelligence reveals itself through the ordering of these physical orders in a new and different way. Something similar is also to be observed at lower levels of the evolutionary process. Thus, when cells work together to form an organism, the main new factor is that certain variations of the behaviour of individual cells previously determined fortuitously by the environment are now ordered intrinsically in the over-all functioning of the organism. Indeed, the evolutionary process can be said, in a certain way, to be leading to ever higher degrees of intrinsic determination of the order of lower orders' actions, so that, at least in this sense, it has a kind of direction of development. Perhaps life could be regarded as an early stage in this process, and intelligence as a later stage. In this connection it may well be that even inanimate matter is evolving (e.g.

26

there may have been a time before which electrons, protons, and neutrons did not exist). Thus, the difference between life and non-life (and between different levels of intelligence) is perhaps not in the process of evolution itself, but rather in the degree and kind of intrinsic order of order which has thus far resulted from the process of evolution.

In discussing the evolutionary process I want to emphasize that the dynamic feature of the process of *ordering* has to be taken as prior to the more static feature of *order*. This process contains two inseparable aspects—the dissolution of an older order and the creation of a new order. Thus, if individual organisms did not die (i.e. undergo a dissolution of their orders) it would be impossible for new orders to come into being in the successive generations. But what is of crucial significance here is not merely the replacement of one order by another. Rather, it is that each change of order (whether in the actions of the individual organism or in the nature of the successive generations of a species) is itself capable of entering into a yet higher order of changes of order. Therefore creation is not just the death of the old organism and the birth of the new one. Rather, its essence is that in it there is scope for the coming into being of ever higher orders of order. And just as order is itself logically and existentially prior to relationships and classes, so the process of ordering is logically and existentially prior to the orders which result from it, and which are created and dissolved in this process.

At this point we come naturally to the question of mechanism. For we are led to ask whether what has been said about the complexity and subtlety of biological, psychological, and evolutionary ordering refers to the basic constitutive differences that make up the order of natural processes, or whether it refers merely to the distinctive differences in these processes, that are perhaps convenient and useful as descriptions, but that have no really fundamental significance.

When one asks such a question one is usually referring tacitly to the argument that on the level treated by physics the whole world is in reality constituted out of some basic kinds of elementary particles, which move mechanically according to certain laws, in a way that is determined by their inertia and by the forces of interaction between them. To be sure, it may well turn out that known particles, such as electrons, protons, and neutrons, are not truly elementary and that they are in fact constituted out of some as yet unknown 'really elementary' particles of a finer nature. Whatever the truth might be in this regard, however, one can assume that the 'really elementary' particles move in a completely determined mechanical way, given in essence by Newton's laws or by some variation of these. This means that the order of motion is assumed to be nothing more than

Some remarks on the notion of order

an automorphism, i.e. a movement in which there is a limited field of possible states, and in which each change corresponds to going from one state to another, within this limited field. Thus, according to Newtonian mechanics, the state of a system of particles is completely specified by the position and velocities of each particle at a given moment of time. The totality of the available states is taken to be the whole set of possible positions and velocities open to the various particles. And the movement consists of a process in which each particle goes from one position and velocity to another position and velocity. The law of movement then determines the order in which this automorphism takes place, for each possible state of the system.

If one accepts such an assumption, it follows that the whole order of behaviour of any system of particles is in reality determined completely by the mechanical order of movement of the constituent particles. To be sure, one may find it convenient to group these particles into systems, such as atoms, molecules, cells, organs, organisms, etc. Because the particles interact with each other, the systems can display a sort of 'collective behaviour' in which they 'work together' in a general over-all way as a kind of relatively stable unit on a higher level. As a result, one can simplify things by abstracting from the basic laws a suitable partial treatment of the order in which these systems move. In effect, this partial treatment ignores the complexities of the deeper structure of the systems in question and approximates them as single systems. But in a more fundamental sense one must regard these systems as in essence nothing more than convenient abstractions for establishing distinctive differences that are useful in the description of the general behaviour of particles when they happen to be aggregated into groups. On the other hand, the fundamental constitutive differences would be only in the particles themselves and in their movements.

The whole comparison of the evolution of the order of orders of movement would then be a sort of figure of speech, like the 'average man' of economics. One knows that the 'average man' is nothing more than a purely conceptual idealization, while the individual man and his groups or aggregates are what actually exist. Similarly, one might say that the hierarchy of orders and orders of orders is, in this point of view, a purely conceptual abstraction, while what really exist are the particles and their groups or aggregates.

In biology, such a point of view is exemplified in the assumption that the *entire* behaviour of cells and organisms can be explained, more or less, mechanically in terms of the properties of molecules, such as DNA, RNA, amino acids, proteins, etc. Of course, one realizes that these are made of smaller particles, such as atoms, which are in turn made of electrons, protons, neutrons, etc. But one assumes that the mechanical laws determining the motions of the fundamental particles are such that through the interactions of these particles there arise groupings, systems, or aggregates (such as D N A molecules) which can be treated in a simplified way as relatively stable units, having certain essentially mechanical properties that are known or that can be discovered by further experiment and observation. And, of course, if this assumption is correct, it follows that the whole development of life, intelligence, society, etc., can in principle eventually be explained by referring it to an ever more complete knowledge of the properties of these basic molecules.

Of course, all these conclusions will follow only if it is true that the basic constitutive order of the universe is indeed that of the supposedly fundamental particles and their supposedly mechanical motions. These conclusions would, however, all fall to the ground if it turned out that natural processes cannot, in general, be reduced to mere automorphisms of mechanical order, and that they contain a really creative movement, in which there appear new orders and orders of orders. Therefore the question of whether the basic laws of physics are in fact mechanical or not is of the utmost potential significance in biology.

Now the fact is that physics has, in the past 50 years or so, been making gigantic strides away from mechanism. In my view, the main steps in this direction have taken place in statistical mechanics (to a relatively small extent) and in quantum theory (where the step is really clear and decisive).

As is well known, statistical mechanics has been able to explain the thermodynamic properties of matter in bulk by means of a statistical treatment of the movements of large aggregates of atoms. But in doing this it has been led into certain very confused questions, having to do with the statistical interpretation of the concept of entropy. Usually the increase of entropy (which takes place irreversibly) has been equated with an increase of what is called 'disorder'. As I have already indicated, however, there can be no such thing as 'disorder'. Therefore all efforts to proceed along this line are bound to end up in confusion. And, indeed, one does discover that proofs of the increase of entropy always encounter contradictions and paradoxes (such as those involved in Boltzmann's H theorem). More and more subtle analyses are made to resolve these paradoxes but one generally finds that these attempts transfer the difficulties to other parts of the theory, where they are rather hard to see (like sweeping the dust under the carpet).

In my view, these contradictions and paradoxes arise because one begins with

29

classical physics, which works solely in terms of a simple mechanical kind of movement generating particle orbits that are curves of *second order* (i.e. curves determined by the differences in two successive steps). On the other hand, the notion of entropy is inseparable from that of probability, which is in turn inseparable from that of randomness. And as has been seen earlier, a random curve (a typical case of which is the orbit of a particle in Brownian motion) is a curve of infinite order which cannot be determined by the differences in two successive steps. So there is an unresolvable conflict between the order of classical mechanics and the order of randomness implied by the concept of entropy. By calling randomness 'disorder' one fails to notice that it is in reality some kind of order, so that one is able to overlook the conflict between the two entirely different kinds of order, and thus one falls into confusion.

This confusion is further confounded by a presently current tacit assumption that in all fundamental laws (having to do with basic constitutive differences rather than distinctive differences) the movement must be either a simple curve of second order (determined by two successive steps) or a random curve of infinite order (treated by the laws of probability). In trying to explain entropy in this way we are led to impose both these extremes at the same time, and thus to attempt the impossible. But the conflict can be avoided if we admit the new concept that the basic laws of physics may involve curves of all orders, from the first or second to infinity. The increase of entropy can be explained as a *change of order* of the orbital curve, from one of lower to one of higher order. This is evidently not a mechanical process. So thermodynamics takes us out of the domain of mechanics. (But, of course, we will not be able to get very far in our inquiry into these processes until we develop a new mathematics of order, just as Newton could not get very far in mechanics until he developed the new mathematics of the differential calculus.)

One of the reasons why this question is hard to put clearly is that, in a random curve, the statistical properties become simpler, and thus in a sense can be said to change from a higher order to a lower one (e.g. as the gas molecules move at random, they produce a practically uniform over-all density of matter). Thus, when the entropy increases, the detailed motions of the individual particles undergo an increase of order, to a state that is less symmetric than before (since a low order of order implies a high degree of symmetry), while statistically the system as a whole moves toward a more symmetric state. The tendency to identify the change of order with nothing more than the statistical large scale properties lies behind the conclusion that the flow of heat leads only to a state of greater symmetry. This tendency is made particularly inevitable by the assumption that the individual molecules move toward disorder, i.e. no order at al . Clearly, from this assumption, it follows that the change of order can only be in the statistical properties. But once we recognize that the individual particles also have an order, which is always tending to increase (with a corresponding decrease of symmetry), then we can avoid the paradoxes that arise when the individual particles are assumed to become 'disordered'.

In the light of the above discussion, it becomes clear that the so-called reversibility of the basic laws of physics is only a simplifying abstraction. That is, if we simplify the real orbit by treating it as a curve of fixed order, then the laws will allow any given motion to be carried out in a reverse order. But if the real orbit is characterized by a continual and generally unidirectional change of order (from one of higher symmetry to one of lower symmetry), then the motion is not really reversible.

To a certain extent, a similar kind of irreversibility may prevail in living matter. For the evolution of life resembles the random curve in that it is a process of at least potentially infinite order. But it is different, in that the random curve merely goes through all the possibilities on a given level, in such a way as to give rise to a long-run tendency to statistical symmetry in its over-all structure. On the other hand, living matter tends to evolve hierarchically in an ever-increasing totality of orders of orders, so that, in the long run, it is always going to higher levels of over-all structure rather than to statistical symmetry. So, even though the increase of entropy involves a change from a lower to a higher order of order on the microlevel, this is very different from the change of order involved in the evolution of life.

In discussing the problems of statistical mechanics we have seen how notions have been introduced into physics that tend to move it away from a mechanistic point of view. However, it is only when we come to quantum theory that we see to full extent how far modern physics has departed from its earlier basically mechanical foundations. This departure involves three new aspects:
1. Process is discrete rather than continuous. Thus, electrons are said to 'jump' from one 'orbit' or quantum state to another without passing through intermediate states. What characterizes the 'jump' is the change of what is called the 'action variable' by an integral number of units (one unit of action being measured by Planck's constant h). However, the continuous change of action that characterizes classical mechanics is recovered with the aid of the correspondence principle, i.e. in the limit where the change of action contains many units, the discrete

changes can be approximated as continuous and are in correspondence with those prescribed by classical laws.

2. The constitutive order of this discrete process is determined by laws of probability, and not by the classical orbit of curves of second order. However, the correspondence principle enables us to recover classical laws, with the aid of the further statement that in the limit where many discrete steps are taken the probabilistic (random) process of quantum law leads, on the average, to the usual orbits of classical physics or a good approximation (i.e. through the statistical law of large numbers).

3. It is found that electrons, etc., can behave under certain conditions like particles and under other conditions like waves. This phenomenon is often referred to as the 'wave-particle duality'. What it means is that not merely have detailed classical laws broken down, but that the *whole order of movement conceived as the displacement of particles from one place to another has been transcended* (e.g. even the Brownian motion curves of infinite order are inadequate to describe the phenomena of the wave-particle duality).

Now, in classical physics it was known that there is a domain of phenomena (electromagnetic fields) which move in an entirely different order, i.e. that of wave motion. But what has been discovered is that, in processes involving only a few quanta of action, phenomena are observed which suggest that in some sense the same entity follows both the particle order of motion and the wave order of motion. However, a further analysis of the situation reveals that there is an inherent vagueness or lack of complete definition in these orders, which is completely foreign to classical concepts. When the particle order is relatively well defined, the wave order becomes correspondingly vague and vice versa. In other words, there is a reciprocal relationship (implied by the so-called uncertainty principle) between the sharpness of definition of the two opposing orders of motion, i.e. wave and particle. Different experimental conditions then determine which of these aspects is sharper and which is less sharp.

Bohr has formulated this behaviour in terms of what he calls the principle of complementarity. To come to this principle, we first note that while wave and particle orders are both necessary, and therefore complement each other in the full description of the phenomena, they also contradict each other when both are completely defined. In this quantum theory this contradiction is, however, avoided with the aid of an assumption of the kind described above, of inherent vagueness of the orders of wave and particle motions. Bohr assumes that this vagueness is not just characteristic of current theories of quantum phenomena, but that it represents a general principle applying universally. And this assumption is, in essence, the principle of complementarity.

While it appears to be possible to formulate Bohr's view in a logically consistent way, I am inclined to favour another approach, suggested by a comparison of the situation to that obtaining in the statistical explanation of entropy, where we are faced with a contradiction between the one extreme of the simple Newtonian order of motion and the other of an infinite random order. We may perhaps surmise that as there is a whole spectrum of new kinds of order between these two extremes, so there is a whole spectrum of new kinds of order between particle motion and wave motion. I have in fact made some progress toward formulating such orders. But because this work is as yet incomplete, I shall not refer to it further in this talk. Rather, I shall try to stay. as far as possible, within the framework of inherent vagueness of orders that is generally accepted by physicists today for the description of the wave-particle dualism.

Even if one remains within this framework, it is clear that the quantum theory implies a genuinely new order in physical law, radically different from the older order of classical mechanics. One may fail to notice this when one thinks of detecting electrons, protons, etc., by their tracks in a cloud chamber. From these tracks one tends to suppose that one practically 'sees' a little ball moving through the chamber, leaving a set of droplets in its path. But actually one sees only the droplets. The 'little ball' is purely an inference. In the classical domain of processes involving many units of action, this inference is well confirmed as a good approximation. But in the quantum domain one discovers that what we have been thinking of as a 'little ball' also moves in an order resembling that of a wave. So it cannot *really* be a 'little ball'. Therefore, in the interests of clarity of thought it would be best if we would drop the idea of the 'little ball' forever (except as a simplification that is approximately valid in the classical domain). The electron is almost infinitely more complex and subtle in its full behaviour than is any implication of such a model.

Some notion of the subtlety of possibilities open to the electron can be obtained with the aid of a more modern form of quantum mechanics, called 'field theory'. In this theory the starting point is the idea of 'quantum state'. This is defined mathematically with the aid of what is called the wave function. Although this function is rather abstract, certain of its properties can be visualized through the notion of order. For example, the places where the wave function is zero define surfaces. If one analyses these functions in some detail, one sees that different quantum states correspond to different orders in which these surfaces are placed.

Some remarks on the notion of order

Thus, a state of well-defined momentum corresponds to a set of planar surfaces in a simple linear order. On the other hand, a state of well-defined angular momentum corresponds to a set of planar surfaces in a cyclical order, going around a common axis. And, more generally, each quantum state is reflected in some such order.

Now, in quantum field theory we have the further basic factor that a transition is described as the annihilation of an existing quantum state and the creation of a new state. But since each state corresponds to a certain order, as described by its wave function, this implies that all movement is being treated as a change of order. And this change of order is in certain crucial ways similar to what happens in biology, where development of the species proceeds through the death (annihilation) of one organism and the creation (birth) of another.

While we do not wish to suggest that the analogy between electrons and living beings is complete, we do wish to emphasize that it goes far enough to show that physics has really totally abandoned its earlier mechanical basis. Its subject matter already, in certain ways, is far more similar to that of biology than it is to that of Newtonian mechanics. It does seem odd, therefore, that just when physics is thus moving away from mechanism, biology and psychology are moving closer to it. If this trend continues, it may well be that scientists will be regarding living and intelligent beings as mechanical, while they suppose that inanimate matter is too complex and subtle to fit into the limited categories of mechanism. But of course, in the long run, such a point of view cannot stand up to critical analysis. For since D N A and other molecules studied by the biologist are constituted of electrons, protons, neutrons, etc., it follows that they too are capable of behaving in a far more complex and subtle way than can be described in terms of the mechanical concepts.

However, probably because biologists are (tacitly if not explicitly) guided by notions resembling the correspondence principle, they are often able to suppose that these quantum mechanical subtleties are of no significance for systems as large as those studied in the inquiry into the genetic structure of cells. Nevertheless, it must be emphasized that, from the point of view of quantum physics, current experiments in this field are extremely crude. Therefore they can be counted on to reveal only the grosser features of what is happening, while finer points that may be of crucial importance can very easily be overlooked.

Even in physics, similar situations have arisen. Thus, one finds in the phenomena of superconductivity and superfluidity how quantum properties can be significant even on the scale directly observable with the naked eye (a scale that is obviously much larger than that involved in the behaviour of D N A molecules). What is striking here is the appearance, even on the large scale, of a kind of stability of certain states of motion, which is made possible only by the fact that action is discrete and quantized. Some physicists (notably Schrödinger) have suggested that the genetic process is made stable because the relevant molecules (such as D N A) are also in well-defined quantum states. On the other hand, without taking such quantum properties into account, it is practically certain (as emphasized by Pattee) that we cannot understand the stability of transmission of genetic characteristics. And if this is the case, it is extremely likely that we will miss certain key aspects of the process when we ignore the quantum properties of the molecules, by treating them as if they were nothing but large-scale classical objects.

With regard to this point, it is important to note that the whole subject is beset with some very serious problems. Thus, the current quantum theory treats the change of quantum states of molecules basically as a random process (of infinite order) with probabilities determined by certain mathematical quantities which are called 'matrix elements' and which can in principle be calculated by solving Schrödinger's equation for the relevant system. Now, on the basis of certain fairly reasonable assumptions about these matrix elements various physicists (notably Wigner) have demonstrated that the probability that a molecule (such as D N A) could engage in a self-replicating process without a basic change of structure is essentially nil.

In a way, this brings us to the same conclusion as that which tends to be held intuitively by biologists, i.e. that in the D N A molecules, quantum mechanical subtleties will be lost in the effects of random motions, so that the system can be treated essentially classically. However, if one reflects on this for a moment, one will see that this leaves us with the serious problem that the underlying physical laws almost certainly imply an unstable behaviour in the replication process, rather than the observed stable behaviour. Does this not lead us to the notion that (as suggested by Wigner), in some way, the laws of physics may themselves be incomplete ?

This brings us to the further question of how universal the current laws of physics actually are. With regard to this question, we know even now that our notion of the nature of the so-called 'elementary' particles is extremely inadequate. Thus, it has been found that these particles can be created, destroyed, and transformed into each other. In addition, a large family of similar, but unstable, particles has been discovered which has been classified phenomenologically in

Some remarks on the notion of order

a complex set of interrelated orders. It already appears to many physicists as if the beginnings of an entirely new order of natural law were being revealed, in which the particles would be like the flowers on a carpet pattern, while there would be something as yet new and unknown, which corresponds to the woven structure of cords that constitute the carpet. So when we analyse the world as if it were made of particles, this might be similar to analysing a carpet as if it were made of the flowers, that can be abstracted from its patterns. Such an analysis might give certain correct results and yet be very misleading when applied too broadly. Similarly, it may well be that the whole structure of physics is inadequate and misleading when extended too far into the processes of living matter.

Of course, we do not at present know whether the new order that must underlie current physical laws will be significant in biology or not. Nevertheless, the fact that the whole order of basic physical law is currently in the process of radical transformation may perhaps suggest to us that none of the inferences drawn from these laws are iron-clad certainties applying with absolute and unlimited universality. Therefore, even before we know just what form these new laws will take, we may perhaps be led to allow ourselves to question the completely universal validity of some of the basic assumptions underlying current physical theories.

When we look at the quantum theory, we see that the assumption that is on the weakest footing is that of the *random order* into which transitions between successive quantum states are supposed to take place.

First of all, there is the fact that the very idea of randomness has been confused by equating it with the impossible and meaningless concept known as 'disorder'. So the first step is to try to clear up our thinking about the randomness of quantum processes.

As in the discussion of classical statistical mechanics, we are led in doing this to assume that in a series of quantum processes (i.e. 'jumping' from one quantum state to another) there may be an order, defined, for example, by a suitable analysis of similar differences in the steps. In a typical case, the process may have a very high order of order, which would, in certain ways, approximate the quality of randomness. The tendency of the degree of order to increase in a unidirectional way would, as in the classical theory, be the explanation of the increase of entropy. And as happened classically, the process would be reversible, only if one simplified the theory, by abstracting the situation in which the order of order was more or less fixed.

It follows from the above discussion that a series of quantum jumps may not have a completely random order (with probabilities given by the matrix elements computed in accordance with the quantum theory). But does this conclusion not contradict the vast body of experimental fact, which has thus far confirmed the inferences drawn from the quantum theory ?

If one looks carefully at this body of fact, one discovers that the randomness of a series of quantum jumps has actually been tested only in a few limited contexts, and then only to a rather limited degree of approximation. A typical test would be to measure some property of a particle and then, a little later, to measure either the same property or another property of this particle. Such observations as have been made thus far involve rather long time intervals between successive measurements. If one reflects a moment, one can see that if one allows a long time to elapse between successive measurements it becomes difficult to distinguish between a process of high order and a random process. To make such a distinction it is necessary to have a series of successive measurements on the same system that follow each other in as short a time as possible.

New experiments are therefore needed here. Recently, preliminary efforts to set up such experiments have been begun by some of my colleagues of Birkbeck College. However, it must be emphasized that the technical difficulties in the way of such a programme are quite considerable, so that one must not expect quick results in this field.

We see, then, that even in physics, quantum processes may not take place in a completely random order, especially as far as very short intervals of time are concerned. But after all, molecules such as DNA are in a continual process of rapid exchange of quanta of energy with their surroundings, so the possibility clearly exists that the current laws of quantum theory (based on the assumption of randomness in *all* quantum processes, whether rapid or slow) may be leading to seriously wrong inferences when applied without limit in the field of biology.

Of course, unless the situation in living matter differs in some way from that in non-living matter, then, although there may be something non-random in the order of physical processes, the quality will nevertheless tend to be lost over long intervals of time. But we have already admitted that, even in inanimate matter, the order of a process is not fixed, and can change, for example, when the entropy increases. It is evidently possible to go further and to assume that, under certain special conditions prevailing in the development of living matter, the order could undergo a further change, so that certain of these non-random features would be continued indefinitely. Thus, there would arise *a new order of process*. The changes in this new order would themselves tend to be ordered in a yet higher order. This would lead not merely to the indefinite continuation of life, but to its indefinite evolution to an everdeveloping hierarchy of higher orders of structure and function. The situation in physics could then be compared to the pattern of sound or 'background noise' which has momentary fragments of simple and symmetric order but a long run tendency to approximate a random order of sounds that gives an over-all impression of statistical symmetry or uniformity. On the other hand, the situation in biology would then be more like the pattern of sound in music, which is ordered in a hierarchy that is in principle capable of further development without limit.

Because of our scientific training, we may at first find it hard to accept such an idea. For we have been heavily conditioned to the belief that the higher orders of nature are determined completely by the lower order mechanical motions of the particles which constitute the complex structures that we meet in everyday life. But, after all, this is merely a rather poorly tested assumption. As yet, it is in fact impossible to exclude the contrary assumption that in some crucial ways the higher order features of natural laws are as 'fundamental' as are those features referring to the movements of electrons, protons, atoms, D N A molecules, etc. Certainly, to fail even to entertain such notions, at least in a provisional way, would constitute a kind of dogmatism. Such dogmatisms could very easily put 'blinkers' on our field of mental vision that could prevent us even from looking for any phenomena that do not fit into current mechanical hypothesis. In particular what it prevents us from looking for are phenomena capable of showing whether, the essence of natural law is the hierarchy of orders that has been described here or whether it is some fundamental level of mechanical order.

What kinds of experiments and observations does this point of view suggest ? Naturally, one can say little in this regard until we have, to some extent, digested the new structure of ideas that it implies. Nevertheless, one can even now make a few preliminary suggestions.

Firstly (as is necessary in physics) one has to inquire very carefully into the randomness of biological evolution, both for the genotype and for the phenotype. Of course, it is recognized by all that natural selection constitutes a non-random feature of the evolutionary process. But the further question is that of whether *all* non-random features are of this type, i.e. essentially external to the basic processes that constitute the living organism itself. Indeed, within such an organism it is now generally assumed that evolution is due entirely to purely random mutations of the genes. The real question here is that of finding whether this process is in fact purely random or not.

David Bohm

One observation that could be relevant would be to trace a series of successive mutations to see if the order of changes is completely random. In the light of what hasbeen said here, it is possible that while a single change (or difference) may be essentially random relative to the previous state of a particular organism, there may be a tendency to establish a series of similar changes (or differences) that would constitute an *internally ordered* process of evolution.

In this regard, it is possible that there is an important difference between an almost stable kind of species, in which mutations tend to be both slow and very nearly random, and a species in the process of transition, in which mutations tend to be fairly rapid and strongly directed in some order, in the way described above. Of course, most species that we can now observe will tend to be almost stable, having passed through their rapid and strongly directed transitional phases long ago. So it is unlikely that a cursory inspection will immediately show up a species that is in an easily observable process of transition ordered in some well-defined direction. Therefore, in most cases that one is likely to meet, the deviations from randomness will very probably be small. However, there may be a few cases of unstable species, in which a series of mutations can be found, which is in an order that is appreciably different from that of randomness.

If such deviation from randomness could be found, this could have very far-reaching consequences. For it would imply that when a given type of change had taken place there is an appreciable tendency in later generations for a series of similar changes to take place. Thus, evolution would tend to get 'committed' to certain general lines of development. But because even order of change can itself change, there could be a higher-order tendency to alter the 'line' of evolution and to start on new lines. Of course, survival in the total natural environment would ultimately decide which lines could be sustained. Nevertheless, an entirely different principle of internal order would determine both the nature of these 'lines' and how they would tend to vary and transform into other 'lines'.

More generally, what is needed, in physics as well as in biology, is to perceive the existing facts anew, in the light of the notion of order, and of a hierarchy of orders of orders (e.g. one may try to see where new orders of order have come into being in evolution). Such perception will evidently tend to lead us to ask new questions in our scientific research. And, as is well known, it is as important to ask the right kind of question as it is to find the answers to the question by observation and experiment. Thinking within a fixed circle of ideas tends to restrict the questions to a limited field. And, if one's questions stay in a limited field, so also do the answers. Thus, nature itself is apparently confirming our

39

Some remarks on the notion of order

assumption that the general framework of current ideas is in principle complete and exhaustive, requiring only detailed development before leading to a full understanding of everything in the universe.

In reality, however, it is only we who thus tether ourselves so that we never wander too far from familiar fields that seem to be safe, secure, and rewarding. Nevertheless, we can, at any moment, cut our tethers, by considering the available phenomena in the light of new general notions, such as those of order. We may fear that we will in this way be led to go out of our familiar rich pastures and to lose ourselves in the unknown immensity of trackless deserts and mountainous wildernesses. But it is not at all unlikely that we can in this way come upon much vaster fertile areas than those which we have known thus far. At least, it seems reasonable that some of us should engage in this kind of exploration, and that not everybody should spend his whole life in extensions of the field of the known.

Further Remarks on Order

David Bohm
University of London

As a result of a very fruitful dialectic resulting from the interchange of points of view at the Bellagio Conference, I now find myself in a position to extend my original remarks on the general notion of order in a considerable number of ways.

▶ *On Metaphysics.* I think the most important aspect of the interchange is the emergence of a common realization that metaphysics is fundamental to every branch of science. Metaphysics is not a well-defined field of study, a single, basic foundation on top of which we erect a towering structure of physics, chemistry, biology, psychology, sociology, and so on; but, rather, something that pervades every field, that conditions each person's thinking in varied and subtle ways, of which we are often not conscious.

Metaphysics is a set of basic assumptions about the general order and structure of existence. Whenever we say '*All* is X' or 'X is *basic*' or 'X is *always* (or *never*) true' we are expressing a metaphysical position. Thus, various ancient Greek philosophers enunciated views, such as '*All* is fire . . . water . . . flux . . . atoms', etc. Scientists later assumed that *all* is a universal mechanism, following Newtonian laws. Then came assumptions, such as '*All* is to be described and predicted by suitable structures of mathematical formulae (such as those of quantum mechanics)'. This Conference heard yet other metaphysical positions exposed, such as: '*All* is to be understood as a structure of automata, expressed in terms of semi-groups' and 'What is *basic* to *all* life is a genetic process, in which changes in the genotype are *always* fortuitously related to the experiences of the phenotype and in which these changes will survive *only* if they are favourable to continued propagation of offspring in the existing environment.'

It seems clear that everybody has got some kind of metaphysics, even if he thinks he hasn't got any. Indeed, the practical 'hard-headed' individual who 'only goes by what he sees' generally has a very dangerous kind of metaphysics, i.e. the kind of which he is unaware (e.g. 'You can *never* change human nature'; 'There must *always* be wars', etc.). Such metaphysics is dangerous because, in it, assumptions and inferences are being mistaken for directly observed facts, with the result that they are effectively riveted in an almost unchangeable way into the structure of thought. What is called for is therefore that each one of us be aware of his metaphysical assumptions, to the extent that this is possible.

Further remarks on order

One of the best ways of a person becoming aware of his own tacit metaphysical assumptions is to be confronted by several other kinds. His first reaction is often of violent disturbance, as views that are very dear are questioned or thrown to the ground. Nevertheless, if he will 'stay with it', rather than escape into anger and unjustified rejection of contrary ideas, he will discover that this disturbance is very beneficial. For now he becomes aware of the assumptive character of a great many previously unquestioned features of his own thinking. This does not necessarily mean that he will reject these assumptions in favour of those of other people. Rather, what is needed is the conscious criticism of one's own metaphysics, leading to changes where appropriate and, ultimately, to the continual creation of new and different kinds. In this way, metaphysics ceases to be the master of a human being and becomes his servant, helping to give an ever changing and evolving order to his overall thinking.

The proper role of metaphysics is as a metaphor which provides an immediate perceptual grasp of the overall order and structure of one's thoughts. It is therefore a kind of poetry. Some 'hard-headed' individuals may object to bringing such 'poetry' into science. But, just as Molière spoke of the man 'who talked prose all his life without knowing it', so even the practical man is 'speaking poetry all his life without knowing it'. The point that I wish to emphasize here is that all of us will think more clearly when we frankly and openly admit that a lot of 'hard-headed common sense' and 'factual science' is actually a kind of poetry, which is indispensable to our general mental functioning.

▶ *The Metaphysics of Process.* The basic metaphysics that I am now considering is one in which *process* is fundamental. I therefore suggest that one entertains the notion 'All is process'. That is to say, 'There is *no thing* in the universe'. Things, objects, entities, are abstractions of what is relatively constant from a process of movement and transformation. They are like the shapes that children like to see in the clouds (e.g. horses, mountains, buildings). Actually, the clouds are an aspect of the movement of air, the condensation of water vapour, and such. The forms that we see in them have only a certain relative stability. Rocks, trees, people, electrons, atoms, planets, galaxies, are also to be taken as the names of centres or foci of vast processes, extending ultimately over the whole universe. Each such centre or focus refers to some aspect of the overall or total process, which is relatively stable (i.e. which has a certain tendency to 'survive'). Some things (like clouds) last a short time, while others last much longer. But fundamentally (i.e. metaphysically) I am assuming that nothing lasts for ever. This assumption cannot be 'proved' definitively. However, it has evidently never yet been falsified in experience, experiment, or observation (i.e. nobody has ever yet discovered anything that is permanent).

Now, the notion of process is evidently practically empty of content until we can say something about its order. Generally speaking, there are three levels in which this order can be discussed. These are:

(i) Quasi-Equilibrium Process
(ii) Dynamic Process
(iii) Creative Process

As a rule, we tend to begin in a situation close to equilibrium, which enables us to recognize certain relatively static, or constant, features of process. We give these features names, and are thus led to regard them as stable objects or entities (e.g. as we can do when looking at clouds). Then, as we see that these features are changing and transforming, we seek to explain their relative stability in terms of a dynamic process of interaction of some basic entities (e.g. the shapes of the clouds are the results of movements of molecules of air and water). Still later, we come to the notion of a creative process, in which there are no basic objects, entities, or substances, but in which all that is to be observed comes into existence as a certain order, remains relatively stable for some time, and then passes out of existence (e.g. as physics now explains the movement of electrons through annihilation of existing orders and creation of new orders). In the metaphysics of process, creation and transformation of order is always taken to be the deepest and most fundamental account of the laws of a process.

▶ *The notion of structure-function.* Although the notion of creative process is taken here as basic, it is necessary to have a language, in which the relatively stable aspects of process can be given a fairly detailed and precise description. Such a language is needed, in order that our communications concerning process shall be capable of being given a certain relatively unambiguous kind of content. It is here that the notion of structure-function becomes relevant.

The notion of structure has already been explicated in my earlier talk as 'Constitutive order of constitutive orders'. To extend the precise specification of structure to the case in which change and transformation are relevant, we are now led to consider that particular kind of process which is known as *function.*

In general terms, one can say that function is a certain kind of ordered change of structure [1]. Each function can be regarded as having an 'input' consisting of a certain range of possible structures. The result of a particular function is to transform any given kind of input structure into some corresponding 'output structure'. A function of the stomach is to transform undigested protein structures into simpler 'digested' structures. A function of the brain is to 'digest' informational structures in a similar but different way.

The above is an extension of the mathematical notion of function : to each value of the 'independent' variable there is a corresponding value of the 'dependent' variable. But, evidently, even this extension of the notion of mathematical function contains only a part of what is more broadly meant by the word 'function'. For, usually, we think of each function as having a role, an aim, or an end, in something that goes beyond the field of that function. For example, suppose that a man is doing a job in the role of a government 'functionary'. This activity evidently has a part to play in some broader set of functions, which evidently extend up to include the whole government. And the function of the whole government is, in some sense, to help regulate the overall order of society.

What is the function of the overall order of society ? Presumably, it is aimed at creating conditions in which individual human beings can live happy lives. But what then is the function of happiness ? Is it to enable people to be 'better citizens' so that they can properly fulfil the functions in their respective jobs ? If this were the case, we would be going around in a circle, in which no sense could ultimately be made of what is happening. The existence of society would then seem to be the mere result of the arbitrary and fortuitous coupling of functions (e.g. as it is said of some : they eat to live and live to eat).

The above discussion illustrates the general feature of function : that it must ultimately be referred to something beyond the field of function if it is to be properly understood. In the case of society, the ultimate aim of all social functions may be taken as happiness, meaning by this the harmonious ordering of life, both individually and collectively. Of course, others may disagree with this and say that the ultimate aim is to express the glory of the state, or of God, or to realize an ideal. But all these are also beyond the field of function.

A similar set of questions arises in the study of biological evolution. In all forms of life we observe highly coordinated, integrated sets of functions. But to understand these we must appeal to what is beyond the field of function. In earlier days, people attributed all this to the creative action of God. Current metaphysics appeals instead to survival value as the trans-functional feature to which all biological function has ultimately to be referred. But when one inquires more carefully, one finds that it is not possible to specify just what is meant by survival value without bringing in something beyond the field of function. For example, the older Darwinian phrase 'survival of the fittest' could have been taken to mean survival of those organisms that best harmonized with (*fitted* into) the prevailing environment. Without some notion equivalent to harmony, fitness, suitability, viability, or the like, the notion of survival becomes largely empty of content, being indeed a mere tautological statement that those forms of life that continue indefinitely to produce offspring from one generation to the next are the ones that will survive.

Such metaphysical difficulties with the notion of function are evidently not restricted to the study of biological evolution or of the organization of society. Indeed, the very word 'organization' contains a tacit reference to what is beyond the field of function. For example, the difference between a mob and a group of people working together on a constructive job is that the latter is so organized that its functions are harmoniously directed to an end determined beyond the whole field of the functioning group, while the former has no such harmonious ordering of its actions.

Physics has traditionally been a science which was supposed to be free of such trans-functional references. But this freedom was always more apparent than real. Thus, even if we restrict ourselves to Newtonian mechanics, we are inevitably led to ask: 'Why do the various parts of the universe function in just the order implied by these laws and not in some other order ?'

That is to say, is the form of Newton's laws a contingency, that depends on something beyond these laws ? If so, we are led back on the endless search for some ultimate ground which must eventually bring us, as has been seen, beyond the field of function. And if, instead of God, it is a 'natural necessity' (i.e. something that could not be otherwise, because of what the world *is*), then we are already appealing to something outside the field of function. Moreover, we are doing so in a very unclear way because there is, in fact, no means of knowing that the world *is* such that Newton's laws are inevitable. Indeed, what such a statement amounts to is that this is what we believe to be true in an absolute sense. And, evidently, believing something to be absolutely true is an act of faith that is outside the field of function. In one way or another, therefore, deep inquiry into the foundations of the laws of Newton must sooner or later carry us beyond the field of function.

The relevance of such trans-functional questions was demonstrated rather sharply when the whole Newtonian structure was overturned by subsequent developments in physics in such a way as to show that the approximate validity of Newton's laws is in fact contingent on the satisfaction of certain conditions (smallness of velocity relative to that of light and largeness of action relative to Planck's constant). Moreover, in the actual development of theories of relativity, cosmology, and quantum mechanics, scientists seem to have been forced to make wide use of further trans-functional criteria, such as 'simplicity' or 'elegance', which have indeed played a crucial role in determining both the form and the content of such theories. In addition, because of the use of probabilistic laws in various fields (statistical mechanics and quantum theory), trans-functional notions implicit in the words 'disorder' or 'randomness' have begun to be brought into the very foundations of physics. As far as one can tell, people mean by the word 'disorder' only that what actually happens fails to harmonize with their tacit conventions concerning what is supposed to constitute order.

Indeed, in all sciences, trans-functional notions are being introduced through largely tacit metaphysical assumptions of various kinds. One of the most common of these assumptions is contained in the distinction between what is fortuitous and what is not. Now the word 'fortuitous' includes the notion of 'randomness' but goes beyond this notion in the sense that two different orders, each perfectly determinate in itself, may have no essential relationship between them. For example, if one assumes that the genotype is fortuitously related to the phenotype, this allows causally determined changes in the genes (e.g. by means of certain chemicals). But these changes are then not significantly related to the basic order and structure of the phenotype.

One cannot fail to note here that the definition of the word 'fortuitous' is inseparable from related words, such as 'essential', 'significant', 'relevant'. All of these words bring in notions going beyond the field of function. At present, our notions of this kind are largely tacit and unconscious. So, in effect, we are introducing all sorts of arbitrary (and indeed fortuitous) elements into our thinking by using the word 'fortuitous' in an uncritical way. It is incumbent on all of us to pay very careful attention to what we really mean by such words. In doing this, we may well open the door to fruitful new lines of research.

▶ *Are harmony and conflict merely subjective judgments ?* As one goes into these questions one cannot fail to notice that trans-functional notions all tend ultimately to involve some kind of tacit distinction between harmony and conflict. But here one may begin to feel uneasy, because of the operation of the widespread belief that this distinction is merely a subjective judgment, whereas it is felt that science ought to be based, as far as possible, on objective fact. Thus physics deals with properties, such as mass, length, time, charge, etc., which are supposed to exist 'out there' independently of human beings, while qualities like harmony and conflict, beauty and ugliness, are supposed to exist only in the eye of the beholder.

I regard it as very important to question this division. Actually, the concepts of length, time, mass, charge, have been created by man. A few thousand years ago, nobody felt that these qualities are what is 'out there'. It is true that they are creations that in some way reflect a reality beyond man's thoughts. Nevertheless, man sees them in nature, as he can see beauty or ugliness.

As a result of our history of technological development, we have many technical words describing all sorts of functions, but few words to describe trans-functional qualities, such as harmony and conflict. Because we can so easily talk about function, and find it so hard to talk about what is behind function, we readily slip into the habit of assuming tacitly that only function is relevant in science. It is therefore necessary to develop the non-functional aspects of our language to correct this imbalance.

In order to do this we can begin by considering a few simple cases where harmony and conflict evidently have more than a purely subjective significance. Thus, suppose that we have motor cars moving along a road. These are co-ordinated in some kind of order, in the sense that they do not collide and destroy each other. Now, if this road intersects another road with a similar stream of cars, the two streams will not in general be coordinated. Rather, they will be only fortuitously related, with the result that they will tend to collide and clash, destroying each other in an order that is not harmony but conflict. With the aid of traffic signals, these two streams can be coordinated harmoniously. Similarly, in an organism, the growth rates of cells have to be coordinated harmoniously. When there is a cancer, this coordination breaks down and there is a state of conflict, which ultimately destroys the organism. In all these cases and many other similar ones it can be seen that we are actually able to distinguish harmony and conflict in a factual way, but that because there is no room for such a distinction in our scientific language we do not believe that it has any significant role to play in the development of our basic theories. (In other words, our tacit metaphysics is that harmony and conflict have no metaphysical relevance whatsoever.)

The importance of these considerations to biology is evident. For example, we can see that survival of an organism depends on two levels of harmony: (a) within the organism itself, and (b) in the relationship between the organism and its environment.

At first sight it may seem that to see this doesn't really help the biologist with his actual work. But, in my opinion, this would be a short-sighted view. For a change in metaphysics can make one alert and attentive to possible new lines of development, which may eventually open up new fields of research.

Then again, there is, as I have indicated earlier, a wide field of investigation to which one can be led by inquiring carefully into the meaning of the word 'fortuitous'. For in the understanding of how this notion is to be limited, one discovers one of the main factors that make real creation possible. (It is not the only one, however.)

Here I can make use of the example from the work of Piaget, in his observations on the development of the intelligence in infants and young children. He cites cases in which children learn to draw by first simply moving the pencil in what appears to be a 'random' or 'fortuitous' way which presumably expresses the ever-changing 'internal' state of the child. But if the child is looking at an object (e.g. a rectangle), this internal state will be influenced by such 'impressions' from the outside. As a result, the line ceases to be 'fortuitous'. Instead it begins to show a vague resemblance to the rectangles. At a certain age the child is able to notice this resemblance and to have a persistent interest in it. As he pays attention to what is happening, the internal 'impression' of the rectangle is probably sharpened up, and this 'expresses' itself by an 'improvement' in the resemblance of the line to the rectangle [2]. At first corners are introduced, and later the lines are straightened out, made perpendicular to each other, and so on. And this is potentially only the beginning of an endless movement, in which can emerge, in principle at least, a very great artist. What is crucial here is that the initially fortuitous relationship of the 'inner' state ('expressed' by the variations in the line) and the 'outer' perceptible structures is steadily transformed. This transformation from fortuitous to necessary is typical of a wide range of creative processes. (There may also be an inverse process in which certain orders, previously necessary, become fortuitous.) In such processes a genuinely new order is brought into existence.

This notion of creation is potentially relevant in all science. In biology, it might help to throw light on the origin of life and of intelligence. In addition, it suggests the need for more careful attention to the relationship between genotype and phenotype (already emphasized by Waddington on other grounds). For, evidently, between the genetic material and the whole organism is a vast hierarchy of structure and process on many levels. As the organism grows, and as it maintains itself, there is room on all these levels for transformations from fortuitous relationships to necessary relationships, and vice versa [3]. Careful attention to these transformations may well be a key to understanding what kind of harmony is really involved in the survival of a species.

However, I would like to extend this notion of the fortuitous and the necessary to the whole of existence. In other words, since even the so-called 'elementary'

48

particles of physics are now known to be created, annihilated, and transformed, there is no known factual reason that would even favour the postulation of an ultimate 'bottom level'. Rather, all that is known seems to fit very beautifully into the metaphysics of process, assumed to be ordered hierarchically, in the general way I discussed in my earlier talk.

Each level or order of the hierarchy is assumed to contain necessary aspects and fortuitous aspects. In the overall movement within this process, the fortuitous can transform into the necessary and the necessary into the fortuitous, thus making possible the creation of new orders (along with the ending of old orders). In this way we come to consider an 'open-ended' universe, with neither 'bottom' nor 'top' levels. Indeed, even the level structure is itself always transforming, although this tends to be more nearly constant than any of the 'sub-features' of the various levels (particles, waves, organisms, etc.). But fundamentally the basic principle is: Nothing is permanent. Change is what is eternal. And each feature of the order of change is itself eternally changing. These changes of change in turn constitute new features, which lead on hierarchically to the totality of the process.

In its totality, this process is too vast and subtle to be encompassed by the measure of man. Man thus necessarily abstracts partial and relatively stable aspects. (But as has been indicated, these too are always changing.) It cannot be too strongly emphasized that, in the first instance, this abstraction is creative. That is to say, the process as a whole makes an 'impression' on the 'inward' side of man and this is somehow 'expressed'. By becoming aware of how the expression is related to new perceptual impressions, man begins to learn to abstract from the practically infinite complexity of the originally 'fortuitous' environment just those features which will ultimately prove to be 'relevant', 'essential', 'significant', 'basic'. Eventually, any particular line of abstraction tends to become fairly habitual and mechanical. Thus one tends to lose sight of the creative origin of our abstractions (both in early childhood of the individual and in the distant past of human society). But real perception in any field (whether scientific or artistic) demands the same kind of fresh creative abstraction that the young infant is carrying out when he is beginning to come into contact with the world.

▶ *Is nature more like an engineer or an artist ?* The whole question of the relationship of the field of function to what transcends function can be brought out very nicely by asking the question: Is nature more like an engineer or an artist ?

Of course, the whole notion of comparing nature in this way is only a metaphor.

Further remarks on order

Nevertheless, it is a useful one. For I submit that, tacitly, most scientists have been treating nature as if it were either an engineer or the work of an engineer.

This trend began to be taken very seriously when, in Newtonian times, the universe was compared with a vast clockwork, originally set in motion by God (who, according to some, also made occasional repairs). Nowadays, it tends to be the computer designer who replaces the clockmaker. But evidently the basic principle is not altered. In all sciences, including biology and psychology, inquiry is founded mainly on an analysis of functions. This is, in essence, the basic approach of the engineer who wants to design a system of functions, with a definite utilitarian end in view. One could perhaps say that in the field covered by biology, nature's utilitarian aim has been assumed to be the production of species of organisms that survive.

Now, if one considers the process of creation as I have been describing it, one sees that it is much more like that of the young child learning to draw and starting on the way to becoming a great artist than it is like that of the engineer who wants to design a system of useful functions. Each 'centre' of process can take in 'impressions' of the rest, and 'expresses' these in its 'environment'. These 'expressions' can now come in once again as new 'impressions', thus starting a cycle of process in which each 'centre' can 'assimilate' its 'environment' and at the same time transform this 'environment' to 'suit' or 'fit' what is 'within'. I suggest that all existence is of this general character, but that living, intelligent beings carry this process to much higher orders than is done in inanimate nature (e.g. the particle, as centre, 'expresses' itself in its environment through its 'field', which also carries into it 'impressions' of the whole universe).

If this comparison is at all valid, might we not also reasonably expect that nature's creations quite generally tend to move towards overall harmony, which is *felt* by us as beauty in our immediate perceptions and which is *analyzed* intellectually as coherent and rationally ordered function? Thus, we are not surprised that almost anything to be found in nature exhibits some kind of beauty both in immediate perception and in intellectual analysis.

Of course, an underlying harmony, first sensed as beauty, tends to make for survival but, in general, does not guarantee such an outcome. It would clarify this issue if we considered nature to be largely neutral on the question of survival of any particular creation. (Indeed, when a real artist has finished a piece of work he doesn't care much what happens to it, as he has already started something else.) Anything that survives will of course have to have a certain harmony and will therefore be felt to have a corresponding beauty. But each type of harmony leads to survival only in limited conditions. Thus, a snowflake melts at room temperature, while a man dies at temperatures low enough to permit a long period of 'survival' for the snowflake.

Such a change of metaphysics becomes particularly significant when it enters the study of human beings. After all, it makes no sense to say that one lives merely in order to function, or even that one lives merely in order to survive, or to produce offspring that survive. A human being cannot do other than seek to live in harmony and beauty, without which survival has no value. Those who cannot do this will seek delusory substitutes in drugs and in exciting stimuli provided by various forms of violence. The same holds true for society. Any society which puts mere survival as the supreme value of life is already on the way to collective decay and the violent ending of life for many of its individual members. Survival makes sense only in a broader context, just as the engineer's organization of functions makes sense only when the ultimate aim of this function is beauty, harmony, and a creative life for all.

▶ *On the self-regulating hierarchy of process.* In the whole discussion, one theme that was common to almost everyone was the study of self-regulating systems in engineering processes, in computer structures, in rhythms of physico-chemical change at many levels of living organisms, in perception, in brain function, and in many other contexts. It therefore seemed worth while to extend the notion of hierarchic order to that of a self-regulating or self-governing process.

In this connection it is useful to recall that '-archy' means 'government'. So the notion of hierarchy is intertwined with that of government in a fundamental way. It is thus hardly accidental that each government tends to be organized hierarchically. In addition, we have to consider the relationships of different 'sovereign' governments which, of course, tend as a rule to approximate anarchy. To avoid the deleterious effects of such anarchy, there are efforts to organize governments of governments (e.g. Federation of the states of the USA, or perhaps ultimate world federation). Thus one obtains a series of levels of parallel governments, organized in super-hierarchies (i.e. hierarchies of hierarchies). Such a hierarchy of hierarchies might in principle extend onward and upward without limit.

All of this is not intended merely as interesting speculation. Rather, I refer to it to emphasize the vast order that is implicit in the full concept of hierarchy. It is something like this sort of order that is probably involved in a living being and even more in an intelligent being. Perhaps even the electrons and protons of an inanimate nature are also organized in some sort of very complex self-regulating hierarchy. The reason I suggest this is that in a metaphysics based on the notion of process we cannot take the continued existence (survival) of any particular aspect for granted. Because the basic order of process is eternal change of everything, we can no longer appeal to the mechanical notion that certain basic objects, entities, etc., 'simply exist' with constant and invariable properties. Rather, the survival of any particular thing, however 'basic' it may be thought to be, demands a complex process of *regulation,* which provides for the stability of this thing, in the face of the eternal change in all that serves to constitute what it is.

As an illustration of how the principle of regulation of hierarchies could operate let us begin by considering the hierarchy in a particular government. As shown in the diagram, a typical government operates on a number of levels, each of which contains various departments. Within a given level the main activity is, generally speaking, the carrying out of appropriate functions (which is done by 'functionaries' in the manner indicated earlier). But between the levels the main activity is that of *abstracting* information about the order of functioning. This abstraction proceeds in two directions. There is an upward movement in which the higher level officials are informed about what is 'essential', 'relevant', 'significant'.

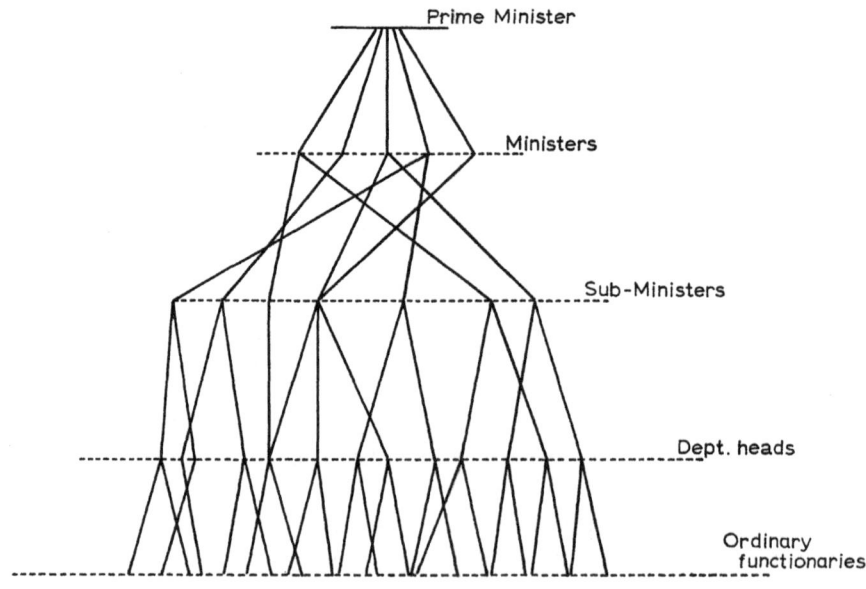

(Naturally, such a selection is necessary, as they would be overwhelmed with a flood of unnecessary data otherwise.) Then there is a downward movement in which the higher level officials inform those lower in the hierarchy how they are to order their actions in the light of the general aims of the government, and in the light of information of all sorts coming from other departments and levels.

Such directives are necessary, since without them the various departments and even various individuals would be likely to engage in conflicting actions. Indeed, even with such information, conflicts are likely to arise from time to time (e.g. the governing bodies of universities may be directed by the Minister of Education to expand, while the Treasury directs them to spend less money). In the upward flow of information, higher officials (in some cases even going up to the Prime Minister) should normally become aware of such conflicts and then give new directives that bring the activities of the departments concerned into a coherent and harmonious relationship.

The proper action of a government therefore depends on a circular process, in which information moves upward and downward all the time. In a properly functioning government there would be no need for any imposition of authority in this circular movement. Rather, each person would want only to do his job; that is, to help regulate the order of society in a harmonious way. The higher officials would always be eager to learn the real state of affairs from the lower officials, who in turn would be eager to consider directives based on the greater range of information available to the higher officials. Actual governments seldom come near to this mode of functioning, for people generally subordinate the real function of their jobs to the effort to extend and glorify the ego, or to make it secure. As long as human beings tend to operate in this way, neither governments nor any other organizations are likely to fulfil their proper functions. Therefore, if we are to use the notion of government as a model of a self-regulating hierarchy, we will have to refer not to actual governments of human beings as they are now, but rather to governments of beings (human or otherwise) who would be free of the ego. (Indeed, scientists find computers so interesting to work on just because of the tacit hope of constructing an artificial intelligence that would not be confused by an ego.)

The above considerations are particularly relevant to the ability of a government to adapt to contingencies. This is evidently dependent on the pattern of the hierarchical organization of the circular flow of information. To help define our terms more precisely, let us refer to this upward and downward flow of information as a 'vertical movement' while the various functions carried out on a particular level will be called a 'horizontal movement'. It is basically the vertical movement that 'regulates' the horizontal movement and thus avoids clashes and conflicts which can arise when unforeseen changes of the general situation cause previously coherent sets of functions to cease to be compatible with each other.

Generally speaking, a particular organization of the governmental hierarchy can meet only some limited range of contingencies. Beyond a certain point, conflicts may arise for which a given hierarchical structure has no means of resolution. At present such a situation tends to give rise to revolution, in which the order and structure of the hierarchy is violently altered. Such violence is generally so destructive that a vast range of new problems is introduced, which may well be even more difficult to solve than were the conflicts that originally gave rise to the revolution. But basically this violence arises because the people involved on all sides of the conflict generally subordinate their functional activities to the advancement of the aims of the ego, either the individual ego or the collective egos of a particular group as opposed to another. In order to use government as a model of a properly self-regulating hierarchy in this context it is therefore necessary to consider an ideal situation in which government officials were interested only in their proper functions but not in any sort of ego, whether individual or collective. When one or more governments are involved in a common crisis, all the people concerned would, regardless of individual or national background, cooperate towards working out a new hierarchical structure and then try to see how it actually functions. If it were still not adequate, they would try again and again until a solution was achieved.

In this kind of situation we would have a fairly clear illustration of the 'timeless' character of the vertical structure of the hierarchy. That is to say, the ordinary functional activities of the hierarchy involve the order of time in an essential way. But changes in the vertical order really do not thus involve the order of time in a correspondingly essential way. (For example, it can be agreed by all that at such and such a moment one vertical order is to be replaced by another.)

Of course, it takes time to work out the consequences of a new vertical order. In the beginning, such a process would go through a complex period of readjustment, but if the new structure is an adequate solution, things would sooner or later 'settle down' to a quasi-equilibrium state that was relatively harmonious once again.

We are thus led to consider the 'timeless' ordering of the vertical structure and its role in the 'time' process of functioning on the horizontal level. In essence, it is the 'timeless' that regulates and dominates the time process. Anything really new and creative has to come in 'timelessly'. (Here one should consider that various parts of the hierarchy can also change their structures in this timeless way.) The order of time is then what carries the creative change forward to complete realization.

Probably the most immediate analogy is between this process and the operation of the human brain. In this connection one should advisedly think of a vast series of hierarchies of hierarchies. This serves to regulate the movements of the organism on all levels so that the latter is able to adapt and meet contingencies from moment to moment. The *functional systems* by which this regulation is achieved will include many kinds of non-linear cycles (having periods from microseconds and less to years and more), some of which were discussed in the Conference.

When an organism meets a new situation that is beyond the capacities of a given organization of cycles to handle, what is called for is a suitable reordering of the vertical structure. On the lower levels of living things this takes place genetically (at least to a considerable extent). That is, most of the individuals concerned die, but a few of them in whom the reordering has 'fortuitously' taken place in the genotype will survive and produce offspring. But at higher levels (including human beings) this reordering can often take place much more rapidly, far up in the hierarchy of the brain and nervous system of the individual. Such a change occurs when the individual is said to have had a 'flash of understanding'. In the point of view that is being suggested here, this kind of 'flash' is a reordering of the hierarchy, which results in a new mode of action of the organism. In essence, the flash takes no time, but it takes time to work out its implications and consequences in terms of new intentions, modes of thinking, and general orders of functioning.

Consider, for example, the set of cycles of body function described by Art Iberall, going from the shorter 'oxygen cycles' of a few minutes to longer 'kidney' cycles of many months. Observations have shown that conditions of stress, such as those commonly met by business executives, can disturb these cycles, to produce illnesses such as high blood pressure. Evidently such men have a pattern of thinking that leads them to engage in activities that are beyond the normal ability of the hierarchy of self-regulating cycles of functioning to cope with indefinitely. Sometimes when these men are told the facts about their situation there is a sudden flash of understanding. It is this flash that reorders the vertical hierarchy of the process of thought, and thus ultimately restructures the entire sets of hierarchies of function.

55

Further remarks on order

Of course, it takes time for the restructuring to take effect. First there is the development of a new intention (to live in a healthier way) along with a new way of thinking about these questions. And then, all the steps needed (e.g. exercises and diet) are taken to implement the new intention. After a year or so, such a man is usually visibly healthier than before. In a way, he has carried out a peaceful revolution in his whole way of life. This revolution started in the flash of understanding that altered the order of his thinking process, in which the crisis had its basic origins. And thus it also altered the organization of the whole of his being.

Is it not possible that analogous processes take place on the biological level? Waddington has brought out how complex is the chain of process between genotype and phenotype. C. Longuet-Higgins has suggested (at least tacitly) that in this chain there is room for 'improvement of programmes', providing a kind of adaptation going beyond that determined merely by a change in the genotype. But is this 'improvement of programmes' not a particular kind of change in the 'timeless' order of the hierarchy?

Now let us come to physics. I want to propose that electrons, protons, etc., are merely the names of aspects of a vast, self-regulating, hierarchical process, operating at the level of inanimate matter. Such a hierarchy would in a natural way show many of the properties that are now attributed to quantum mechanical systems. For example, consider a decision made on a given higher level of the government. As the effects of this decision percolate downwards, they spread out, somewhat like a wave. But then, from the lower levels, information is abstracted in an ever more 'concentrated' way, so that it may come back to another department of the original level at a given 'point'. If one looks only at that single level, it seems that the decision has 'moved through space' from one department to another. In such a view one could perhaps attribute all these decisions to a hypothetical 'little ball' that passed from one government building to another (this ball could perhaps be called the 'decidon'). But then, sociologists investigating the phenomena from this standpoint would be puzzled at the wave-like properties of the decidon. The analogy to the electron is quite evident. May it not be that the word electron is actually only the name of a certain complex and relatively stable mode of action of some universal hierarchy of material process?

Such a notion gives a very good model of 'quantum transitions'. Thus each 'quantum state' would correspond to a quasi-equilibrium order of the hierarchy. When physical conditions were such that a given hierarchical order could no longer adapt its function to the actual situation, there would be a sudden 'timeless change of vertical order, resulting in a new quantum state. After a short period of readjustment, this would settle down to a new quasi-equilibrium mode of functioning. The change of order would be discontinuous while the function would be continuous. But because the complex details of function are not abstracted into the higher levels of information defining the structure of the hierarchy, it is impossible from a knowledge of this structure (i.e. the quantum state) to predict, in each individual case, exactly when the revolutionary change of structure will take place. At most, this can be done statistically, on the average, for an ensemble of systems with similar hierarchical structures.

This brings us to an interesting new concept, i.e. the *law* of the vertical order of the hierarchy. Now, in the horizontal mechanical order of functioning, the law of motion is known (in classical physics at least) to be some sort of second order differential equation. This can be expressed as a 'second order' discrete process, in which the basic law is the similarity of successive differences. That is, if A denotes the first state of the system, B the state after a short time interval later, and C the state after a corresponding interval still later, then we can say that B is to A as C is to B. This law then determines the whole time sequence (i.e. D is to C as C is to B, etc.).

Let us now, for the sake of exploration, suppose that in the movements thus far studied in physics the vertical order has a similar law. In other words, the differences of adjacent levels are similar. So we could write: L_2 is to L_1 as L_3 is to L_2, etc. (where L_n is the nth level). This law would determine an ever-changing level structure in the timeless order of the hierarchy. If we introduce an order parameter τ, this law would have some formal resemblance to the mechanical law determining the horizontal movements in terms of a time parameter t. Indeed, the timeless order parameter τ would even have some kind of vague and general statistical relationship to the time order of process, measured in terms of the time parameter t. (This is because it takes some time for information at one level of the timeless order to reach another level where its effects would be felt in the mechanical order.) Nevertheless, it would be of crucial importance to distinguish the two kinds of order, i.e. timeless and time.

I suggest that, in quantum mechanics, what has been called time (t) in Schrödinger's equation should actually refer to the timeless order parameter τ if one is to understand what it really means. In other words, Schrödinger's equation should not be regarded as determining the actual movement of things in time. Rather, it determines the 'timeless order' which is really the order of the hierarchical structure, and therefore the 'quantum state', in the sense defined earlier. This quantum state is then only statistically related to the mechanical time order of t. Real time involves a change of order, i.e. a change of quantum state, or a 'quantum jump'.

If these general notions are valid, then it follows that current quantum mechanics is incomplete, in the sense that it leaves out altogether (except in a crude and vague way) the treatment of the actual time process of function and change of state. This suggestion then points to a possible new direction of research, i.e. development of laws of the time order of physical process. I am now working on these with some success, but lack of space prevents me from giving further details on this point.

▶ *On the separation of the observer and the observed*. When we consider the notion that information is what is moving up and down in the vertical order of the hierarchy, we come to the yet deeper question of the observer and the observed (subject and object). It has generally been accepted in science that observer and observed are separate and distinct entities. Such an assumption (which is evidently a basic part of our generally accepted metaphysics) has led to a veritable hornets' nest of entangled and confused problems, growing out of the effort to understand how subject and object are related. But in the metaphysics of process, observer and observed cannot be taken as separate entities. Rather, they are only names of aspects of the total process. And, indeed, we see in the notion of the hierarchical structure that since information is moving upward and downward at all levels, there is no need for a separate 'subject' who would be 'doing the observing'. Rather, inanimate nature *is* both observer and observed.

The same holds true for living beings, which are also hierarchical structures, built out of higher order hierarchies of the hierarchies that constitute inanimate nature. Thus the vertical movement of information continues onward into living beings. In human beings this movement extends further into the hierarchies that constitute awareness, intelligent perception, and understanding. Therefore, at no stage is there a separate observer or subject. The observer is the totality of all that exists, and this is also the observed. Indeed, the movement of information is the dominant order in the whole process, from which the mechanical order is abstracted as 'subordinate'.

Of course, we always have to discuss some partial aspect, abstracted from this totality. In this aspect there are always the two streams—'factual' information coming upward and 'directive' information going downward. It is the upward stream going on through our own sense perception into consciousness that

58

makes our own observation possible, while the downward stream makes possible our participation in the total process. Thus, as indicated earlier in the discussion of Piaget's observations on the young child who is learning to draw, perception and action are always two sides of one circular process (whether in the electronic level or in the brain). This is indeed the basic metaphysical position to which I would like to call your attention here.

Finally, I would like to consider the question raised at the Conference: Is the total hierarchy finite or at least denumerably infinite ? In my view, this is an inappropriate question based on an inadequate understanding of the general metaphysics that is under discussion here. In the metaphysics of process we have to start with the notion that the totality is vast and beyond the measure of man, not only quantitatively but in the potential and actual richness of its qualities. Indeed, we *always* begin with 'what is', which is the unknown. From this, mankind can, at any point in its history, abstract a certain knowledge, having partial, relative, and limited validity. This abstraction is basically creative, i.e. the outcome of a movement of 'outward' participation and an 'inward' perception of this participation. In this 'circular' movement, mankind creates its percepts, concepts, language, and so on. What has been created up to a given time provides the very terms of communication available at that time. These terms enable us to focus on some limited aspects of the vast field of what is to be perceived directly with the senses, and to treat these as 'basic', 'essential', 'significant', 'relevant', while other far more numerous and richer features are treated as 'fortuitous' and 'irrelevant'. Experience guided by these notions is then always showing us the limits of validity of the corresponding abstractions. These limits indicate the need for extending the 'circular' process to create newer abstractions leading to new percepts, concepts, language, etc.

Indeed, even the notions that I have discussed here are only a part of this unending process, in which nothing is fixed, not even the terms of discussion of the total metaphysical context. So it would not make sense to postulate a finite or denumerably infinite hierarchy, since the process metaphysics implies that, as our experience is extended, the whole notion of hierarchy will very probably have to change radically and fundamentally. Even the notion of process metaphysics itself may well change beyond recognition, for all that we can now know. But all that I am suggesting is that this metaphysics may be relevant in the present stage of development of human knowledge.

In this metaphysics, man is a part of the vast totality of all process. At the same time, it is man who has created the abstractions, with the aid of which he

Further remarks on order

is able to recognize certain features of the process in perception, remember them, communicate them, etc. By means of these abstractions, man is able to assimilate the world within his consciousness while he is also participating creatively in the world, to help transform it, so that it will be more suitable to his needs (which in turn are changing in this process). These two movements (assimilation and creative participation) are two inseparable sides of one 'circular' process.

Notes and References

1. These notions of function have been explained in an article: D. Bohm, *Proceedings of the International Conference on Elementary Particles*, Kyoto, 1965.
2. Is this not like the 'improvement' of computer programmes, suggested by C. Longuet-Higgins?

3. It is the development of a quasifortuitous relationship between genotype and phenotype which permits the former to be nearly 'insulated' from the latter, thus making possible a certain reliability of hereditary transmission, as emphasized in Neo-Darwinism.

Bohm's Metaphysics and Biology

Marjorie Grene
University of Texas

1 A philosophical observer at a scientific conference is a kind of ethologist (or epistemenologist?) watching the conceptual behaviour of the other animals. The Second Serbelloni Symposium was outstanding not only because it brought together a group of extremely ingenious and well-trained performers, but also, and above all, because it produced a confrontation of two different conceptual patterns or, to borrow Kuhn's term, two paradigms, one orthodox and relatively restricted (and restricting) in its scope, the other heterodox and comprehensive. The result was not, as often happens in cases of deep-seated conceptual disagreement, simply the clash of two sub-groups. Rather, the first was literally comprehended—that is, described and explained—by the second, though its members were, with some exceptions, unaware that this is what had occurred. It was—I hope—a case of evolution in action: where the species doomed to extinction, innocently unconscious of its lack of 'fitness', continues happily to perform its traditional rites. The spectacle was instructive, but difficult to report, for two reasons. On the one hand, David Bohm in his original paper and in his 'Further Remarks' has himself indicated plainly how his 'metaphysic of process' assimilates and explains the truncated metaphysics of orthodox biology (and physics and computer science and psychology, etc.). Yet on the other hand, most of the contributors to and probably most of the readers of this volume, subscribing as they do to the still current orthodoxy, which as a matter of fact flourishes exceedingly at present, rather like the horns of the Irish elk, are unlikely to see the pertinence of Bohm's metaphysical remarks to their own methodology, and so are unlikelier still to see anything but *im*pertinence in my remarks on these remarks. The poor best I can offer in these circumstances is to try to put Bohm's speculations and, by implication, the metaphysics of the orthodox majority, into their historical context in terms of the major development of philosophical thought in the past three centuries or so.

2 It is otiose, yet necessary, to point out once more that the major trend of modern thought has been held captive by the brilliant success of the scientific revolution of the seventeenth century. The revolutions of the twentieth century have occasioned *some* fundamental rethinking of basic principles, and may yet—if Bohm's predictions are correct—have more far-reaching effects than they have,

61

explicitly, had so far. But the chief model of 'scientific method' is still that of the Galilean-Newtonian philosophy. And there was something deeply incoherent about this philosophy from the start. Bohm indicates the source of this incoherence when he points out that the acceptance of Newtonian mechanics depends in the last analysis on an act of faith. Why should our mathematicizing be true of nature ? There is no intrinsic reason, for example, why Newton's geometrical proof of Kepler's second law should demonstrate anything about what goes on in the sky. For an Augustinian, what we think, when we think clearly, and what there is in nature, both, if at different levels of perfection, express God's being. Descartes, with his sharp and simple dichotomy of cogitation (= mathematicizing) mind over against extended matter, has to invoke God *ex machina* to hold the two together. But he is still sufficiently an unquestioning Augustinian, so that, for him, the invocation works. With the secularization of thought (metaphysical as well as scientific : as Bohm quite correctly states, they are never wholly unrelated, since 'metaphysics' is just the most comprehensive range of anybody's thought, whether he knows it or not), the Cartesian dichotomy becomes unstable. Its uneasy synthesis in Kant, with nature reduced to phenomenon and mind to moral will, depends, still, on Kant's undoubting pietism for its ultimate support. That gone, it is only a short, inevitable step on the one hand to the *Nullpunktsexistenz* of the Sartrean for-itself, which frankly lives by contradiction, and on the other to the fruitless and equally self-contradictory objectivism of the contemporary philosophy of science (and of many scientists). (For the fruitlessness of the latter, see the outcome of Wittgenstein's *Tractatus,* for its self-contradiction, the argument of Russell's *Human Knowledge;* or see also the critique of E. Straus in *Vom Sinn der Sinne,* or of course Whitehead.) Of course each lingering remnant of the divided cosmology tries to account for the whole : Sartre's 'dialectical reason' serves up a caricature of nature ; modern epiphenomenalism, a caricature of mind. Along these lines there is just nowhere further to go—and there never was ; but it has taken us three hundred years, and indeed may take still longer, to find this out.

The incoherence lies not just in mind-body dualism, which has long since given way to a belief in matter-in-motion as the sole reality, but, as Bohm emphasizes, in the deep-lying divisiveness of our conceptual framework along a number of related lines. To cut off mind from nature is to cut off subject from object, so sharply that science itself (the product, after all, of subjects) becomes irrational and reality meaningless. Science becomes computation-for-the-sake-of-prediction-for-the-sake-of-computation-for-the-sake-of-prediction . . . .

'understanding' a merely subjective addendum, and 'truth' a dirty word, dropped in weak moments, like words with one less letter, but decently avoided for the most part in polite society. And the world so known ? It used to be, and, as Bohm points out, for many biologists still is, the seemingly solid one-level nature of Democritean atomism, where faith that God made and keeps united our thoughts and their objectives gives way to the equally, if not still more, irrational faith that more complex orders *must* be explained out of, and exhausted by, those that are simplest, and ultimately out of the one 'real' order of matter in motion. Taking subject and object together we have, in Whitehead's words, 'a mystic chant over an unintelligible universe'.

For what the subject-object dichotomy entails is a separation of order from the ordered, of meaning—which shrinks to a game with meaningless counters—from what is meant—which shrinks to an infinite aggregate of equally meaningless data. In philosophical jargon, it entails, as the literature (and indeed, literature) richly shows, a radical division between value and fact: in Bohm's terms, harmony becomes a little secret preference of our own, and beauty a private vagary, rather than, as it is, the criterion of our access to reality. In contrast, Bohm's linking of understanding, beauty, and the timeless orders that govern emergent process may herald, in my view, a comprehensive alternative to the self-denying ordinances of modern thought: self-denying in that it alleges itself to be only the compulsive outcome of its own neural processes. (Waddington is right, of course, in calling Bohm's view Whiteheadian, but it may prove more viable than Whitehead's own cosmology, since it can be developed, I believe, without recourse to a doctrine of eternal objects: a radical *in*coherence, I feel, in Whitehead's system.)

True, there are some other signs also of relaxation in the cramping cosmology that still governs most scientists' minds. On the 'subject' side, books like Hanson's *Patterns of Discovery* or Kuhn's *Nature of Scientific Revolutions* may help. The slowly growing influence of Polanyi's *Personal Knowledge* is a hopeful sign, although unfortunately many of those who profess to accept its conclusion have grasped only the most superficial theme of its complex argument. On the 'object' side, not only Bohm has argued (and Waddington in his summary accepts this) that physics itself has come to the end of its Democritean chapter, and so a new and richer synthesis may be in sight. But for one thing Bohm's position on physics still appears to most professionals as extreme heterodoxy, and on the other hand authoritative biologists, especially in molecular biology—both Crick and Watson, for example—still argue that a complete one-level, particulate

Bohm's metaphysics and biology

ontology is, if not already here, just around the corner. And often, I believe, even when they seem to moderate their position by pointing to the richer complexities of modern physics, biologists are just concealing, to themselves and others, their real reductivism behind a screen of Gibbs statistics, Volterra-Lotka equations and the like. There is still a long way to go before we are out of the woods.

3 Meantime there is a further fundamental disability in the Cartesian-Newtonian world view which was exhibited most beautifully at Bellagio, both in its representatives and in Bohm's manner of transcending it. This is a disability also stressed by Whitehead: the incapacity to develop an adequate conception of life.

Not only Descartes' *bête-machine* but the principal thrust of Kant's transcendental analytic make it plain that in terms of the chief modern tradition this must be so. At the very start of the first Critique Kant distinguishes between acts which I perform and so have as things in themselves, but cannot know, and intuitions (*Anschauungen*) which are passively present to me, and which alone can supply content for my knowledge. The knowable, in other words, is passive; agency cannot be known. But as Whitehead argued, as Suzanne Langer argues, as Bohm argues, what we know as alive we know precisely in and through its *activity*. Bohm adduces here Piaget's account of child development; one could cite also von Weiszäcker's *Gestaltkreis* or Goldstein's concept of preferred behaviour. This is also, I believe, the fundamental (if sometimes concealed) import of Waddington's stress on epigenesis: organisms *are* not simply, they *act*. And acts are, in Bohm's terms, *creative* processes: they bring into being an order that was not. But thought in the Cartesian tradition has restricted itself, in viewing the object of knowledge, to its passivity: to what Bohm calls equilibrium and dynamic processes, excluding creative process. In the beginning was God's creation, thereafter a 'clockwork' universe, but no creative creature. This is again a consequence, if you will, of the subject-object dichotomy, of which the logical outcome on the subjective side is the Sartrean for-itself, and on the object side Crickian reductivism. Bohm argues, of course, that even in physics a concept of creative process has been found indispensable; my point here is simply that without a concept of act, that is, of creative change of order, there can be no adequate concept of life.

But, it will be objected, that is just what the conference did move to develop: we had brilliant applications of automata theory, statistical dynamics, chemical engineering, and so on, to biological problems. Of course there is no limit to such applications, and in the appropriate context

64

they are all to the good. The question remains: what *is* the appropriate context ?

The context within which the most articulate participants at the Symposium approached their problem was thoroughly functional. Almost everyone used, constantly and as self-evident, the term 'biological' as synonymous with 'functional', 'adaptive', 'conducive to survival'—strictly, in evolutionary terms, conducive to leaving descendants—or 'produced by Natural Selection'. The best, indeed the perfect, specimens of this breed of thought were Longuet-Higgins, Gregory, and Maynard Smith; even Waddington, though not *quite* orthodox, appears in his summarizing notes as an interesting mutant of the same species.

The fundamental principles of this reigning form of biological thought are two: first, uniformitarianism extrapolated to the faith that ultimately all explanation is one-levelled in terms of least particulars. (It's all, after all, physics and chemistry— see Arbib's note on L.-H., p. 336) Secondly, that the only allowed supplementation of such a monolithic materialism is the reference to adaptation personified in the concept of Natural Selection, that is, of the 'mechanism' by which the less adapted are eliminated in favour of the better adapted. Now naturally, I hasten to say, I too cross myself when speaking of Natural Selection. Yet for all my efforts I am still unable to understand quite what it really means, especially when I hear from Dobzhansky about 'selection *sub specie aeternitatis*' or from Simpson that 'whatever we see, selection sees much more', and so on. Let me try once again.

Look at Longuet-Higgins' summarizing statement: 'The secret of life is the ability of living creatures to improve their programs.' 'Improve', when challenged, he altered to 'adapt'. Now this is plainly the neo-Darwinian *credo* using the contemporary tools of population genetics and information theory. Whatever computer techniques it embodies, what it adds to the Democritean platform—the ultimate reduction of all process to its simplest level—is simply the axiom of adaptivity: the thesis that living things are adaptation machines. This thesis is then supposedly sufficient to generate evolution. What does it mean and how does it work ?

Medawar recently stated the core of contemporary evolutionary theory as consisting in the two propositions: (i) that the terrestrial populations existing at a given future time will differ statistically in some degree from those existing today; and (ii) that the genetic constitution of those populations will have some connection with their changed phenotypic characters. We may add to these Waddington's emphasis on the role of the phenotype in the selective process controlling (i) and (ii). These are all perfectly harmless statements which no one, philosopher or otherwise, would want to challenge. But they tell us nothing at all about the epistemic relation between discourse about least particulars: gene pools or populations of gene pools, and discourse about cells or organ systems or organisms, nor about the ontological import of such discourse, all of which, in Longuet-Higgins' 'Of course in the end it's all physics and chemistry', they presuppose. Nor do they tell us how from a time when there were no living cells or organ systems or organisms there came to be such entities. When we try to go further here, we find a peculiar muddle, a muddle which needs to be disentangled before an adequate theory of evolution, emergent or otherwise, can be formulated.

I have tried to analyse this peculiar muddle a number of times elsewhere and so, of course, have numerous other people. Langer in her new book has some good arguments; and indeed Bohm's argument on the necessity of the transfunctional seems to me absolutely conclusive. But perhaps in my less imaginative way I may briefly make one more attempt.

Behind Medawar's harmless statement, the principal presupposition of modern evolutionary theory is the thesis formulated in 1932 by R. A. Fisher: 'Evolution *is* progressive adaptation, and consists in nothing else', the thesis echoed in Longuet-Higgins' 'self-adapting programs'. Yet adaptation on its own, or progressive adaptation on its own, as neo-Darwinism takes it, is by no means self-explanatory, or indeed explicable. For one thing, the concept of evolution as identified with progressive adaptation is basically ambiguous. On the one hand, such adaptations are supposed to be mechanically self-generating—through mutation, natural selection, recombination and isolation—and so to entail no teleological reference. Yet on the other hand, adaptation, like its Victorian twin, 'utility', is itself a teleological concept: it is adjustment of something to something for some end. To interpret organisms as adaptation machines, as neo-Darwinism does, therefore, is to interpret them as complexes of means for ends, and therefore teleologically. Secondly, if the means-end reference is admitted, one must ask further: means to what end? But the 'end', for Darwinism, dare not be some 'higher' form of life, the next 'level' to which evolution aspires: that would be to reintroduce a forbidden version of 'unscientific' teleology. It must then, and is usually said to, be *survival* that adaptation is 'for'. But in that case the whole 'theory' becomes a tautology, a complicated way of saying simply that what survives survives. (And Medawar's two propositions, which were stated in response to the accusation of tautology, simply open out the tautology a little way, only to let it form again when we try to interpret those propositions in their full theoretical context.) Finally, and fundamentally, the trouble is, really, that the concept of adaptation is essentially a relative one—relative to the existence of two things: the organism to be adapted and the *environment* to which it is to be adapted. Helmuth Plessner makes a similar point in an early work, *Die Einheit der Sinne*, in the context of an analysis of sense perception. Referring to the physiological investigation of perception, he points out that perception cannot be explained wholly in terms of adaptation, since there must be *something there already to become adapted. Adaptability*, which is a potential relation between organ, medium, and object of perception, must precede adaptation. Thus the adaptable entity, the organism which *can* achieve adaptation, must be assessed in its own right, by its appropriate norm, before the detailed conditions of its adaptation can be specified. The same is true in the context of evolutionary theory, or indeed wherever the attempt is made to rely on utility or adaptation as a principle of ultimate explanation. Adaptation is a crypto-teleological concept, but teleology even when explicit is itself dependent on the prior evaluation of the ends evoked. And in the case of evolutionary biology, the end is not simply survival but the survival of—a type, a mode of living, an order of orders, in Bohm's language, adjudged as significant in itself. That is the only judgment that can fill in the tautology of survival-for-survival-for-survival.

4 Such a judgment, however, you are all, or nearly all, unwilling to make, or to admit that you make, and that despite the undeniable force of Bohm's argument on function/transfunction. Why not?

There are temperamental reasons: the passion for model making, and social reasons: the prestige and power granted to machine makers in our society. But the fundamental reason lies, I believe, in the basic ambiguity of the concept of 'mechanism' itself. The world of the seventeenth century's 'new mechanical philosophy' was, apart from its Infinite Designer, a one-level universe, whose laws would ultimately be specifiable in terms of its 'hard, impenetrable' least parts. Its laws are, in Bohm's phrase, automorphic, expressing the simplest order of matter-in-motion. Take away the Designer and you have a self-regulating system, just what, three hundred years later, automata theorists have triumphantly learned to produce. But a machine, as Polanyi has demonstrated (both in *Personal Knowledge* and recently, August 1967, in *Chemical and Engineering News*) is, essentially, not a one-level, but a hierarchical system. It demands operating principles ordinally complementary to, i.e. depending on, but controlling, the laws of physics and chemistry which govern the behaviour of its parts. Because, however, we have put it together out of discrete parts which we control and

67

can specify, we can easily neglect this two-level structure and hold that physics and chemistry alone 'produce' and 'explain' the machine. The clockwork seems to make the clock. Now of course the clock needs its clockwork. But you cannot in terms of physics and chemistry alone say anything about telling time. You cannot in terms of physics and chemistry alone distinguish any message, whether the time of day or the hereditary program of an organism, from a noise.

Why is *this* message so hard to put across ? Because of the compulsion of Democritean thinking times the self-deception of 'utility'. Machine thinking, as I have argued for the case of evolutionary theory, is crypto-teleological. But when you make the end explicit, in engineering terms it is still a means; and you can keep running, like Yellow-Dog Dingo, without ever stopping to face the fact that *some* intrinsic value, some harmony, in Bohm's language, some timeless order, is the controlling principle which all the while governs your unending course. Or if you're pushed, it is still the minimal order, the maximally meaningless order, survival, or motion-for-the-sake-of-motion, that turns out, allegedly, to be your goal.

Bohm asks : Is nature more like an engineer or an artist ? Admittedly, the two have much in common. The artist must control his material craftsmanly : he is also an engineer. And the great engineer also achieves beauty. Moreover, the emphasis on engineering concepts as against *mere* physics and chemistry (see for example Gregory's *The Brain as an Engineering Problem*) is in fact an advance from a one- to a two-level ontology, and that is all to the good. But the difference in ends, and therefore in the logical structure (the order of orders) of the two enterprises is what matters here. The engineer makes artefacts, without intrinsic significance or intrinsic reality. They are in essence means to the perpetuation of what is. The artist makes new realities, richer orders that never were on land or sea, dependent of course on conditions specifiable in terms of lower levels, but neither predictable nor explicable in terms of them. Such harmonies, such emergent orders, have to be apprehended, not through manipulation of means for means for going on going, but through understanding : understanding, again in Bohm's terms, as the union of observer and observed in the presence of beauty.

To the physics or engineering minded, such formulations may well appear absurd, wildly metaphysical, 'subjective', 'irrational'. Of course, so did Galileo to the good Aristotelians. I can only point, in conclusion, to convergences in other contemporary writers with Bohm's cosmology. The three principal orders he specifies in his paper are identical with those distinguished by Merleau-Ponty in *The Structure of Behaviour,* and Merleau-Ponty like Bohm points to artistic creation as the paradigm on which we ought to lean if we would understand our own way of being (the order of intelligence). And of course the analogy of nature and art (not artefaction, or invention, but creation, artistic discovery) forms the central theme of Suzanne Langer's recent work. There is no doubt, in my view, that we must acknowledge and implement philosophically Peirce's insight: that while logic is subordinate to ethics, in the sense of practical or engineering knowhow, ethics *sive* engineering is subordinate to aesthetics; our sense of beauty, of the intrinsically meaningful, dominates, whether we will or no, our grasp of what is real, of what is worth making real or allowing to perish. We all seek, in our own way, as Plato saw, to achieve immortality, to find a timeless order, through begetting on the beautiful. And only he who has seen the beautiful itself 'can breed true virtue, since he alone is in contact not with illusion but with truth'.

Comments by C. H. Waddington

I want to make two brief comments on metaphysical points—but certainly not to attempt any full discussion of all the metaphysical issues raised.

1 A remark in passing to Marjorie Grene. I know that the fashion of the time is very much against notions which have such transcendental-sounding names as Whitehead's 'eternal objects'; but I shall require convincing that David Bohm's 'timeless orders' are either very different, or much more likely to prove acceptable to the philosophical establishment.

2 A more serious comment, to David Bohm. I agree with the argument, on page 44, that function implies the trans-functional, i.e. must ultimately be referred to something beyond the field of function. The question is, how is this trans-functional to be conceived of ? There is, I think, a danger that people may think that it must be not only trans-functional but also transcendental. David seems to me to come near to implying that in his discussion of 'vicious circles' ('they eat to live and live to eat'). Now I do not think all such 'circles' are vicious; and I doubt if we can find, and am reluctant to believe that we need, anything beyond such circular and therefore self-sufficient systems. In fact the title of a chapter in one of my books (*The Ethical Animal,* 1960) was 'The Shape of Biological Thought, or the Virtues of Vicious Circles'. What is unsatisfactory about David's 'Eat to live' example is not the circularity, but that the circle is too small, eat is inadequate to act as the complement to live. David seems to get nearer to my outlook when he writes (p. 45) that the reference 'beyond the field of function'

may be to a 'natural necessity', that is, to the fact that the world just is structured in that particular way.

I think the extent of my agreement with Bohm, and the directions in which I should like further clarification from him, may be best expressed if I quote two paragraphs from my discussion of 'function' in *The Ethical Animal*.

'When we assign a function to something, we in fact assert two propositions about it. First, that it forms part of a causal network; and secondly, that the results of the causal network, when observed over the range in which they are expressed, exhibit some general property. Another way of expressing the latter point is to say that the causal network is organized. The concept of function is in fact very closely connected with that of organization, and can be regarded as a derivative of that more general notion. Is organization, then, an illegitimate concept? It probably is so in terms of a crudely mechanical materialist picture of the world i.e. a picture in which we consider that all existing things can without loss be reduced to the movements and interactions of some ultimate constituent particles. But such a picture has never been more than a theoretical aspiration in biology, and is at present out of date even in the physical sciences. There are now, I think, few scientists who would consider it illegitimate to conclude that groups of elementary constituents may, by entering into close relationships with one another, build up complex entities which then enter into further causal inter-actions with one another as units. It is this fact, of the integration of groups of constituents into complexes, which in certain respects operate as units, which is spoken of as organization. In so far as it occurs, the concept of function is a legitimate one. If we have some complex entity A which acts as a unit, we can regard it as exhibiting organization of its constituent elements. Suppose that within A we can discern certain sub-units, P, Q, R, then the function of P within the organized system A is the contribution which P makes towards those types of behaviour in which the unitary character of A is exhibited' (p. 62).

The question whether the trans-functional, Bohm's 'timeless order', is also transcendental is of course of particular importance and has a particular urgency in the ethical connection with which I was concerned. I quote again: 'Adopting the usual terminology of biology, we can say that the function of ethicizing is to make possible human evolution in the socio-genetic mode. Now, once we have assigned the function to a general type of activity we have a rational criterion against which to judge any particular example of that activity' (p. 29).

'It is as well to consider what is implied by such a mode of approach by taking as an example some aspect of human activity which is less emotionally loaded than ethical beliefs. Consider for example the activity of eating. The human newborn infant has first to develop into the sort of creature that goes in for eating. In this development innate factors probably play a much greater role, and extrinsic factors a lesser one, than they do in the development of the infant into an ethicizing being, but this alteration in the relative importance of the two types of factors is of minor consequence in the present context. Next the child will acquire certain specific food habits, becoming accustomed to and accepting a particular diet. This is a process analogous to the development of specific formulated ethical beliefs. In order to find a basis for criticizing these food habits we have first to enquire what is the function of eating. We find that it is to make possible the growth of the body. Inspecting the growth of human beings on a wide basis, we discover that it manifests a general character which we describe as health. We can then ask of any particular food habit or diet how effective it is in bringing about healthy growth. The criterion we are applying here is one of general accordance with the nature of the world as we observe it. If any individual approaches a nutritionist and says that he prefers to grow in an abnormal and unhealthy manner, the nutritionist can do no more than tell him that if he does so he will be out of step with nature. The criteria, of biological wisdom in the case of ethics, or healthy growth in the case of eating, which can be derived in this way, are immanent in nature as we find it, not superimposed on it from outside. However, even if one considers that there is some overriding supernatural being from whom our ethical standards are ultimately derived, it is surely blasphemous to suppose that the nature he has created is such as to deceive us as to his true wishes. Thus, even an immanent criterion, if we have discerned it aright, would not contradict, though it might of course fall short of, a supernatural one' (p. 30).

The Practical Consequences of Metaphysical Beliefs on a Biologist's Work: an Autobiographical Note

C. H. Waddington
University of Edinburgh

Several of the more 'hard-headed' characters at the Second Symposium expressed from time to time, at cocktails or after dinner, a suspicion that metaphysical considerations of the kind introduced by David Bohm have ultimately no real impact on the directions in which science advances. They suggested that they were merely part of the froth churned up while the theoretical physicists flounder and thrash about trying to find a firm footing in the deep and dangerous waters of quantum theory, sub-nuclear particles, and the like ; and that when this footing has been found the froth will settle down and disappear. It is software, they suggested, so soft as to be deliquescent and ultimately evanescent.

I do not agree with this. I should like to argue that a scientist's metaphysical beliefs are not mere epiphenomena, but have a definite and ascertainable influence on the work he produces, by reminiscing for a moment about my own career. I am quite sure that many of the two hundred or so experimental papers I produced have been definitely affected by consciously held metaphysical beliefs, both in the types of problems I set myself and the manner in which I tried to solve them. I do not want to argue now—though I'll do so later if anybody wants me to— that these were really the most interesting problems, and that I set about them in the right way. Maybe my metaphysics was leading me up the garden path (though I don't think so) ; but the point I want to make now is that it was leading me somewhere and was therefore something more than a set of decorative flourishes on the proscenium arch, giving on to the stage in which the real action takes place.

As David Bohm points out, metaphysics is a sort of poetry ; and we are all talking poetry all the time—if you doubt it, look again at what Richard Gregory has to say about such a basic activity as perception. As poetry, metaphysics can be absorbed through communication-channels other than extended rational exposition. So, to begin this metaphysical-experimentalist's autobiography, I will mention two (or perhaps three) notions which infiltrated into my thinking at a very early stage, without much benefit of academic dignity, and which have remained there ever since.

When I was a schoolboy there was a peculiar period after you had taken your entrance or scholarship examinations to the university in December of year n, and before you actually went to the university in October of year n + 1. This was the time when your schoolmasters could really have fun with you. I had been in the classical sixth form, learning mainly Latin and Greek and a certain amount of mathematics; but when it dawned on me that all my friends were leaving and I had better leave too and go to the university, I decided I had better try to get in on the basis of chemistry. Fortunately we had a chemistry master, E. J. Holmyard, who was something of a genius of a teacher. During one summer holiday and autumn term, he taught me the whole of chemistry, at least enough to push me into the lowest grade of assisted entry to the university, an Exhibition. After that he could really break loose and teach me what he was interested in. His passionate interest happened to be the functions of the Alexandrian Gnostics, and the Arabic Alchemists derived from them, in transmitting both the philosophical ideas and the technical knowhow from the Greek civilization which expired around A D 200, to the European one which began to come alive at the Quattrocento. So he made me learn a smattering of Arabic and look at a large number of very odd late Hellenistic documents. Two ideas stuck:

▶ *The world egg* 'Things' are essentially eggs—pregnant with God-knows-what. You look at them and they appear simple enough, with a bland definite shape, rather impenetrable. You glance away for a bit and when you look back what you find is that they have turned into a fluffy yellow chick, actively running about and all set to get imprinted on you if you will give it half a chance. Unsettling, even perhaps a bit sinister. But one strand of Gnostic thought asserted that *everything* is like that.

▶ *The Ouroboros*, the snake eating its tail. This famous symbol, which is as well known in ancient China as in Alexandria, expressed the whole gist of feedback control almost two millenia before Norbert Wiener started 'creating' about the subject at M I T and invented the term 'cybernetics'. Here is a drawing of an ouroboros which I made for an essay which I wrote while I was still at school, presumably around 1923. I reproduce it because you will see that inscribed within the ouroboros is a third subsidiary notion; the slogan 'εν το παν,' 'hen to pan', 'the one, the all', a phrase which implies (in a cybernetic context, be it remembered) that any one entity incorporates into itself in some sense all other entities in the universe (Fig. 1).

Before these highly poetic metaphysics had any practical influence on my

73

Practical consequences of metaphysical beliefs

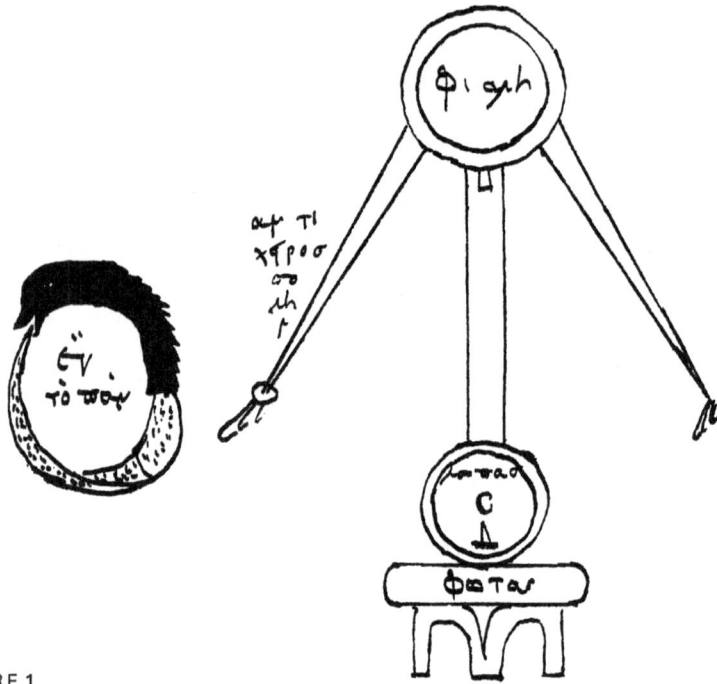

FIGURE 1
The Ouroboros, together with an alembic (distillation vessel) redrawn from an alchemical document known as the Chrysopeus (i.e. gold-maker) of Cleopatra.

scientific work, there was added to them a large body of much more explicitly rationalized thinking; in the first place that of Whitehead, to whose writings I paid much more attention during the last two years of my undergraduate career than I did to the textbooks in the subjects on which I was going to take my exams. Later this was joined by some infusions of thought which claimed to be materialist—either 'fancy' (dialectical), which preceded Whitehead and seemed to me to be in the main left behind by him; or 'crude', the prime example being Morgan and his school, who insisted that the gene is not just a logical construct from Mendelian ratios [cf. Woodger's definition

$$\text{mend} =_{Df} Aeq'\hat{x}\hat{y} (x, y \varepsilon \text{ whz.} \quad Apr'Zyg'x \uparrow K \upharpoonright Apr'Zyg'y \varepsilon 1 \rightarrow 1)],$$

but is just a simple lump of stuff. But one was anyway surrounded by materialists, and the whole of science was dominated by essentially Newtonian conceptions

74

of billiard-ball atoms existing at durationless instants in an otherwise empty three-dimensional space. It was, for me, Whitehead who suggested new lines of thought.

What was this Whiteheadian metaphysics ? I will sketch very briefly what were the salient features in my eyes.

1. The raw materials from which we start to do science—or with which we finish the scientific testing of a theory—are 'occasions of experience'.

2. An occasion of experience has a duration in time (cf. David Bohm, 'there are no things, only processes').

3. An occasion of experience is essentially a unity. Any attempt to analyse it into component parts 'injures' it in some way. Yet we cannot do anything with it unless we do analyse it. Our first step towards an analysis is to dissect the unity into an experiencing subject and an experienced object. The dividing line between these two is both arbitrary and artificial. It can be drawn through various positions, and wherever it is drawn it is never anything more than a convenience.

4. The content of any occasion of experience is essentially infinite and undenumerable. Moreover, wherever we draw the line between experiencing subject and experienced object, the latter will always remain undenumerable. If this were not so, we should merely have to describe the totality of the content of the experience, and the experience itself would be created.

5. The experience, which Whitehead refers to as an 'event', has, however, some definite characteristics. Whitehead refers to these as 'objects'—the word which is usually used, and has indeed been used above, in a quite different sense. Definiteness of the Whiteheadian objects in an event implies that, although the event has some relation to everything else past or present in the universe, these relations are brought together and tied up with one another in some particular and specific way characteristic of that event. (Whitehead was writing *before* quantum mechanics became a dominant influence in our thought, but compare these notions with the idea that a particle must also be thought of as a wave function extending throughout the whole of space time.) For this tying-together of universal references into knots with individual character, Whitehead used various different phrases at different periods in the development of his thought. For present purposes I am content to stay with the least far-reaching of these, when he spoke of the coming together of the constituent factors in an event as a 'concrescence'. Later he described the way in which an event here and now incorporates into itself some reference to everything else in the universe as a 'prehension' of these relations by the event in accordance with its own 'subjective

Practical consequences of metaphysical beliefs

feeling'. This is a metaphysics very close to that advocated by David Bohm when he speaks of creativity, and argues that nature is more like an artist than an engineer. Privately my own thought runs along similar lines ; and I think they may be extremely important to the way in which one behaves in one's whole personal life ; but I do not see that they have had any direct influence on the way in which I have conducted experimental work, which is the subject which we are discussing here. As far as scientific practice is concerned, the lessons to be learned from Whitehead were not so much derived from his discussions of experiences, but rather from his replacement of 'things' by processes which have an individual character which depends on the 'concrescence' into a unity of very many relations with other processes.

So, without going into further metaphysical sophistication, let's get down to brass tacks. What did I actually do as a practising biologist, and how was this influenced by this metaphysical background ?

I began work as a palaeontologist, studying the evolution of certain groups of fossils. And I chose, as my main interest, a group which forces on one's attention the Whiteheadian point that the organisms undergoing the process of evolution are themselves processes. The Ammonites were cephalopods, related to squids and the Nautilus, which laid down spiral shells. The animal occupied only the latest-formed part of the shell and from time to time moved forward a little, leaving behind it the part it had previously inhabited. Thus following the whorls of the spiral shell outwards, from the centre to the periphery, one has a record of the whole life-history of the animal ; it never appears just as an adult whose juvenile stages have vanished. The whole developmental process is preserved so that one cannot avoid examining it. And the process is, of course, complex, with many facets. On the surface of the shell there are 'ornaments'—ribs, knobs, tubercles, etc.—which change as development proceeds ; the cross-sectional shape of the tube may change, and so may the closeness with which it is coiled ; and behind each living-chamber a partition is laid down which meets the outer shell in a complex 'suture' which is one of the most characteristic features of a species (Fig. 2).

In most types of animal, in which the adult form of the individual replaces the younger stages, the only way to study the developmental history is to collect a large population containing juvenile as well as adult stages. Age can usually not be determined directly, but one can take some measure of size as an indication of it. Fig. 3 shows a graph I made, around 1927, of the variation in ratio of breadth to length in a collection of fossil shells (a Brachiopod, *Terebratula*).

76

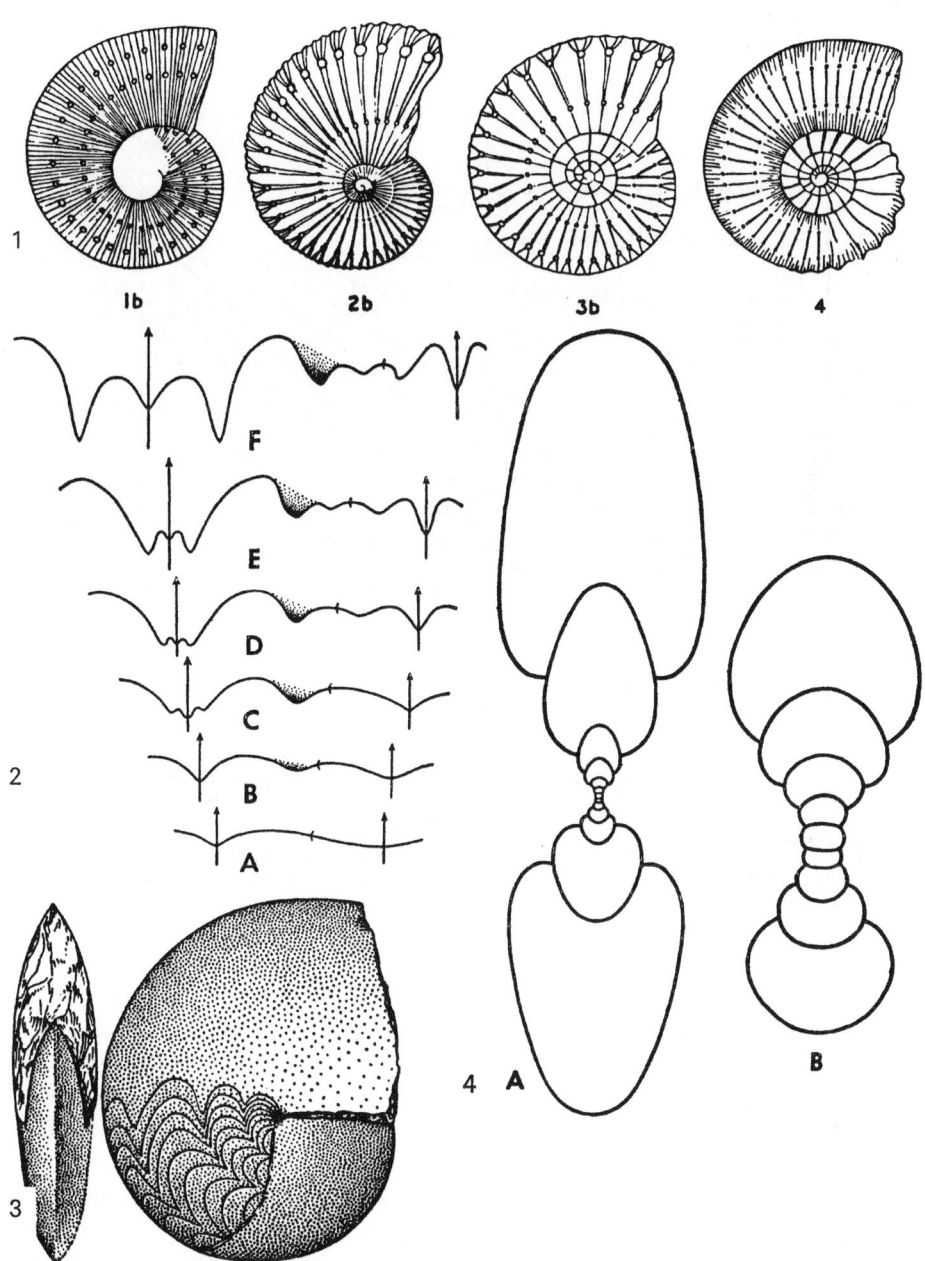

FIGURE 2
Ammonites. In 1 (above) 1b, 2b, 3b and 4 are an evolutionary sequence of species, to illustrate changes in the form of the ribs and knobs. 2 shows the development of the suture in an early species, from the young form A to the adult condition F. 3 illustrates a very tightly wound form, and shows the sutures. 4 is a cross-section of a loosely coiled species, to illustrate the change in the shape of the spiral tube, the younger stages being shown enlarged at B. (From *L. Mollusca.* In (ed. Moore) *Treatise on Invertebrate Palaeontology,* vol. 4. Geol. Soc. America and Univ. Kansas Press).

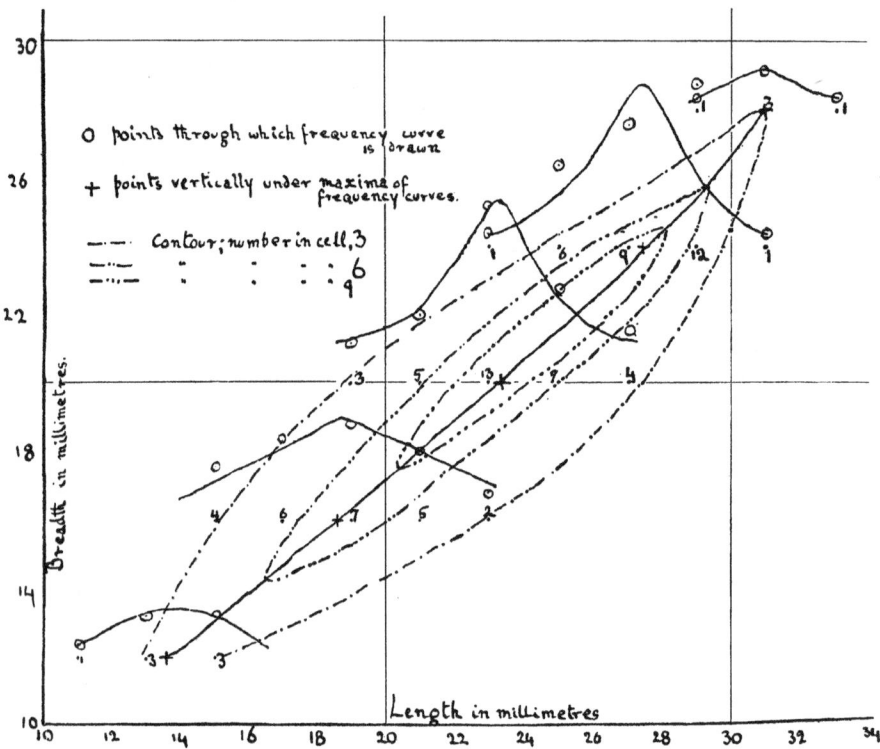

FIGURE 3
Frequency distribution of length and breadth in a population of fossil *Terebratula*, shown as a contour map with superposed frequency curves at five intervals. (From an unpublished exercise of my own, used in 1927 by my friend Reuben Heffer to try his hand as a printer in preparation for joining the family book shop.)

These early exercises left me with a deeply ingrained conviction that the evolution of organisms must really be regarded as the evolution of developmental systems—which was the title of one of the first articles I wrote about evolution when I returned to the subject some years later. It is of course related to such old ideas as Haeckel's 'biogenetic law'—phylogeny repeats ontogeny—but I took it also as a guiding principle in population genetics. I still think that when modern population geneticists express the variation in a population by means of a timeless frequency curve, which deals only with the adults, this is a simplification which needs justification—which of course it may often at least partially possess, for instance when one is dealing with the imagos of an insect like *Drosophila*,

78

which metamorphoses suddenly into an adult form which thereafter does not change. But when I started doing experiments on *Drosophila* evolution, in the forties and fifties, I treated even that insect as a developmental system, and by manipulating the environment in which it developed was able to uncover the rather novel process of genetic assimilation. Thus my particular slant on evolution — a most unfashionable emphasis on the importance of the developing phenotype — is a fairly direct derivative from Whiteheadian-type metaphysics.

In my early career there was a considerable period in which I was concerned directly with developmental systems themselves, rather than with their evolution. My approach to experimental epigenetics was again strongly influenced by Whiteheadian metaphysics, but also by genetics. In fact, when I decided that I wanted to do something more experimental than is possible in palaeontology, I first tried to become a geneticist. My first two published papers were, one in plant genetics, and another a collaboration with J. B. S. Haldane along classical neo-Darwinist lines, in which we studied the effect of inbreeding on the segregation of linked genes, writing down, and trying rather unsuccessfully to deal with, great series — up to 27 members, I think — of finite difference equations. So I had my taste of the thin gruel of mathematical formalism, as well as of the strong nourishing soup of time-extended populations of fossilized developmental systems. But my attempt to become a geneticist was a failure, because at that time in Britain there simply was no way in which one could earn a living at the subject. So I became an experimental embryologist. For some years my most immediate interest was in solving the technical problems of carrying out meaningful experiments on types of embryo which no one else had tackled successfully, such as birds and mammals, or of doing biochemical work on the very small tissue-fragments isolated from various regions of an egg.

However, the theoretical structure of the subject also needed a good deal of attention, and this is where the Whiteheadian approach came in. In the twenties, T. H. Morgan had argued that epigenesis should be considered in terms of the activities of genes, but he had little effect on the workers in this field. Embryological theories involved such notions as the 'segregation' or 'differential dichotomy' of 'potencies' to develop into organs such as the nervous system, kidneys, gut, etc.; or, at a more advanced but still very imprecise level, notions such as 'the organizer', 'induction', and the like. I wanted to return to Morgan's idea that the only 'potencies' it is meaningful to talk about are the potential activities of genes. So did several people who were primarily geneticists, but who had become interested in development without having actually worked on it very much.

Practical consequences of metaphysical beliefs

Two rather radically different lines of approach were followed. The one which was—and in fact still is—favoured by most geneticists depended on what I think may be called an 'atomistic' metaphysics. It set out from the assumption of the existence of single genes, and it asked, at first, what does A do, and, later, what controls whether gene A is active or not ? There is only space to mention, just to remind you, some of the key figures. Goldschmidt concluded that genes control rates of processes ; Muller that they manufacture more, or less, or none of some substance, or sometimes an anti-substance, or even a quite new substance. Garrod (on human metabolism), followed by Haldane (on flower pigments), led on to the identification by Beadle and Ephrussi, in the early 1940s, of these substances as enzymes. Again in the forties, Hadorn in *Drosophila* and Grüneberg in mice studied the manifold developmental consequences which may follow an alteration in a single gene. Finally, about 1960, Jacob and Monod produced their story of the mechanism which controls the activity of single genes (or small operon groups) in prokaryotic organisms in which the chromosome normally lacks protein.

Clearly, this line of approach has paid off very well. But to my mind it does not really deal with the questions which the epigeneticist faces. For one thing, there is the difficulty that the Jacob-Monod control system could scarcely work, without considerable modification, in cells with proteinaceous chromosomes. But the problem is deeper. In cells of higher organisms we are not usually, if ever, confronted by the switching on or off of single genes. What we find is a whole complex cell becoming either a nerve or a kidney or a muscle cell In the late thirties I began developing the Whiteheadian notion that the process of becoming (say) a nerve cell should be regarded as the result of the activities of large numbers of genes, which interact together to form a unified 'concrescence'. This line of thought had several ramifications. For instance, just before the Beadle-Ephrussi era I showed in detail how the development of the wing of *Drosophila* is affected by the activities of some forty different genes. Again a few years earlier it had become apparent that the 'gene-concrescence' itself undergoes processes of change ; at one embryonic period a given concrescence is in a phase of 'competence' and may be switched into one or other of a small number of alternative pathways of further change—but the competence later disappears and if you've missed the bus the switch won't work. My main pre-occupation, however, was with the nature of the switches, which was the subject of the experimental biochemical work I was doing, with Needham and others, on 'embryonic induction'. Influenced—probably over-influenced—by genetics. I insisted that the switch must have sufficient specificity to recognize particular genes. We showed that, in these terms, the specificity resides inside the cells which react to induction—we called it 'the masked evocator'. This is very similar to the situation discovered by Jacob and Monod many years later in bacteria, where again the specific repressor molecules are internal to the cells which react to enzyme-inducing substances. If I had been more consistently Whiteheadian, I would probably have realized that the 'specificity' involved does not need to lie in the switch at all, but may be a property of the 'concrescence' and the ways in which it can change. Because of course what I have been calling by the Whiteheadian term 'concrescence' is what I have later called a *chreod*, a notion which René Thom has explicated; and the switches are Thom's *catastrophes*. The specificity *need* not be in what precipitates the catastrophe, but could reside only in the possible stable regimes (limit cycles in the simplest case) into which the system could be flipped.

Let us return to the beginning. The whole of this, I am afraid rather long-winded, exposition has not been aimed at showing that my line of thought, which I have derived from Whitehead, is the correct or best one. I should in fact admit that it has not so obviously paid off as the 'what does a single gene do ?' line ; although I also continue to feel that it tackles deeper, because more embracing, problems. However, the point was to illustrate the fact that metaphysical presuppositions may have a definite influence on the way in which scientific research proceeds. And that point I have, I think, established, even if you feel that my metaphysics has led me up the garden path. And, after all, I am a biologist; it is plants and animals that I'm interested in, not clever exercises in algebra or even chemistry. The garden path has its attractions for the likes of us, and all of us who want to understand living systems in their more complex and richer forms are fated to look like suckers to our colleagues who are content to make a quick (scientific) buck wherever they can build up a dead-sure pay-off.

Since I am an unagressive character, and was living in an agressively anti-metaphysical period, I chose not to expound publicly these philosophical views. An essay I wrote around 1928 on 'The Vitalist-Mechanism Controversy and the Process of Abstraction' was never published. Instead I tried to put the Whiteheadian outlook to actual use in particular experimental situations. So biologists uninterested in metaphysics do not notice what lies behind—though they usually react as though they feel obscurely uneasy—and philosophers like Marjorie Grene may get so far as to conclude (p. 65) that I am not a wholly orthodox mechanical materialist.

The Status of Neo-Darwinism

J. Maynard Smith
University of Sussex

Introduction. By Darwinism is meant the idea that evolution is the result of natural selection. Neo-Darwinism adds to this idea a theory of heredity. In its most general form, the theory of heredity is Weismannism, i.e. it is the theory that changes in the hereditary material are in some sense independent of changes in the body or 'soma'. In particular, the theory of heredity is Mendelian, i.e. it assumes that heredity is atomic, and obeys either Mendel's laws or some modification of them explicable in terms of the behaviour of chromosomes (e.g. linkage, polyploidy).

There are two reasons for discussing neo-Darwinism at this conference. The first is that only in the study of evolution is there a body of biological theory in any way comparable to the theories of physics; a conference on theoretical biology can hardly refrain from discussing it. The second is that the theory in at least some formulations is tautological. 'The survival of the fittest' appears to mean merely that survivors survive. There seems little point in trying to explain evolution by a tautology.

In this article, therefore, I shall attempt first to formulate the theory of a non-tautological form. I shall then discuss what types of observation might refute it, because this is the best way of seeing whether the attempt at a non-tautological formulation has been a success. Finally, at a less philosophical level, I shall discuss what problems the theory can cope with, what problems are at present unsolved because of a lack of data or of adequate mathematical tools, and what problems seem at present inaccessible to solution without introducing new concepts,

▶ *The formulation of neo-Darwinism.* The main task of any theory of evolution is to explain adaptive complexity, i.e. to explain the same set of facts which Paley used as evidence of a Creator. Thus if we look at an organism, we find that it is composed of organs which are at the same time of great complexity and of a kind which ensures the survival and/or reproduction of their possessor. Evolution theory must explain the origin of such adaptations.

At the outset we are faced with a difficulty: we have no way of measuring the degree of complexity of a structure. Thus although most of us would readily agree that the organs of a man are more complex than those of an amoeba, and those of an amoeba more complex than those of a bacterium, we have no agreed

82

criteria on which to base this decision, and no way of deciding by how much one organism is more complex than another.

It may therefore seem odd to start formulating a theory of evolution by introducing a term which cannot be fully defined. However, I see no escape from doing so. If organisms were not both complicated and adapted, living matter would not differ from dead matter, and evolution theory would have nothing to explain.

Evolution is explained in terms of three properties, multiplication, heredity and variation, which organisms can be observed to possess. They will be considered in turn.

(i) *Multiplication.* All living organisms are capable of increasing in numbers in at least some environment. Multiplication is necessary because without it natural selection is impossible; you cannot cull a herd which can only just maintain its numbers. It is a corollary of this condition that life must consist of individuals and not of a continuum if it is to evolve.

(ii) *Heredity.* Briefly, like must beget like. More precisely, before we can say that entities have heredity, a number of different kinds of entities, A, B, C, etc., must exist, and each must tend to produce offspring like itself. Thus fire, if supplied with fuel, will multiply; but it does not have heredity, because the nature of a fire is determined by the fuel it is burning, and not by the nature of the fire from which it was lit.

(iii) *Variation.* If heredity were perfect, evolution would be impossible. Occasionally an offspring must differ from its parent. Viewed in this light, variation is merely the unreliability of heredity, and as Pattee [1] has emphasized, the problem is to explain why the reliability of replication is so high, not why mistakes are sometimes made.

However, this is not the whole story. If variation is to lead to evolution, then some variations must alter 'fitness', and at least some of these must increase fitness. By fitness is simply meant the probability of survival and reproduction. A melanistic moth is by definition fitter if it is more likely to survive, and a myopic man may be fitter if his myopia enables him to escape the draft. [Much confusion has arisen because 'fit' is not used in this sense in the phrase 'the survival of the fittest'. If it were so used, the phrase would indeed be tautological. A more precise though less elegant (and hence less 'fit') phrase would be 'the survival of the adaptively complex', i.e. organisms are adaptively complex or, as Bohm might say, 'harmonious', because such organisms survive better than less harmonious ones.] It follows from this definition that fitnesses can only be compared in a specified environment or range of environments.

Given entities with those properties, variants of higher fitness will replace their less fit ancestors: according to neo-Darwinism, this replacement constitutes evolution. Very early in evolution there arose a distinction between 'genotype' and 'phenotype', because those genotypes which gave rise to a phenotype were fitter than those which did not. By 'genotype' I mean that part of an organism which is replicated; by 'phenotype' I mean a structure or sequence of structures developing under the instructions of the genotype, and whose function it is to ensure the replication of the genotype. (These are not quite the accepted meanings of the words in genetics, but I would rather misuse words than invent new ones.) To paraphrase Butler's unexpectedly perceptive remark that the chicken is the egg's means of ensuring the production of another egg, the phenotype is the genotype's way of ensuring the production of another genotype. Once there is a distinction between phenotype and genotype, there is a process of epigenesis, and a process of decoding whereby the instructions in the genotype are translated into the structure of the phenotype.

Something must now be said about the origin of new variations, i.e. of mutation. It has been said that mutation is 'random'. Apart from the difficulty of defining the word, the statement is in one sense untrue, because different mutagenic agents produce different kinds of change in the genetic material.

Observation suggests that two things are in fact true about mutation: (i) most mutations lower fitness. If this were not so, evolution would proceed without natural selection; (ii) if a variant phenotype arises because development occurs in a changed environment, this will not produce corresponding changes in the genotype, such as to give rise in the next generation to the variant phenotype. This is the Weismannist assumption, expressed colloquially by saying that acquired characters are not inherited. Note that it does *not* say that changes in the phenotype cannot cause mutations, because of course they can. The apparent randomness of mutation arises because genotype and phenotype are connected by an arbitrary code.

The Weismannist assumption is expressed in molecular terms in the 'central dogma', which states that information can pass from DNA to protein, but not from protein to DNA; more precisely, if a new kind of protein is introduced into a cell, this cannot direct the synthesis of a new DNA molecule able to direct the synthesis of more of the new protein.

So far I have been describing a set of properties of organisms or, more precisely, a set of properties which neo-Darwinism assumes all organisms to have. This is not by itself a theory of evolution. The theory of neo-Darwinism states that these properties are necessary and sufficient to account for the evolution of life on this planet to date.

The limitations of time and place are important: of time, because in future we shall doubtless control our own evolution and that of our domestic animals and plants by the direct biochemical manipulation of DNA; of place, because we have as yet no grounds for asserting that if evolution has occurred elsewhere in the universe, it has done so by neo-Darwinist processes, although I would be willing to conjecture that it has.

It may help to clarify my position if I say that I accept Bohm's (this symposium) argument that to understand biological function we must appeal to what is beyond function, and also his statement that 'current metaphysics' (= neo-Darwinism) appeals to survival value as the trans-functional feature. Where I think he goes wrong is in regarding this procedure as tautological. He has been misled by the phrase 'the survival of the fittest'. Of course Darwinism contains tautological features: any scientific theory containing two lines of algebra does so. That it is not tautological *in toto* is best demonstrated by showing that it can be falsified, as I will now try to do.

▶ *Possibilities of refuting neo-Darwinism.* If this formulation of neo-Darwinism is not tautological, it must be possible to suggest observations which would refute it. Such observations could take two forms: (i) it could be shown that the assumptions made by neo-Darwinism are not in fact true of all organisms; (ii) patterns of evolution may occur which are inexplicable on the neo-Darwinist assumptions.

The possibilities will be considered in turn. It seems unlikely that we can show that organisms do not multiply or do not vary. However, the assumptions about heredity and about the origin of new variation could readily be disproved if they are false. Thus it should be possible to demonstrate Lamarckist effects if they occur, or 'inertial' effects whereby if one mutation in a given direction has occurred the next mutation is more likely to be in the same direction, or even 'teleological' effects whereby a succession of mutations occur which are individually non-adaptive but which together adapt the organism to a new environment.

By and large, such types of mutational event seem not to happen, and it is difficult to see in molecular terms how they could happen. It is impossible to *prove* that they do not, just as it is impossible to prove that heat never flows from a cold body to a hot one. All one can do is to assume they don't until someone demonstrates that they do.

The status of neo-Darwinism

I will turn now to the possibility that patterns of evolution occur which cannot be explained by neo-Darwinism. The first possibility is that evolutionary changes occur more rapidly than can be explained by neo-Darwinism. This would be quite easy to demonstrate if it occurred on a small scale in the laboratory. Thus suppose for example a population of fruit flies were kept at an unusually high temperature. By measuring the genetic variance of temperature tolerance in the population before starting, it would be possible to predict the maximum rate at which temperature tolerance would increase. If in fact it increased faster than this, then the population would have evolved by a mechanism other than neo-Darwinism.

However, most critics of neo-Darwinism accept that the theory works at the level of laboratory experimentation. They suggest instead that there are large-scale features of evolution which call for additional types of explanation. Here we are up against the difficulty that we do not understand epigenesis, and we therefore do not know how many mutations would be necessary, for example, in the genotype of a small dinosaur to turn it into a bird. Therefore we do not know how many generations of selection would be needed to produce the change. This difficulty, combined with the imperfections of the fossil record, mean that we are unlikely to be able to disprove neo-Darwinism by showing from an examination of the fossil record that evolution has proceeded too rapidly. All one can say is that where we do have a reasonably continuous record, the observed rates of change are many orders of magnitude slower than those which can be produced in the laboratory.

If, however, neo-Darwinism were false, one would expect to be able to demonstrate its falsity by examining the end-products, i.e. existing organisms. Thus it follows from neo-Darwinism that if we find an adaptively complex organ, then the organ will contribute to the survival or reproduction of its possessor. One apparent exception arises in cases such as the worker bee, which have organs favouring the survival of their close relatives; but since their close relatives share many of their genes, this is explicable on the grounds that the phenotype of the worker bee ensures the multiplication of its own genotype.

If one invents counter-examples, they seem absurd. Thus if someone discovers a deep-sea fish with varying numbers of luminous dots on its tail, the number at any one time having the property of being always a prime number, I should regard this as rather strong evidence against neo-Darwinism. And if the dots took up in turn the exact configuration of the various heavenly constellations, I should regard it as an adequate disproof. The apparent absurdity of these examples only shows that what we know about existing organisms is consistent with neo-Darwinism. It is of course true that there are complex organs whose function is not known. But if it were not the case that most organs can readily be understood as contributing to survival or reproduction, Darwinism would never have been accepted by biologists in the first place.

Thus there are conceivable observations in the fields of genetics, of evolutionary changes in the laboratory, and of physiology, which could disprove neo-Darwinism. In palaeontology, although there is perhaps no possibility of a formal disproof because in our present state of ignorance about epigenesis it would always be possible to argue that a sudden evolutionary change was due to a single mutation, there are in practice many conceivable observations which would throw grave doubts on the theory. It therefore seems to me absurd to argue that the theory is tautological, though I readily admit that it is often formulated tautologically.

At present there are in my opinion no adequate observational grounds for abandoning the theory. This is of course no reason for not seeking for such grounds—I am all for people looking for Lamarckian effects, or for exceptions to the central dogma. But in the meanwhile, the theory explains so much that it is impossible to operate in biology without accepting it, just as it is impossible to operate in physics without accepting Newtonian mechanics, or some other theory which subsumes Newtonian mechanics as a special case.

▶ *The successes and failures of neo-Darwinism.* I have suggested that neo-Darwinism has not as yet been refuted. But is it of any interest ? Does it tell us anything not immediately obvious ? Does it solve problems ? Since the ability to solve problems seems to me one of the essential characteristics of a scientific theory, these questions are important. Perhaps the easiest way of answering them is to list some of the problems which can be thought about within the context of neo-Darwinism, and which would be unanswerable in any other context. This is not to say that all these problems have been solved—I would say that the first four have been largely solved, the fifth only partly solved, that we lack essential data for the solution of the sixth, that there are accepted but partially erroneous solutions to the seventh and eighth, and that both conceptual difficulties and lack of information prevent the solution of the last:

(i) How rapidly will gene frequencies change under selection ?

(ii) How can one predict the effects of selection for a continuously varying character?

(iii) What processes are responsible for the genetic variability of sexually reproducing species ?

(iv) How many selective deaths are needed to replace one gene in a population by another ?

(v) Will selection bring genes affecting the same character on to the same chromosome ?

(vi) Can selection be responsible for the evolution of characters favourable for the species but not to the individual ?

(vii) Can one species divide into two without being separated by a barrier to migration ?

(viii) In what circumstances will sexual reproduction accelerate evolutionary change ?

(ix) Has there been time since the pre-Cambrian for selection to program the length of DNA known to exist in man ?

These problems—except for the last—illustrate the field within which neo-Darwinism has been successful. Even when problems are unsolved, this is because of a lack of data or of mathematical technique rather than of concepts.

The failures of neo-Darwinism arise because of the absence of theories in the adjacent fields of epigenesis and of ecology. Lacking a theory of epigenesis, we cannot say how many gene substitutions are required to convert a fin into a leg, or a monkey's brain into a human one. Consequently we cannot say how many generations of selection of what intensity were needed to produce those changes. There is one exception to this statement of ignorance. Knowing the genetic code, we also know how many mutational steps are needed to convert one protein into another. Consequently we can speak with more precision about the evolution of proteins than about the evolution of legs or brains.

The difficulties which arise from our ignorance of ecology can best be illustrated by discussing the related problem of whether neo-Darwinism can explain the evolution of increasing complexity. Neo-Darwinism predicts that *in the short term* individuals will change in such a way as to increase their fitness in the environment or range of environments existing at the time. This may lead to an increase or a decrease in complexity. Sometimes, as in the evolution of tapeworms or viruses, it has led in the direction of decreasing complexity, albeit in an increasingly complex environment.

Thus there is nothing in neo-Darwinism which enables us to predict a long-term increase in complexity. All one can say is that since the first living organisms were presumably very simple, then if any large change in complexity has occurred in any evolutionary lineage, it must have been in the direction of increasing

88

complexity; as Thomas Hood might have said, 'Nowhere to go but up'. But why should there have been any striking change in complexity ? It is conceivable that the first living thing, although simple, was more complex than was strictly necessary to survive in the primitive soup, and that evolution of greater fitness meant the evolution of still simpler forms.

Intuitively, one feels that the answer to this is that life soon became differentiated into various forms, living in different ways, and that within such a complex ecosystem there would always be *some* way of life open which called for a more complex phenotype. This would be a self-perpetuating process. With the evolution of new species, further ecological niches would open up, and the complexity of the most complex species would increase. But this is intuition, not reason. It is equally easy to imagine that the first living organism promptly consumed all the available food and then became extinct.

What we need therefore is first a theory of ecological permanence, and then a theory of evolutionary ecology. The former would tell us what must be the relationships between the species composing an ecosystem if it is to be 'permanent', i.e. if all species are to survive, either in a static equilibrium or a limit cycle. In such a theory, the effects of each species on its own reproduction and on that of other species would be represented by a constant or constants. We want to know what criteria these constants must satisfy if the system is to be permanent. A start on this problem has been made by Kerner [2] and Leigh [3].

In evolutionary ecology these constants become variables, but with a relaxation time large compared to the ecological time scale. Each species would evolve so as to maximize the fitness of its members. If so, a permanent ecosystem might evolve into an impermanent one. For example, a predator-prey system might be permanent because the prey could burrow and so escape total extinction. But if the predator evolved the capacity to burrow too, the ecosystem would become impermanent.

What then are the criteria to be satisfied if an ecosystem is not only to be permanent, but is to give rise by evolution to permanent ecosystems of greater species diversity ? We have no idea. But the first living organism, with its food supply, had to comprise such an ecosystem if evolution was to lead to increasing complexity.

References

1. Pattee, H. H. *Towards a Theoretical Biology 1* (Edinburgh University Press 1968).
2. Kerner, E. H. *Bull. Math. Biophys. 19* (1957) 121–46.
3. Leigh, E. G. *Proc. Nat. Acad. Sci. 53* (1965) 777–83.

Addendum on Order and neo-Darwinism by David Bohm

After finishing my remarks on order for the Bellagio Conference I remembered a very interesting conversation with Maynard Smith, the significance of which suddenly struck me. So it seemed appropriate to add a brief discussion of what we talked about, because it so nicely brings out what I want to say about the effects of tacit metaphysical assumptions on the whole course of one's thinking.

In this discussion I shall *underline* metaphysical words such as *all, only, always, never, basic, relevant, significant,* etc.

The conversation in question began when I recalled an article that I had read about population limitation by birds. In this article it was stated that birds (and other animals) had somehow evolved methods of controlling population so as not to outrun their food supply. In particular it was noted that not all the birds mated, and that when the flock was large, the fraction that mated was correspondingly reduced. To explain this behaviour it was proposed that it was related to the swarming of the entire flock, which occurred before mating. The suggested explanation of population limitation was, then, that when the birds swarmed, each one could note how large the flock was. The perception would somehow influence the propensity of each bird to mate, in such a way that when the flock was large the fraction would decrease.

This explanation was very severely criticised by Maynard Smith on the basis of neo-Darwinist conceptions of the evolutionary process. In essence, the argument was that those birds that tended not to mate in given circumstances (e.g. the perception of a large swarm) would not pass on this characteristic to succeeding generations. On the other hand, those birds that did tend to mate would pass on the tendency to mate. So in the long run all birds would tend to mate, no matter how large the swarm. Therefore the explanation cannot be accepted as a possible one.

In saying this, Maynard Smith emphasized very strongly that hereditary characteristics are passed on *only* in the individual organism, and *never* by any collective entity that could be called the 'species' or the 'group'. Indeed, the notion that any given characteristic has survival value for the species or the group as a whole was implied to have no *direct relevance* for giving an account of how the genetic process actually takes place, because there is no way *at all* in this process for the state of the species or the group to play a causative role. After all, one would hardly want to assume, for example, that each individual

Addendum by David Bohm

bird has some 'altruistic' tendency to put the welfare of the flock first, when it comes to determining whether he will mate or not. Such behaviour might conceivably occur in individual human beings, who can understand abstractions such as 'the community as a whole', but almost certainly, individual birds cannot formulate such abstractions.

At the time, I found Maynard Smith's arguments very convincing. But now I can see that they are based on a number of tacit metaphysical assumptions which could perhaps usefully be questioned. In essence, these assumptions are: (1) Changes in the genotype are *always* related to the changing experiences of the phenotype in a *completely fortuitous* way; (2) Systematic changes in the behaviour of the phenotype are *always* the necessary results of changes in the genotype, and *never* occur without such changes.

In explanation of assumption (2) it should be stated that, of course, no one would do something so absurd as to suppose that *every* change in the phenotype has to be caused by a corresponding change in the genotype. What is assumed in neo-Darwinism is, as far as I have been able to tell, that changes in the phenotype not thus caused by changes in the genotype have a *completely fortuitous* relationship to the transmission of hereditary characteristics.

Thus, if different birds had different tendencies to refrain from mating when the flock is perceived by them to be large, these tendencies would either have to correspond *only* to genetic differences between the individual birds, or else they would have to be *totally fortuitous* in the way in which it entered into the genetic process as a whole. If the tendency is determined *only* by the genotype of the individual bird, then, as has been seen, that genotype cannot survive, and if it is *entirely fortuitous,* then by hypothesis, there is no means by which it could be transmitted.

Let us now see whether we cannot get outside of these rather limited metaphysical possibilities, in a way that does not violate our feelings as to what seems reasonable or possible, in the light of all that has been learned over the past few hundred years. To do this, let us begin by supposing, for the sake of argument, that the genetic constitution of the birds is such that the tendency to mate in each individual is not completely determined by the genetic structure alone. That is to say, we allow room for variations or fluctuations in this tendency, even among birds that all have the same genes. However, instead of supposing that the variations or fluctuations in the behaviour of individuals is subject to an *absolute* randomness (disorder, lawlessness, etc.) we propose that they are *only* 'relatively' fortuitous. That is to say, they are fortuitous in relationship to the genetic constitution of the individual birds, but not in relationship to the actual situation of these birds in their overall environment.

Let us assume, for example, that the propensity for mating of a given bird depends not only on the size of the flock when swarming took place, but also on his position in the flock, as well as on myriads of other factors, such as what food he has eaten, what is the state of the wind as it ruffles his feathers, etc. So even in birds of the same genetic constitution, there would inevitably be 'scatter' in mating behaviour, which would however vary systematically in such a way that a large flock would tend to reduce its rate of mating.

Of course, one could use the theory of probability, with its 'normal distributions' to *describe* such a 'scatter'. But at present we conventionally refer also to *laws* of probability, implying thereby a kind of 'ironbound' necessity for *complete randomness, disorder,* and *lawlessness* in the variations and fluctuations of behaviour of individual birds that make up the 'scatter'. Once we realize that probability theory is *only* a *conventional* (as well as approximate) *description* and not a *law,* we are free to consider limits in the degree and kinds of fortuitousness of this behaviour, which lead to radical changes in the meaning of the whole picture.

One of these changes is that hereditary characteristics can now belong to the group or species without being analytically deducible from properties of the individual considered in isolation from the group. For the genetic constitution of the individual may determine *only* a range of possible behaviour. The actual behaviour of each individual is fortuitously dependent on the *total* environment, and this latter can include the *actual* group. For example, the tendency of birds to refrain from mating may prevent the group from exhausting its food supply. Because *all* the individuals have, in this regard, the same genetic constitution, those that mate do not thereby propagate to their offspring any more tendency to mate than is possessed by those birds that refrained from mating.

In thus noting the possibility of going outside current metaphysical assumptions I do not wish to blame or criticize those scientists who hold particular views, such as neo-Darwinism. Indeed, such views have evidently been very fruitful, and have helped lead people to many useful and interesting discoveries. But now it may well be necessary to give attention to various tacit restrictions that could be getting in the way of further progress.

In this regard, the situation in biology is not really basically different from that prevailing in other sciences, such as physics. For example, quantum theory has certainly been a gigantic step forward. Nevertheless, it contains the tacit metaphysical assumption that when the quantum state (i.e. the wave function) is determined, then the fluctuation in behaviour of individual atoms is *completely random*, and must remain so, no matter what question the physicist may come to inquire into, nor what conditions may come to be established for these atoms. So, in effect, there is an assumption of a *law of lawlessness* (which is evidently in some ways an inherently self-contradictory notion). Would it not be more reasonable to suppose instead that the behaviour of individual atoms is fortuitously related to the quantum state, and that it might therefore be non-fortuitously related to other things (some of which may perhaps be thought of only in the future) ? Then, both for the birds and for the atoms, we could consider that some of the factors responsible for fortuitous variations could change systematically in certain (generally as yet unknown) contexts. It is clear that it will be useful to remain alert to these extended possibilities rather than to close our minds with tacit assumptions of whose metaphysical character we are in general not even aware.

Comment on Bohm's Addendum by J. Maynard Smith

The example which Bohm discusses serves very well to bring out the difference between us. The theory of population regulation which Bohm mentions is due to Wynne-Edwards [1]. My own interest in the theory was aroused because on the one hand it appeared, for the reasons given by Bohm, to contradict neo-Darwinism, but on the other hand it provided an explanation for certain phenomena not otherwise easily explicable. I am enough of a Popperian to think that a proponent of neo-Darwinism should pay particular attention to observations or ideas which appear to contradict his theory.

In thinking about Wynne-Edwards' ideas I was led to make almost precisely the assumption about the effects of genotype and environment on breeding behaviour suggested by Bohm. I considered [2] the evolution of a species divided into reproductively isolated groups, each group consisting of genetically identical individuals. Some groups consisted of individuals which would refrain from breeding in certain circumstances ; thus when a group grew too large, some members would refrain from breeding, but which members refrained would depend, not on genetic constitution which was the same for all members of the group, but on such things as age, position in the flock, etc. It was supposed that groups consisting of such individuals were at a selective advantage, because less likely to outrun their food supply, over groups consisting of genetically different individuals which continued to breed irrespective of circumstances.

93

The status of neo-Darwinism

Thus although I considered the possibility mentioned by Bohm, I also took into account the possibility that there would be genetic as well as environmental differences in readiness to breed. I think Bohm makes the tacit metaphysical assumption that all individuals are genetically identical. The justification for taking genetic differences into account is that I know of no phenotypic character for which there is not a genetic component of the variance. The crucial question is therefore this: granted that there are genetic differences (e.g. in the density which must be reached before an individual in given circumstances ceases to breed), how will the species evolve? The answer depends, among other things, on the size of the groups and the amount of migration and interbreeding between them. As it happens, we have little information about populations from this point of view, and it was for this reason that in my article on 'the status of neo-Darwinism' (written before I saw Bohm's addendum) I listed the problem of the evolution of 'altruistic' traits as one for whose solution we lack factual information.

Now there is nothing in all this which runs contrary to neo-Darwinism. It may well be that the particular model I chose for the evolution of an altruistic trait was the wrong one. What is important is that it is possible within the framework of neo-Darwinism to analyse the evolution of population regulation, of altruistic traits, and of the related phenomenon of sexual reproduction, and to make fairly precise statements about the conditions which must be satisfied if particular characteristics are to evolve. But if the Weismannist assumption were relaxed, it would not be possible to say anything of comparable precision. It follows that if experimental evidence were to oblige us to abandon Weismannism, our theory of evolution would lose most of its power to make precise and testable statements.

Thus while I agree with Bohm that we should make our assumptions explicit, that we should question them, and that if the evidence requires it we should alter or abandon them, I cannot see why he should think that 'it may well be necessary to give attention to various tacit restrictions, that could be getting in the way of further progress'. I think science progresses by making assumptions more restrictive, not less so. It would be as easy to abandon Weismannism in genetics as to abandon the law of conservation of momentum, but the result in both cases would be a loss and not a gain in explanatory power.

References

1. Wynne-Edwards, V. C. *Animal Dispersion in relation to Social Behaviour* (Oliver and Boyd 1962).

2. Maynard Smith, J. *Nature, 201* (1964) 1145.

Note on Bohm's Addendum by C. H. Waddington

Note on Bohm's Addendum by C. H. Waddington

If the conversation between Bohm and Maynard Smith really took the form reported here, it seems to me to have been rather unsatisfactory. It is made to appear that Maynard Smith adopted the most narrowly orthodox neo-Darwinian line, implying that selection acts directly on genotypes, so that the selection against birds which fail to mate in dense populations is bound to be effective. Bohm rightly makes the point, which I have been insisting on in several places in these essays, that selection acts on phenotypes, and that there are many steps between the phenotype and the genotype. This point is actually often admitted in some neo-Darwinian genetics which can be considered relatively orthodox, though more frequently in the context of artificial selection; it is less often that its full implications for the theory of evolution in nature are considered (see my article on The Paradigm for an Evolutionary Process).

One does not, however, have to be very unorthodox to admit that selection may sometimes be ineffective, so that the failure of certain birds in a population to have offspring may have negligible or even no genetic consequences. The subject of 'selection limits' is of great importance to practical breeders concerned, for instance, with increasing the numbers of eggs laid by hens, where there would at first sight seem to be obvious selective advantage under natural conditions, let alone artificial ones. It is well known that natural selection does not in practice lead to a continual increase in the number of eggs produced by an individual; and poultry breeders find that, practising artificial selection, they can push up the number quite a way, but then come to a standstill. Geneticists concerned with the fundamental mechanisms of animal breeding have made considerable studies on the problem, e.g. from my own laboratory [1–4].

There are a number of mechanisms which can be appealed to: (i) The individuals in the population may be genetically identical as far as the selected character is concerned. (More strictly, the additive genetic variance for this character may be exhausted.) This is the explanation put forward by David Bohm. It is likely to occur only after the selection has been effective over a considerable number of generations. (ii) The phenotypes selected may be determined by heterozygous genotypes. In the simplest case, if, in a population consisting of AA, Aa and aa individuals, the Aa ones breed less (or more) than both the AA and aa, the situation settles to an equilibrium, after which no effect is produced on the frequency of the genes in the next generation; (iii) The phenotype selected for by one criterion may, for physiological or other reasons which are very difficult to circumvent, be selected against by some other criterion, which will probably operate at some other point in the life history. This seems to be usually the most powerful determinant of a selection limit; (iv) It is possible that there are inherent physiological limits, which random mutation has not succeeded in finding a way round. For instance, the number of eggs laid by a bird may be limited by the energy-expenditure required to digest enough food to have a surplus to lay down as stores in eggs. But it is doubtful whether one can find any cases of this which cannot more usefully be considered as examples of (iii).

Finally, there is a mechanism of a quite different order which has been invoked. Suppose that there are no selection limits of the kind just described which restrain the progressive increase in birds which will mate however high the population-density rises; it is suggested that a flock which has gone far along this line may, in a bad season, find it eats itself out of a living, and thus leaves an empty space into which there expands some nearby flock in which the deterioration of the control of population numbers had not gone so far. The reality and power of such 'inter-population' or 'inter-deme' selection is at present a highly controversial issue. The true-blue orthodox neo-Mendelians won't have it at any price, maintaining either that it can't or doesn't actually occur in nature, or that, if it did, it would do no more than slow up the processes that would occur without it; the more unorthodox argue that it might slow them so much as to deprive them of any practical validity; or, going further, that it would lead to the formation of gene-pools in which the control system was built in, for instance by producing populations in which the available genes fix a selection limit by one of the mechanisms (i) to (iii) above.

The only point of listing all these technicalities of selection-genetics is to show that we know what the form is sufficiently to put them on one side and come down to what, I take it, David was really getting at.

He used what he understood Maynard Smith to say (whether he really did so or not) to make the following main points:
1 To argue that, if selection favours individuals with a certain character, there will be a genotype for that character, and the frequency of that genotype will increase, is to indulge in metaphysics. And, moreover, in a metaphysics of a 'thing' kind rather than a 'process' kind. A genotype, i.e. a hereditary memory-store, is a static thing, a phenotype—the results of epigenesis—is a process. This is a point that I also have been making, *ad instantum* rather than *in abstracto*, throughout these meetings. The point to be emphasized is that much orthodox conventional scientific theorizing is full of unacknowledged metaphysical assumptions, which it is as well to have out into daylight.

96

Notes by Marjorie Grene

2 He wants to point out that if an entity has characteristics which are strictly determined in contexts in which you have earlier been interested (organisms with a specified genotype, atoms with a determined quantum state), they may exhibit properties in other contexts, and these properties need not be *completely random* or *lawless*, but may be related in very interesting ways to other variables in the situation which have previously not attracted attention but which may repay study. I think no biologist would deny that the phenotype variations which can be produced among a group of genotypically identical individuals are not completely random or lawless, but are related, in interesting ways, to the environmental circumstances in which epigenesis occurs (cf. the studies on identical twins reared apart). Here the main point would seem to be that, even when we have 'scientific' answers, we need to realize that they are partial answers to only *some* of the questions worth asking.

References

1. Falconer, D. S. and King, J. W. B. *J. Genet.,* *51* (1953) 561.
2. Falconer, D. S. *Introduction to Quantitative*
Genetics (Oliver and Boyd, Edinburgh and London, 1960).
3. Roberts, R. C. *Genet. Res., 8* (1966) 347.
4. Robertson, A. *Proc. Roy. Soc. B*, 153, 234.

Notes on Maynard Smith's 'Status of neo-Darwinism' by Marjorie Grene

Darwin demonstrated that given multiplication, heredity, and variation, natural selection follows. The first three were taken as given facts which together entail the fourth. (So powerful was his argument that it now seems natural selection is what we have, and multiplication what we infer from it: 'multiplication is necessary because without it natural selection is impossible'!) But Darwin was aware, at least sometimes, of the limitations of his theory in a way in which many of his twentieth century successors are not. First, he recognized that natural selection controls only adaptive characters and that some of the complexities of biological phenomena may not be adaptations. (They cannot survive if they are maladaptive, but natural selection may leave room for viable alternatives.) True, he was inclined to think of organisms primarily in Paleyan terms, that is, as mechanisms for self-maintenance and survival, but he did admit the possibility that not all characters might be explicable in these terms. And secondly, he recognized, at least occasionally, that natural selection theory does not offer any explanation of increasing complexity. This point is made whimsically in Kingsley's legend of a land where animals 'progress' backwards—from man to amoeba, as it were.

The status of neo-Darwinism

Maynard Smith's statement seems to me to illustrate admirably the confusion of contemporary theorists with respect to these two limitations. Of course, as long as natural selection theory remains a theory for studying natural selection it provides a powerful framework for experimentation on such questions as Maynard Smith enumerates. But what happens when he tries to show that this framework is adequate to explain evolution as a whole ? He wants to show that the theory is not tautological : that it is not just a device for measuring the survival of what survives. To do this he has to distinguish the 'fittest' as the 'harmonious' or the 'adaptively complex' from the genetically 'fit', i.e. those most likely to leave descendants. But either, as he sometimes argues, such 'complexities' are again to be assessed as superior in adaptivity, that is, in genetical fitness, and so the theory again collapses into tautology. Or 'complexity' is not reducible to survival value, and then, as he also admits, Darwinism or neo-Darwinism has no theory of increasing complexity at all. But increasing complexity is an essential feature of evolution. So neo-Darwinism provides an excellent and ingenious set of analytical instruments for measuring selective phenomena, plus either a vacuous theory of evolution or none at all.

Further insights may come, Maynard Smith suggests, from theories of epigenesis and of ecology. Bohm suggests that they may come from a study of creative process in non-living as well as living nature. Such theories would by no means undermine natural selection theory in so far as it limits itself to its proper range : but they would supplement it by explaining evolution in terms of laws and processes for which selection provides the necessary, but not the sufficient, conditions. Sufficient conditions it seems to provide only when it (1) extrapolates adaptivity to all biological phenomena, thus pretending to make the functional self-sufficient in a manner in which it cannot be so (see Bohm's notes and my comments on them) and (2) surreptitiously assimilates the concept of complexity, or of an increasing order of order, to that of adaptation, so extrapolated.

Some comments on Maynard Smith's contributions by David Bohm

Maynard Smith in his *Status of neo-Darwinism* has made what I regard as a very useful contribution to the clarification of the metaphysical assumptions underlying this subject. As indicated in my earlier articles, the main point about anyone's metaphysics is not to criticize it (though this can often be an appropriate

Further comments by David Bohm

thing to do). Rather is it to bring out what the metaphysics is, to make clear what is fact, and what assumption. I think that, in general, Maynard Smith has done just this with regard to neo-Darwinism. However, it seems to me that in a number of significant respects he is still overlooking metaphysical assumptions by tacitly accepting as fact certain notions that are in reality either suppositions or conventional definitions of terms.

The question at issue concerns Maynard Smith's reply to the frequently-voiced criticism that the basic conceptions of neo-Darwinism are tautological. He answers this on two levels, which I shall discuss in succession.

First, he says that there is a clear and unambiguous (i.e. tautology-free) definition of fitness, which is simply 'the probability of survival and reproduction'. (Adding also that this notion is more precise and elegant than that of 'the survival of the adaptively complex' or of my own notion of 'harmony'.) Now, in my view, the words 'probability of survival and reproduction' have little or no meaning, except in certain very narrow and strictly limited contexts, which could well fail to exist, for the most part in typical natural environments.

It should, however, be emphasized here that I am questioning not merely the metaphysics of biology, but also that involved in the use of probability theory in physics, and, generally, in all applications of statistics. The basic point, the significance of which has, in my opinion, been almost universally overlooked, is that probability theory has meaning only when a situation of randomness prevails. As indicated in my earlier articles, the problem of randomness has never been clearly defined, and is indeed full of sources of confusion and contradiction. Nevertheless, one sees that mathematicians and scientists do have at least certain (more or less tacit) notions in mind when they apply such ideas. Consider, for example, a series of coin throws. Not only does the average frequency of heads or tails tend to come near to a half in a long series of throws. Equally important, if we select arbitrary sub-sequences in which any particular results have already been obtained (e.g. ten heads in a row), then the frequency of heads still tends to come near to a half in a long series of throws. This is (at least in part) what is meant by randomness of the sequence. If this requirement is not satisfied, then the application of probability theory can lead to confused and erroneous results.

Now, evidently, the above condition for randomness is satisfied in coin throws. It is also satisfied in many applications in physics, chemistry, biology, and social statistics. But it is not clear that wherever there is a 'scatter' or variations of results, it is always possible to define a meaningful probability of occurrence

99

of a certain kind of event. For this 'scatter' may, in certain interesting and significant ways, fail to be random.

Of course, statisticians are well aware of this problem. To try to deal with it they have introduced the notion of probabilities of correlation of various classes of events (e.g. in a non-random sequence, the probability that B follows A is different from the probability of B in the whole sequence). But this depends on the (usually tacit) assumption of a yet higher order of randomness which applies to the distribution of correlated sets of events. In other words, the definition of probabilities of correlation does not remove the problem of establishing randomness; it only pushes it into the more obscure problem of what happens to higher order combinations of correlated events. For reasons similar to those given in connection with simple sequences of events, it follows, therefore, that the failure of randomness cannot always be comprehended in terms of probabilities of correlation of various classes of events.

Consider, for example, the Wynne-Edwards theory of limitation of bird populations, discussed in my 'Addendum' and in Maynard Smith's 'A Comment on Bohm's Addendum'. In the 'Addendum' I proposed that a given set of birds may have a common genetic constitution, with, however, a variable propensity for mating, that depends on various factors in the environment. (Of course, Maynard Smith is right to point out in his 'Comment' that the consideration of a fixed genetic constitution is also an assumption. However, it was my intention to propose this merely as a convenient basis for argument, and not as a fixed and generally valid metaphysics.) Now, these environmental factors are in general changing in a complex and non-repeating way. Not only is this happening in the external environment. Even more, it is being proposed that the flock or group of birds is itself a key part of the environment of each bird, in the sense that the propensity for mating of such a bird depends on the size of the flock that he perceived in swarming; where he was, in the flock, when he perceived it; and on many other factors, most of them not precisely or clearly specifiable.

Of course, one may, if one wishes, assume that all these factors, specifiable and unspecifiable, fluctuate at random in a way that has no *significant** relationship to the survival of the individual or the group. However, the theory of Wynne-Edwards specifically proposed that (among other factors) the propensity for mating depends on the size of the flock. This is therefore not a 'randomly' fluctuating variable. On the contrary, it is (in this theory) a key causal factor

* In this article I am using the convention of italicizing metaphysical words.

on which depends the whole process of reproduction and survival of the flock.

Now, consider the situation of a biologist who is trying to discover what it means to assume a 'probability of survival and reproduction'. Of course, in the relatively strictly-controlled conditions prevailing either in a laboratory or in the practical work of animal or plant breeders, the *relevant* factors may for the most part either be fixed or else fluctuate 'at random'. In these cases, such probabilities can be meaningfully defined. But then, in nature, there is a vast range of processes that vary in a way that is neither known nor controlled. Under natural conditions these processes may sometimes be random, but not *always*. It is pure metaphysics to suppose that assumptions of randomness that are valid under artificial or carefully selected natural conditions will continue to hold for *all* or even for *typical* natural conditions. The theory of Wynne-Edwards is only one example of how, under certain natural conditions, new variables (the size of the flock) may appear which are significant for the whole process of reproduction, and which are neither fixed nor 'randomly fluctuating' in relationship to the process. But as far as anybody can tell *a priori*, there is a potential infinity of such as yet unknown variables. Whenever one assumes that there exists a 'probability of survival and reproduction', one is assuming tacitly that the *relevant* effects of *all* that is unknown are *always* either fixed or else fluctuate at 'random' in relationship to the process of survival and reproduction.

Even if one assumes a 'probability of correlation of the process of survival and reproduction to this potential infinity of unknown variables', one does not solve this problem. For, as has been already indicated, such a probability has meaning only if the corresponding yet higher order combinations of variables are either fixed or fluctuate 'at random'. But after all, any particular natural process of evolution covers a limited period of time. If there is an essentially unlimited number of fluctuating 'variables' of very many kinds, and only a limited number of individuals and of generations of individuals, it is not at all unreasonable to suppose that the total set of events is not in general large enough to satisfy the necessary conditions for the meaningful application of the concept of 'probability of survival and reproduction'.

Unless the conditions for randomness described above are actually satisfied, the neo-Darwinist conception of 'probability of survival and reproduction' becomes confused. At best, it can be regarded as a tautology and, at worst, as a definitely self-contradictory notion. If one says that 'probability of survival and reproduction' means the actual relative frequency of creatures of a certain kind that did in fact survive in a specified historical process, then it is evidently a tautology. If it is taken to mean more than this, in the sense that one uses the probability concept to draw further inferences, that process is not logically justifiable, and may in fact lead to erroneous and contradictory results.

In my view, René Thom and Waddington have indicated a possible way out of this difficulty. Instead of thinking of the random fluctuations implied by the concept of probability, one thinks of a 'chreod', describing a vast range of possible ordered developments having certain relatively stable features. In this 'chreodic' process, it is essential to emphasize that selection operates on the phenotype rather than on the genotype. The main role of the genotype is to contribute to the conditions of selection. Thus, as Waddington has often emphasized, in general, a particular gene does not determine either survival or probability of survival (except possibly under certain limited conditions). Rather, like any other factor in the overall environment, it makes a certain contribution that is relevant in the total process of life that may lead either to survival or to non-survival. Whether there is survival or not depends in fact on the degree and kind of harmony that exists between phenotype and environment. Changes of the genotype will in general change this harmony in certain ways (either favourable or unfavourable). Changes of the environment will likewise also alter the harmony. But it is all one process in which effects of genotype and environment on the phenotype are interwoven in an inseparable way.

The tendency of neo-Darwinism to treat genotype and phenotype as two quasi-independent kinds of processes is based on tacit assumptions of randomness that can be expected to have in general only a limited applicability. So while neo-Darwinism may well lead to correct results in certain fields of study, it seems necessary to note that the extension of this theory to the *whole* of natural evolution is a kind of metaphysical assumption that there is good reason to question.

Now I come to Maynard Smith's second level of answer to the criticism that neo-Darwinism is tautological. This is that one can show the non-tautological character of neo-Darwinism by suggesting observations that would refute it.

The difficulty with such a response is that if the basic concepts of neo-Darwinism should (as seems more likely than not) turn out to be inapplicable or confused in their meanings outside of certain rather limited kinds of contexts that are not typical of natural evolution, it is hard to see why the refutability of the theory in its limited context of applicability would answer the criticism of tautology in broader contexts. Here it is surely significant that Maynard Smith attributes what he calls the 'failures of neo-Darwinism' to the absence of theories in the adjacent fields of epigenesis and ecology. As I have already indicated, it is just in these domains that one is led to question seriously whether concepts like 'probability of survival' have any real meaning (because conditions for randomness are not satisfied). Indeed, it may well be that the efforts of scientists to think in terms of such concepts is a principal barrier to progress in these fields. Rather, concepts like 'chreod' and 'harmony' may be much more relevant here.

One sees here a characteristic difficulty in finding evidence that could really refute a general and basic set of metaphysical assumptions. It is generally admitted that each theory has its 'failures'. But only very rarely does one regard these failures as evidence tending to refute the metaphysics. Rather, one places the fault in particular failures to solve certain 'problems' that arise within the framework of the theory. But if one is not fully aware of the metaphysical assumptions, he may not see that these problems could actually be inherently and basically insoluble, because his metaphysics makes assumptions whose meanings are confused when extended out of certain limited areas, where they may work fairly well.

It is here that it is relevant to consider Maynard Smith's 'Comment' on my 'Addendum'. Towards the end he says that science progresses by making assumptions more restrictive, not less so. I wonder whether this is *always* true, or whether it is only sometimes true. If we accept it as *always* true, we have made ourselves a particular kind of metaphysics concerning the nature of scientific research. Of course, modern science did in fact grow up, at least in part, in a struggle against the vague, speculative, and untestable features of mediaeval metaphysics. But does it not now tend to overemphasize sharpness of definition of concepts and testability (or refutability), thus jumping to the error of the opposite extreme ?

I think that Popper's emphasis on refutability has been wrongly understood, in many ways. It seems to me almost as if many scientists would give the highest *value* to any theory, merely because it happens to be easily refutable, while vague notions of potentially very great scope are frequently brushed aside, as having little or no *significance*. Here I would like to stress that the mere restrictiveness and refutability of a theory is not *always* the main consideration determining its *value*. Consider, for example, Ohm's law in physics, which is wonderfully precisely defined, restrictive, and refutable, but which is not generally regarded as having a deep and broad *significance*. On the other hand, early in this century there were vague notions, such as that light shining on metals will liberate electricity. It was very hard at that time to make precise and restrictive assumptions about

103

this phenomenon. Yet, in time, the study of it led to the photo-electric effect, and helped lead to the quantum theory. So I would like to emphasize, not the immediate restrictiveness and refutability of a theory, but rather that these are features which we can reasonably aim to achieve ultimately. It does not, therefore, follow that a theory that is immediately more restrictive than another is *always* a better one for the progress of science.

Perhaps one could here usefully recall the notion that Nature is more like an artist than an engineer (which was discussed in my 'Further Remarks on Order' paper). Therefore it requires a basically artistic attitude to understand it. Now, when an artist wants to learn to make a picture, for example of a man, he begins by indicating the general form first in a vague way and then step by step learns with the aid of careful observation how to articulate this picture in more detail. The attempt to draw an extremely detailed picture at too early a stage inevitably leads to confusion, because until one has drawn the general outlines one generally does not know what is the next level of articulation of structure that it is appropriate to look at. Similarly I would propose here that in science it is generally necessary to begin with vague general notions at an early stage, and to articulate these notions later as the subject develops. The metaphysical assumption that a more restrictive and more easily refutable theory is *always* a better one is therefore liable to lead at best to superficiality (as the accidentally defined details of the theory are given basic *significance*) or to confusion (as its concepts are extended too far beyond the limited experimental domain in which they have been demonstrated to work).

With regard to biology, I want to say that in my view, at least, this subject is now in an extremely early stage of development. Therefore somewhat vague notions like chreod, harmony, etc., may well have a key part to play. Rather than *base* everything on precisely defined ideas like 'genotype determines probability of survival and reproduction' (which could very easily have no meaning in typical areas of the domain of natural evolutionary process), it seems to me that it is more *relevant* to consider both genotype and external environment as different but closely interwoven aspects of what may be called the 'total environment' or 'total set of conditions' of the phenotype.

These considerations on falsifiability and on the question of vagueness *v.* restrictively precise definiteness of a theory help, I think, to explain a certain misunderstanding between Maynard Smith and myself that comes out in the 'Addendum'. When I received Maynard Smith's 'Comments' on this 'Addendum', I was surprised to learn that as far as I could see he had no very strong objections

Further comments by David Bohm

to the Wynne-Edwards theory of regulation of bird populations, and that indeed
he had himself tried to work out a model of such a theory. How, I asked, was it
possible for such a misunderstanding to develop? The answer that occurred to
me is that it was due to our deep metaphysical differences. To me, nothing that
Maynard Smith wrote seems to constitute any real evidence against the theory
of Wynne-Edwards. But as far as I can see, the fact that this theory has not yet
been formulated in a precise, restricted, and easily refutable way seems to mean
to Maynard Smith that it is not conducive to the progress of science. Since I
did not, at the time, understand his point of view very well, I tacitly assumed
that his severe criticism of the theory implied that he had much stronger arguments
against the theory than those that he actually wanted to make (i.e. that it was
not restrictive enough).

Paradigm for an evolutionary Process

C. H. Waddington
University of Edinburgh

The theory of evolution has passed through two main phases. In the first
('Darwinian'), the essence of the process was held to be natural selection
operating on a hereditary system characterized by 'blending' inheritance, in
which new hereditary variation was brought into being by some unknown
mechanism. In the second ('Post-Darwinian'), the hereditary system on which
natural selection acts had the Mendelian properties of dependence on discrete
alternative states of the hereditary factors and the production of new variation
by mutation. The enormous advances made in our understanding of evolution
by calling on the resources of Mendelian genetics need no emphasis.

The formulation of the logical structure of the typical, or paradigm, process of
evolution, assuming Mendelian heredity, was initially carried out with perhaps
greater attention to refuting certain loudly expressed objections to the new
outlook than to developing fully its own inherent character. Of the great
triumvirate who laid the first foundations for a fully logical, i.e. mathematical,
formulation of Mendelian evolution, Haldane and Fisher were English—and
England had just seen one of the most ferocious (and silly) of all academic
battles, in which the anti-Mendelians, led by Pearson and Wheldon, had gone
down to defeat at the hands of the believers in Mendelism and Discontinuous
Variation, whose champion was Bateson [1]. In the heat of the fight there had
been some corrosion, on both sides, of trust in Darwinian mechanisms of evolution.
Bateson, emphasizing discontinuity not only in heredity but also in phenotypic
variation, toyed with ideas of Mutationism allied to those of de Vries, while
believers in various types of over-riding evolutionary forces, such as orthogenesis,
found that the vagueness of the material basis for the hereditary system postulated
by Wheldon and Pearson offered scope for their own equally nebulous ideas.
Thus the two English progenitors of what came to be called 'Neo-Darwinism'
considered that one of their main tasks was to establish, as clearly as possible,
that Darwinian natural selection would, after all, 'work' in Mendelian populations.
The other main pioneer, the American Sewall Wright, was less affected by these
predominantly British squabbles. His mathematical formulation is far less drastically

This essay is dedicated, with respect and affection, to my good friend Theodosius Dobzhansky,
in celebration of his seventieth birthday.

simplified for polemical purposes; but its richer intellectual content calls for subtler and more difficult mathematics, and until recently it has been less influential than those of Haldane and Fisher.

Before attempting to re-formulate the essential logical features of an evolutionary process, it will be as well to remind ourselves of the formulations which were given by the pioneers. The first and simplest of them was Haldane, beginning in a series of papers in the obscurity of the Proceedings of the Cambridge *Philosophical Society* from 1924 onwards. This work was summarized in his book *Causes of Evolution,* 1932. An indication of the atmosphere in which it was written can be found from the quotation he chose as the motto for the Introduction: 'Darwinism is dead.—Any sermon'. On page 20 we find the statement, surprisingly apologetic to modern eyes: 'But I propose to anticipate my future argument to the extent of stating my belief that, in spite of the above criticisms, which are all perfectly valid, natural selection is an important cause of evolution.' The argument, when it comes, is related to systems of which the paradigm case is described as follows (pp. 180, 181): 'In a random-mating group a population composed of the three genotypes in the ratio $u^2AA : 2uAa : 1aa$ is stable in the absence of selection, and any group whatever reaches this stable equilibrium after a single generation of random mating. . . . Now after selection the population $u_n^2AA : 2u_nAa : 1aa$ is reduced to $u_n^2AA : 2u_nAa : (1-k)aa.$'

Haldane developed this paradigm mainly by studying the rates of change of gene frequency, for genes of various kinds, in populations with different types of mating system.

The most drastic simplifications involved in this paradigm are:
(i) The system essentially implies an equilibrium, in which the frequency of a gene selected against is reduced to zero, or to the level at which it is maintained by recurrent mutation; but the paradigm assumes that the initial conditions are not at equilibrium. Nothing is explicitly stated about why this should be so; it might be, for instance, because a change in environmental conditions has altered selective values, or because a totally new gene has occurred by mutation.
(ii) There is no explicit mention of the phenotype, and certainly no hint that phenotypes can be affected by environments as well as by genotypes.
(iii) There is no mention of the fact that the effect of a given gene is influenced by the rest of the genotype. In some of the later developments Haldane does discuss specific interactive effects between two or more genes, but he leaves on one side the pervasive effects of 'the genotypic milieu', to use the terminology of that time.

107

Paradigm for an evolutionary process

Fisher's paradigm avoids the last of these implications, but otherwise differs more in mathematical technique than in logical structure. Instead of measuring selection coefficients by linear coefficents, such as Haldane's k, he uses 'Malthusian parameters' expressed in exponential terms, which state the numbers of offspring produced by organisms of the genotype in question. In his paradigm, the difference in the Malthusian parameters for two alleles enters an expression which also involves the frequencies of the alleles and a parameter which expresses the phenotypic difference produced by altering one allele to the other in the actual population, taking into account its breeding structure (e.g. amount of inbreeding or assortative mating) and all the other genes present. He develops his paradigm initially into 'The Fundamental Theorem of Natural Selection', in the form: *'The rate of increase in fitness of any organism at any time is equal to its genetic variance in fitness at that time.'*

This is a statement which has proved extremely difficult to interpret. In the first place, it is clear that the word 'organism' must be shorthand for 'population of organisms'; but though this makes it easy to attach a meaning to 'genetic variance in fitness', it does nothing to elucidate what may be meant by an 'increase in fitness of an organism (= population)'. It is usually held to imply, if not to be synonymous with, an increase in the numbers of the population; and since genetic variance is essentially a positive quantity, this would lead to the conclusion that all animal populations must always increase in numbers, which they do not. Some way has to be found to get around this difficulty. We will return to it later in connection with Maynard Smith's remarks. Here I would only point out that Fisher's paradigm still involves the first two simplifications characteristic of Haldane's.

It is not so easy to disentangle and describe any particular situation as the paradigm adopted by Sewall Wright. He was not concerned to demonstrate the point on which Haldane felt he had to make a bold assertion: 'that natural selection is an important cause of evolution'. He took this for granted, and was more interested in the circumstances in which the operations of natural selection are mitigated or even overcome by other factors. However, his basic picture of an evolving population has a further element of inclusiveness and flexibility over and above that introduced by Fisher. Wright deals in the selective values of whole genotypes, considered as combinations of alleles at large numbers of loci. These values are envisaged in terms of a hyper-surface in a space in which fitness provides one dimension, while the others express the vast number of possible gene combinations. In his earlier papers at least, Wright was mainly concerned with an initial situation whose non-equilibrium character he carefully defines. He conceives of the fitness hyper-surface as comparable to a rough piece of country, with many hills and valleys; and he sets out to consider the mechanisms by which a population, which for some contingent reasons finds itself at the top of one hill, may travel across a valley and thus reach the top of some other, possibly higher, hill in the neighbourhood. Much of his work is therefore concerned with what might be called quantization processes in evolution; and the mathematical tools he uses deal largely with changes in the frequency distribution of gene frequencies in populations, particularly with processes which lead to certain alleles becoming 'fixed' at frequencies of 0 or 100%.

Wright's treatment began by avoiding simplification (i), i.e. failure to specify an initial state of disequilibrium, but only by invoking a rather special case, one in which a population has, by chance, got into a metastable position. His formulation can without too great difficulty be modified to deal with other types of initial non-equilibrium conditions, such as heterogeneity of the environment in space or time, but it remains true that these are not explicitly incorporated into any general paradigm. Further, he makes little more open reference to phenotypes than do Haldane or Fisher, and he does not incorporate into his scheme any suggestion that the phenotypes on which selection acts are affected by environments. He therefore employs simplification (ii) just as the others do.

Now, on the face of it, the two great problems of the Theory of Evolution— once we have granted that natural selection is an effective agent—would seem to involve just those points omitted by simplifications (i) and (ii). One problem is adaptation, and the focus of the long-continuing debate about Lamarckism is precisely that organisms so often exhibit adaptations which *look as though* they were responses to the environment, but which turn out not to be so in any direct way. Any paradigm which omits the effects of environments in altering phenotypes would seem to make it difficult, if not impossible, to deal with this (leaving it to 'random mutation' is not dealing with it). It was only by taking this factor into account that a solution could be found, in the form of genetic assimilation. Again, the second main problem is that of speciation. Here again everything leads us to the conclusion that diversity of the environment in space and/or time is of the essence, and that a paradigm which implies that the fitness of a genotype is single-valued is likely to prove inadequate

▶ *Maynard Smith's defence of neo-Darwinism.* The comments made above do not in any way imply that we should abandon neo-Darwinism*; they only

* Hereafter I shall contract this to neo-D which I shall also treat as an adjective.

suggest that some of the simplifications on which the mathematical theory has been based have outlived their usefulness and should be revised. The nature of the problem involved may be better appreciated if one looks at the article by Maynard Smith in this volume, in which he tries to demonstrate that neo-D is not a mere tautology by enquiring how one might refute it. I shall argue that none of his suggested 'refutations' would really require us to abandon the theory. (As a side-issue, I should like to remark, as I have done earlier [*Ethical Animal,* p. 151], that I have never been convinced by Popper's argument that, while hypotheses cannot be proved, they can be disproved ; in practice they can always be suitably amended to deal with the objections raised. Popper encourages a fashionable current of thought in the philosophy of science, which states that the thing to do with a hypothesis is to try to refute it. This is the treatment Maynard applies here, and in my opinion the result suggests that it is not a very useful line of approach—searching for an improved paradigm, as suggested for instance by Kuhn, may prove more rewarding.)

Let us consider, in turn, the various possible 'refutations' which Maynard Smith describes.

The first, which he dismisses undiscussed, would arise if 'we can show that organisms do not multiply', and multiplication he has defined as 'increasing in numbers in at least some environment'. Now it is perhaps a quibble, but it might be pointed out that a once-numerous species which had lost the power of increasing in numbers in any environment might still undergo neo-D evolution for some period while its numbers were declining ; it is not unlikely that the Sequoias, for instance, are in this situation. But in general I agree that no serious line of attack is likely to emerge in this connection. I also agree that it would be fatal to neo-D if we could show that organisms do not vary ; but it would be fatal to the idea of evolution in general, not only to the neo-D version of it.

Maynard Smith then speaks of 'the assumptions about heredity and the origin of variation', and he has stated that a refutation would occur if 'it could be shown that the assumptions made by neo-D are not in fact true of all organisms'. Now, of course, many different sorts of non-Mendelian heredity have been demonstrated in a variety of organisms—episomes in bacteria ; chloroplastal, mitochondrial or more general types of non-chromosomal genes, such as Sager's ; organelles in the cortex of Ciliates ; and so on. Some types of bacterial transformation, or episomal heredity, could even be interpreted as examples of Lamarckian phenomena. And there are certain rather weak, 'inertial' effects of the kinds he mentions ; for instance the occurrence of a duplication of a locus makes possible

110

the evolution of a protein dimer with two related polypeptide chains. If these were really refutations of neo-D it would have been refuted already. But in fact, of course, they are regarded as mere details and special cases. The main body of the theory is not noticeably weakened, though it has to give up the claim—which its more enthusiastic protagonists sometimes announce—to be the sole and sufficient explanation of all evolutionary phenomena.

All these attacks on neo-D are, in fact, attacks simply on Mendelism, not on anything which neo-D has added to Mendelism. To refute neo-D in this manner requires no more, and no less, than a refutation of Mendelism.

Maynard Smith then turns to some suggested refutations based on rates of change, but points out, justly in my view, that we cannot make any quantitative predictions in this field, and that therefore no refutations are possible.

Finally he turns to 'examining the end-products—the existing organisms'. He tries to invent animals whose organs exhibit an order which is clear enough to be undeniable but which it is implausible to attribute to any form of adaptation for reproductive efficiency. I think he fails to realize that he has come into a region where there are also epigenetic rules and types of organization to be considered. For instance, his first example is: 'If someone discovers a deep-sea fish with varying numbers of luminous dots on its tail, the number at any time having the property of being always a prime number, I should regard this as rather strong evidence against neo-D.' I should not draw such a conclusion so quickly. Which prime numbers? If they were 1, 3, 5, 7, 11, 13, 17, 19, I should suppose that the spot-producing mechanisms worked with some threshold-type action, so that at a low level it could produce 1 spot, and at higher levels added more spots two at a time until it got to 7, then the next effective jump gave an extra 4. We might explain 13 by saying that going from 7 through 9 to 11 put us well above the relevant threshold, and therefore the next jump goes back to being only a 2-jump; and after that, of course, we would get back to a 4-jump and reach 17 (Fig. 1). Then we would deal with 19 as we did with 13, and 23 as with 17. Then we'd go up another notch to a 6-jump, and get 29. After that I'd be willing to pass it back to Maynard Smith and remind him of his article in the 'Prolegomena' about epigenetic mechanisms for counting large numbers.

He goes on: 'And if the dots took up in turn the exact configuration of the various heavenly constellations, I should regard it as an adequate disproof.' If we are to take 'exact' quite literally, this might carry some weight; but if we allowed some latitude in the configurations, the observation, if made, would in my opinion refute not neo-D, but some as yet unformulated theory of epigenetic

Paradigm for an evolutionary process

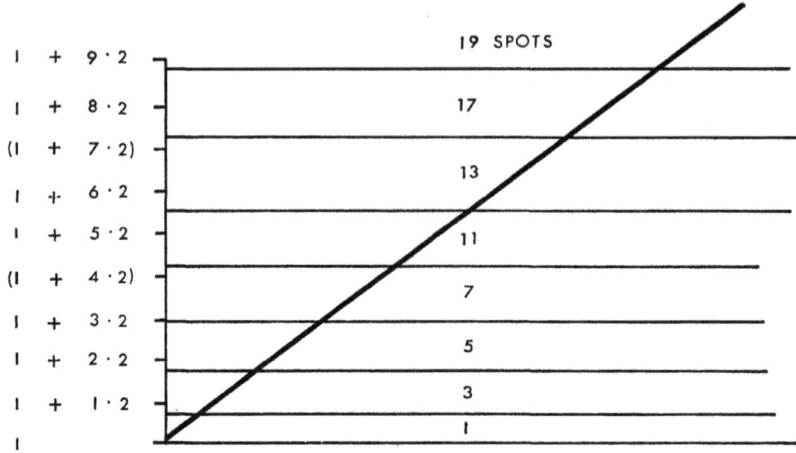

FIGURE 1

Hypothesis for explaining prime numbers of spots. One spot appears in any case. Further spots are added, two at a time, according to the concentration of a substance produced under the control of the genotype, which varies within the population. Each phenotype has a certain range of stability with regard to this substance, these ranges (shown on the graph) gradually increasing at higher concentrations of the substance. The adjustment of these stability-ranges to get the first eight prime numbers is not wildly implausible.

mechanisms. All neo-D can say about the configuration of spots is that they are useful; there is no *a priori* way of telling whether the fact that they look like something else is relevant or not; if they look like something that seems unlikely to be relevant, the first hypothesis is that the epigenetic system imposes limitations which make it impossible to obtain the useful effects without this associated surprising side-effect.

I should like to give two concrete examples in this connection. There is a family of Lepidoptera known popularly as 'Lantern Bugs'. Many of the species have large prolongations on the front of the head. In the species *Laternaria lucifera* this prolongation bears a number of spots and patches which give it an extraordinary resemblance to a crocodile's head. Now the prolongation is only a few millimetres long; it is difficult to believe that any animal which has been predated by crocodiles sufficiently to have evolved avoiding mechanisms will ever mistake the moth's head for a real reptile. What adaptive value can have led to its evolution? Is not this almost as odd as Maynard Smith's 'absurd example' of a fish with spots on its tail arranged like the stars of the Great Bear? I cannot explain it, but I do not propose to abandon neo-D on that account.

My other example is, please believe me, not wholly frivolous. When those neo-D stalwarts J. B. S. Haldane and Julian Huxley were young men, they co-operated in writing an extremely good elementary book on *Animal Biology*. For the frontispiece they went, like Maynard Smith, to the fauna of the deep sea. And they picked out illustrations of two species which no one could fail to see as very recognizable caricatures of the two authors (Fig. 2). Now Haldane and Huxley could certainly be formidable adversaries to run across unexpectedly in the darkness of some deep argument; but no, that cannot have been the neo-D reason for the evolution of these fish. The resemblances are as unexplained as would be the constellation-spots. One can say little more than that, if enough

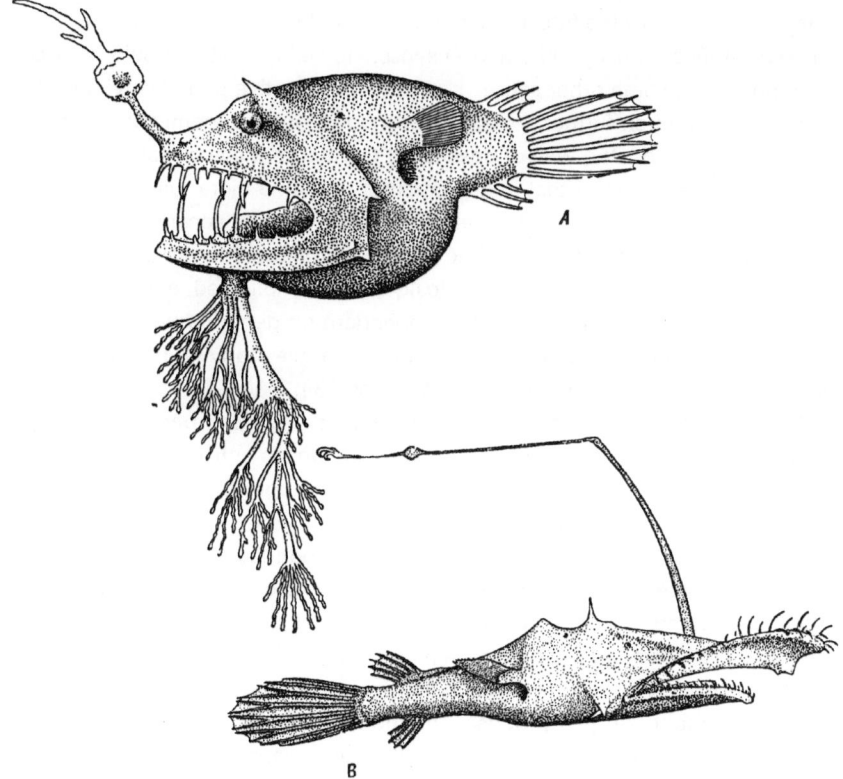

FIGURE 2
Two deep-sea angler-fishes, to show their extraordinary structure and their adaptations to their mode of life.

species go on evolving long enough, a lot of funny things are likely to turn up.

The point that is non-frivolous about this example is that the deep-sea fauna is one of those which presents in dramatic form one of the great problems of evolution. How did it come about that any animals ever went into such an uninviting and difficult environment? The standard neo-D paradigm, particularly in the Fisher-Haldane version, has altogether too little to say about the colonization of new habitats, especially of habitats which are poorer than the original home of the species, as the abysses are poorer than other parts of the sea.

▶ *Towards a post-neo-Darwinian paradigm* As a matter of fact, Maynard Smith actually gave away the whole story about 'refuting' neo-D before he started to show how he thought this might be done. On page 83 he admits that the phrase 'the survival of the fittest' *is* a tautology (and therefore irrefutable), if the word fitness is used in its neo-D sense. He claims that it should be used, in that phrase, in some other sense: he suggests 'adaptive complexity', but does not discuss this much further, except for the statement that complexity cannot be precisely defined. One way of describing the aim of a search for a new paradigm would be to say that it is an attempt to define more fully what this concept of 'fitness' should be. It is, after all, very generally accepted that fitness is a difficult and obscure concept. The simple Haldane-Fisher fitness has to be modified in situations which are technically more complicated, e.g. with non-random mating, or selection intensities dependent on gene frequencies [2]. More radical modifications have been made by people studying heterogeneous environments or more complex genetic systems [3].

A few authors have tried to formulate a concept of general parameter which will always change in one direction during evolution, as entropy always increases in physical systems, and as Fisher seems to have thought that his fitness would always increase.

If such a parameter could be defined, one could deduce from it the nature of the 'evolutionary force', which keeps evolutionary processes on the move in the face of the many factors which tend to bring it to a halt at some position of equilibrium (see page 115). Among possible parameters, MacArthur [4] has suggested that one measuring efficiency of using limited resources would always increase in evolution; Slobotkin [5] proposes that homeostasis is always increased. But both these authors, in my opinion, began their arguments from too close to the conventional neo-D paradigm, and did not pay enough attention either to the epigenetic effects of the environment or to changing distributions of organisms throughout a heterogeneous environment. I have myself [6] suggested that the parameter which is continually increased in evolution would have to express ability of members of the system to find some way or other of keeping alive and leaving offspring. This is perhaps not very far from MacArthur's suggestion, but involves the possibility that the organisms may not simply become more efficient at using available resources, but may begin exploiting new resources. The way to give more precision or more penetration to such ideas is, I suggest, a new investigation of the logical structure of the evolutionary process.

▶ *The post-neo-Darwinian paradigm* 1. Suppose you have a material structure P with a characteristic Q such that the presence of P with Q produces Q in a range of materials P_i under circumstances E_j. Then you could have 'natural selection' to increase the range P_i and E_j. (For instance, a crystal dislocation in material P, where Q would be replicated in a variety of materials P_i in several environments E_j.) But the whole system would not qualify as 'life' because, in my original phrase, it is not interesting enough: in rather more objective language, because this set of postulates makes no provision that Q has an effect on E. But Q can only affect E if it is not merely a memory-store to be replicated, but is also an operator. To say that Q becomes an operator is the same as saying that Q becomes a phenotype. This might, logically, involve no, or very little, or very much, translation of the memory-store Q into the effective operator phenotype Q^*. But in practice—and perhaps because of a profound law of action-reaction—it is difficult (impossible?) to find a Q which is stable enough to be an efficient store and at the same time reactive enough to be an effective operator. Thus a considerable translation from Q to Q^* is characteristic, and may be necessary, for all systems which can be accorded the name of 'living'.

2. Having now got a system in which natural selection will evolve something effective (or interesting) enough to be called 'life', we need to specify conditions under which evolution will continue. If the range of conditions E_j is single-valued, evolution will produce a single optimum phenotype Q^* (probably, but not necessarily, depending on a single optimum genotype Q) and then stop. Similarly, if E_j has only a finite number of values, evolution will eventually reach a stable end-state, with a number of different Q^*s, each optimum for one value of E_j. The paradigm situation therefore demands two further conditions: (i) that E_j is an infinite-numbered set; and (ii) that there are sufficient Qs to provide Q^*s suitable for an infinite sub-set of E_js.

The infinity of E_js is ensured by the fact that Q^*s are components of E_js. Thus any evolution of a particular Q^* into Q'^* automatically changes a number of E_js in which this Q^* is a component. In more biological language, the environment which exerts selection on one organism is influenced by the presence of other organisms; and as the other organisms change in evolution, so the environment of the first organism is altered, and it must evolve too. In a still cruder formulation, we can say that the evolutionary appearance of a new species automatically creates potential new environmental niches ready for exploitation by some further new form. ('Big fleas have lesser fleas, upon their back to bite 'em, and lesser fleas have smaller fleas, and so *ad infinitum.*') [7]

The second requirement, that the available genotypes must be capable of producing phenotypes which can exploit the new environments, requires some special provision of a means of creating new genetic variation. It might at first sight seem simplest to ensure this by a mechanism in which the new environment would itself stimulate the production of new appropriate hereditary variation. This is the Lamarckian hypothesis. It is indeed a fact that new environments can often produce new appropriate phenotypes (E'_j produces Q'^*). But except in very special circumstances E'_j does *not* produce Q'. Instead, what we normally encounter are systems in which Q is continually giving rise to a range of altered forms ($Q'_1, Q'_2 \ldots$) by a process which is 'random' in the sense that it is unrelated to any selectively-effective E_j (though it must certainly be controlled by rules of occurrence not related to some E_j).

The relative importance of these two 'evolutionary forces'—new environments, new genes—needs some consideration. The classical neo-D position is that all gene mutations occur with a specifiable frequency, so that there is no question of genuinely novel, first-time-ever mutated genes; and that mutation frequencies are so low (10^{-5} or less) that they can almost never overcome the selection pressures. Such considerations gave rise to a dogma of the ineffectiveness of mutation as an evolutionary force; it merely provides some raw material on which other more effective forces might act. But this dogma grew up in the pre-molecular period of genetics, when one was thinking of, e.g. the mutation rate of the plus allele of the 'white eye locus' in Drosophila to an allele which gave a white eye phenotype. We have to rethink the situation now that we realize that the protein produced by the w locus (or group of proteins, from a group of loci [8]—but that is an irrelevant complication from the present point of view) can be reduced to complete ineffectiveness by changes in any one of a large number of different peptides. If one characterizes mutations, not by such crude criteria as 'producing white-eyed rather than red-eyed phenotypes', but in molecular terms as substituting amino-acid P for Q at position W in a polypeptide, then the rate of particular mutational steps is reduced, probably by several orders of magnitude. This means that they are even less able to overcome the effects of selection by sheer frequency of occurrence; but it also raises the possibility that some mutation rates may be so low that in comparison with the time-scale of the relevant evolutionary events, it may not be justifiable to consider them in terms of continuous rates at all. We may have to bring back into our thinking the possibility of radically new, never-seen-before, mutations—considerations which were in the forefront for the earliest Mendelian evolutionists, such as Bateson and de Vries in the first decade or so of this century.

Before one can evaluate the relative importance in evolution of (i) changes in environment, as against (ii) mutations-as-alterations-of-molecules or mutations-as-determinants-of-epigenetically-complex-phenotypes, one needs to realize the range of spectrum of levels of organization in which evolution may occur. At one end of the spectrum is:

(a) the purely macromolecular. Is it (selection-wise) a 'good thing' to substitute valine for glutamic acid at position 6 from one end of the chain of haemoglobin (producing the sickle-cell character)? In higher organisms there probably is a gene for doing this floating around in the population anyway—or you could find some other way of dealing with the environmental selective pressure. The question is more important in relation to the very earliest phases of evolution, when the living systems comprised little more than a bare minimum of molecules which could just keep going. How did one improve the very first, barely effective, DNA polymerases, oxidative enzymes, or chlorophylls? The course followed by evolution may, at such stages, really have been influenced by the nature of the polypeptide substitutions that mutation threw up.

(b) From this, there is a complete continuous spectrum of increasing levels of complexity—through the 'hypermolecular', which is what classical neo-D concerns itself with—to the 'ecosystem' evolution involved in such a question as: how did the London sparrow cope with the success of the petrol engine in driving off the streets all those horses, with their offerings of dung full of delicious seeds? How did the rabbit of the gentle English fields evolve a method of flourishing in the harsh Australian outback, with not a single species of its normal food plants, or the red deer of the Scottish heather-covered hills find its evolutionary way into a set-up in which, ecologically married up with the Australian wombat, it dominates much of the wilder country of New Zealand and, incidentally, grows to about twice its normal weight in its home country? [9]. It is extremely unlikely that any of these evolutionary episodes had to wait for any new polypeptide sequences. All the Qs necessary to produce the appropriate

Paradigm for an evolutionary process

Q^*s were, one imagines, already present in the population, and the effective 'evolutionary force' was the occurrence of the new environments (E_js).

The relative importance of the two 'evolutionary forces' arising from new random mutations and new environments must therefore change as we pass from the macromolecular end of the evolutionary spectrum through the hypermolecular to the ecosystem end, but wherever we are within this range, each of these 'forces' has *some* importance.

It is important to emphasize that the new genetic variation must not only be novel, but must include variations which make possible the exploitation of environments which the population previously did not utilize. In Longuet-Higgins' terminology, it is not sufficient to produce new mutations which merely insert new parameters into existing programmes; they must actually be able to rewrite the programme. Essentially the same point was made by Pattee. In a letter (12 February 1968) discussing tactic copolymer growth as an elementary genetic system, which he raised again in the *Prolegomena*, he writes: 'The fundamental reason that evolution is limited in such single tactic copolymers is that the rule for monomer addition or the code is intimately dependent on some fixed length of the description. If evolution is to lead to unlimited complexity, then the code must accept a description of indefinite length and the description itself must be able to grow indefinitely without changing the code.'

The predominance of 'random mutation' over Lamarckian mechanisms in tellurian ('this-earthly') biological systems does not arise because the latter could not, in theory, support evolutionary processes. It is presumably related to the problem described at the end of paragraph 1, how to combine a store which is unreactive enough to be reliable, with something which interacts with the environment sufficiently actively to be 'interesting'. To ensure reliability the store must be rather unreactive; the solution has been generally adopted by tellurian systems to provide the necessary variability by a random process not depending on reaction with the environment rather by endowing the storage material with some residual reactivity [10].

3. We need also to provide that the variant Q's actually come in contact with the variety of new E_js. This may be brought about simply by arranging for very wide but quite indiscriminate geographical dispersion of newly fertilized zygotes; and this is the mechanism on which the whole plant kingdom relies. The world of non-sessile, motile animals has a more economical method of achieving this result by employing behavioural mechanisms, which lead animals either to explore situations more or less at random (which is little better than plants

can do) or, more typically, to choose those environments in which they can most easily earn their daily bread and escape their enemies. This is an elaboration, of a rather Lamarckian character in that it involves a reaction to the immediate circumstances, which is over and above the absolute necessities of the evolutionary paradigm; but one which is very important in animal evolution.

4. In order to accommodate the numbers of altered or mutated Q's required to evolve into the new environments E'_j, the paradigm situation must be one describable only in terms of populations, not of individuals.

5. The new environments E'_j are, as we have seen, essentially complex entities, being functions of a number of variables, among which are, for instance, a number of phenotypes Q^*s of various species, as well as various physical quantities, etc. The evolution of an optimum phenotype for a particular new environment, $Q^*_{E'_j}$ will therefore in many cases involve combination of a number of different variant genotypes, $Q'_p, Q'_q, Q'_r, \ldots$, etc. It is therefore not surprising to find that most evolving systems have developed some method of encouraging the production of appropriate combinations of Qs. Provision for this is not an absolutely necessary constituent of the evolutionary paradigm, since some organisms get away with nothing more radical than very large numbers of short life-cycles, though there are few, even of the viruses and bacteria, who cannot do anything better than this. Nearly all organisms have evolved mechanisms which facilitate recombination of variant Qs, usually by some sexual or parasexual process.

6. In so far as the necessary heterogeneity of the environment comes into existence as a consequence of interactions between existing organisms (i.e. the E'_js are functions of Q's), two consequences will follow: (i) There will be a continual increase in the number of species which are optimum in some or other of the ever-increasing number of E'_j. Thus the diversity of the organic world will continually expand; (ii) Since a new, attainable E_j may be a function of an indefinite number of existing phenotypes, there will be a tendency for the evolution of even more complex phenotypes, capable of operating optimally in environments of increasing complexity. When there are both water-snails and water-visiting mammals, evolution can produce parasites with life-cycles involving interactions with both hosts; when there are night-flying insects, then, but not before, evolution can produce larger night-flying predators, such as bats, which need a sonar system to prevent them running into obstacles or to locate their prey.

7. The necessary heterogeneity of environment has another consequence, which cannot be omitted from the paradigm situation. Selection operates on phenotypes, and phenotypes are affected by environments as well as by genotypes. Further,

119

since there is a necessity for mobility (whether passive or active) to ensure that new environments are explored, it will not in general be true that the environmental influences which contribute to the formation of the phenotype are identical with those which exert the most important selection on it. (Consider, for instance, an annual plant; the phenotype of a plant and its seeds will be influenced by the weather, etc., in the year n, while the main selection may be exerted by the weather in year $n + 1$, when the seeds germinate.) Thus the paradigm situation must incorporate two theoretically separable effects of any environment E_j: on the development of a Q into a $Q^*_{E_j}$, and on the selective values of the variety of Q^*_x which arrive within or pass through it.

The complete paradigm must therefore include the following items: A genetic system whose items (Qs) are not mere information, but are algorithms or programs which produce phenotypes (Q^*s). There must be a mechanism for producing an indefinite variety of new Q'^*s, some of which must act in a radical way which can be described as 'rewriting the program'. There must also be an indefinite number of environments, and this is assured by the fact that the evolving phenotypes are components of environments for their own or other species. Further, some at least of the species in the evolving biosystem must have means of dispersal, passive or active, which will bring them into contact with the new environments (under these circumstances, other species may have the new environments brought to them). These environments will not only exert selective pressure on the phenotypes, but will also act as items in programs, modifying the epigenetic processes with which the Qs become worked out into $Q's$.

From the standpoint of this paradigm, how should we envisage the classical evolutionary concepts of adaptation and fitness? I confess I have not sufficient mastery of the sophisticated mathematics which would be necessary for a rigorous exposition. Instead of attempting that, I will suggest, for adaptation, an allegorical illustration which I originally applied to man in his world (motto to Chap. 4 of *Behind Appearances*), but which I think applies almost equally well to any organism: *Man in the world is like a caterpillar weaving its cocoon. The cocoon is made of threads extruded by the caterpillar itself, and is woven to a shape in which the caterpillar fits comfortably. But it also has to be fitted to the thorny twigs—the external world—which supports it. A puppy going to sleep on a stony beach—a 'joggle-fit', the puppy wriggles some stones out of the way, and curves himself in between those too heavy to shift—that is the operational method of science (and of the evolution of biological systems).*

120

C. H. Waddington

As regards fitness, we have to define that concept in terms which allow for the existence of heterogeneous and evolving environments, and of organisms capable of active or passive dispersal through a range of environments which are acting both as agents of selection and as subsidiary programs affecting the development of the phenotypes which are selected. The fitness even of a single phenotype cannot, therefore, be represented by a single-valued coefficient, but only by a matrix, or a continuous distribution of values, which specifies also the variety of environments, in which selection may occur. If we wish to attach 'fitness' to a single, multigenic *genotype*, we should have to increase the dimensionality of the matrix so that it could take account also of the epigenetic-programming aspects of the environments; and if we wished to emulate the classical neo-Darwinists and speak of the fitness of single genes, we should have to increase the matrix again to incorporate all the various genetic combinations in which it might occur in the population in question (it seems highly dubious to me that any process of averaging over all these combinations, as advocated, e.g. by Fisher, has any biological validity).

In contrast to the difficulties of conceiving of 'the fitness of a gene' from this point of view, the concept of the fitness of a population comes close to giving one an opportunity to use the adjective 'perspicuous' in the precise sense given to it by that pedant Fowler (*Modern English Usage*)—'means (the being) easy to get a clear idea of'. The fitness of a population is the degree to which its gene pool gives it the ability to find some way or other of leaving offspring in the temporarily and spatially heterogeneous range of environments which its dispersion mechanisms offer to it.

But, even if not too difficult to formulate, this concept has a joker in it—which is what makes it interesting and challenging. How to know what the 'temporarily heterogeneous range of environments' may bring? A new Ice Age, a new virus, a new predator? The gene pool can, of course, preserve for some time genes which turned out to be useful in relation to critical situations in the more or less recent past: Lewontin [11] has discussed these possibilities in a very stimulating way, and shown that, unless the environment behaves itself—is not capricious—they are rather limited. From a more general point of view, one can envisage a number of possible strategies. As (i) to canalize your development, build up well-buffered chreods, and insist on developing into some good all-purpose almost invariant form in spite of whatever environmental effects get thrown into the epigenetic programs—as mice concede little more than a fractional elongation of the tail to the difference between growing up in a hothouse rather than a

121

cold-storage depot [12]; or (ii) acquiring a gene pool which allows an extreme flexibility in the end-results of development—good examples are some small crustacea, such as *Daphnia* and *Artemia*, in which the minute physico-chemical variations between every pond of water are reflected in the shape of the adults, or plants such as the 'water arrowroot' *Sagittaria sagittifolia,* in which the leaves of one and the same plant have radically different forms when they are growing wholly under water, on the surface, or in the air. To make a success of this gambit, it is of course necessary to ensure that the plasticity of epigenesis is in general, or at least in really crucial situations, such that the environmentally-modified forms are selectively useful in the environments that produced them. It is no use allowing your muscular development to be influenced by the environmental demands on the use of muscles if the dependence takes the form that using muscles causes them to be consumed and to wither away; (iii) another ploy is to develop some general defence mechanisms which do not have to know in advance what they will have to defend against. 'Random' mutation provides this facility to some extent, but only over periods of many generations of selection. If the population can afford to wait that long, as bacteria can, to add a general mechanism for rapidly spreading a good defence, once it has been acquired, by such systems as Infective Resistance Factors, is obviously a useful second stage. Perhaps the best example of a generalized defence is the vertebrate antibody-production system, which seems able, even within one lifetime, to protect against an enormous range of invading foreign substances.

The systematic exploration of the evolutionary strategies in facing an unknown, but usually not wholly unforecastable, future would take us into a realm of thought which is the most challenging and very characteristic of the basic problems of biology. The main issue in evolution is how populations deal with unknown futures; is this problem so different from that described by Gregory, when he says that 'perception involves the continual solution of a series of puzzles'? In epigenesis we find systems which will develop into perfectly good lenses or livers, even when there is something non-standard in the conditions which normally guide the cell's synthetic machinery into those paths. In all these cases we are forced to consider the nature of mechanisms which can operate effectively on the basis of inadequate information. This seems to be one of the central general problems of Theoretical Biology. Life might be defined as the art of getting away with it; and Theoretical Biology as the attempt painstakingly to explicate just how it is done.

C. H. Waddington

Notes and References

---

1. Wheldon's polemical journal, *Questions of the Day and of the Fray*, did not cease publication till 1924 (?); and the anti-Mendelian influence was strong enough to ensure that Fisher's first major genetical paper, arguing amongst other things that continuous variation is explicable on Mendelian principles, was rejected by the Royal Society of London, and published by the less prestigeful Royal Society of Edinburgh, in 1918.

2. Wright, S. Stochastic processes in evolution, in *Stochastic models in Biology and Medicine* (ed. Garland), (Univ. Wisconsin Press 1964) p. 199.

3. Levins, R. *Amer. Nat. 96* (1962) 361; *Amer. Nat. 97* (1963) 75; *J. Theoret. Biol. 7* (1964) 224; *Evolution 18* (1965) 635; *Genetics 52* (1965) 891.

4. MacArthur, H. H. *Proc. Nat. Acad. Sci. Wash. 48* (1962) 1893.

5. Slobotkin, L. B. *Am. Sci. 52* (1964) 343.

6. Waddington, C. H. *The Nature of Life* (Allen and Unwin 1961) p. 109, *Prolegomena* (1967) p. 22.

7. Several ecologists argue that there will be, for ecological reasons, a selection pressure towards the development of more complex ecosystems, involving an ever-increasing number of species. The fundamental selection is towards increasing the stability of the ecosystem; when there are only a small number of interacting species, there is a tendency for violent fluctuations in numbers, which may lead to the extinction of some species and the complete collapse of the whole system. See, for example, Hutchinson, G. E. *Amer. Nat. 93* (1959) 145; MacArthur, R. H. *Ecology 36* (1955) 533; Dunbar, M. J. *Ecological Development in Polar Regions* (Prentice Hall 1968).

8. Green, M. M. *Proc. 11th Int. Cong. Genet. 2* (1965) 37

9. For this and other examples see *The Genetics of Colonising Species*, I U B S Symposium (ed. Baker and Stebbins) (Acad. Press 1965).

10. In connection with the nature of the mutational events which may play a part in evolution I would like to advance some ideas, which are, I freely admit, so speculative that I have relegated them to a note rather than the body of the text; but many biologists will consider them so outrageously heterodox that they may refuse to consider them at all—and these I should beg to think again.

One of the major problems of evolution theory is to understand how the sharp discontinuities between major taxonomic groups—Phyla, Families, Species-Groups, and so on—have come into being. A simple-minded empirical inspection of the facts would suggest, as it did for instance to Goldschmidt when he wrote his *The Material Basis of Evolution* 1940 and *Theoretical Genetics* 1955, that it might be profitable to contemplate the possibility of the very occasional occurrence of what Goldschmidt called 'systemic mutations', which result in a complete restructuring of the genome, achieved either in a single step, or at least in rather few generations. When Goldschmidt wrote, no clear-cut example could be given in which the occurrence of such a process had been observed. The orderly minded orthodox biological world closed its ranks against this suggestion that revolutionary processes may happen. It became accepted that the only respectable doctrine is that evolution never involves anything but step-by-step Fabian gradualism, plodding along a weary way similar to that by which the annual milk yield of dairy cows or egg yield of hens is slowly improved—the occurrence of a little allopolyploidy or rearrangement of chromosomes by two or three

123

breaks could be admitted, but would only push the basic philosophy from Bourgeois-Liberal to right-wing Social Democrat.

It is still impossible—so far as I know—to quote a compelling instance in which a systemic mutational event has been observed in an evolving multicellular organism. But events which appear to be essential of this kind are becoming well known in the field of cell culture. It is a common experience that cells isolated from vertebrate tissues usually grow in culture for a fairly restricted number of cell generations—a hundred or two—and then die out, *unless* they undergo some sort of change which brings into being cells capable of forming an 'established line', which can then be sub-cultured in perpetuity. The nature of the change from a 'strain' to an 'established line' is highly obscure, but it often involves what looks like a complete restructuring of the genome; there may be a considerable reduction in number of chromosomes, accompanied in some cases by considerable changes in chromosome morphology. (For a recent review see *The mammalian cell as a differentiated microorganism* by Howard Green and George J. Todaro, *Ann, Rev. Microbiol. 21* (1967) pp. 574—600.)

The fact that cells in culture can throw up, within at most a few cell generations, new types of cells capable of giving rise to 'established lines', and that the change may involve a very drastic reshuffling of the genome (usually, in the cases observed, with a loss rather than a gain of chromosomal material),

is evidence that something like a 'genetic revolution' or 'systemic mutation' can occur. It is, of course, more difficult to see how such an event in an evolving population could be propagated so as to affect the future, but if such events are not ruled out of court by the nature of genetic processes it seems silly to close one's mind to the possibility that evolution has found some way of making use of them.

An example of an evolutionary phenomenon which suggests a very radical reorganization of the genome between nearly related species is the astonishing difference found by Forbes Robertson (in press, *Genet. Res.*) in the DNA sequences of *Drosophila melanogaster* and *simulans*, as tested by molecular hybridization. RNA manufactured *in vitro* on a *melanogaster* DNA template hybridizes only one-third as well with *simulans* DNA as does the RNA made on the *simulans* template; the results of a reciprocal experiment are very similar. Although it seems certain that the hybridizations only occur between substances related to the highly reiterated stretches of DNA, this is still strongly suggestive that the differences between these two species involve much more radical and pervasive alterations of base sequences than were contemplated a few years ago, when it seemed that all that was involved was a few inversions, and translocations of large sections of the chromosomes.

11. Lewontin, R. C. *Bioscience 16* (1966) 25.

12. Barnet, S. A. *Biol. Rev. 40* (1965) 5.

Some comments on Waddington's paradigm by J. Maynard Smith

1. Neo-Darwinism and Mendelism. Wad argues that 'to refute neo-D requires no more, and no less, than a refutation of Mendelism'. This of course depends on how you define neo-D. In my 'defence' of neo-D I made Weismannism rather than Mendelism the central assumption. Otherwise the refutation of neo-D is trivial : bacteria do not Mendelize but they do evolve. But the Weismannist assumption (roughly, if the phenotype of an individual is altered by an altered environment, this will not cause that individual to produce offspring with the new phenotype) is not disproved by the types of non-Mendelian heredity mentioned by Wad.

Nevertheless, I agree with the main point Wad is making – the most direct way (but not the only way – see below) of refuting neo-D is to show that its genetic assumptions are wrong.

2. Refutation by the end-products of evolution. I argued that complex structures which did not contribute to the survival of their possessors would refute neo-D. Wad argues that since development is based on algorithms, it can lead to inexplicably complicated (=funny) results. I think that I am right, but I agree that no single example could be decisive. My point is that when biologists are confronted by a structure, they analyse its function in terms of its contribution to survival, and this method of analysis usually works. If it didn't – i.e. if it often turned out that an organ when studied in detail could not be interpreted as contributing to survival – then biologists would have abandoned Darwinism long ago. But it is of course true that there are at any one time plenty of structures, usually ones which have been little studied, which cannot be interpreted adaptively.

3. 'The survival of the fittest' and Waddington's paradigm. It is perhaps a pity that this phrase was ever introduced into biology, because it is a standing invitation to philosophers to argue that Darwinism is a tautology. There seem to be three ways of treating the phrase:

(a) Assume 'fittest' means 'most likely to survive', and you have a boring tautology.

(b) Replace 'fittest' by some more sophisticated definition of survival capacity, and you may have an interesting tautology. Thus a conclusion may follow necessarily from certain assumptions, but still be interesting, e.g. the conclusion that planetary orbits are elliptical, given Newton's laws of motion and gravitation. This is particularly likely to be the case if the conclusions are more easily tested than the assumptions, as is the case in the Newtonian example.

Thus one might try to deduce from the laws of heredity some property which will be maximized. This is what Fisher's 'fundamental theorem' does, and what Wad does in his article. So far the approach seems to me to have been unfruitful, mainly because the assumptions (i.e. of genetics) are easier to test than the conclusions (i.e. the course of evolution).

If Fisher's 'fundamental theorem' is interpreted to mean that evolution must lead to an increase in the rate of growth in numbers (i.e. 'fitness') of a population, then the theorem is simply false. If, alternatively, one interprets fitness as a

125

mathematical function of the frequencies of genotypes in a population and of their *relative* probabilities of survival, then the theorem is, with certain qualifications, true, but, as Wad points out, difficult to apply.

The snag with Wad's less mathematical attempt to find something which increases in evolution is that it leads to a false conclusion. Thus he concludes 'the fitness of a population is the degree to which its gene pool gives it the ability to find some way or other of leaving offspring in the temporarily or spatially heterogeneous range of environments which its dispersion mechanisms offer to it'. Now fitness in this sense is not necessarily maximized. For example, plant species commonly and animal species occasionally lose the capacity for sexual reproduction. Such a change usually leads to extinction. It is a lowering of fitness in Wad's sense, yet I see no reason to doubt that the change occurs by natural selection. (c) The third approach, which I adopted in my 'defence', is to reformulate the phrase 'the survival of the fittest' in a non-tautological way, by taking fitness to refer not to some function, sophisticated or otherwise, of survival capacity, but to the properties of 'adaptive complexity' or 'harmoniousness' or what have you — i.e. to those properties of living organisms, and sometimes of their artifacts, which distinguish them from inanimate matter, and which call for an explanation.

I do not think approaches (b) and (c) are mutually exclusive (although obviously confusion arises if 'fitness' is used in two senses). The difference between Wad and myself lies not so much in our views about the mechanism of evolution, which are rather similar, but in what we were trying to do. In my article on 'the status of neo-Darwinism' I was trying to defend the present orthodoxy from criticisms of a philosophical and fundamentally Lamarckist type. Wad, perhaps rightly, regards this argument as no longer very interesting, or as something for molecular biologists to worry about, and has therefore been trying to say something new about evolution.

Reply by C. H. Waddington

To my way of thinking, John's comments introduce some confusions into this discussion, which it may be well to try to clear up. They are largely terminological, and connected with the fact that when I offer a criticism of neo-D, John tends to reply with an impassioned defence of Darwinism or Weismannism or some other well-accepted historical precursor of the views I was discussing. Let us first, then, agree on what we mean, at least roughly, by the doctrines attached to these various names.

Reply by C. H. Waddington

By *Darwinism*, I mean the theory that organisms come into existence by a process which involves material heredity from their progenitors under the control of natural selection. This is certainly *not* a tautologous statement, since there is an alternative to it—which was in fact generally accepted before Darwin, namely that organisms are brought into being by 'Special Creation', or something of the kind. What John refers to as 'refutation by the end-products of evolution' would be refutation of Darwinism itself, not merely of neo-D.

By *Weismannism*, I mean the same as John does, according to the statement in his first paragraph. This again certainly is *not* tautologous: It is a necessary but not a sufficient condition that evolution should be of the neo-D type. I am not attacking it. Perhaps John is right in thinking that it still needs defending in any company in which philosophers are present, but I don't think we need waste much time on rehashing the old arguments in this meeting.

By *Mendelism* I mean the theory that heredity is transmitted in the form of discrete factors which can segregate and recombine. I think it is confusing to introduce the verb 'to Mendelise', a piece of lab jargon dating from the days of Bateson and Punnett, when it had the meaning 'to exhibit the phenomenon of segregation into classes with one or other of the classical 'Mendelian ratios', i.e., 3 : 1, 9 : 3 : 3 : 1, etc.' Many types of organisms (e.g. polyploids) do not Mendelise in that sense, but no one would deny that they exhibit Mendelian heredity. I think the same is true even of bacteria, and do not accept John's contention that their behaviour refutes Mendelism.

By *neo-D* I mean the view that Weismann's doctrine—that there is no influence of the phenotype on the genotype—can be transferred from the individual level to the population level, and that an adequate theory of evolution can be formulated in which 'fitnesses' are attributed to genotypes. John slides altogether too easily between the Weismannist point that the environment of an individual does not affect the heredity he transmits, and the quite different argument that the environment of a population does not affect what they transmit. I maintain that a population's environment does influence, quantitatively, what they transmit, because natural selection acts on phenotypes which are partially environment-dependent. John also gives away too much in his para. 3(a), where he suggests that if we define fitness as 'most likely to survive' we have only a boring tautology, and that we need to define it in a more sophisticated way to get a tautology as interesting as Newtonian mechanics. I have never wished to deny that the results obtained by classical neo-D were as interesting and valuable in their field as Newton's conclusion that planetary orbits are elliptical—and just as unavailable to commonsense unassisted by algebra. The point I am making is comparable to the criticism which might be offered against Newtonian mechanics—that it deals with point-masses, frictionless surfaces, an unresisting medium. It is good mathematics, but as science it is good in outer space, but poor in the sticky conditions on earth. Now, as soon as we have phenotypes we are in a realm whose correlate in the physical world would be that of friction, turbulence, bodies occupying volumes, and all the other complexities which have to be added to the Newtonian picture before it can be actually used. This is what we now need to do to neo-D.

But we do not, in my opinion, need to give up the basic point that fitness is essentially 'survival capacity' (i.e. capacity to leave offspring). I was surprised to find John, in his para 3(c) ready to allow this position to be overrun by the enemy. But it is the fundamental strategic strong-point of the whole of Darwinism. Once you concede that the thing that survives, i.e. that contributes most to evolution, is something other than the thing which leaves most offspring, then you might as well go straight back to Special Creation and have done with it. The point is not to compromise on the issue that the only way to contribute to evolution is to leave offspring, but to ask more sophisticated questions about just which organisms *do* leave more offspring.

Finally, I do not follow the last paragraph in John's section 3(b). Of course some plant and animal species sometimes have lost the capacity for certain types of reproductive performance, including sexual reproduction in general ; but I see no reason to doubt that this loss is produced by natural selection of those individuals with the greatest capacity to leave offspring in the environments immediately available. Of course, again, selection, which operates on the differences in fitness between contemporary individuals, may push a population into an evolutionary situation in which it is unable to cope with changes in its environment, and so becomes extinct. But to state that a certain property is being maximised within a population does not imply that it is always getting greater. Natural selection will pull up fitness as far as it can, but that may still not be far enough to ensure the survival of the population. The point which John is making here against my concept of 'fitness' is the same as that which he advanced to David Bohm against Wynne-Edwards concept (p. 90) ; and I feel justified in defending myself even more strongly than I defended Wynne-Edwards on p. 95.

Gibbs Ensemble and Biological Ensemble

Reprinted from the Annals of the New York Academy of Sciences
Volume 96, Article 4, pages 975–84, 2 March 1962

Edward H. Kerner
University of Buffalo

With apologies to practising biologists, especially ecologists, who labour first-hand with perhaps the hardest observational material in all science, I am going to trespass a bit in difficult and unfamiliar terrain and venture some ideas at second-hand.

My proposal is to consider seriously as a first step in a general theory of ecology the scheme of population dynamics advanced by a great mathematician, Vito Volterra (1931), almost 30 years ago. The dynamics can be written down in one line:

$$\frac{dN_r}{dt} = \epsilon_r N_r + \frac{1}{\beta_r} \sum_s a_{sr} N_s N_r$$

(N_r: population of rth species).

The meaning may be simply stated: beyond an intrinsic growth or decay rate leading to the Malthusian $N_r(t) = N_r(0)e^{\epsilon_r t}$ (ϵ of either sign) if each species were isolated from all others, the growth rate of each is conditioned further only by binary (N_r, N_s) interactions between species (a_{sr} measuring the strength of interaction) in a purely reciprocal or predator-prey manner ($a_{sr} = -a_{rs}$).

It would seem rather odd to suggest so simplistic a scheme as a guide to understanding what occurs in the real ecological world. Objections abound. Since birth, death, and growth are complicated stochastic processes and population numbers are discrete, not continuous variables, how can deterministic differential equations be useful? What of the age distributions of different species? Where is account taken of the complex of breeding habits and peculiarities of the life cycles of the different species? Where is any description given of populations in space as well as time, of immigration currents and herding instincts, or of season, climate, geography, and topography? What about symbiotic and other kinds of interactions besides the antagonistic ones?

I believe that these and other questions are answerable only by considering what one may expect of a theory or of an initial theoretical attempt. In ecology we are faced with a stunning, even paralysing, array of variables and factual data. For theoretical purposes is it not a matter of trying to sift out of the array some few

129

of the most basic elements, the grossest realities, and of trying to see what organization amongst them alone may first be found ? In short, should we not try to build that first rudimentary model that captures some of the distilled, if distorted essence of the reality ?

Let us imagine that our knowledge of atomic phenomena had preceded that of celestial mechanics and that our empirical data in physics were wholly at what we now call the quantum level, our objects of study being atoms, molecules, and quanta. Things would be difficult. Such theories as might be attempted would likely deal directly in probabilities, the data being rather lawless and experiments being imperfectly reproducible. It would be at some stage, we may guess, a far-fetched speculation to suggest, for instance, that in a collision of two large molecules we forget the empirically visible complicated internal characteristics of each, forget the myriad electrons and nuclei and the massive data about them ever before our eyes, and treat the collision as a *two*-body collision, inventing a— to us—gross fiction of some kind of 'billiard ball' representation of the molecules. Nevertheless, let us trust that the invention would be seen to be useful for explaining *some* things, although surely not everything. The further perception by a latter-day Newton of some simple limiting kind of deterministic 'laws' that could *for certain purposes* be used to help interpret the discretenesses, discontinuities, and probabilities of our daily experimental fare would not be without merit. To be sure, we should not be fooled by such laws, knowing full well that reality was not so simple.

I suggest that the Volterra generalized predator-prey scheme (1) grasps a root matter of macroecology, (2) has the necessary generality and simplicity to allow a wide assortment of observationally testable results to be made, that is, has real theoretical viability, and (3) admits appreciable elaboration, both in respect of the primitiveness of the predator-prey scheme itself and of comprehension of population variations in space as well as time, and also, in principle, of effects of physical factors such as temperature and radiation intensity. Its crudity is not its weakness but its strength. At the least I think it illustrates the *kind* of thing a general theory can possibly do in ecology, although the theory itself may be wrong or incomplete. By its generality is meant its capacity to view many species in association, without recourse to the abstracting of one or a few species from the rest of the biological world.

The root matter referred to is simply this : cells eat cells and thereby beget cells, but some classes of cells, instead, live off nonvital matter and energy (the interconnection of the nonvital and vital elements may at first be set to one side). This

130

Edward H. Kerner

has a peculiarly biological character, apart from the specific meaning of 'cell', without very much counterpart elsewhere in the observable world (an exception may be found in certain classes of biochemical reactions *in vivo* where chemical concentrations may be coupled by Volterra-type equations [Hinshelwood, 1951; see also Lotka, 1920]; more generally, the equations of chemical kinetics are not unrelated to those of Volterra). It is universal (or trite, if you like, but surely not trivial); it is perhaps the single overmastering phenomenon in the ecosphere, without which any comprehensive view of the large biological world is bound to be vacuous. Volterra's dynamics is but a mathematical paraphrase of it.

Let us not forget that at least in *some* microecological cases (figure 1) the Volterra scheme speaks the essential truth, and that on the macroscopic scale it answers, as none other does, to possibly the grossest fact of general observation: that of the unremitting fluctuations in populations beyond the incidents of season and migration.

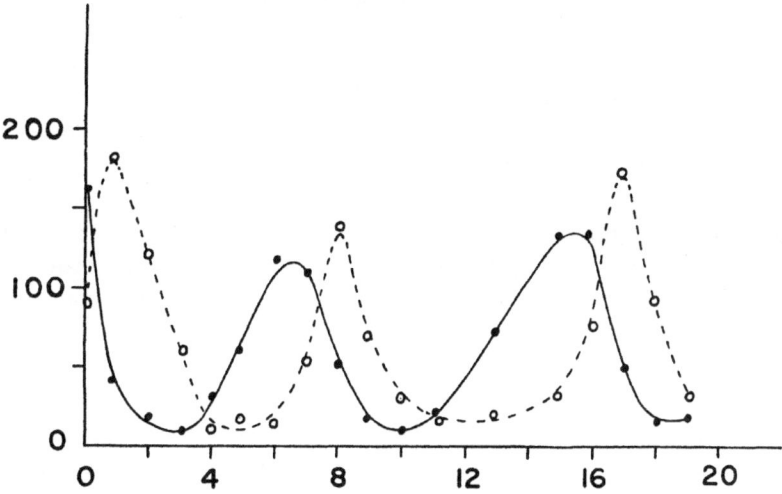

FIGURE 1
Population variations of *Paramecium aurelia* (*broken curve*) and *Saccharomyces exiguus* (*full curve*). Ordinate shows the number of individuals in 0·2 gm samples; abscissa is time in days. Adapted from Gause in D'Ancona (1954)

In thus advocating the blurred rather than the sharp view of the biological scene, toward the end of encompassing dominant effects, there is of course no assertion of the literal validity of the Volterra mechanics, any more than that it is claimed that

131

complex molecules are billiard balls and that classical theory can describe quantum phenomena. Rather the position set forth is that in the long term and over the large view of a big multicomponent biological association, the association operates in some measure *as if* actuated first by the rough biological rule incorporated in the Volterra dynamics, observed population oscillations being describable as effectively of Volterra type. 'Predator' and 'prey' of course need not have only literal meaning but, additionally, the significance that one species' losses add up somehow to another's gain, without precise specification of just how.

It is certainly true that the behaviour of many laboratory microcosms indeed shows Volterra's scheme to be nugatory as a general microscopic proposition ; characteristically in some of these cases it is probabilistic rather than deterministic results that are manifested : for example, the survival of one of two competing species, but not always the same one. Behind any deterministic scheme such as Volterra's there are surely more basic stochastic elements. This in itself is no bar to useful deterministic theorizing. One may even surmise that the useful deterministic theory, granting its usefulness, must stand as some deducible limit within the complete stochastic theory, a touchstone and a clue.

I have mentioned the simplicity of Volterra's mechanics. Actually the coupled differential equations admit of complete analysis only for two interacting species. Wherein is their general simplicity ? It is in two conservation laws : (1) they admit one integral to view and none other is apparent, the quantity

$$G \equiv \Sigma q_i \beta_i (e^{v_i} - v_i)$$

is conserved throughout the motion, v being log (N/q), q the average population level $\bar{N} = (1/T) \int_0^T N(t) dt$; (2) they admit a Liouville's theorem in the many-dimensional space $v_1, v_2 \ldots$ ('phase space'), to the effect that a bag containing a cloud of phase points (each corresponding to different sets of initial values of the N_i or v_i and thereafter being propelled by the dynamical laws) keeps its volume but not its shape unchanged throughout the course of its motion. The parallel with physics is striking : the two-body problem is amenable to full discussion, the many-body problem is not ; yet from energy conservation and Liouville's theorem (in the physical phase space of coordinates and momenta of the many particles) progress can be made. The name of this progress is statistical mechanics, its chief author is J. Willard Gibbs. The above-mentioned cloud is a Gibbs ensemble and is the mental instrument for statistically assaying the vastly complicated and practically unknowable motion of any single phase point ; it is exactly the amassing of ignorance in the cloud by which knowledge is gained. As one may see, statistical mechanics has in essence nothing to do with mechanics, but is really a statistical theory of differential equations; it simply happens that the Volterra equations, as well as Newton-Hamilton's and possibly others, fit into the scheme.

The statistical-mechanical idea is that despite (or even because of) our inability to unravel the details of motion of very large numbers of variables, we can still draw statistical conclusions about them and can describe through system-type quantities, such as temperature and entropy, important macroscopic features of the whole (for example, entropy tends to increase, and energy tends to flow downward along temperature gradients). Moreover, if we had full knowledge of the details it would be an embarrassment against the necessary poverty of data obtained from observation: necessary because there would not even be enough data books for recording information about all the variables. Thus the full knowledge would come to be reduced to manageable proportions via averaging procedures dictated by the meager data available; statistical mechanics of some type would be invented. The statistical overlay of dynamics, in short, is very nearly compulsory when the dynamics covers large numbers of degrees of freedom. If very accurate laws of population dynamics were known, whether deterministic or stochastic, some means of surveying the mass of them statistically would still be essential to understanding macroscopic observations. It could be argued from this standpoint that the accuracy of the laws destined in part for a statistical hopper might not have to be so great, especially if the observations themselves were not very fine.

The conservative character of Volterra's dynamics, if it have some general meaning, then stands as a curious feature of a system that physically is energetically and materially open and nonconservative. It is an interesting question how this may be comprehensible in the roughest conceptual terms from the underlying biochemical ultimates; one wonders how understandable it is in principle (Elsasser, 1958).

The role of any macroscopic parameter—such as temperature—is somewhat different biologically from what it is physically. Whereas temperature as a system property is directly observed physically, it is only indirectly observable in the biological association, for the association is observed fundamentally through population-time curves of one or a few selected species. The physically comparable observation is that of the displacement along an axis, as a function of time, of a Brownian particle. Both curves bear the stamp of the operation of complex environs; neither is predictable in detail but only in respect to statistical properties such as mean amplitude or mean frequency of oscillation. In short, both are kinds (different kinds) of random noise and, as such, they function as 'thermometers'

Gibbs ensemble and biological ensemble

or other thermodynamic indicators as well as anything else could. The macrofeatures of the large system cannot but reveal themselves through the fluctuations of the microobservables. To say, for instance, that the whole system is in thermodynamic equilibrium is only to say that different long strips of the microcurves exhibit always the same statistical qualities : the strips of noise, taken from different time intervals, look 'the same'. The macroecological world is in some ways bound to be a strange one for us, for we are submicroscopic elements in it—somewhat like electrons in a molecule in a great sea of molecules—and therefore conscious firstly of the fine particularities that impinge on us. Nevertheless, how does the biological world look to an observer far removed from local detail and unable to distinguish even the discrete individuals comprising a population, but seeing things only grossly, as the chemist sees the colour change of an indicator in the course of a titration without so much as a glance at its atomistic structure ? Our hyperfine powers of observation and our too-intimate engagement with what we are observing may perhaps be hindrances to a certain order of understanding.

The basic rules of statistical mechanics tell what distribution of phase points in phase space is proper to the description of thermodynamic equilibrium ; it is the canonical distribution

$$P(v_1, v_2, \dots)\, dv_1\, dv_2 \dots = e^{(\Psi - G/\theta)}\, dv_1\, dv_2 \dots$$

giving the probability that the v_i are found in the intervals v_i, $v_i + dv_i$. Here θ means 'temperature'', Ψ 'free energy'', the factor $\exp(\Psi/\theta)$ being a normalization constant. If one replaces G by the Hamiltonian of a physical system (that is, the energy written as a function of coordinates and momenta of the component particles) one obtains the chance for finding the coordinates and momenta of all particles in any infinitesimmal range of their possible values. For a gas of weakly interacting atoms this leads to the Maxwell-Boltzmann law

$$P(V_x, V_y, V_z)\, dV_x dV_y dV_z \sim \exp - \left(\frac{\frac{1}{2}M(V_x^2 + V_y^2 + V_z^2)}{kT} \right) dV_x dV_y dV_z$$

for the probability that any one of them has a velocity in V_x, $V_x + dV_x$; V_y, $V_y + dV_y$; V_z, $V_z + dV_z$ (here θ stands for kT = Boltzmann's constant $\times$ absolute temperature).

In completely parallel fashion for the biologic case, if one at random opens the door on an association and peers in at some species, for example the kth, one has a chance

$$P(v_k)\, dv_k \sim \exp - \frac{q_k \beta_k}{\theta} (e^{v_k} - v_k)\, dv_k$$

of finding

$$v_k \equiv \log (N_k/q_k) \equiv \log n_k$$

in the range

$$v_k, v_k + dv_k$$

Or, in n-language, one's chance for finding

$$n_k \text{ in } n_k, n_k + dn_k$$

is

$$P(n_k) \, dn_k \sim n_k^{x_k-1} e^{-x_k n_k} dn_k$$

x_k being short for $q_k \beta_k / \theta$. Substantially this same law was invoked by Corbet *et al.* (1943) for interpreting butterfly- and moth-catch data, with x_k taken as some small parameter (meaning here $\theta \gg q_k \beta_k$).

The way is not yet clear for using the apparatus of Gibbs ensembles to interpret the basic observational datum, the single population-time curve that was likened before to a Brownian particle's displacement-time curve. The answer here is that (ergodic hypothesis)

time averages = ensemble averages

more exactly that the time average of any function of population numbers, $f(N_1(t), N_2(t), \ldots)$, translated into a function of phase variables, $F(v_1(t), v_2(t), \ldots)$ is

$$\frac{1}{T} \int_0^T F(v_1(t), v_2(t), \ldots) \, dt = \int_{\text{(all phase space)}} F(v_1, v_2 \ldots) e^{(\psi - G/\theta)} dv_1 \, dv_2 \ldots$$

where we assume that the association has been let run a long time and has attained equilibrium, and where T is long enough to cover very many cycles of population oscillation. Strict proof of ergodicity is difficult to ascertain, but indirect arguments and the weight of statistical-mechanical experience point to its validity.

A principal result stemming from this understanding of averages is that (bars meaning time- or ensemble-averages)

$$\frac{\theta}{q_k \beta_k} = \frac{\overline{(N_k - \bar{N}_k)^2}}{\bar{N}_k^2} = \overline{\left(\frac{N_k}{\bar{N}_k} - 1\right) \log \frac{N_k}{\bar{N}_k}}$$

Thus the meaning of 'temperature' is in effect that it measures the amplitude of oscillation (in two ways) away from the average population level $\bar{N}_k$. Zero temperature is the completely quiescent state $N_k = \bar{N}_k$. We can think of 'heat' exchange (that is, G-exchange) between two associations at different temperatures in a meaningful way; the high-θ association cools off — all its species experiencing a fall in amplitude of fluctuation — as the low-θ association warms up, all its members increasing their amplitudes of fluctuation. A heat capacity can indeed be defined analogously to the physical one and, curiously, it has a temperature dependence much like common physical ones. Similarly entropy can be defined; it shows a monotonic rise with rising temperature. However, calorimetry is not thermodynamics. The concept of *work* is missing. To introduce it, the variations of external parameters (such as volume in the physical case, or radiation intensity or physical temperature in the biological case), as they induce changes in the association, need to be considered; the general nature of such variations remains to be studied.

Plainly the calculation of time averages relevant to the observationally basic population variations of one species will be vitally controlled by the statistical parameter x_k. More elaborate analysis brings out qualitatively that: (1) at small x

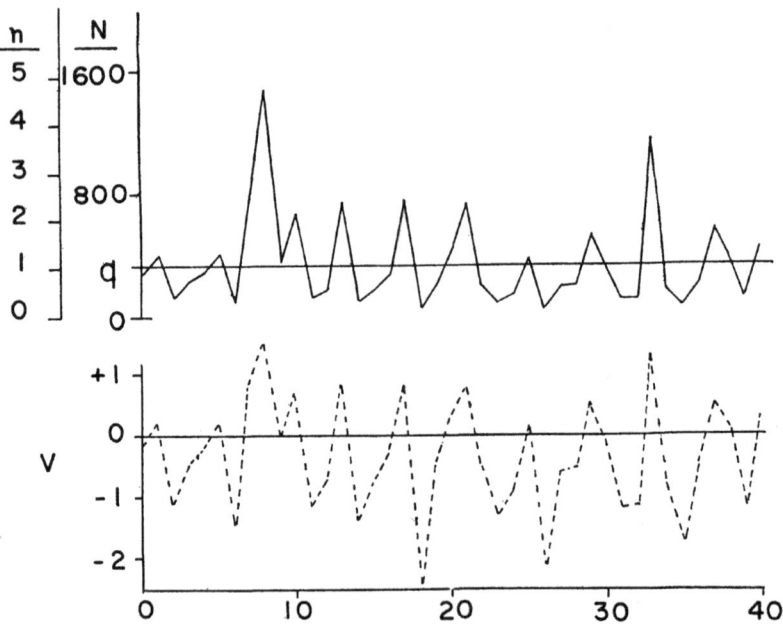

FIGURE 2
Labrador fox-catch for the first 40 years of a 90-year period. Ordinate N, upper curve, shows the catch, q being the average catch. Ordinate n is the catch measured in units of q. According to Elton, it is reasonable to assume that the catch is proportional to the population. The lower curve shows the reduced variable $v = \log n$. In the statistical theory, such a population-time curve as this is taken to represent a kind of 'noise'. Adapted from Kerner (1959); fox-catch data from Elton (1942)

136

(high θ) the population spends most of its time at below-average levels, oscillating below in long shallow troughs and above in short high peaks; (2) at large x (low θ) the frequency of oscillation is highest (but is altogether a slowly varying function of x) and the upward swings away from the mean level are closely comparable with the downward swings; (3) the amplitudes of oscillation above and below the mean level, taken separately, are not rapidly varying with x except at small x (for example, $x < 1$) where the upward amplitude alone increases rapidly with decreasing x; and (4) the mean amounts of time spent above and below the average level are similarly slowly dependent on x except at smaller x.

The important question is: Can the single parameter x describe the host of statistics that can be ferreted out of the population-time curve? A whisper of an answer is given in figures 2 and 3 and table I, where Elton's (1942) Labrador fox-catch data is analysed and a comparison with theory made (original sources

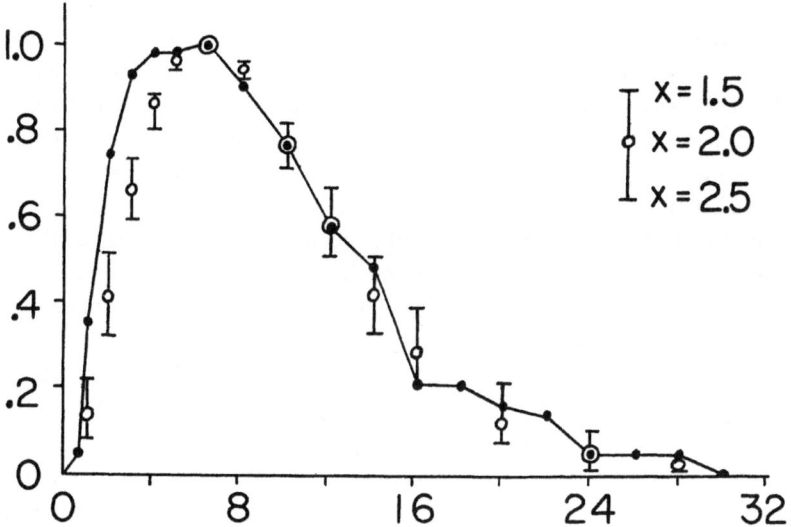

FIGURE 3
Comparison of observed and theoretically expected frequencies with which the fox-catch curve crosses a given axis. In figure 2 draw horizontal lines at intervals of 50 fox catches; the ratio of crossings of the curve on these lines to the crossings on the line q is here plotted as the ordinate against the ordinal number of the line as abscissa. Solid dots are observed values; circles are expected values corresponding to the choice $x=2\cdot0$ of the parameter x; for comparison, expected values for $x=1\cdot5$ and $2\cdot5$ are also shown. The statistical errors of the observed values are appreciable, possibly 40 per cent or more. Adapted from Kerner (1959)

137

Gibbs ensemble and biological ensemble

TABLE I
Comparison of observed and theoretically expected statistical data for the fox-catch curve of figure 2

Quantity averaged	Time average	x	x range
Fraction of time spent at below-average levels	0·595	1·90	0·70—15·0
Amplitude of oscillation of n for $n < 1$	0·445	2·10	1·00—2·75
Amplitude of oscillation of n for $n > 1$	0·729	1·71	1·42—2·05
$(n-1)^2$	0·536	1·87	1·69—2·07
$(n-1)\log n$	0·489	2·04	1·86—2·27
$\log n$	−0·258	2·08	2·31—1·91

The question at issue is whether the *single* theoretical parameter x can account for a variety of statistics. The first column gives the observed time average, using directly the polygonal line of figure 2; the second column gives the theoretical x resulting in the same value; and the third column gives the range of x corresponding to a ±10 per cent alteration of the observed values. Adapted from Kerner (1959)

may be consulted for details). The whispered answer seems to be, possibly yes.

A final point remains to be mentioned about Volterra's model. Its conservative character stems in a basic way from the evenness of the number of interacting species. This has quite a dubious and artificial air about it. Its general import, however, may not be without real biological meaning, since, when the number of species is odd, what occurs is a tendency of decay into an even association with exhaustion of at least one species. In dynamical terms, the system exhibits some dissipative characteristics. It has been part of ecological and evolutionary thought for a century or more that closely similar species cannot both occupy the same ecological niche indefinitely, but that the slightly more 'successful' species eventually completely supplants the other (Volterra-Lotka principle, see Hardin, 1960). It turns out to be feasible to describe such extinction-domination struggles within Volterra's theory as an odd → even decay. The advantage over the usual description is (1) the avoidance of the unrealistic element, in the latter, of an ultimately static population of the successful species, (2) an accounting of the strugglers in immersion in a larger biological world. Under the hypothesis of close similarity of the competitors, the dissipation still can be viewed through the lens of the conservative theory; a species of secularly drifting statistical mechanics can be constructed; and the short-term oscillations of the strugglers may be seen against the backdrop of the controlling long-term rise of one and fall of the other. A description of the workings of the creation and annihilation of species is not beyond the range of the theory.

138

Edward H. Kerner

References

Corbet, A. S., R. A. Fisher & C. B. Williams. 1943. The relation between the number of species and the number of individuals in a random sample of an animal population. *J. Animal Ecol.*, **12** : 42–58.

D'Ancona, U. 1954. The Struggle for Existence (E. J. Brill, Leiden, Germany).

Elsasser, W. 1958. *The Physical Foundation of Biology* (Pergamon Press, New York, N. Y).

Elton, C. 1942. *Voles, Mice, and Lemmings* (Clarendon Press, Oxford, England).

Hardin, G. 1960. The competitive exclusion principle. *Science*, **131** : 1292–7.

Hinshelwood, C. 1951. Decline and death of bacterial populations. *Nature,* **167** : 666–9.

Kerner, E. H. 1957. A statistical mechanics of interacting biological species. *Bull. Math. Biophys.*, **19** : 121–46.

Kerner, E. H. 1959. Further considerations on the statistical mechanics of biological associations. *Ibid.*, **21** : 217–55.

Kerner, E. H. 1961. On the Volterra-Lotka principle. *Ibid.* To be published.

Lotka, A. J. 1920. Undamped oscillations derived from the law of mass action. *J. Am. Chem. Soc.*, **42** : 1595–8.

Volterra, V. 1931. *Leçons sur la Théorie Mathématique de la Lutte pour la Vie* (Gauthier-Villars, Paris, France).

Volterra, V. 1937. Principes de Biologie Mathématique. *Acta Biotheoretica*, **3** : 1–36.

139

A Statistical Mechanics of Temporal Organization in Cells

Reprinted from Society for Experimental Biology Symposium No. 18, 1965, pp. 301–26

Brian C. Goodwin
Massachusetts Institute of Technology

Running through science there is a duality in the description of systems, which has resulted in some of the deepest problems of scientific method and analysis. This duality arises from the analytical procedure which reduced all complex, macroscopic systems to a set of simpler, microscopic components. Natural processes are thus regarded as having two sides : a macroscopic or phenomenological one, which is observed directly ; and a microscopic one, which may not be directly observable but which is postulated to underlie and in some sense explain the macroscopic process.

Cell biology is currently going through a molecular revolution in which almost daily discoveries are laying bare the molecular organization of cellular processes. However, the macroscopic aspects of cell behaviour, which form the content of cell physiology, are left largely unexplained by these developments. Thus, for example, recent studies in molecular biology have shown that negative feedback control mechanisms operate at the molecular level in cells, but detailed analysis of these control circuits fails to explain the integrated, higher order properties of cells, such as homeostasis, competence, and circadian organization. There remains a gap between molecular biology and cell physiology, which is a direct consequence of the analytical method of science.

The problem of bridging such a gap, of resolving microstructure and macrostructure in systems, is not new in science. It arose first in connexion with classical thermodynamics, which is a phenomenological or macroscopic science, and the kinetic theory of gases, which is a molecular or microscopic theory. The man who set himself the task of deducing thermodynamic laws from the dynamic properties of molecules was Boltzmann. He and his successors, notably Willard Gibbs, succeeded in developing a theory which allows one to do just this. It is called statistical mechanics, and in considerable measure it succeeds in overcoming the duality inherent in scientific analysis, although in so doing it has posed some theoretical problems which to this day remain unresolved.

The purpose of this paper is to study the possibility of forging a direct link between molecular biology and cell physiology by using essentially the same formal, logical procedures which were used in the development of classical

140

Brian C. Goodwin

statistical mechanics. The starting point will be a study of the dynamic behaviour of molecular feedback control circuits in cells, which will play, in our theory, a role similar to the kinetic theory of gases in statistical thermodynamics; and the goal is the derivation of certain macroscopic variables which are formally analogous to thermodynamic quantities: temperature, entropy, free energy, etc. These macroscopic variables will then be used to describe and analyse the coordinated or organizational properties of cells, such as homeostasis and adaptation. One of our tasks will be to discover the observational or experimental significance of these new thermodynamic-like variables, and to see how they can be used to give analytical precision and quantitative content to our intuitive ideas about the nature of cellular organization.

THE CONTROL CIRCUITS

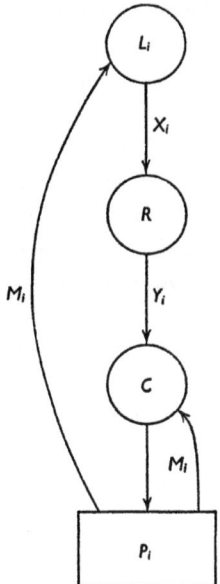

FIGURE 1

The kinetic theory of gases had a ready-made dynamic foundation in Newtonian mechanics. Unfortunately there is no parallel to this for the dynamics of molecular control mechanisms. However, on the basis of the purely qualitative features of

141

cellular control circuits, and the behaviour of negative feedback devices studied in engineering, it is possible to derive a set of differential equations which describe, albeit in very crude form, the dynamics of a class of control mechanisms which on circumstantial grounds seem to have a high probability of occurring in cells.

There are at least two types of feedback mechanism operating in cells to control macromolecular activities. They are known as feedback inhibition and feedback repression, and according to current theory they have the characteristics shown diagrammatically in figure 1. Here L_i represents a genetic locus which produces messenger ribonucleic acid (mRNA) of the ith informational species in quantities represented by X_i. This specific 'signal' encounters a cellular structure, R (a ribosome), where its activity results in the synthesis of informationally homologous protein in quantities denoted by Y_i. The protein then travels to some cellular locus, C, where it exerts an influence upon the metabolic state either by enzyme action or by some other means. (We will concentrate in the following on the case where Y_i is an enzyme.) The result of this activity by the enzyme is the generation of a metabolic species in quantity M_i, which enters a metabolic pool, P_i.

Two different closed control loops are now obtained by a feedback of the metabolite either to the site of enzyme activity, C, or to the genetic locus, L_i. In the first case the metabolite reduces the activity of the enzyme ; this is feedback inhibition (Umbarger, 1956 ; Magasanik, 1958). In the second case the metabolite reduces the rate of deoxyribonucleic acid (DNA)-mediated synthesis of mRNA at the genetic locus, and this is called feedback repression (Vogel, 1957). There are a great many details and problems which this description leaves out. For example, there is the question of how M_i exerts its effect upon the enzyme (cf. Gerhardt & Pardee, 1962) ; and the nature of the actual genetic repressor remains unknown, although there is evidence that it is a complex between M_i and some macromolecular species (Monod & Jacob, 1961). Furthermore, there is the possibility that a third level of control exists, wherein the metabolite feeds back to the ribosome-mRNA complex and reduces the rate of protein synthesis.

In the following analysis we will consider in detail only the dynamics of feedback repression, involving the control loop from the gene, through mRNA and protein to metabolite, and back to the gene. The control sequence in this circuit is believed to be the following : X_i controls the rate of synthesis of informationally homologous protein ; Y_i controls the rate of production of the ith metabolite ; and M_i controls the rate of synthesis of mRNA as well as taking part in metabolic reactions. These quantities will therefore be regarded as the essential control variables in the respective biochemical reactions. This involves the assumption

Brian C. Goodwin

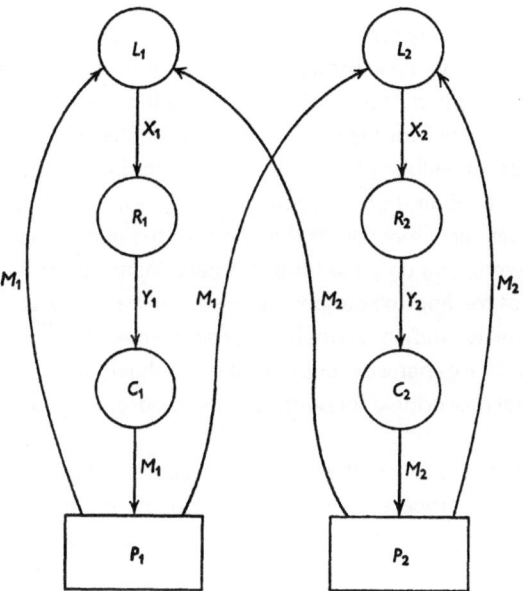

FIGURE 2

that other possible rate-limiting factors, such as size of nutrient and energy pools,
concentrations of aporepressors (i.e. the macromolecule postulated to act in
conjunction with the metabolite to form the effective repressor complex), etc.,
can be treated as parameters of the 'motion' of the variables, X_i, Y_i and M_i. That is
to say, these quantities are considered to remain constant or to change very slowly
compared with the dynamic motion of the control variables. It is possible to
extend this control scheme to cover more complicated circuits, such as when
several enzymes form a biosynthetic sequence and are controlled simultaneously
by one single metabolic species, the end-product of the sequence (Gorini, 1963).
We will not, however, consider these modifications here.

More complicated situations will also be analysed, in which repression occurs
between different control components, so that we get the situation shown
in figure 2. Here a metabolite controlled by Y_i interacts by repression with
another genetic locus, L_2, while a reciprocal interaction occurs from L_2 to L_1,
through M_2.

143

Temporal organization in cells

We will now establish a set of very coarse functional relationships between the variables X_i, Y_i, and M_i, which describe the essential dynamic features of the type of control system which we have in mind. If the differential equations thus obtained can serve as the dynamic basis for the construction of a statistical mechanics and 'thermodynamics' of cellular processes, then the crudeness of the initial equations will not be a severe limitation in the macroscopic analysis. This is because it is the nature of a statistical mechanics that many of the microscopic details are smoothed out, as it were, and only the fundamental dynamic properties are retained. Therefore in spite of the approximations necessary in the microscopic description of the control circuits, we may nevertheless get some idea of the macroscopic quantities relevant to a general description of cell behaviour, and some suggestion of experimental procedures for controlling and observing these macroscopic quantities.

Consider first an equation for the rate of synthesis of the ith species of protein, when the controlling variable in this process is X_i, the concentration of the corresponding species of mRNA. The general form of the equations to be considered is

$$\frac{dY_i}{dt} = f_i(X_i, Y_i, M_i) - g_i(Y_i, M_i) \qquad \qquad \text{....(1)}$$

where $f_i(X_i, Y_i, M_i)$ is a function describing the rate of synthesis of protein, while $g_i(Y_i, M_i)$ relates to the rate of its degradation. The simplest conceivable functions which satisfy the control requirements of our model are given by:

$$f_i(X_i, Y_i, M_i) = a_i X_i, \quad g_i(Y_i, M_i) = \beta_i$$

Then $\dfrac{dY_i}{dt} = a_i X_i - \beta_i$ $\qquad \qquad \text{....(2)}$

Here $a_i X_i$ represents the rate of mRNA-controlled protein synthesis, while β_i is the rate of protein degradation, assumed to be a constant. Since protein synthesis is an almost completely irreversible process, no terms for the reverse reaction on the template are included.

In equation (2) the constant a_i is a composite parameter containing a rate constant for the template synthesis of the ith species of protein and concentration terms for activated amino acids, the precursors of protein synthesis. The simplification involved in using such a representation for what is clearly a very complex biochemical process may seem to invalidate our analysis from the start. The one feature of the real process which is incorporated in this equation is control of protein synthesis by messenger RNA. Since it is the dynamic consequences of this control which we seek to investigate, we will go no further than equation (2) at present. The more general form of equation (1) allows certain modifications to be included if they prove to be essential.

CONTROL EQUATION FOR mRNA SYNTHESIS

The equations which we consider for messenger RNA synthesis will be of the general form

$$\frac{dX_i}{dt} = \phi_i(X_i, Y_i, M_i) - \psi_i(X_i, Y_i, M_i) \qquad \qquad ....(3)$$

Here $\phi_i(X_i, Y_i, M_i)$ is a function describing mRNA synthesis, and $\psi_i(X_i, Y_i, M_i)$ represents the rate of its degradation. It is assumed that the kinetics of repression of mRNA synthesis by the metabolite M_i are essentially the same as those of enzyme inhibition. This means that we are dealing with a surface-binding phenomenon wherein the repressing molecule or complex combines reversibly with the DNA template and so interferes with its synthetic activity. The template also combines reversibly with the precursors for RNA synthesis.

There are many details which must be considered before even a crude expression for control of mRNA synthesis by metabolite can be obtained. These include the kinetics of the reaction between DNA templates and activated nucleotides on the one hand and repressors on the other, and the relationship of the repressing complex to the metabolite concentration. It also requires certain assumptions about the 'storage capacity' of the metabolic pool. When the pool size is very small, one would expect very little or no metabolite to 'spill out' of the pool and repress the activity of the genetic locus. But when the pool is large, then a considerable fraction of the metabolite would be expected to serve a repressive function. These and other considerations are treated in detail elsewhere (Goodwin, 1963), and unfortunately there is too little space to present them here. I will therefore pass directly to the simplest possible equation which can be derived on the basis of our assumptions for the control of mRNA synthesis by metabolite, which is

$$\frac{dX_i}{dt} = \frac{a_i}{A_i + m_i M_i} - b_i \qquad \qquad ....(4)$$

Here the rate of degradation of mRNA, b_i, is assumed to be a constant. The other parameters, a_i, A_i, and m_i are quite complicated functions of more elementary constants. The storage capacity of the metabolic pool is an important quantity in determining the size of A_i and m_i.

In equation (4), the main characteristic is the appearance of M_i in the denominator of the expression for mRNA synthesis. This is a consequence of the assumption that repression is a surface adsorption phenomenon. Similar expressions were obtained by Szilard (1960) in his very interesting studies of control processes in cells.

CONTROL EQUATION FOR METABOLITE SYNTHESIS

There is one more step to take before the equations reach their final form. Since the concentration of metabolite was assumed to be controlled by the concentration of enzyme, Y_i, we can write down one more control equation showing how these variables are related. The simplest kinetic scheme for the synthesis of M_i is an equation of the form

$$\frac{dM_i}{dt} = r_i Y_i - s_i M_i \qquad \qquad \dots (5)$$

The parameter r_i represents a composite constant which includes the rate constant for the enzyme, the concentration of its substrate, and the Michaelis constant for the reaction whose product is M_i. The term $s_i M_i$, s_i a constant, implies that M_i is drawn off from the pool at a rate dependent upon its own concentration, i.e. that this process is primarily metabolite-controlled.

At this point in the argument we use a device which is frequently employed in kinetic studies. Because enzyme-catalysed reactions are much faster than template-directed macromolecular synthesis, their ratio being about 1000 : 1 (cf. Goodwin, 1963), it can be assumed that the variable M_i will be always at, or very close to, a steady-state value in relation to the comparatively slow rates of change in the variables X_i and Y_i, which are macromolecular quantities. We can therefore write

$$\frac{dM_i}{dt} = r_i Y_i - s_i M_i = 0 \qquad \qquad \dots (6)$$

and then solve for M_i in terms of Y_i. This puts the equation (4) into the final form

$$\frac{dX_i}{dt} = \frac{a_i}{A_i + k_i Y_i} - b_i \qquad \qquad \dots (7)$$

where $k_i = \dfrac{m_i r_i}{s_i}$.

We thus end up with a pair of differential equations in X_i and Y_i, (2) and (7), which represent the dynamics of a particular type of closed feedback control circuit. In the derivation of these equations many assumptions have been made which may prove to be invalid, and an extremely complex system has been reduced to one of almost absurd simplicity. However, the hope is that the major qualitative features of the control system have been included in the model; and it can be

146

Brian C. Goodwin

made much more complicated and sophisticated without altering its essential dynamic properties.

In all this we have not mentioned the units of the variables, which are in some sense concentrations. The units which have actually proved to be most useful are rather unorthodox; they are simply molecules per cell. The reason for this choice is that the usual chemical units for homogeneous solutions, moles per litre, have little meaning for macromolecular populations in the heterogeneous conditions of the cell interior. Furthermore, it has been estimated that for mRNA, the size of the populations can be very small in certain cells, amounting to perhaps 100 molecules of a particular informational species. There are many other questions which cannot be considered in detail here, and I must refer to a more complete treatment in my book *Temporal Organization in Cells*.

THE DYNAMICS OF THE CONTROL SYSTEM

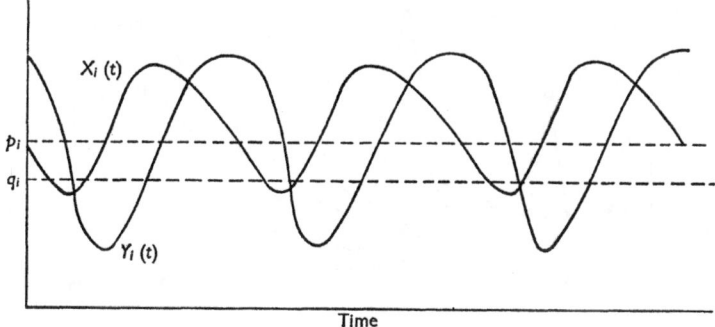

FIGURE 3

Figure 3 shows the dynamic behaviour of the variables X_i and Y_i in the control circuit of figure 1, whose motion is governed by the equations

$$\left.\begin{aligned}\frac{dX_i}{dt} &= \frac{a_i}{A_i + k_i Y_i} - b_i \\[2mm] \frac{dY_i}{dt} &= a_i X_i - \beta_i\end{aligned}\right\} \qquad \dots (8)$$

The variables undergo continuing oscillations as functions of time, varying about fixed, steady-state values p_i and q_i, which values are obtained by setting the equations equal to zero and solving for X_i and Y_i. Thus the major dynamic characteristic of the control circuit is the occurrence of oscillations in system

147

variables. It is a common experience among engineers that negative feedback control systems have a strong tendency to oscillate. Engineers usually try to avoid these 'parasitic' oscillations, as they are called, and design their control systems to prevent, if possible, the oscillations occurring. It is a central assumption however, in this study that, in the course of evolution, cells have not selected against dynamic oscillations in their control circuits, but have made use of them to organize the staggering complexity of cellular dynamics into a well-ordered, rhythmic sequence of biochemical processes. The oscillations are thus regarded as the dynamic basis of temporal organization in cells.

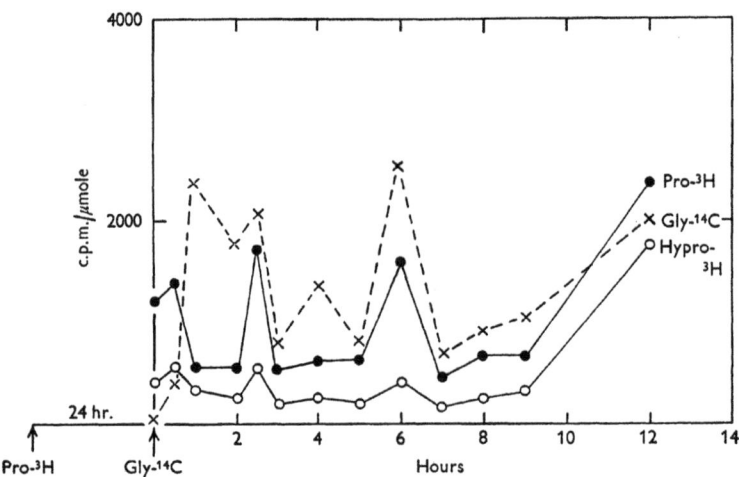

FIGURE 4
Dynamics of the appearance of proline, hydroxyproline, and glycine in extractable collagen of embryonic chick bone tissue. (After Tanzer & Gross, 1963)

There is much indirect evidence to support the idea that oscillations are an intrinsic feature of molecular dynamics in cells, most of which has arisen in connexion with the study of rhythmic phenomena, particularly circadian rhythms (cf. Harker, 1958; Pittendrigh, 1960, 1961; Halberg, 1960). Recently, however, some very remarkable experimental observations have been made on the dynamics of proline incorporation into collagen in embryonic chick cells by Tanzer & Gross (1963) and by Jackson (1963), which provide direct evidence for this belief. These investigators have observed very marked fluctuations in the specific activity of the proline pool of chick embryos after injection of radioactive proline, and synchronous fluctuations in the specific activity of labelled collagen. In figures

148

Brian C. Goodwin

4 and 5 some typical results are reproduced, with the kind permission of the authors. Figure 4 shows the dynamics of the appearance of three different amino acids, proline, hydroxyproline, and glycine, in extractable collagen of embryonic chick bone tissue after administration of labelled amino acid to the embryo. Jackson's observations on free proline in the chick embryo are shown in figure 5. Here we have observations on a system undergoing large and fairly regular fluctuations, which can hardly be ascribed to random noise in the embryo or the experimental procedure.

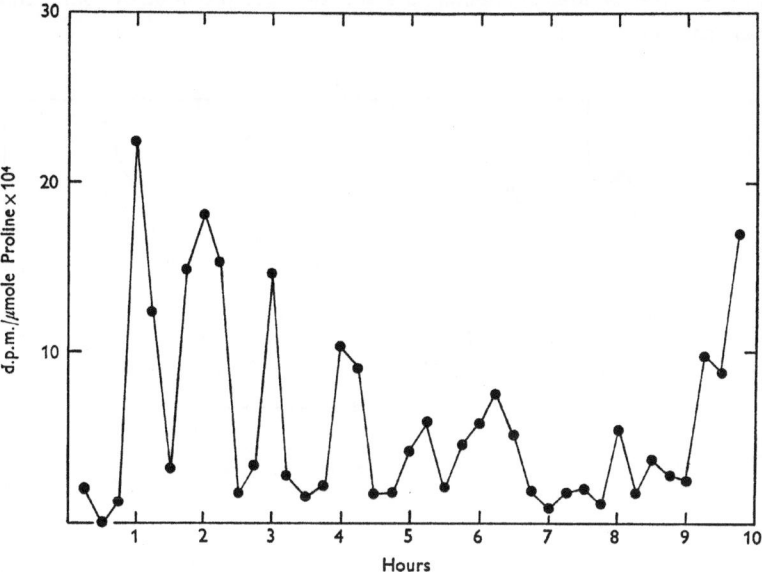

FIGURE 5
Free proline. (After Jackson, 1963)

Further evidence that the observed variations represent a non-random cellular variable is given in figure 6. It is possible to eliminate the fluctuations by a well-defined modification in the experimental procedure. The addition of a chaser of cold proline to the chick embryo after the hot injection causes the variations to disappear, and the dynamics of the system are smoothed out. This result finds a ready interpretation in the present theory of oscillating control circuits. Proline, a non-essential amino acid, is synthesized in chick cells from precursors via a

149

biosynthetic sequence which according to our assumptions is controlled by the end-product, proline. Only as long as the amino acid is being produced by this endogenous, self-regulating system will oscillations occur. (That the variable M_i, in this case taken to be proline, undergoes oscillations is an immediate consequence of equation (6), since Y_i oscillates.) If proline is added exogenously in quantities which saturate the control system, then the oscillations will cease, and in fact the endogenous production of proline will be shut off. The theory predicts that any oscillating circuit should be damped out in this manner by adding saturating quantities of the feedback molecule, the end-product, provided of course that the cells are permeable to it.

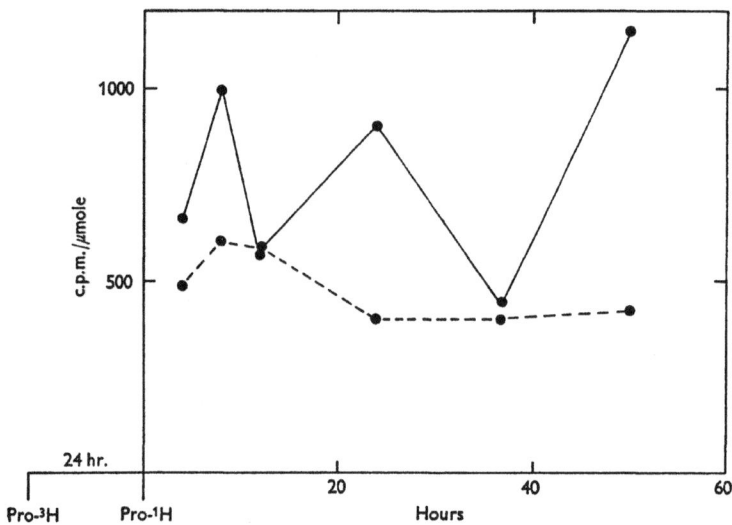

FIGURE 6
Hydroxyproline: specific activity. (After Tanzer & Gross, 1963)

Furthermore, the level at which the cut-off occurs should be at maximum pool size, i.e. at or near the bottom of the oscillating specific activity curve. This is also seen in figure 6. More work is required before it can be definitely concluded that what is being observed in these studies is a periodic variation arising from oscillations of the kind described in this paper. The absence of strict periodicity can be explained by the fact that not all the chicks used for the study are in synchrony at

150

Brian C. Goodwin

the level of intracellular metabolic oscillators. Indeed, it is remarkable that periodicity is even suggested by the investigations, since this implies a high degree of synchrony between cells synthesizing collagen, and between chicks of the same developmental age. These studies do suggest, however, that oscillatory behaviour is a characteristic of molecular dynamics in cells, and they provide a direct experimental approach to the study of temporal organization in cellular control systems.

THE INVARIANT INTEGRAL

The property of equations (8) which allows us to construct a statistical mechanics is the fact that they have an integral. This we see by combining the two equations into one and writing

$$(a_iX_i - \beta_i)\frac{dX_i}{dt} + \left(b_i - \frac{a_i}{A_i + k_iY_i}\right)\frac{dY_i}{dt} = 0$$

This expression can now be integrated, with the result that we get a constant of the motion (an invariant integral):

$$G_i(X_i, Y_i) \equiv \frac{a_iX^2}{2} - \beta_iX_i + b_iY_i - \frac{a_i}{k_i}\log (A_i + k_iY_i) = \text{constant} \qquad \dots (9)$$

This integral is analogous to the energy integral in physics, and it allows us to proceed to a thermodynamic-like description of cell behaviour, arising from the oscillating dynamics of molecular control mechanisms.

It is convenient to transform to new variables defined by

$$x_i = X_i - p_i$$
$$1 + y_i = 1/Q_i(A_i + k_iY_i)$$

where $Q_i = A_i + k_iq_i$, and p_i, q_i are the steady-state values of X_i and Y_i which make dX_i/dt and dY_i/dt simultaneously zero. The equations thus become

$$\left.\begin{array}{l}\dfrac{dx_i}{dt} = b_i\left(\dfrac{1}{1+y_i} - 1\right) \\[2em] \dfrac{dy_i}{dt} = c_ix_i\end{array}\right\} \qquad \dots (10)$$

where $c_i = \dfrac{a_ik_i}{Q_i}$.

The integral now takes the form

$$G_i(x_i, y_i) \equiv \frac{c_ix_i^2}{2} + b_i[y_i - \log (1 + y_i)] = \text{constant} \qquad \dots (11)$$

151

Temporal organization in cells

Now as a first approximation we can regard the cell as being made up of a large number of single control circuits of the type shown in figure 1. All these units are coupled together biochemically within the cell because they draw upon common metabolic pools of activated precursors for macromolecular synthesis. We call this type of coupling 'weak interaction', a concept which plays a very important role in the statistical mechanics. The very complex array of interactions occurring in the biochemical space in which the deterministic control components function is not explicitly defined in our theory, for we are largely ignorant of its details. However, we cannot ignore these interactions, for they constitute an integral part of cell structure. In the statistical mechanics these undefined interactions are represented dynamically as a 'noisy' biochemical space in which specific control processes take place, and through which dynamic motion is transmitted from one oscillator to another. It is thus that random processes and distribution theorems enter the dynamics of the control systems, one consequence of which is a maximum 'entropy' theorem.

The crude model which we thus obtain for the cell is a large number, say n, of individual, self-regulating control circuits immersed in a biochemical space which supports their existence. Each control circuit has an integral of the same type, and so for the whole system there is a general integral

$$G(x_1, x_2, \ldots, x_n; y_1, \ldots, y_n) \equiv \sum_{i=1}^{n} G_i(x_i, y_i) = \text{constant}. \qquad \ldots (12)$$

This is the integral which will play a role in the statistical mechanics similar to the Hamiltonian or total energy integral in physics. The dynamic system made up of the set of n control circuits with a total of $2n$ variables ($x_i, y_i; i = 1, 2, \ldots, n$) will be referred to as the epigenetic system of a cell.

When we consider coupled control circuits of the type shown in figure 2, a somewhat altered set of equations is obtained by using essentially the same assumptions and procedure which led to equation (8). The equations are:

$$\left. \begin{aligned} \frac{dX_1}{dt} &= \frac{a_1}{A_1 + k_{11}Y_1 + k_{12}Y_2} - b_1 \\ \frac{dX_2}{dt} &= \frac{a_2}{A_2 + k_{21}Y_1 + k_{22}Y_2} - b_2 \\ \frac{dY_1}{dt} &= a_1 X_1 - \beta_1 \\ \frac{dY_2}{dt} &= a_2 X_2 - \beta_2 \end{aligned} \right\} \qquad \ldots (13)$$

Brian C. Goodwin

These lead to a new integral wherein quadratic terms appear, i.e. terms like $X_1 X_2$. These terms reflect the presence of what are called 'strong interactions' in the system, and they lead to complicated dynamic effects characteristic of coupled non-linear oscillators. It is these strong interactions which lead to complex but stable relationships between oscillators, so that a definite temporal ordering of biochemical processes can be achieved. These effects include synchronous locking or entrainment between oscillators, wherein two coupled components lock together at the same frequency; and subharmonic resonance or frequency demultiplication, wherein a stable but asymmetric relation arises between two coupled oscillators, so that in a certain time-interval they complete a number of oscillations which is a rational fraction of those completed by the uncoupled (free-running) components. Both these phenomena appear to occur in cells or organisms with diurnal rhythms (cf. Pittendrigh & Bruce, 1957). The properties of the molecular control circuits studied in this paper are regarded as providing the dynamic foundation for such behaviour in cells.

THE STATISTICAL MECHANICS
The central mathematical content of a statistical mechanics is a probability distribution which allows one to use powerful statistical techniques for evaluating mean values of various functions which can be used to study the general, macroscopic properties of the complex, integrated system. The most useful distribution for our present purposes is that defining a construct known as the canonical ensemble. Once again lack of space prevents a full discussion of the significance of this theoretical concept, and reference must be made to a fuller treatment (Goodwin, 1963). The probability distribution is defined as

$$\rho = e^{(\psi - G)/\theta} \qquad \dots (14)$$

where G is the general integral of the system, defined by equation (11) in the case where there is no strong interaction in the system, and by a more complicated integral for a system with coupling between control components. In this expression θ is a quantity analogous to temperature in physical systems, while Ψ is analogous to free energy. These thermodynamic expressions will be prefixed in the following by the adjective talandic, deriving from the Greek ταλαντωσις, meaning oscillation. Thus we will speak of talandic temperature, talandic free energy, etc., in referring to the macroscopic quantities which define the 'thermodynamic' properties of oscillating control systems of the type considered.

By using the procedures of statistical mechanics, the following result is readily obtained:

153

$$c_i \overline{X_i(X_i - p_i)} = \theta = \frac{b_i k_i^2}{Q_i} \overline{\frac{Y_i(Y_i - q_i)}{A_i + k_i Y_i}} \qquad \qquad \dots (15)$$

In these expressions the bar signifies that the mean value of the quantity is referred to.

This shows us that $\theta = 0$ when $X_i = p_i$ and $Y_i = q_i$, i.e. when the variables are at their steady-state values and no oscillations occur in the system. Thus the point of zero talandic temperature is where there is no dynamic motion, analogous to the zero point of temperature in physics, i.e. where there is no kinetic motion of molecules. It may prove to be the case that this point of absolute zero with no oscillating motion in the control circuits is just as unattainable in living cells as its analogue is in physical systems, and that it is reached only with the death of the cell.

The condition of large θ corresponds to a very excited state in the oscillators, with the variables undergoing oscillations of large amplitude. It is of fundamental significance in this analysis that when θ is small, the oscillations are small and have the character of sinusoidal oscillations, i.e. they are effectively linear. However, as θ increases, the oscillations get larger in amplitude and progressively more non-linear. In the system with strong coupling, it can be shown that the complex interactions characteristic of non-linear oscillators, such as frequency demultiplication and synchronous locking, cannot occur when θ is very small, and can arise only at elevated θ-values. Thus the talandic temperature of the system is found to be directly related to its degree of non-linearity, and may be regarded as providing in some sense a measure of the organizational capacity of the system in the time domain. We will shortly discuss this question in greater detail.

It is interesting to consider the actual units of θ and the order of magnitude required for significant non-linearity in the dynamics of the control circuits. In equation (14) the parameter c_i is $a_i k_i / Q_i$. Here a_i has the units of a rate constant, $1/\text{time} = 1/T$. The units of k_i, an equilibrium constant, are $1/\text{concentration} = 1/C$. Q_i has no units, and b_i is a rate, C/T. Thus we get from (15):

$$\frac{1}{T} \cdot \frac{1}{C} \cdot C^2 = \frac{C}{T} = \frac{C}{T} \cdot \frac{1}{C^2} \cdot C^2$$

The talandic temperature scale thus has the units C/T.

The magnitude of θ in the epigenetic system is determined by the microscopic parameters of the control circuits as well as by their oscillatory motion. Some of these parameters can be calculated on the basis of current estimates for macro-molecular synthetic rates, but other parameters can only be guessed at. Thus the

154

studies of Loftfield & Eigner (1958) and Dintzis (1961) show that the protein synthetic time in the cells of higher organisms is of the order of a very few minutes; and investigations by Penman, Scherrer, Becher & Darnell (1963) suggest that the mean lifetime of mRNA in He La cells is several hours. These and other results allow one to estimate the values of the parameters a_i, β_i, and b_i. However, a quantity such as the storage capacity of a cell for a particular metabolic species can only be guessed at. It is nevertheless possible to make some reasonable suggestions for the values of the microscopic parameters in the equations (8), and the results obtained are (cf. Goodwin, 1963):

$$a_i = 200, b_i = 5/12, a_i = \tfrac{1}{5}, \beta_i = 20, k_i = 24, Q_i = 480$$

These are to be regarded as order-of-magnitude estimates for a control circuit in the cell of a higher organism.

Using these values, it has been calculated that an oscillation of considerable magnitude, with the population of mRNA of a particular species varying about a steady-state value of 100 molecules/cell with an amplitude of roughly 50 molecules, requires that θ has a value of approximately 100. The period of such an oscillation is about 5 hours. In general the estimates suggest that θ will be in the range from 25 to 10^3, giving oscillations with periods in the range of 2–14 hr. With θ less than about 1/10, the system approaches linearity and higher-order dynamic behaviour characteristic of non-linear oscillators is unlikely to occur.

THE CELL AS A RESONATING SYSTEM

In general this analysis of intracellular dynamics leads one to the point of view that the cell is a kind of resonating system which cycles constantly through a set of states. It is not a quiescent, passive system which moves or changes state only in response to stimuli from the environment, but rather is a vibrant entity with intrinsic dynamic activity which is oscillatory in nature. This oscillatory activity is a type of biological energy, and the suggestion made here is that cells should be considered as resonant systems with properties somewhat analogous to those which characterize conjugated biochemical species such as the purines and pyrimidines and confer upon them a position of unique biochemical importance. The many resonant modes which, according to this analysis, arise from the oscillatory behaviour of intracellular negative feedback control circuits and their interaction could explain the remarkable intrinsic coherence and stability of biochemical processes in cells. They could also account for the great lability which cells have for adaptive response to the physical environment or to other cells. The introduction of frequency and resonance considerations into the analysis of cell-cell interactions in developing embryos provides a new dimension for the investigation of such phenomena as embryonic induction, competence, and individuation (i.e. the self-organizing features of interacting embryonic cell populations).

The particular class of control circuits which has been considered in this analysis and called 'epigenetic' could account for relatively slow cellular rhythms, with periods of several hours. However, shorter control loops such as those arising from feedback inhibition would be expected to result in oscillations with shorter periods, and one should be prepared to find a wide spectrum of frequencies occurring in cells. Furthermore, the phenomenon of subharmonic resonance, well known as a consequence of interactions between non-linear oscillators, greatly extends the frequency spectrum over that of the free-running or uncoupled control circuits. Thus, for example, if one of the control circuits of figure 2 shows, separately, oscillations with a period of about 3 hours and the other a period of about 6 hours, then the whole system when appropriately coupled can show oscillations with periods ranging from 3 to 24 hr, and greater. As the order of the subharmonic increases, the amplitude of the oscillation also increases, so that relatively small free-running amplitudes can be greatly magnified when oscillators are coupled. Thus, for example, cells and organisms can generate slow rhythms, such as tidal and circadian rhythm, by the proper interaction of control processes whose fundamental frequencies are considerably greater than 1 or 2 oscillations per day. There is evidence (Pittendrigh and Bruce, 1957) that this is in fact how organisms generate frequencies which keep them in time to terrestrial periodicities.

Another phenomenon commonly observed in non-linear systems is for two oscillators with different frequencies to get locked together at the same frequency when they interact. This is known as synchronous locking of oscillators, and the common frequency can be at or anywhere between the free-running frequencies of the interacting components, depending upon the coupling. The phase relations between locked oscillators are also widely variable with the parameter values. It seems likely that stable temporal relations of this kind between different control processes in cells could play a very important role in keeping different biochemical activities always in a correct phase or time relationship with one another (Halberg, 1960).

The stability of the temporal relationships which can arise from the interaction of non-linear oscillators is strongly dependent upon the 'amount' of non-linearity in the system. In the statistical mechanics which has been described briefly in this paper, the general parameter θ is, in a rather direct sense, a measure of the non-linearity present in the system. The talandic temperature level is therefore a major macroscopic quantity in determining the stability of the time structure which can emerge in strongly interacting control processes. Thus the frequency spectrum of a cell and its resonance characteristics, in the sense suggested by this study, are intimately related to θ, which emerges as a general system parameter of central importance for the analysis of temporal organization in cells.

TALANDIC TEMPERATURE AND HOMEOSTASIS

The type of stability with which we are concerned in this study is a stability of temporal relations between biochemical processes. The above considerations on synchronous locking and subharmonic resonance allow one to glimpse the richness of time structure which can arise in the dynamics of oscillating control systems when strong interactions between control units are introduced, compared with the relatively simple dynamics of uncoupled oscillators. They also emphasize the importance of θ in relation to temporal ordering between the interacting components. Even if there is a rich pattern of strong interactions in the system, they may result in very little temporal organization if θ is small. Only at elevated talandic energy levels (large θ) can the dynamic consequences of non-linear interactions emerge.

We can thus begin to see how talandic temperature relates to such physiological notions as homeostasis. When θ is large, small disturbances will not cause synchronously locked oscillators to drift out of phase, for example, because they will spontaneously lock again after the disturbance has ceased. But when θ is small, the forces of dynamic interaction between oscillators are greatly reduced, and the oscillators will not be ordered by stable relations to one another. Random disturbances will cause the components to drift without any mutual coherence, and any time structure in the system will rapidly decay. The importance of θ as a major system parameter in relation to the homeostatic capacity of cells in time thus becomes evident.

ADAPTIVE ASPECTS OF TALANDIC PHENOMENA

The emphasis implicit in the present theory is on rhythmic properties of cells which are stationary in time, and are neither growing nor differentiating, but simply maintaining themselves. The homeostatic properties of coupled oscillators discussed briefly above relate to an internal stability of relations in strictly periodic systems, which cycle through the same state at regular intervals. These ideas are readily extended to include the relations which arise between cells and periodic behaviour in environmental variables. Periodicity is a dominant feature of a terrestrial

157

environment, so that adaptation to such an environment involves rhythmic variation in the biochemical activities of cells. The most obvious rhythms in organisms are those which are linked to the light - dark cycle of the planet, and the homeostatic aspects of these diurnal rhythms have been discussed by Harker (1964). We may observe that the phenomenon of entrainment is extremely important in ordering cellular activities relative to environmental variables. Thus once again the talandic temperature of oscillating control systems becomes an important variable in considerations of the capacity which an epigenetic system has for adaptive and stable response to environmental signals. A cell in a very low talandic energy stage can respond only passively to a periodic environmental signal ; it has very little 'oscillatory energy' of its own to use in adapting its activities to environmental cycles, so that its response depends largely upon the intensity of the exogenous stimulus. However, a cell in a high talandic energy state has a high capacity for adaptive response to periodic environmental signals by means of nonlinear interactions. It can also organize its biochemical activities with greater autonomy since the exogenous stimulus need be used only as a clue to set the endogenous oscillators relative to the environment, the oscillators then maintaining a coherent rhythmic state by means of the high talandic energy of the system.

TALANDIC FORCE AND TALANDIC WORK

So far we have considered only stationary or conservative behaviour in control systems, and that is the natural province of the present theory. The equilibrium condition of the epigenetic system occurs when all the steady-state values $(p_i, q_i ; i = 1, \ldots, n)$ are fixed and the talandic energy is equally distributed over all components so that the talandic temperature is uniform throughout the system and equations (13) are satisfied. However, it is possible to move away from this equilibrium point and to consider the forces which are at work to change the state of the epigenetic system. The correct quantitative approach to this question is through the analogues of force, work and free energy in the present theory. The concept of force enters a statistical mechanics in association with quantities which are called external parameters. These are the environmental variables which act upon the epigenetic system, such as light, temperature, inducers, mutagens, etc. These stimuli cause the system to move to new states by altering the values of the microscopic parameters. Considering for simplicity the case of control systems without strong interactions, the generalized force conjugate to an external parameter, s_r, is defined as

$$F_r \equiv -\frac{\partial G}{\partial s_r} = -\sum_{i=1}^{n} \left\{ \frac{\partial c_i}{\partial s_r} \frac{x_i^2}{2} + \frac{\partial b_i}{\partial s_r} [y_i - \log(1+y_i)] \right\}$$

The canonical mean of this force can be shown to be equal to

$$\bar{F}_r = -\sum_{i=1}^{n} \left\{ \frac{\partial \Psi_{pi}}{\partial c_i} \frac{\partial c_i}{\partial s_r} + \frac{\partial \Psi_{qi}}{\partial b_i} \frac{\partial b_i}{\partial s_r} \right\} \qquad \ldots (18)$$

where Ψ_{pi} and Ψ_{qi} are the talandic free energy functions for the variables x_i and y_i, respectively. The work done by the stimulus, acting over a period t_0 to t_1, is defined as

$$W = \int_{t_0}^{t_1} \bar{F}_r \frac{ds_r}{dt} dt$$

$$= -\sum_{i=1}^{n} \int_{t_0}^{t_1} \left\{ \frac{\partial \Psi_{pi}}{\partial c_i} \frac{\partial c_i}{\partial s_r} + \frac{\partial \Psi_{qi}}{\partial b_i} \frac{\partial b_i}{\partial s_r} \right\} \frac{ds_r}{dt} dt$$

If now θ is held constant and the process is a reversible one, we can write

$$W = -\sum_{i=1}^{n} \left\{ \int_{c_i(t_0)}^{c_i(t_1)} \frac{\partial \Psi_{pi}}{\partial c_i} dc_i + \int_{b_i(t_0)}^{b_i(t_1)} \frac{\partial \Psi_{qi}}{\partial b_i} db_i \right\}$$

$$= \Psi_0 - \Psi_1$$

Thus we see that the amount of talandic work done in a reversible process is equal to the change in the talandic free energy of the epigenetic system. This is strictly analogous to the situation in classical thermodynamics.

TALANDIC TEMPERATURE AND GROWTH

Now the direction of 'spontaneous' change in a thermodynamic system is that defined by a decrease in the free energy function. Figure 7 shows the family of curves obtained when Ψ_{pi} is plotted as a function of c_i, for various values of θ. We see immediately that Ψ_{pi} decreases as c_i decreases. (Similar curves are obtained for Ψ_{qi} as a function of b_i.) Since $c_i = (a_i k_i)/Q_i$, $Q_i = A_i + k_i q_i$, and $b_i = a_i/Q_i$, a decrease in c_i and b_i means generally that the steady-state values p_i and q_i increase. The direction of spontaneous change in the system is therefore an expansion: the macromolecular populations tend to increase if the constraints in the system are such that some movement away from equilibrium is possible. Thus we may say that the dynamic system we have been studying has a spontaneous tendency to grow.

Another property evident from figure 7 is that when θ is large there is a much greater tendency for this expansion to occur than when θ is very small, for the talandic force causing the system to expand is directly proportional to the slope of the talandic free energy curve, as we see from (18). When θ is small the slope

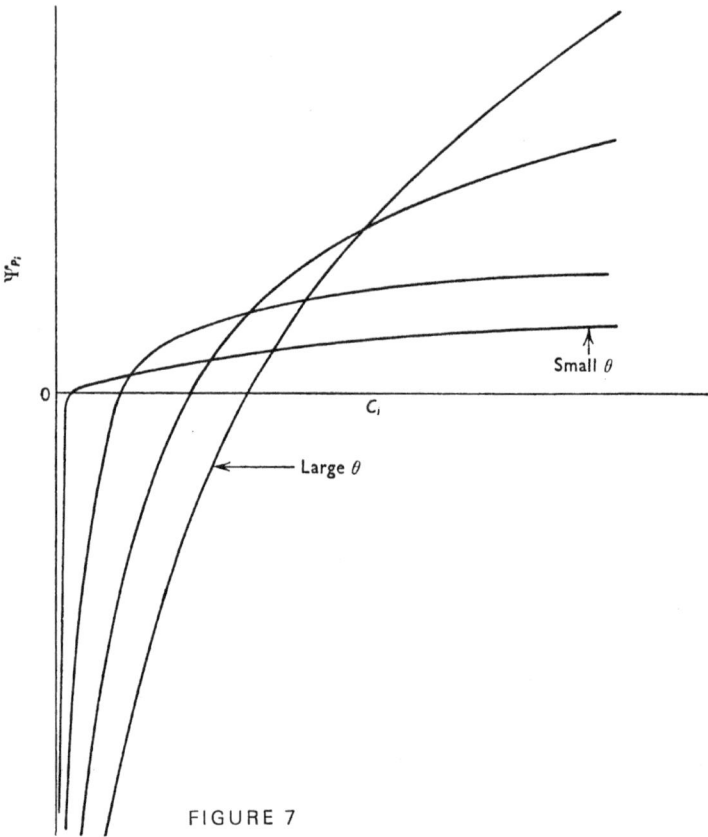

FIGURE 7

of the free energy curve in figure 7 is small over most of the range c_i; but with θ large, the slope is uniformly large, and the system will change its state rapidly when parametric constraints are loosened. These considerations suggest how one may understand cell growth rates, and lay a quantitative foundation for their analysis in terms of the dynamic properties of cellular control systems. We also have here a possible basis for analysing the adaptive responses of cells to external stimuli such as inducers, in terms of the concepts of talandic force and work.

TALANDIC PROPERTIES OF EMBRYONIC SYSTEMS
Turning briefly now to an embryological problem, the ideas of the present theory

may have some relevance in relation to the property of competence in embryonic cells. There is good evidence that the length of time during which a cell or tissue is competent to respond to a particular inductive stimulus is determined by processes within the cell, and is to a considerable extent independent of its environment (Waddington, 1934, 1936; Holtfreter, 1938). Some kind of timing mechanism appears to be operating in embryonic cells which fixes their period of competence, and the notion immediately presents itself that the fundamental dynamic property which underlies such 'clocks' is once again oscillatory behaviour in cellular control systems. One might visualize the development of a particular competence and its duration in analogy with the setting of an alarm clock. At particular developmental stages in the differentiation of a cell, biochemical clocks might be started which go off at some later time. The ringing of the clock would be the attainment of a certain level of activity at certain genetic loci, with the occurrence of particular protein species at critical concentrations in the cell. The alarm would ring for a while, and then stop; the high activity would last for some time, and then drop off again, as occurs in cells with circadian time-structure. However, these postulated clocks in embryonic cells would not be truly cyclic, in the same way that an alarm clock is not cyclic; it goes off once, and then has to be set again. Embryonic processes are manifestly non-cyclic and irreversible in this sense, and this creates a difficulty for a theory in which the equilibrium condition is defined by steady states. There appear to be ways of extending the theory to cover irreversible processes in the embryological sense. One could then approach the problems of developmental canalization and homeorhesis (Waddington, 1957), questions which are still outside the province of the present theory. The suggestion we would like to make here, however, is that the phenomenon of embryological competence may well find an explanation in terms of cellular timing mechanisms which are generated by non-linear interactions between biochemical oscillators of the type considered in this study.

There is a further aspect in which considerations of temporal organization in cells may be of significance in relation to embryonic processes. Periodicities in time can generate periodicities in space by leaving a physical trace, as it were, of the rhythmic activity. In connective tissue there occurs a rather remarkable ortho-gonal layering of collagen fibres, one layer of fibres lying at 90° to the next. It is very difficult to imagine how this could occur during embryonic development if there is a continuous, steady release of collagen from chondrocytes. However, if collagen is released from these cells periodically, with a well-defined rhythm, then one has a new variable with which to explain the physical periodicity. We have

161

already seen that there is experimental evidence for oscillations in the synthesis of collagen in the results of Tanzer & Gross (1963) and Jackson (1963), shown in figures 4–6. Gross (1961) has discussed the structural layering of collagen fibres in some detail from this point of view, and he suggests that oscillatory behaviour in embryonic cells may frequently underlie spatial periodicities in the morphology of the adult. The possibility of using time structure to generate space structure gives to the developing organism a further dimension for morphogenesis. The homeostatic properties of interacting biochemical oscillators in cells then apply directly to connective tissue formation as a homeorhetic process.

EXPERIMENTAL ASPECTS OF THE THEORY

A theory of biological organization has little value if it cannot be given an experimental foundation. In this last section we will consider how talandic temperature, the central macroscopic variable to have arisen in this study, may be controlled and measured experimentally in cells. Successful experimental control of θ would amount to an explicit test of the theory; whereas if such control cannot be realized, then the theory would have to be abandoned, for it would then have practically no predictive value.

The main idea which has arisen in this study is that a system organized micro-scopically in the manner described by equations (8) can exist in many different talandic energy states without any change occurring in the steady-state values of the system variables, p_i and q_i. In terms of the epigenetic system of cells, this means that with all microscopic parameters fixed so that the steady-state levels of the different species of molecular and macromolecular populations do not change, it should be possible for the cell to be in any one of a large number of different states in the sense of its level of oscillatory activity or excitation. The question of how to control θ thus becomes the problem of exciting or damping the oscillators experimentally and observing these changes of talandic energy in some way.

One procedure might be as follows: consider a culture of tissue with well-defined rhythmic behaviour, such as the circadian respiratory periodicity in cultured adrenal glands reported by Andrews & Folk (1963). Let the culture be maintained on a defined medium, in which there is a limiting quantity of an essential amino acid, so that protein synthesis is controlled by the concentration of this component. Suppose that the culture is given a small pulse of the limiting amino acid, so that protein synthesis is briefly stimulated. The pulse should be small enough for the added quantity to be used up in, say $\frac{1}{2}$–1 hr after it is

162

added ; or else there should be a flow of medium through the culture vessel such that the concentration of amino acid can be constantly regulated, and pulse times closely controlled.

The effect of the pulse should be a temporary stimulus of protein synthesis. This one would anticipate, both because the protein synthetic system will be temporarily relieved of amino acid limitation, and because of the inductive effect which amino acids appear to have on mRNA synthesis (Stent & Brenner, 1961). The pulse should therefore cause a transient shift in the oscillating trajectories of the different protein species. The direction in which such a disturbance will shift an oscillatory trajectory, whether towards or away from the steady-state value, will depend upon what part of the trajectory the system is on when the disturbance begins. If it is on that part of the trajectory which lies below the steady state, then the stimulus will shift the system 'up' towards it, thus decreasing the amplitude of the oscillation. But if the protein concentration is above the steady-state axis when the stimulus begins, then the shift is away from the axis and the amplitude of oscillation will be increased.

We come now to an observation which is critical to the argument. The oscillations occurring in the feedback control mechanisms studied in this work are distinctly asymmetrical. In figure 3 it can be seen that the variables undergo greater excursions above their steady-state values than below, and it can be shown that this asymmetry increases as θ increases (Goodwin, 1963). The asymmetry is particularly marked in the case of oscillations in protein concentrations, and the fundamental dynamic reason for this behaviour is the occurrence of Y_i in the denominator of the expression for the rate of mRNA synthesis. Our assumptions about the dynamics of feedback control of macromolecular synthesis in cells thus have an important consequence : protein and mRNA concentrations spend more time above the steady-state axis than below. It follows therefore that a transient increase in protein and mRNA synthesis will be more likely to occur on that part of the cycle which is above the steady-state axis, and hence should cause the trajectories to move away from this axis. That is to say, the small pulse of limiting amino acid will have a greater probability of increasing the amplitude of the oscillation, hence increasing the talandic temperature, than of decreasing it.

However, a single pulse is not likely to cause a permanent change in θ. What is required to bring about such a change is a repetition of the pulse at intervals of perhaps 2–3 hr over a fairly long period of time, say 2 days. The idea is that the pulsing should not cause any permanent change in the microscopic state of the system ; the pool sizes, synthetic rates, etc., returning to their initial values after the pulse ceases.

Temporal organization in cells

We now suggest that the way to observe changes in θ is through the rhythmic behaviour of the cells. It can be shown in the theory that the period of the oscillations is directly proportional to the square root of the talandic temperature when θ is fairly large (giving well-defined temporal behaviour in cells, such as circadian rhythms). A basic assumption in this study is that circadian periodicity is directly related to oscillations in biochemical control circuits. We are thus led to the conclusion that the above-described pulsing treatment of cells with well-defined rhythmic behaviour should result in a slowing down of the biological clock, since the periods of the primary oscillations in the cells should increase. As the pulsing treatment continues, the prediction is that the clock period will get progressively longer.

There are many complications which may be expected to arise in such an experimental design. However, the basic principle suggested for the experimental control and observation of θ is periodic, transient disturbances of a rhythmic system, designed to excite the dynamics of the control circuits without permanently changing the microscopic parameters. It is clearly essential that some experimental procedure be devised whereby θ can be controlled and observed, otherwise it can hardly serve a useful purpose in the study and analysis of cell behaviour. A more direct observation of oscillatory motion, and hence a more direct measurement of talandic temperature in cellular control systems than via biological clocks, would certainly be desirable. The experimental methods of Tanzer & Gross (1963) and Jackson (1963) seem to approach this goal, but in this case continuous observation is not possible since the cells are destroyed during the analytical procedure. It is possible that further refinements in such optical techniques as microspectrophotometry and fluorometry will allow one to make measurements on the 'motion' (in the sense of changing concentrations) of molecular populations in single, living cells, and thus measure θ directly. Some experimental foundation is clearly essential for a theory which seeks to replace the intuitive notions which we have concerning cellular organization by precise, quantitative concepts. The analysis presented here is still in a very exploratory state, but the hope is that it may lead to a theory which has real content and can contribute to the development of a theoretical biology which is firmly grounded in experiment.

References

Andrews, R. V. and Folk, G. E. (1963). Respiratory circadian periodicity in cultured hamster adrenal glands. Fed. Proc., 22, 382.

Dintzis, H. M. (1961). Assembly of the peptide

chains of hemoglobin. *Proc. nat. Acad. Sci., Wash.*, **47**, 247–61.

Gerhart, J. C. and Pardee, A. B. (1962). The enzymology of control by feedback inhibition. *J. biol. Chem.*, **237**, 891–6.

Goodwin, B. C. (1963). *Temporal Organization in Cells* (London: Academic Press).

Gorini, L. (1963). Control by repression of a biochemical pathway. *Bact. Rev.*, **27**, 182–90.

Gross, J. (1961). Collagen. *Sci. Amer.*, **204**, (5), 120–30.

Halberg, F. (1960). Temporal coordination of physiologic function. *Cold Spr. Harb. Symp. quant. Biol.*, **25**, 289–310.

Harker, J. E. (1958). Diurnal rhythms in the animal kingdom. *Biol. Rev.*, **33**, 1–52.

Harker, J. E. (1964). Diurnal rhythms and homeostatic mechanisms. *Symp. Soc. exp. Biol.*, **18**, 283–300.

Holtfreter, J. (1938). Veränderungen der Reaktionsweise in alternden isolierten Gastrulaektoderm. *Roux. Arch. EntwMech. Organ.*, **138**, 163–96.

Jackson, D. S. (1963). Personal communication.

Loftfield, R. B. and Eigner, E. A. (1958). The time required for the synthesis of a ferritin molecule in rat liver. *J. biol. Chem.*, **231**, 925–43.

Magasanik, B. (1958). The metabolic regulation of purine interconversions and of histidine biosynthesis. In *A Symposium on the Chemical Basis of Development*, pp. 485–94. (Ed. McElroy, W. D. and Glass, B.) (Baltimore: Johns Hopkins Press.)

Monod, J. and Jacob, F. (1961). General conclusions: Teleonomic mechanisms in cellular metabolism, growth, and differentiation. *Cold Spr. Harb. Symp. quant. Biol.*, **26**, 389–401.

Penman, S., Scherrer, K., Becher, Y. and Darnell, J. E. (1963). Polyribosomes in normal and poliovirus-infected He La cells and their relationship to messenger-RNA. *Proc. nat. Acad. Sci., Wash.*, **49**, 654–62.

Pittendrigh, C. S. (1960). Circadian rhythms and the circadian organization of living systems. *Cold Spr. Harb. Symp. quant. Biol.*, **25**, 159–84.

Pittendrigh, C. S. (1961). On temporal organization in living cells. *The Harvey Lectures*, Series 56, p. 93 (New York: Academic Press).

Pittendrigh, C. S. and Bruce, V. G. (1957). In *Rhythmic and Synthetic Processes in Growth.* (Ed. Rudnick, D.) (Princeton: University Press.)

Stent, G. S. and Brenner, S. (1961). A genetic locus for the regulation of ribonucleic acid synthesis. *Proc. nat. Acad. Sci., Wash.*, **47**, 2005–14.

Szilard, L. (1960). The control of the formation of specific proteins in bacteria and in animal cells. *Proc. nat. Acad. Sci., Wash.*, **46**, 277–92.

Tanzer, M. L. and Gross, J. (1963). Collagen metabolism in the normal and lathyritic chick. *J. exp. Med.* (In the press.)

Umbarger, H. E. (1956). Evidence for a negative-feedback mechanism in the biosynthesis of isoleucine. *Science*, **123**, 848.

Vogel, H. J. (1957). Repressed and induced enzyme formation: A unified hypothesis. *Proc. nat. Acad. Sci., Wash.*, **43**, 491–6.

Waddington, C. H. (1934). Experiments on embryonic induction. I. The competence of the extra-embryonic ectoderm in the chick. *J. exp. Biol.*, **11**, 211–17.

Waddington, C. H. (1936). The origin of competence for lens formation in the amphibia. *J. exp. Biol.*, **13**, 86–91.

Waddington, C. H. (1957). *The Strategy of the Genes* (London: Allen and Unwin).

165

New Thoughts on Bio Control[1]

A. S. Iberall

General Technical Services Inc., Pennsylvania

My colleagues, notably Dr Cardon, and I have been pursuing a closely coordinated novel view of the complex biological system. This is contained in nine basic reports written for NASA and a number of derived papers that we have given or reported on at various meetings [1]. I will attempt to give you some sketchy outline of our central ideas up to date. (It is only fair to the reader and ourselves to state that Dr Cardon and I are physical scientists only recently concerned with biology.)

Starting in our first report—available as a N.Y. Academy of Science review of September 1964—we sought to define the regulating and control functions of the system as a whole.

First we sought the central theme of the biological system and found it well enunciated in the biological literature in the form of Bernard's and, later, Sechenov's and Cannon's [2–4] insistence on the primacy of the interior milieu and of its regulation—homeostasis—as the significant theme of operation of the complex system. We then found and accepted H. Smith's view of the kidney [5] as the master regulator of the system, for in its function, more than anywhere else, does the preservation of the character of the watery internal milieu lie. As physical scientists—basically in the tradition of the modern chemical engineer with a view of flow charts, unit processes, and regulating and control functions—we thus had a core model of chemicals in a watery environment with a master flush valve at the exit to preserve the watery environment and regulate concentration contents.

It was clear, next, that ever since Wiener [6] the internal electric system had to be related to possible regulating and control functions. However, within the spirit of discovery that we were seeking, it appeared not so much that Wiener had developed key knowledge about the electrical system—Adrian and Sherrington [7, 8] were much more influential and important from a biological point of view—but that he called attention to a control (cybernetic) point of view for the system.

At this point it was necessary to put some mechanisms, systems, and processes into the bath. Here we had to start with what would be specialized and peculiar to physical theory and its dynamic analysis. If one were to look for regulating and control chains, it would be first necessary to dispose of any self-sustained

[1] This report was substantially presented, by invitation, for the local Biomedical engineering section—IEEE—Phila., Sept. 1966; Dept. Physiology—U. of Pittsburgh, Dec. 1966; Serbelloni conference on Theoretical Biology, Lake Como, Aug. 1967.

oscillatory or so-called limit cycle phenomena. These are indicative of unstable linear chains that have unknown causality and 'purpose' to begin with in the analysis of a new unknown system. We proposed to tease these out and lay them aside and then to continue to look for regulators and controllers.

To those who might consider this idea vague, it really consists of the following: what is unique in the description of the dynamics of a system—whether by equations, physical chains, or verbal chains—is the set of singular states of motion of a system and the kinds of operating stability that one finds around these singularities. It is thus that the actual 'stable' dynamic motion of a system is cast. This is the topological overview of non-linear mechanics in considering the dynamics of systems. As Minorsky [9] puts it so well, the non-linear limit cycles of Poincaré are the stable non-linear steady states of operation of a system, just as the damped oscillation is the stable linear steady state of operation of a system. Whether a system be linearly stable (at rest) or non-linearly stable (in limit cycle oscillation) depends upon the specific character of its singularities. In an *a priori* unknown system, they can only be discovered experimentally.

We therefore went searching, and began to tease out a broad spectrum of oscillators, which may also be called engine cycles, rhythms, DC to AC converters, or non-linear limit cycles. In fact, after a while, it dawned on us that we had become keepers of biological macrospectroscopy, the dynamic spectrum of biological effects, in which the chains are biochemical, bioelectric, and bio-mechanical. (When we were introduced to Brian Goodwin's work [10], we regarded him as a keeper of microspectroscopy.) We found so many such chains that we were led then to believe a key idea: that what the biological system consisted of was a great collection of limit cycle oscillators; basic regulating and control functions were to be found in the mediation of the stability of these oscillators, mainly by inhibition; that the slowly varying DC parameters of the milieu mediate the operating point—in amplitude, or frequency, or in actual stability margin—of these oscillators so as to achieve desired regulation.

I can name some of these 'major' physiological oscillators for clarity and suggest their nominal time scale.

bioelectric nervous waves (spikes, etc.)	0·1 sec. per cycle
heart beat complex	1 sec.
ventilation	4 sec.
blood circuit flow	10 sec.
blood flow oscillations	30 sec.
metabolic oscillations	100 sec.

New thoughts on bio control

vasomotor oscillations	400 sec.
many fast endocrine oscillations	300–1000 sec.
gas exchange oscillations	2000 sec.
metabolic fuel oscillations	5000 sec.
heat balance oscillations	3 hours
circadian rhythms	24 hours
water cycles	$3\frac{1}{2}$ days
many longer range endocrine rhythms	20–40 days

Since Cannon's concept of homeostasis, on careful reading, was a concept of a regulating process or chain, but with the implication (in his examples) of quasi-static regulation [4], it seemed appropriate to propose a change in name to indicate a more dynamic nature of the regulation. For this, the term homeokinesis was offered, for it is by manipulation of the kinematic variables—of space and time alone—that this dynamic regulation takes place.

By our fourth report [1] we were then prepared to propose a new operational definition of life.

'Thus life is tentatively defined as any compact system containing a complex of sustaining non-linear limit cycle oscillators, and a similar system of algorithmic guiding mechanisms, that is capable of regulating its interior conditions for a considerable range of ambient environmental conditions so as to permit its own satisfactory preservative operation; that is capable of seeking out in the environment and transferring and receiving those fluxes of mass and energy that can be internally adapted to its own satisfactory preservative operation; that is capable of performing these preservative functions for a reasonably long period of time commensurate with the "life" of its mechanical-physical-chemical elements (i.e. clocks made of parts that should wear for hundreds of years that run for two seconds are not crickets); and—likely as a luxury part of the definition—that are capable of recreating their own system out of materials and equipment at hand (one will have to note in the future how much of the biological system can be rebuilt, or one can recall the story of the amorous young bridegroom who has watched his bride remove most of her apparent charms before his eyes—teeth, eyes, hair, wooden leg, etc.).

'The purpose of this definition is to guide the physical search for "explanations" of the operation of the biological system; and to leave the physical scientist closer to some physical base by which he can model, "build", or assess systems that resemble naturally "living" systems by suitable operational definitions; and a clue that "life" does not have to be explained by only one mechanistic scheme per system, but may involve many possible types of successful operation.'

We view the function of such a definition, in an operational Bridgman sense, that it should give us a hunting licence in the form of a basic concept of what we are hunting for. In this definition it is declared to be dynamic regulating chains. These may be fully unstable and oscillatory, or may be marginally unstable and aperiodic. We would hope that by this means we can have forced the intrusion of thinking about dynamic processes into biology. The significance of dynamic AC analysis of systems does not have to be stressed to any group who has been influenced by modern electrical engineers.

Further, to give the system its primary functional keynote, its 'purpose,' we have proposed the following catch-phrase:

The biological system is an intermittently self-actuated motor system operating in both short and long term, that seeks to sustain the metabolic reaction: fuel + oxygen → carbon dioxide + water, plus sufficient power so that it can continue its self-actuated motor activity to seek to sustain its metabolic activity.

Shorter still: *the system eats and moves about, so that it can continue to eat and move about.*

Thus, for example, in our second summary paper at the 1965 IFAC Symposium in Tokyo, we could represent our overview of the biological system two years after our N.Y. Academy paper. As an excerpt:

'The experimental survey of a complex biological system like the human discloses a large number of autonomous oscillators continuously operating in the system. A partial frequency spectrum in man would consist of primary neural frequencies in the range of 5 to 50 cps, a muscle motor unit frequency at about 10 cps, a heart beat about 1 cps, breathing rate about 1 cycle per 4 seconds, several eating cycles per day, circadian rhythms of 1 cycle per day, sex urge approximately 1 cycle per few days and menstruation about 1 cycle per 30 days. To these may be added cycles, demonstrated within the past few years, in ventilation rate, local skin temperature, and metabolism of approximately 100 seconds, 400 seconds, 30 minutes, and 3 hours. The first appears to be an engine cycle, primarily in skeletal muscles, in which the major heat production takes place. The 400 second cycle appears to represent a vasomotor action which partitions blood flow among the major organ circulations. The 30 minute cycle may be a gas exchange cycle, likely representing a total body carbon dioxide equilibrium, and the 3 hour cycle is probably an overall thermal balance. A 3-day cycle has been found in body weight. It has been tentatively identified as a water balance cycle.

New thoughts in bio control

'It has become increasingly apparent that the many oscillators in the biological system are not incidental characteristics of the system, but represent the working components of the system. In summation, they are the biological system. In accordance with this view, it is proposed that homeostasis, Cannon's organizing biological concept of a complex regulation characteristic of the system, is obtained as a result of shifting the stability of these intrinsic non-linear oscillators. The oscillators are likely modulated or shifted in operating point by electrical and chemical signals. In our view such action is an illustration of dynamic regulation. It is possible that the biological system is not able to operate in any other way. In fact, it is likely that the same type of instability mediation is the foundation for all automatic control theory. However, the more general thesis is beyond the scope of discussion in this paper.

'To illustrate the mechanisms of homeostasis, a number of biological systems will be discussed. A first case in point is that of heat production in mammals, which is of interest as part of the system of human thermoregulation. Local skin temperatures, ventilation rate, and oxygen consumption rate all show sustained limit cycle oscillations with a prominent component near 100 seconds. We have postulated that this is a heat engine cycle ; that the engine consists essentially of skeletal muscles ; that the level of operation of local muscle engines is regulated by an oxygen choke which limits the oxygen supply and thus the rate of local oxidation ; that regulation of oxygen rate is achieved in a 100 second cycle of capillary red blood cell flow, mediated by the vasodilatory effect of adrenaline. That adrenaline is the hormone mediator of the cycle is suggested by its calorigenic action, its vasodilatory effect primarily in skeletal muscles, its effect of increasing muscle activity and a time of action for transient effects in a range between one and two minutes.

'The mechanism of the adrenaline action is not known.'

The essence of the matter, as this last thought intends to imply, is that having now proposed all of these grandiose systems views of the complex mammal, the human, the burden of proof is now upon us to make the description stick. This, to us, has meant that we must identify and run down these dynamic regulating chains. This is what we are now beginning to try to do. However, there are so many of them that we have to choose carefully and for the most hard-hitting value. Our resources are really quite limited. However, to add some more flavour to the presentation, I think I can add what we had in addition grasped more recently and expressed in our sixth report, August 1966 [1].

170

In the metabolic reaction:

Time elements	Fuel	+	A matrix Oxygen	→	CO_2	+	H_2O
1							
2							
3							
4							
5							

we basically consider that there seems to be a matrix of regulation levels and effects, a hierarchy of regulation functions that seem to pile up under each item; the regulation of fuel, oxygen, CO_2 and H_2O. Many of these are quite independent and tend to appear redundant. What we expressed in the past, coming more nearly into focus in this sixth report, is that when the ecology provides cueing sensitivities of a temporal type which may have survival value, and there is a suitable cyclic time fracturing (or phase locking) commensurate with this period provided by possible physical-chemical chains, these chains will become a dominant dynamic regulating element in the system. Thus, the spectrum of effects we have identified seem to fit this scheme. However, beyond this, we have finally vaguely seen that the spectrum tends to be even more limited. We have further said in our sixth report:

'The apparent competition between autonomous physiological oscillators and the environmental cues (day, month, year) is to be resolved by recognizing that cycles must fit and thus be entrainable as small numbers with all such cues. We eat, defecate, urinate, sleep a few times per day. These are useful adaptations. An animal (call him a dinosaur) that must chase food for too long to make one meal can't make it. He will not satisfactorily entrain. Time is against him. When time is ample, then the system instabilities will lead to an orbital entrainment in such cycles that can fit the time comfortably. This is one added thought of how the patterning richness is regularly reinforced. The second thought emerges even more strongly from endocrine considerations.

'Each endocrine gland seems to put forth a spectrum of hormones, and it has been a little discouraging to attempt to exhaust, by long lists, their apparent multiplicity of functions. Yet major functions emerge, and the spectrum covers functions that are usefully adaptive. However, it also emerges that more than one gland may put out hormone components that collaborate at common functions. In reviewing time constants, it appeared fascinating that, while the spectrum is rich, it tends to appear finite in number. Thus the concept of time fracturing of physical-chemical chains seems to roost on a more specific perch. Any particular ductless gland may contribute hormonal elements. However, most of them will be involved in chains with a small number of time constant ranges. (For example, we might propose some typical numbers, 0·05–0·1 second, 0·5–1 second, 5–10 seconds, 20–40 seconds, 60–120 seconds, 300–500 seconds, 20–40 minutes, 3–4 hours, 20–8 hours, 3–4 days, 20–40 days.) It is not the case that they must be involved in such time ranges. However, those that can be entrained in chains that have considerable adaptive value have greater survival value for their species. Thus the hormone patterns are interlocked—from gland to gland— to form a matrix of chain function (sugar metabolism, for example) and time. It is this highly locked-in matrix that exists in the endocrine system, the nervous system, etc., that provides the underlying functional structure of the system. The patterns then form and wrap themselves around these more permanent poles. The final ingredient, of course, is that these salient polar times fit, by small numbers, and interact with the A C cueing of the environment. The night–day cycle is a most powerful polarizer. The breakdown of time ultimately to the 1–2 minute or the 0·05–0·1 second cycle time is of course more subtle. The 28-day menstrual cycle drive is quite real, and the breakdown into 4 weeks is reasonable. It is disconcerting to consider cues for a $3\frac{1}{2}$ day water cycle. Which is causal for which ?' (That is, the week for the water, or the water for the week ?)

'The existence of such time cycles does not mean that all animals must share them. Development in each species can have occurred independently. It is likely that similar chemical cycles may be arrived at, but this arises because the bio-chemical chains, in general, are not so specific, although particular ones may be sharply deterministic.'

Thus far, we have sketched out our scheme for the system via its more 'in-voluntary' physiological oscillator chains. Now I will discuss a little our proposal that this is extendible to the less 'involuntary' psychological or behavioural system.

We have postulated that what is most characteristic of behaviour is an orbital synchronization with various elements in the milieu. This model has been sub-consciously influenced by a quantum mechanical model of free electron conduction in metals. In Frenkel's *Wave Mechanics* [11] there is a delightful modelling of the electron motions as follows: The electrical field causes an accelerated motion of free electrons with a relaxational phase on collisions. The free electron path is thus defined as a sequence of mean free path relaxations and mean relaxation time between collisions with what are basically not crystal lattice points but lattice dislocations. However, at any point the electron is not locked up for one oscillation but it makes a number of pinwheel rotations. The human—it is proposed—acts similarly. A signal DC or AC in the milieu—for example, the flick of the skirt of a passing comely female, or the pulsations in the stomach, or a painful nerve signal, etc., each mediate the oscillator complex, and lock the system 'posture' for that 'moment' into an orbital synchronous path. The command algorithm (an extremely large code book) was not all determined at birth, or did not emerge at various maturational points. It was partially 'learned' and developed out of the matrix of experience. Such development, I was taught a long time ago, represents an epigenetic process.

To give behaviour some picturable substance, consider the bank of physiological oscillators as not being rigidly closed but with partially open inputs, and quite a few in number. Let the oscillators resemble the various resonators of a piano or organ, each of which can have its pitch, amplitude, and timbre mediated. Now let the input winds of the milieu—both internal and external—play over the bank of open inputs. They will make melodic patterns in the organ. Thus far, we only have an instrument like a passive Aeolian harp—even if the individual resonators are active elements.

However, let these oscillations also drive the motor system into space-time orbits. Then the system will be urged to drift down the path of life, however not continuously. It will lock up, circle quite a few times, pass on, etc. There are open portions in this path motion.

Continuity of behaviour is provided by the learned part of the algorithm. Mother, in the mother-child relation, teaches the child various patterns that are fairly adequate to provide the range of needs that will saturate the physiological oscillators over time.

To make this more comprehensible, there is likely a two-state oscillator complex— this is Freudian derived—representing a satisfying or 'euphoric' state, and an 'anxious' or 'dysphoric' state. In this view we find Henry Stack Sullivan's formulations [12] most nearly fitting what we were looking for. Any first infant experience may be satisfactory or not. However, all that can emerge in time is an alternation of such states. In the infant brain, the major signalling complex is the oral one. We consider that this signalling complex, pounding into the plastic brain, develops correlates for satisfying and unsatisfying states. Later, as the nervous system develops and the more detailed outlining of the system, descending toward the anal and genital regions, gradually helps set up a conflict of signalling into the brain, there is a tendency to separate out and segregate these signalling interfaces into the brain, becoming represented as analogue imprints of these system interfaces in the brain. The summation of these signalling complexes, as represented by their imprints in the more primitive structures in the brain, we think of as representing the structure of Freud's id (or *es,* the *it* of the system, as he named it in his own language). It is this measure which somehow is to lead us to anxiety-provoking or euphoric states. We then visualize that the coordinated summation of this oscillator complex, projected into higher centres, represents the ego sum of the system. A more nearly unitary measure of the buzz of information to ego centres is provided by a part sexual flux of libido. Its state presents a measure of the existing, satisfying, operating state of the system. Guided by this measure, system operators fall into an anxious or euphoric state, represented as a two-state operative complex.

As another integrative measure taken by the brain, unification of discreet signal information into an integrated analogue of the information represents the gestalt of the analogue. The brain is thus compacted with a large number of unified analogues to various signalling complexes and motor responses. It is such analogue packages that the mother helps develop in the maturing child together with a program routine of a tolerably satisfying nature that threads through the system hungers.

The missing ingredient is Freud's super-ego, the ego ideal, which represents the image of a satisfying ego state. This, added to the developmental algorithmic content on how to achieve a satisfying ego state, provides the motion for the behavioural patterning that emerges. The conflicts from all the signalling interfaces are not to be satisfyingly resolved by moving in some direct path. The super-ego is developed as a rule book on what motor actions will preponderately move in the direction of a more satisfying state. The speed of motion will also depend on the oscillatory two-state operator system—whether overall anxious or euphoric.

Thus modulated by the two-state oscillator, under the driving guidance of the super-ego algorithm, the system will hurl itself into motor activity that will synchronize in orbital paths with specific oscillator patterns.

In the larger developmental picture there are a number of large-scale flaring instabilities of a maturational nature. Two of these we can be certain of physiologically, and the others, if not yet clear physiologically, are certainly clear observationally, even under comparative anthropological study.

The first is the oral interface of infancy. The primacy of food seeking and of establishing routines for food acquisition may be fairly detailed.

The second is the great flaring instability of adolescence. The physical characteristic changes are, of course, noticeable. However, equally noticeable is the interest in the opposite sex and the preparedness of the system behaviourally and physiologically for mating—the maintenance of the metabolic reaction through reproduction of the species. The adaptive value of this aperiodic flaring of an entire sexual cycle is obvious.

The third newly-emergent flare is Sullivan's chum or peer stage in which the child is integrated sufficiently for it to suddenly break its tie to the mother's orbit and discover its own kind, its peers and chums.

Each of these periods is marked by the continuing development of the guiding algorithm, within the framework of interpersonal relations and the surrounding milieu. A useful thought is that man is an instrument to be used by and for man. It carries with it the idea that the system gain at zero frequency is indeterminate. (This means that the system is unstable at rest.)

The basic command algorithm for behaviour can be considered to be a continued adaptation to make the ego image agree with the ego ideal by means of the sequence of practised analogues that are stored within the brain.

The marginal stability of the system is such that environmental signalling is always blowing over the inputs and putting the analogue systems into continued melodic lines of response. The system is always practising its repertoire. It is this continued response that makes a follower-type characteristic melodic line appear for each and every questing and questioning environment.

In a successful biological species there is a satisfactory patterning of behaviour which threads all of the system hungers. In an unsuccessful individual it may lock up or become too wild. I have presented these ideas less sketchingly in our eighth report as a joint effort with Warren McCulloch. The report became available just at the time of this conference. An additional ingredient that emerged in that report is the foundation for interpersonal forces, the keynote of behaviour stressed by Sullivan. If, in the interior, there is this well-coordinated body image of both inside and outside the system, with its many vector dimensions, then upon meeting a member of a like (or similar) species in a reacting situation, there is an exchange of body image. The individual imagines the nature of his body image projected into the other individual as the other person's apparent actions emerge. If the image and actions are concordant, then an 'empathy', a binding 'force', emerges. The basic two elements of concordance, likely, are complementarity and congruence.

By these means, roughly, we have attempted to bring a dynamic concordance through the biological system by its spectrum of dynamic oscillator chains. Our problem now is to demonstrate the reality of these chains.

We can provide a more rudimentary illustration. We have taken a large motor system—the hind limb of the complex mammalian animal—as a suitable laboratory to test some of these ideas. For example, we are attempting to follow the detailed metabolic reaction in that system.

So far, in the gross animal we have demonstrated appreciable dynamic cycles in :
  respiration rate
  heat production rate
  temperature.
From their temperature concordance, we have postulated that their operative dynamics are to be associated with pulsing dynamics in the cardiovascular system at the level of the microcirculation. We have demonstrated corollary dynamics in the femoral system (for the first 3) and in capillaries (for the 4th) :
  blood sugar oscillation
  blood oxygen oscillations
  blood CO_2 oscillation
  red blood cell oscillations.
We have traced a line of evidence from Krogh, Sir Thomas Lewis, among modern microcirculationists, and others.

To illustrate briefly, we postulated that there should be a chemical engine cycle faster than a thermal relaxation time of three hours. We found it in thermal, metabolic, and ventilatory cycles at 100 seconds, 400 seconds, and greater. We have chosen to highlight the 100 second cycle as the engine cycle because of its large magnitude of variation. For example, we have found a running variation of near two to one in metabolism and ventilation at the 100 seconds level, ranging as much as peaks that are 5–10 times larger than minima in a 5 hour observation period [13].

We have decided that the large cycles must stem from major mesodermic organ systems such as muscle, liver, heart, brain, etc. We have sought and found allusion to such time scales in the microcirculation and expected to find it in capillary opening and closing. Instead we found the cycle in the red blood cells flow. We postulated that the engine cycle could either be run by metering of fuel, oxygen, or a combustion by-product. The high level of regulation posed for fuel and the rapid follower regulation and storage capacity for CO_2 ruled these two out, whereas the more limited storage of oxygen suggested that the engine cycle must be run by an oxygen choke. In this view, the muscles were considered to be an unstable system capable of metabolic conversion at any level to which they were supplied with oxygen. The fluctuation of red blood cells was an additional step in the evidence for this concept. Early in 1966 another step was found in the experiments of another investigator (Whelan), who showed that oxygen tension in the tissue was low and oscillatory—in the range of 0–5 mm Hg, with a period of the order of 100 seconds.

We are continuing in our effort to show an adrenaline involvement in setting this local system into its oscillatory chain. We carry with us the provocative thought that the brain sets a motional pattern, but not the resisting load. Then it must implement its choice by providing the needed oxygen carburettor to run the muscle engine.

In any case, the story and proof of such chains is exciting business. Yet it is beset by the following situation.

We have measured and found ventilation to be oscillatory time and time again. It has been verified by a few other investigators, for example, Goodman [14], now at N I H ; more recently, Lenfent [15]. Yet we can find no other ventilation physiologists who report these results.

We have found normal heart rate to be oscillatory. This has essentially not been reported by other observers.

We have found that Anderson [16] and others (for example Hansen [17]) have found blood sugar to be considerably oscillatory and we have verified these findings ; yet many physiologists deny these findings.

Thus, many of the observations of fundamental dynamic findings are themselves in doubt. We have sufficient confidence in our experimental findings to make even the consideration of the validity of our theses to hang on the validity of our dynamic observation. If the others are right and we are wrong in measurements, then it is dubious that our dynamic concepts should receive much attention. However, if measurements in normal unanaesthetised animals held normally quiescent or normally active demonstrate such dynamic oscillations, then we feel that our ideas may warrant the attention that we believe they deserve.

177

New thoughts in bio control

References

1. A. Iberall and S. Cardon. (1) *Ann. N. Y. Acad. Sci., 117* (1964) 445.

A. Iberall and S. Cardon. (2) N A S A C R-141, Jan. 1965.

A. Iberall and S. Cardon. (3) N A S A C R-219, May 1965.

A. Iberall and S. Cardon. (4) N A S A C R 129, Oct. 1964.

A. Iberall, S. Cardon and T. Jayne. (5) Dec. 1965, Interim Report.

A. Iberall, M. Ehrenberg and S. Cardon. (6) Sixth Report to N A S A, Aug. 1966.

A. Iberall. (7) *Math. Biosciences, 1* (1967), 375.

A. Iberall and W. McCulloch. (8) *Currents Mod. Bio., 1* (1968), 337.

E. Young. (9) N A S A C R-990, Dec. 1967.

A. Iberall and S. Cardon. In *Proc. I F A C Symp. on Syst. Eng. for Control Syst. Design* (Sci. Council Japan, Toyko, 1965).

E. Young, A. Iberall, M. Ehrenberg and S. Cardon. *Proc. Ann. Conf. Eng. in Med., Bio.,* Vol. 8, 1966.

M. Ehrenberg, C. Oestermeyer, E. Bloch and S. Cardon. *Microvasc. Res.* (forthcoming).

A. Iberall, M. Ehrenberg, S. Cardon and M. Simenhoff. *Metab.* (forthcoming).

2. C. Bernard. *An Introduction to the Study of Experimental Medicine,* Schuman (N. Y., 1949).

3. I. Sechenov. *Selected Physiological and Psychological Works* (For. Lang. Pub. House, Moscow, 1952).

4. W. Cannon. *The Wisdom of the Body* (Norton, N. Y., 1939).

5. H. Smith. *From Fish to Philosopher* (Ciba, N. J., 1959).

6. N. Wiener. *Cybernetics* (Wiley, N. Y., 1961).

7. E. Adrian. *The Mechanism of Nervous Action* (U. Penn. Press, Pa., 1932).

8. C. Sherrington. *The Integrative Action of the Nervous System* (Yale U., Conn., 1961).

9. N. Minorsky. *Nonlinear Oscillations* (Van Nostrand, N. Y., 1962).

10. B. Goodwin. *Temporal Organization in Cells* (Academic, London, 1963).

11. J. Frenkel. *Wave Mechanics* (Dover Repr., N. Y., 1950).

12. H. Sullivan. *Collected Books* (Norton, N. Y., 1953).

13. A. Iberall. Trans. A S M E Series D, J. Basic Eng., *82* (1960) 92, 103, 513.

14. L. Goodman. Trans. I E E E, Biomed. Eng., *13* (1966) 67.

15. C. Lenfent. *J. App. Physiol., 22* (1967) , 675.

16. G. Anderson, Y. Kologlu, C. Papadopoulos. Metabolism *16* (1967), 586.

17. K. Hansen. *Acta. Med. Scand. Suppl., 4* (1923), 27.

Cellular Oscillations and Development
Comment on the papers of Iberall and Goodwin

by C.H. Waddington
University of Edinburgh

The argument that a fundamental characteristic of cellular biosynthesis is a tendency to oscillation and the establishment of limit-cycles, opens a new theoretical
approach to many classical problems, including those of differentiation. This is an
aspect of biology in which our understanding is, in my opinion, at a very much
less satisfactory level than many optimistic devotees of molecular biology pretend.
The Jacob-Monod account of how single structural genes are turned on in
prokaryotic organisms goes only a minimal distance towards explaining the phenomena of eukaryotic cellular differentiation ; and an appeal to histones as a *deus ex
machina* is as imprecise theoretically as it is shaky experimentally. In these circumstances it is, I think, excusable, and may even prove profitable, to publish
speculative ideas which still lack the empirical confirmatory evidence which would
be demanded in an area in which we had a firmer grasp of the outlines of the
logical structure within which effective theories must be framed. In this spirit—of
throwing bread upon the waters—I should like to make some comments on the
'Goodwin-oscillator, limit-cycle' model of biological systems. I have already
briefly alluded to some possible applications of these new ideas [1].
▶ *Differentiation of oscillating systems* There I suggested that we can now contemplate a previously unsuspected type of inductive interaction between cells or
tissues, in which one side of the reaction acts as a 'temporal template' and entrains
the oscillations of other elements to its own frequencies and phases. I should like
now to add the point that one of the major difficulties in the theory of differentiation would be greatly simplified if we could interpret embryonic determination in
terms of a fixation of certain frequencies in an oscillatory system. This is the
difficulty which arises from the fact that during development general characteristics
become determined earlier than specific and detailed ones. These general
characteristics are often of a kind which it is very difficult to interpret in terms of the
activities of single genes. This, which is the part of the situation that offers the real
challenge to understanding, does not, of course, always occur. For instance, it is
well known that in the early development of the amphibia, part of the dorsal ectoderm of the gastrula may, by induction, become determined as neural tissue, when
it will proceed to differentiate into some part or other of the nervous system, though

179

it is left for later processes, such as interaction with the neighbouring mesoderm, to determine whether this will be part of the fore-, mid-, or hind-brain, the spinal cord, or peripheral nerves. At the time when the cells are determined simply as 'neural tissue of some kind' the only changes known to occur are the visible alterations of cell shape involved in the formation of the columnar epithelium of the neural plate. It is not wholly implausible to suggest that these changes are due to the coming into activity of one or a few genetic loci ; and one might argue that the first step in the differentiation of the neural tissue was therefore the 'activation' of a few genes, which would be followed later by activation of others, which convert the cells into the various parts of the brain, spinal cord, etc.

But consider another case. It is a well-established fact (see for instance [2]) that small groups of cells in the limb buds of the chick embryo are determined as fore-limb as opposed to hind-limb (or vice versa) well before they are determined as proximal versus distal (thigh versus toe, or upper arm versus fingers), let alone determined as muscle, cartilage, bone, etc. For instance, Saunders transplanted a small piece of tissue from the region which would develop into the thigh into a position near the future wing tip. The transplant developed into toes and claws ! That is, it retained its character of 'something to do with the hind-limb', but was still flexible as regards its distal-proximal character, and, finding itself in the distal region of a limb, developed in the parts appropriate to the distal region of a hind-limb.

I have always found it extremely difficult to see how to comprehend results like this in terms of gene activation. It seems to me totally implausible to suggest that there is a gene, or even a small group of genes, 'for' hind-limb, and another one 'for' fore-limb. We could only make such a supposition if we could envisage there being some substance (gene-produced protein) which characterized all tissues in the hind-limb and distinguished them from those of the fore-limb.

A picture which is intellectually much easier to accept (though experimentally it might be difficult to approach) would emerge if we could consider that the characteristic feature of these early determined states of developing cells is a certain pattern of oscillations. We would not have 'a neural plate substance, a fore-limb substance, a hind-limb substance', etc., but neural plate, fore-limb or hind-limb oscillatory patterns, which could be regarded as analogous to musical themes or chord sequences. The later phases of differentiation, in which the cells of the fore-limb bud become differentiated into the various cartilages, bones, muscles, etc., must certainly involve the 'activation' of different structural genes controlling the proteins in these different sorts of cells ; but we could interpret these changes as similar to the development of the initial theme according to the conventions of some school of classical musical composition—I suppose the analogue of what jazz musicians do to a chord sequence in a jam session would be some sort of cancer!

I confess that I have not yet seen a way in which one could *prove* that frequency of oscillation is involved in embryonic differentiation. But most embryological phenomena seem easily interpretable in such terms. For instance, in the determination of the neural system in amphibia it is generally agreed that two processes are involved: one which induces competent ectoderm cells to become neural tissue, in the first instance of a type characteristic of the fore-brain; and another which interacts with the former in such a way as to produce the more posterior parts of the nervous system, such as spinal cord. There is still argument whether the former process must act earlier and the latter later, or whether the temporal order of them is unimportant (see [3]), but both schools of thought agree that when you combine 'fore-brain' with 'spinal cord' you find that you get all the intermediates between these geographical extremes, such as hind-brain, ear region, and so on. This is compatible with an interaction of frequencies; but also with several other theories, e.g. of gradients of chemical concentrations. What we need is an experimentally decidable question which could discriminate between the *a priori* plausible possibilities.

▶ *Morphogenesis in oscillatory systems.* Several new theories of morphogenetic processes could be developed in relation to systems consisting of numbers of mutually interacting oscillators. For instance, at the Third Symposium Wolpert and Cohen sketched out a theory in which the relation between a transient peak and some other oscillation, which served as a time base, was invoked as a possible way of encoding the 'positional information', which, Wolpert argued, is necessary to specify the behaviour of a given cell within a mass of tissue. The type of theory I suggested in 1965 was quite different. I was thinking of the formation of fields of vibration within delimited areas. The well-known 'Chladni figures', in which the nodes of such fields are exhibited by strewing the vibrating surface with light powder, often have a character which reminds one strongly of biological forms— for instance, in the degree of precision with which the curves are drawn, the degree of geometrical complexity of the curves, and the way in which a given pattern may suffer a slight, graded overall distortion when one of the parameters is changed. Although the elementary theory of simple examples of such vibratory fields has been well explored mathematically, systems in which there are variations in the nature of the material, or in which a stochastic element plays a part, are difficult

181

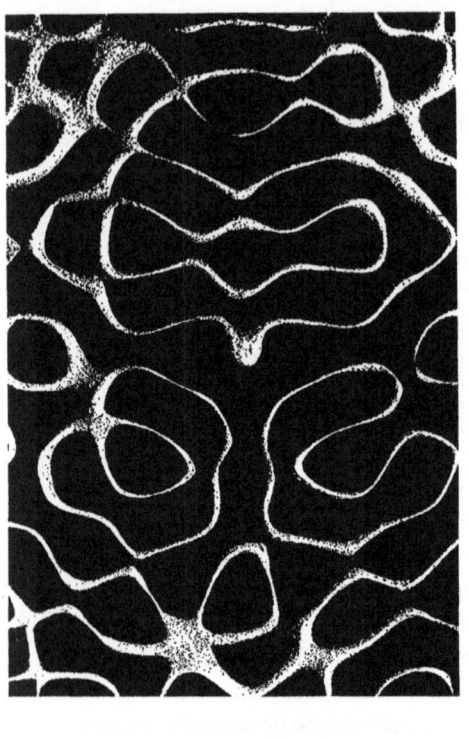

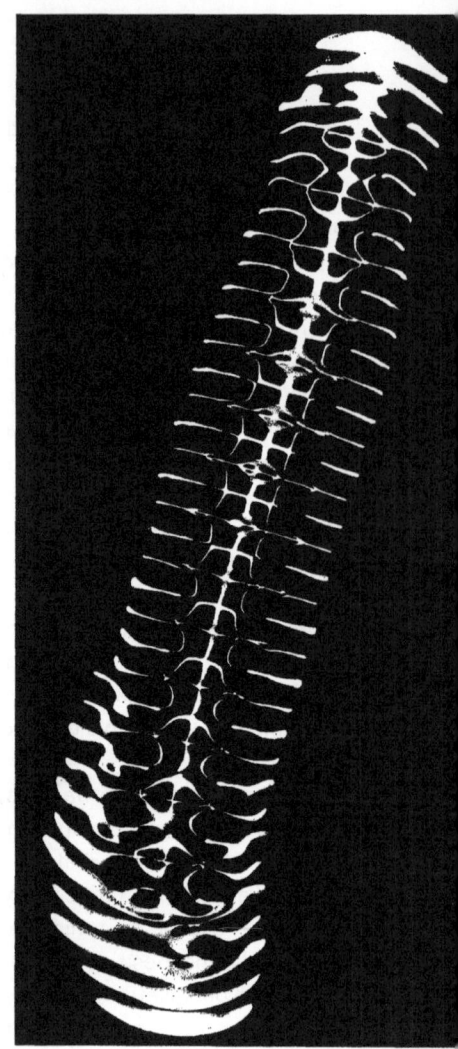

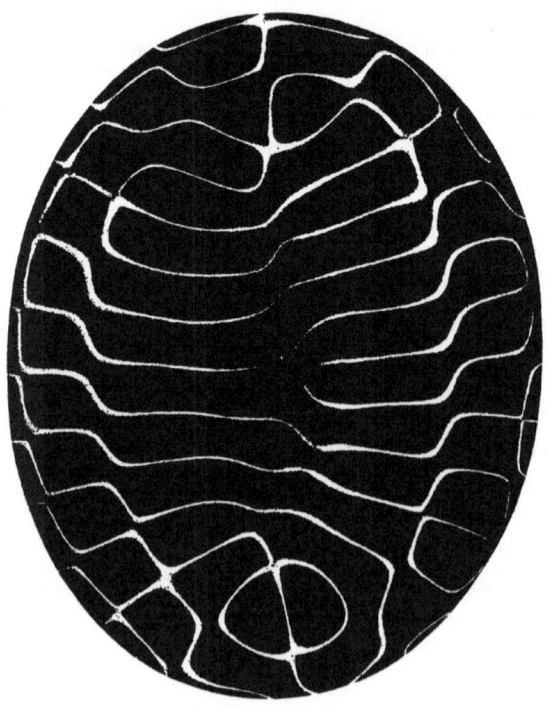

Three examples of the patterns of nodal lines
on vibrating metal plates (from [4])

C. H. Waddington

to handle analytically. A doctor living near Basel, Hans Jenny, has made a hobby
of exploring this field, which he calls Cymatics, by a variety of experimental procedures. To illustrate the 'biological look' of the forms produced, I have copied
three figures selected from the many in his recent, very beautiful book [4].

References

1. Waddington, C. H. *J. Theoret. Biol. 8* (1965)
367.

2. Saunders, J. W., Jnr., Cairns, J. M. and
Gasseling, M. T. *J. exp. Zool. 135* (1957) 503.

3. Yamada, T. *Adv. Morphogen. 1* (1961) 1.

4. Jenny, Hans. *Cymatics* (Basilius Presse,
Basel, 1967).

183

A Physicochemical Basis for Pattern and Rhythm

*A shorter version reprinted from Intracellular Transport © 1966
Academic Press Inc., New York*

John I. Gmitro and L. E. Scriven

Department of Chemical Engineering, Institute of Technology,
University of Minnesota, Minneapolis, Minnesota

INTRODUCTION

Any general principles governing the origins of regular patterns in space and rhythmic oscillations in time seem very likely to find application, directly or in-directly, at many levels of biological science—beginning with the intracellular, supramolecular level. A basic problem in the physical sciences which is increasingly attracting mathematicians and engineering scientists is the explanation of how specific dynamic patterns and rhythms can arise in spatially uniform, steady-state situations. Of course, macroscopic systems always suffer some sort of low-level noise, but how can chaotic, weak disturbances have no effect in some circum-stances, yet trigger development of strong, regular pattern and rhythm in others?

The genesis of dynamic patterns depends on the coupled effects of transport processes and transformation processes. The study of both lies at the heart of engineering science today. Their application to multicomponent, chemically reactive systems is the special concern of chemical engineers. This is one area in which chemical engineering and cell physiology run parallel; probably both could profit from closer communication and perhaps even active collaboration. The report that follows is offered as an example of current research in engineering science which may be of interest in connection not only with intracellular transport but also with other biological phenomena that may be better known to the reader than the authors.

We begin by mentioning a few strictly physical examples of dynamic pattern and rhythm. From these we attempt to abstract the key factors and to identify a set of specific problems which can be precisely formulated from the viewpoint of physical science—that is, in mathematical terms. The first several of these are then formulated and solved for a prototype class of situations involving simultaneous diffusion and chemical reaction in a variety of geometric configurations. The bear-ing of the results on signal propagation, pattern and rhythm generation, and mechanical movement is discussed in more qualitative terms which we earnestly

184

John I. Gmitro and L. E. Scriven

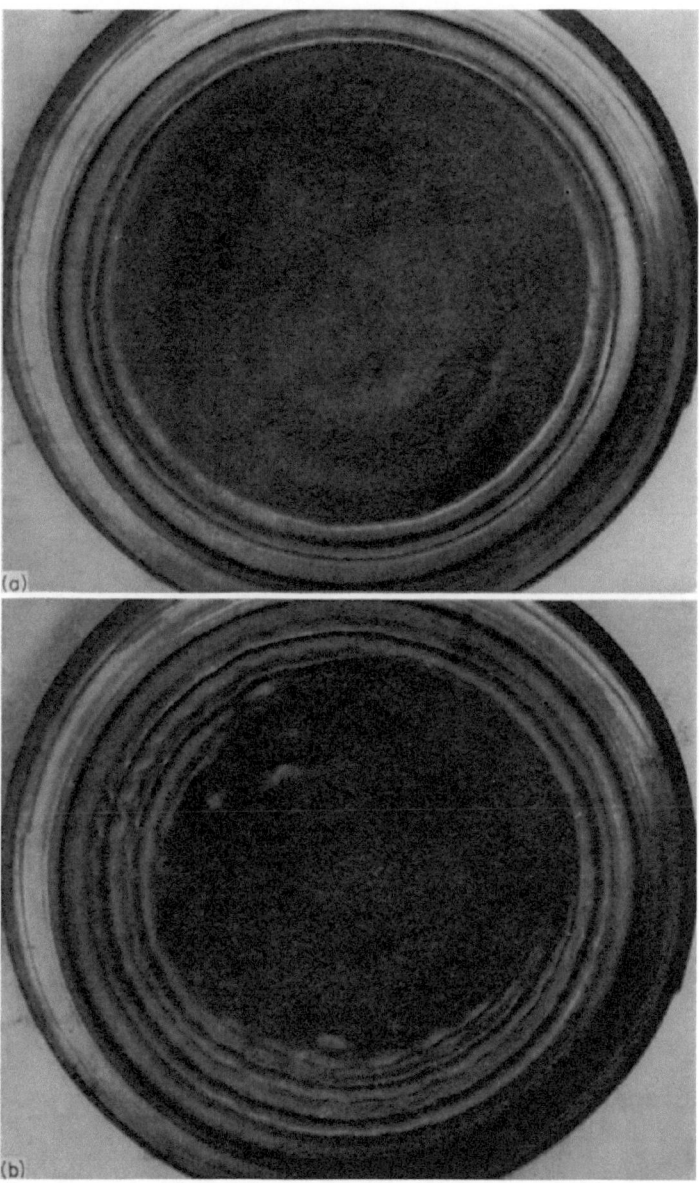

FIGURE 1
Development of Bénard cells in a dish of liquid heated uniformly from beneath. (Photographs courtesy of E. L. Koschmieder, Harvard University)

185

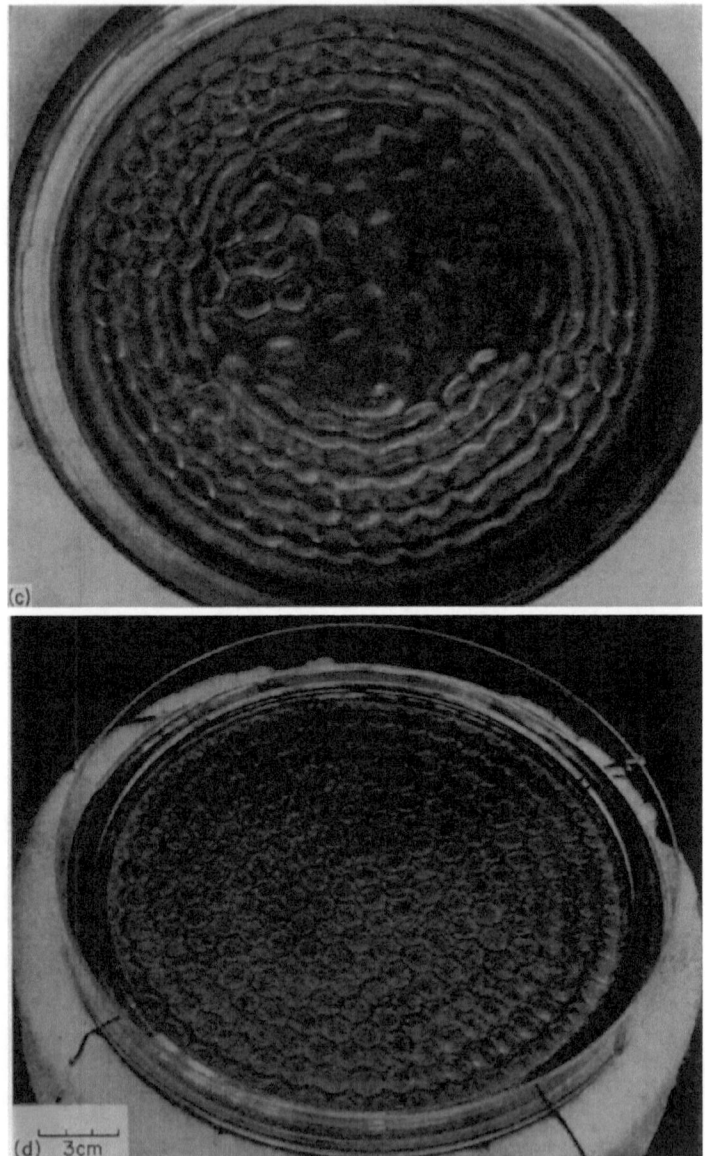

FIGURE 1 (continued)
Development of Bénard cells

hope will be informative to the reader who chooses to skip over the mathematical language of the sections on Formulation, and Instability and Wave Propagation.

Figures 1 (a–d) are photographs showing the development of a dynamic flow structure known as Bénard cells in a shallow dish of ordinary liquid that is being uniformly heated over its bottom side. Almost the same flow patterns can be brought about by two different physical mechanisms, one stemming from the dependence of surface tension on temperature, the other from the dependence of density on temperature (Scriven and Sternling [14]). In the latter case, the hotter, buoyant fluid at the bottom of the dish tends to rise and the colder, denser fluid at the top to sink. Such a turnover would lower the potential energy of the system and render it more stable until the heating from below re-established the unstable density profile. In the turnover itself, hot rising columns would necessarily exist somewhere alongside cold sinking columns, and in this situation of lateral velocity and temperature gradients there would be viscous forces opposing the flow and heat conduction reducing the buoyancy differences responsible for the flow. Thus there are two competing tendencies: one toward establishment of dynamic pattern, the other toward its destruction once it is formed (Sani and Scriven [13]). In fact, the rate of heating from beneath must exceed a certain critical value before the anabolic process can surpass the catabolic process sufficiently to establish flow, which tends to settle down in the steady Bénard-cell pattern if the critical value is not too greatly exceeded. The photographs indicate that the presence of the side of the dish favors concentric ring cells at first; these ultimately break up to give the strikingly hexagonal planform of Bénard

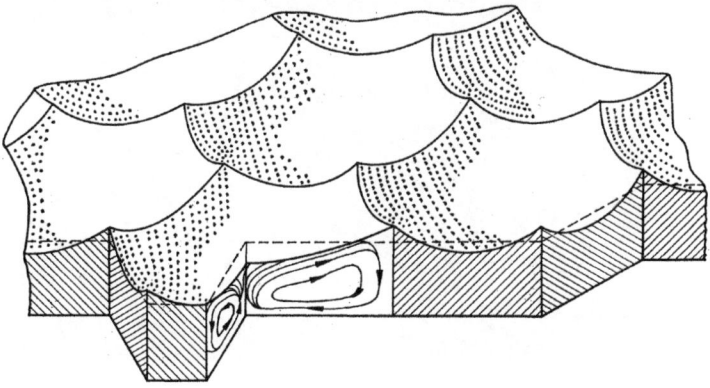

FIGURE 2
Schematic diagram of Bénard cells, showing streamlines of flow within a cell

cells, provided the dish diameter is much larger than the natural cell size. Flow within a cell is diagramed in figure 2, where streamlines are shown. The boundaries between cells are simply symmetry planes across which there is no flow of fluid. They are purely dynamic.

Bénard cells remain fixed in location and the flow within them is steady. They are an example of *stationary* convection. If a dish of liquid mercury is spun fast enough about its axis, a second type of convection, called *oscillatory*, occurs. The cellular planform becomes a little more complicated although still basically hexagonal, while the flow within cells may diminish and reverse periodically or the cellular pattern itself may translate through the liquid. In either case the net result is an oscillating, or rhythmical, flow at each point in the liquid.

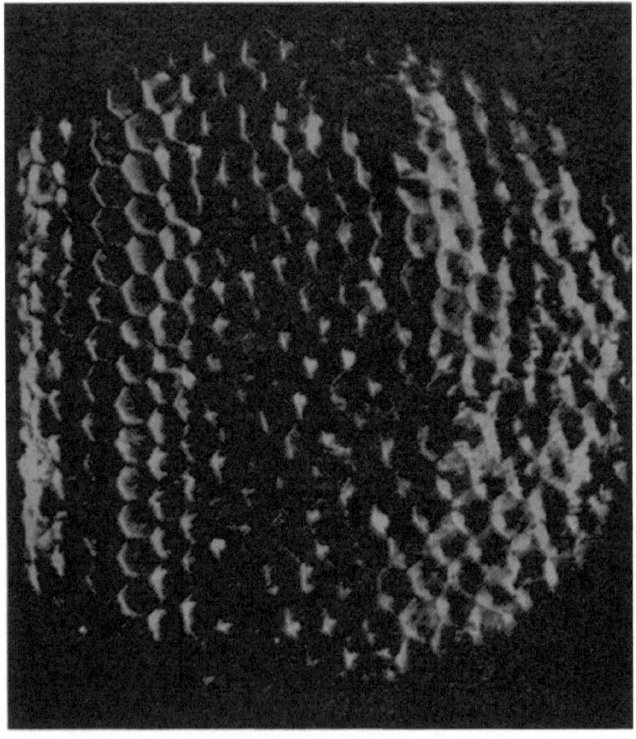

FIGURE 3
Top view of the decanted surface of a solidification front with a hexagonal tessellation.
(Photograph courtesy of John Wiley & Sons Inc.)

John I. Gmitro and L. E. Scriven

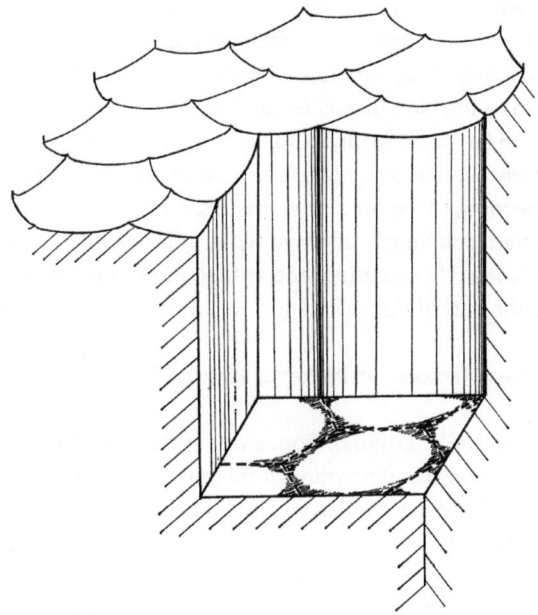

FIGURE 4
Concentration patterns frozen into the solid as hexagonal columns behind a solidification front. Broken lines indicate boundaries of the unit pattern. Regions of high solute concentration are shaded

The values of the critical heating rate for the onset of flow have been very successfully predicted by the theory of convective instability (cf. Chandrasekhar [7]). The analysis below is patterned after that theory.

Another remarkable instance of dynamic structure occurs under certain conditions of freezing of solid out of molten solution. The transformation process is solidification with its accompanying heat release. Both heat and solute are transported. There is a tendency for solute to be redistributed in a regular hexagonal pattern at the freezing face of the solid. This tendency, which stems from local supercooling in the melt, is opposed by thermal transport and reduced by diffusional transport. When these catabolic processes do not prevail, a regular concentration pattern may form over the freezing face ; this results in the tessellated profile pictured in figure 3 and a permanent, hexagonal-column concentration pattern frozen into the solid behind as diagramed in figure 4 (Chalmers [6]). The static structure is merely a partial record of the dynamic processes by which it has been produced—this is the important point.

189

A physicochemical basis for pattern and rhythm

Many other well-studied examples of dynamic structure in physical systems could be cited from fluid mechanics, meteorology, geophysics, and astrophysics. As for biological examples, we prefer to leave these and the biological implications of what follows to the biologically expert reader.

In all cases the key factors appear to be three. First, transformation processes: changes in physical state, as by phase transition, or in chemical state, as by chemical reaction. Second, transport processes: changes in location, as by convection or diffusion. Third, coupling of the two types of processes together: both must proceed simultaneously and affect each other. For our present purposes chemical reactions play the part of transformation while simple diffusion plays that of transport.

Analysis of these factors can be logically organized around the following set of problems:

1. Origin of pattern and rhythm from a uniform and steady state of transformation in systems in which departures from uniformity give rise to transport processes (equilibrium systems are a particular case). A natural adjunct of this item turns out to be

2. Signal transmission by propagation of small local disturbances in an initially uniform and steady-state system. With solutions to these problems, one can study

3. Control of pattern size, rhythmic period, propagation velocity, and wavelength, especially the dependence of possibilities on the complexity of the system—here, the number of participating chemical species. Beyond this lie more difficult questions of

4. Evolution and stability of particular patterns, rhythmic variations, and waveforms, in which so-called nonlinear effects are likely to be dominant. Before answering these questions one can investigate

5. Effect of pre-existing pattern on spontaneously developing pattern and rhythm and disturbance propagation, as most simply exemplified by these processes in homogeneously compartmentalized systems. If at the outset chemical effects alone are considered, as is done here, a parallel problem is

6. Coupling of chemical patterns and waves to electrical and mechanical stress fields in the material and thereby to forces, accelerations, and movements. Ultimately this coupling and the accompanying convective transport should be included in the first problem.

This is a large undertaking and we restrict ourselves here to the first three and last items, focusing on the physicochemical side of the overall problem. For a variety of reasons, some of which may become evident, it has seemed desirable to study pattern and rhythm in *surfaces* or *membranes*, and *lines*, or *fibres*.

FORMULATION

The basic system under consideration is diagramed in figure 5. It consists of a membrane or thread, uniform across its thickness, within which various chemical species are reacting and diffusing along its length. At the same time some or all of the participating species are exchanging with the surrounding media. The number of participating species is left open, for an important question to be answered is what influence the number of species has on pattern and rhythm. Here reaction is the only transformation process and diffusion the only transport process inside the system; modifications necessary to account for convective, electrical, and other effects can be made subsequently.

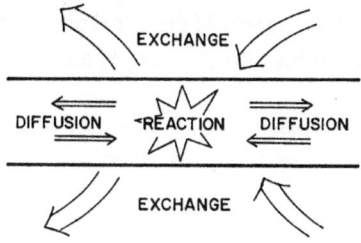

FIGURE 5
Diagram of basic system considered in the text. Reaction and diffusion take place within or on a membrane or thread. There is also exchange with the surrounding media

Equations of change

At every location within the system each chemical species obeys the conservation equation,

Rate of accumulation within the system	net rate of production by = chemical reaction within the system	net rate of influx by + diffusion along the system	net rate of input by exchange + with the surroundings ....(1)*

In mathematical symbols the equation becomes

$$\frac{\partial c_i}{\partial t} = R_i + J_i + Q_i \qquad \qquad ....(2)$$

* For the purposes of this publication the mathematical treatment has been reduced to a brief series of notes, which it is hoped will provide some indication of the approach used. Fuller details are printed in the original article. Numbers of equations, tables and figures have been taken from the original.

191

A physicochemical basis for pattern and rhythm

where c_i is the local concentration of the ith species and t is time. The system is not really described until constitutive relations specifying the rate processes present have been substituted for the rate symbols R_i, J_i, and Q_i.

. . . .

Excursions from uniform steady state

At steady state the concentrations of participating species are constant in time, by definition. If the state is uniform as well, there are no gradients in the system, and the equation of change reduces to a statement that rates of production and output of each component must just cancel:

$$0 = R_i{}^s - \sum_{j=1}^{N} H_{ij}{}^s(c_j{}^s - c_j{}^\circ) \qquad \qquad \dots(7)$$

Here $c_j{}^s$ stands for the concentration of the jth component in the steady state. The corresponding values of the rates of chemical production and of the mass-transfer coefficients are $R_i{}^s = R_i(c_1{}^s, c_2{}^s, \dots, c_N{}^s)$ and $H_{ij}{}^s = H_{ij}(c_1{}^s, c_2{}^s, \dots, c_N{}^s)$ respectively.

Unsteady states can be represented by concentrations that are sums of steady-state values and excursions from the steady state:

$$c_i = c_i{}^s + x_i \qquad \qquad \dots(8)$$

. . . .

With matrix notation and the rules for matrix multiplication the equation of change of excursions, obtained by substituting Eq. (8) and Eq. (7) in Eq. (2), takes the compact form of a standard linear partial differential matrix equation:

$$\frac{\partial [x]}{\partial t} = [K][x] + [D]\nabla^2[x] \qquad \qquad \dots(12)$$

Equation (12) states that if a steady state of the system is slightly perturbed, the rate at which the excursion grows or decays is controlled by the competition between (a) first-order processes depending on chemical reaction within the system and mass exchange with the surroundings, and (b) diffusive processes governed by 'bumpiness' of the concentration distributions within the system.

Whether an excursion grows, whether it periodically oscillates, and what size of pattern is likely to emerge as it develops, are questions that can be answered by solving Eq. (12).

Solution in terms of elementary patterns

Equation (12) can be solved by the methods of harmonic analysis, which rest on

192

a remarkable theorem going back to Fourier: Any spatial pattern may be expressed as a weighted sum of members of a suitably chosen set of *elementary patterns* (Bell [2]; Tolstov [18]). Mathematically, the elementary patterns are a complete set of characteristic functions, or eigenfunctions, appropriate to the geometric configuration of the system of interest.

. . . . .

These are eigenfunctions of the Laplacian operator and satisfy the equation

$$\nabla^2 F_k(r) = -k^2 F_k(r) \qquad \qquad ....(14)$$

The constant k is known as the characteristic parameter, or eigenvalue, corresponding to the eigenfunction F_k. Geometrically, k describes the mean size, or wavelength, of the corresponding elementary pattern. In fact, the mean pattern size, l, is $2\pi/k$. (If k is complex, as for wave propagation, then $l = 2\pi/k_r$; the imaginary part k_i gives the rate of attenuation of pattern or wave with distance.)

. . . . .

Standard methods are available for determining eigenvalues and eigenvectors of a given matrix (Frazer *et al.* [8]). The eigenvalues are themselves the roots of the determinantal equation

$$\det([K] - k^2[D] - \lambda[I]) = 0 \qquad \qquad ....(18)$$

where [I] is the identity matrix. For every value of k there are as many eigenvalues, $\lambda_{k,1}, \lambda_{k,2}$, etc., as participating chemical species, in general.

A solution of the equation of change in an excursion, Eq. (12), is

$$[x] = \sum_{k=0}^{\infty}\left(\sum_{n=1}^{N} a_{k,n}[A]_{k,n}e^{\lambda_{k,n}t}\right)F_k(r) \qquad \qquad ....(19)$$

Before taking up the interpretation of this result it should be pointed out that there may be other solutions, solutions that amount to propagating waves in particular. For example, the function representing simple harmonic waves propagating in the x-direction, viz.*

$$[x] = [a]e^{i(k_r x - \omega t)}e^{-k_i x} \qquad \qquad ....(20)$$

is found by substitution to satisfy Eq. (12) provided the wave-number k ($= k_r + ik_i$) and frequency factor ω ($= 2\pi f$) together satisfy the determinantal Eq. (18) with λ replaced by $-i\omega$.

Equation (18) for the eigenvalues of the transformation-and-transport matrix is the pivot on which the whole theoretical analysis turns. . . .

---

* The excursion [x] is a set of real quantities which may be taken equally well as the real part or as the imaginary part of the right-hand side.

A physicochemical basis for pattern and rhythm

To sum up with regard to instability: Any uniform, steady-state system of chemical reaction and diffusion suffers low-level noise, either transmitted from surroundings or arising internally in molecular fluctuations. Appearing more or less continually and randomly throughout the system as small concentration perturbations, the disturbances are without effect or trigger development of pattern and rhythm, according to the nature of sets of coefficients that characterize reaction, exchange, and diffusion. What matters are the eigenvalues of matrices comprising these coefficients; in particular, the eigenvalue containing the largest growth factor. Because the eigenvalues through the coefficients may depend, as remarked above, on factors other than the participating species, it is entirely possible that a previously stable steady state may turn unstable owing to changes in such factors. Factors other than the participating species can also control whether the ensuing instability is stationary or oscillatory, how rapidly it develops, and the pattern size and rhythmic frequency. Once the number of participating species is known and values of the coefficients are available, one can determine all of these features from the solutions of Eq. (18) for every admissible value of the size factor, k.

. . . .

EXAMPLES

In the preceding sections we have shown how the origin of regular patterns and rhythms in the type of system under consideration, and chemical signal transmission as well, can all be thought of as immediate consequences of Eq. (18), which involves sets of parameters that describe chemical reaction, exchange with the surroundings, and internal diffusion in small excursions from the uniform steady state originally present. We also outlined the derivation of the pivotal equation from basic physicochemical principles. Our object now is to show by means of examples what some of the implications of the analysis are, particularly in regard to control of pattern size, rhythmic period or frequency, propagation speed, and wavelength.

To do this we examine cases of one, two, and three participating species. In the first two we can write the characteristic equation for the eigenvalues, $\lambda_{k,n}$, explicitly in terms of the system parameters and the pattern factor k; in the third case it is scarcely worth while attempting to do so (a cubic equation must be solved). From the characteristic equation, whether or not in explicit form, the relationships can be found which the system parameters must satisfy, as functions of k, in order for a given type of instability to occur—stationary, oscillatory, marginal. The

194

Routh-Hurwitz criteria for eigenvalues of matrices and Routh's algorithm for polynomials with real coefficients are very useful (Gantmacher [9]).

. . . .

DISCUSSION

Without examining thoroughly all of the new possibilities of control of pattern size, rhythmic period, growth factor, wave propagation speed, attenuation factor, and wavelength which accompany a third participating species, we can see that richer ranges of possibilities come with each additional compound and the reactions into which it enters. Instability behaviour just possible with fewer compounds can be realized in a variety of ways ; entirely new behaviour can be produced. The same is true with regard to propagation of chemical waves when the number of participating species is increased.

Chemical concentration waves could provide large numbers of parallel signal-transmission channels. In small-scale systems these might for some purposes be competitive with electrical transmission means. A steady-state reaction system can be arranged to propagate signals at far faster speeds than diffusional movements, and without the attenuation that also limits purely diffusional processes, as shown above. Concentrations of participating species can be exceedingly low ; indeed, concentration could be reinterpreted as the probability density of finding a molecule of the species in a given locale.

In an engineering sense these observations mean that more species broaden the range of alternatives and may permit a better optimum solution to any particular design problem. Indeed, a new alternative may provide a solution where no acceptable solution was available before. The point is that increased numbers of species in a multicomponent system of transport and transformation may have decisive influence on the appearance and nature of pattern and rhythm.

Linearized stability theory illuminates the origin of pattern and rhythm as well as the propagation of small disturbances. But although its predictions of dominant pattern size or wavelength are likely to be close to the values for resultant steady states or limit cycles, the particular pattern shape and variation of amplitude with time or waveform which evolve are invariably determined by nonlinear effects that have not been considered here. Until an analysis of these effects has been completed, many patterns can be put forward as candidates.

Representative geometric configurations and patterns
The analysis and examples that have been presented actually pertain to a whole gamut of line-like and surface-like configurations. All that is necessary is to place

appropriate interpretations on the elementary pattern functions, $F_k(r)$, and the pattern-size factor, k. The most important cases are listed in Table II; representative examples are shown in figure 11. Some mechanical deformations that might be produced by the chemical patterns are also shown, and will be discussed.

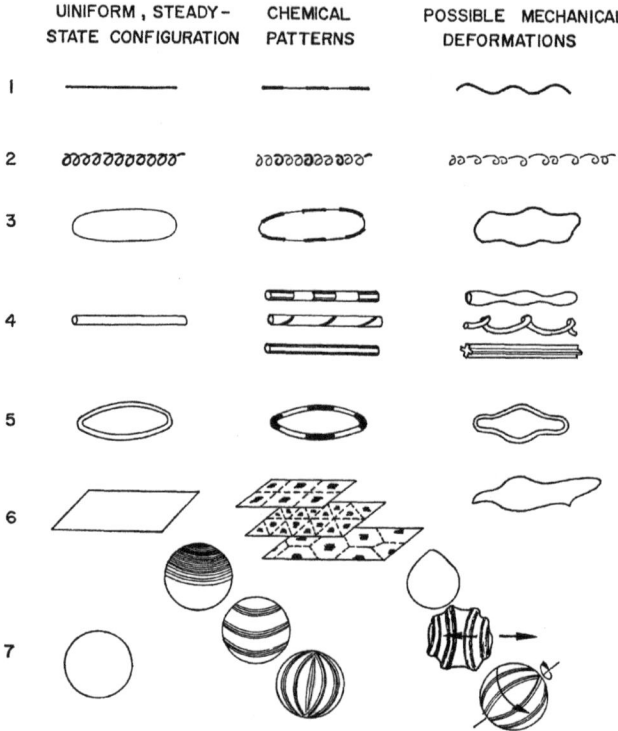

FIGURE 11
Representative geometric configurations with some regular chemical patterns that can arise spontaneously. Concentration dependent stress could produce the corresponding mechanical deformations

If the system is unbounded, or at least very extensive compared to significant pattern sizes, then continuous ranges of pattern size l and the factor k must be considered. If, on the other hand, the system is closed, as are loops, rings, and spheres, then these factors can take on only certain discrete values. The reason is that the basic unit of pattern, whatever it is, must repeat itself in the system an integral number of times. Rarely can the fastest-growing among all pattern sizes be

John I. Gmitro and L. E. Scriven

TABLE II
Elementary pattern functions for representative geometric configurations*

Configuration	Coordinates	Elementary pattern functions	k^2
Line	x	$e^{i2\pi x/L}$	$4\pi^2/L^2$
Circle	ϕ	$e^{im\phi}$	m^2/R^2
Cylinder	x, ϕ	$e^{i(2\pi x/L+m\phi)}$	$\dfrac{m^2}{R^2}+\dfrac{4\pi^2}{L^2}$
Ring	θ, ϕ	$e^{i(n\theta+m\phi)}$	$\dfrac{m}{R_1^2}+\dfrac{n^2}{R_2^2}$
Plane	Rectangular: x, y	$e^{i(2\pi x/L_x)+2\pi y/L_y)}$	$4\pi^2\left(\dfrac{1}{L_x^2}+\dfrac{1}{L_y^2}\right)$
	Polar: r, ϕ	$J_m(\alpha r)e^{im\phi}$	α^2
Sphere	θ, ϕ	$P_n^m(\cos\theta)e^{im\phi}$	$\dfrac{n(n+1)}{R^2},\ -n\leqslant m\leqslant n$

* Position is measured by standard length and angle coordinates.

so accommodated, in which case the nearest admissible pattern size is likely to be dominant.

A cylindrical surface is two-dimensional and can support circumferential as well as longitudinal wave patterns. A rich assortment of combinations can be made, e.g. the helically wound concentration pattern diagramed in the fourth row of figure 11. Moving patterns might provide means for facilitating diffusion of selected species along a thread, or along a membrane for that matter.

Of all the concentration patterns that can exist on flat surfaces (and certain others) only three are strictly regular—those in which the basic unit is a square, triangle, or hexagon, as sketched in the sixth row of figure 11. Pattern functions that are combinations of elementary pattern functions are known for all three (Chandrasekhar [7]). There is no way of predicting from the linearized analysis whether one of these three or some less regular candidates having the same pattern size will be established when a large expanse becomes unstable. The presence of a boundary or edge nearby may favour one pattern over another in a fairly predictable way, however. The configuration of the source in cases of wave propagation may also favour one pattern over another. Figure 12 serves as a reminder that a point source produces circular waves on a homogeneous and isotropic plane; a line source, lineal waves.* Many, many other wave patterns can of course be generated by interference of multiple sources.

*These are described by functions listed for the plane in Table II, with $m=0$ and $L_y=0$ respectively.

197

A physicochemical basis for pattern and rhythm

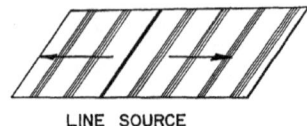

LINE SOURCE

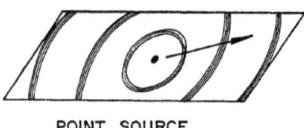

POINT SOURCE

FIGURE 12
Lineal and circular waves propagating on a plane

In the case of the closed spherical surface, linear theory provides a little more information. Before describing it we should point out some of the patterns possible on a sphere: two axially symmetric zonal patterns ($n=1$ and $n=4$; $m=0$) and one sectorial pattern ($n=10$, $m=10$) are shown in the seventh row of figure 11. There are only five strictly regular patterns: these correspond to the five regular polyhedra, the Platonic solids, as shown in figure 13. Of these, the cube and

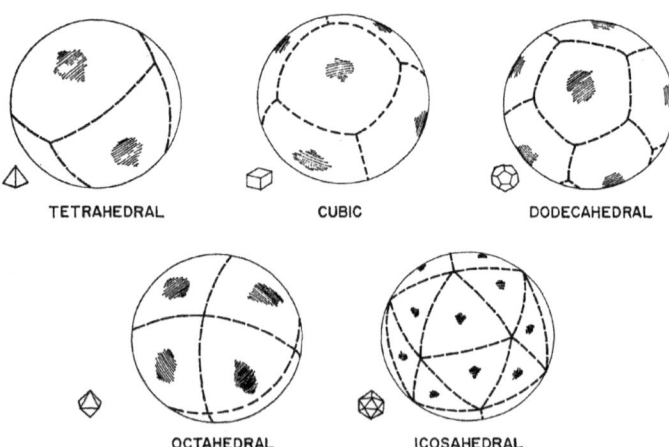

TETRAHEDRAL　　　CUBIC　　　DODECAHEDRAL

OCTAHEDRAL　　　ICOSAHEDRAL

FIGURE 13
Strictly regular patterns on the surface of a sphere, corresponding to the five regular polyhedra

198

octahedron are a conjugate pair, and so are the dodecahedron and icosahedron ;
for if the concentration maxima form one pattern, the minima must form the other
(the same is true for the triangle-hexagon pair on simple surfaces). Further
distinctions can be drawn on the basis of multicomponent concentration patterns,
but we need not go into these here. Pattern functions that are combinations of
elementary spherical harmonics are known for all five.

Because the surface of a sphere is closed, the pattern-size factor, k, can take on
only certain discrete values, which happen to be given by $k^2 = n(n+1)/R^2$, where
n is any non-negative integer. Types of pattern corresponding to values of n from
0 to 10 are listed in Table III. It is significant that for $n = 1$ there is a unique pattern

TABLE III

Spherical patterns for $n=0$ to $n=10$*

n	Type of pattern
0	Completely uniform
1	One axisymmetric pattern, nonregular
2	One biaxial pattern, nonregular
3	Only strictly regular pattern corresponds to the tetrahedron
4	Only strictly regular pattern corresponds to the cube and octahedron
5	No regular patterns
6	Only strictly regular pattern corresponds to the dodecahedron and icosahedron ; one complex polyhedral pattern with cubic/octahedral symmetry
7	One complex polyhedral pattern with tetrahedral symmetry
8	One complex polyhedral pattern with cubic/octahedral symmetry
9	One complex polyhedral pattern with tetrahedral symmetry
10	One complex polyhedral pattern with cubic/octahedral symmetry ; one complex polyhedral pattern with dodecahedral/icosahedral symmetry

* $k^2 = n(n+1)/R^2$.

and likewise for $n = 2$. Among the basic patterns for which $n = 3$ the only one that
is strictly regular corresponds to the tetrahedron. Those corresponding to the
cube-octahedron pair and dodecahedron-icosahedron pair enjoy the same
distinction at $n = 4$ and $n = 6$ respectively. Now, since the dominant value of k, and
hence n, is under the control of reaction, exchange, and diffusion, the linear theory
indicates clearly how by adjusting the rate coefficients of these processes (a)
a simple axisymmetric pattern can be selected and (b) one of the strictly regular
patterns could be selected if some tendency for regularity existed. Whether a
strictly regular pattern could actually be obtained depends on nonlinear effects
under study.

199

A physicochemical basis for pattern and rhythm

Chemical patterns on a closed loop and, cursorily, on a sphere were considered by the mathematician Turing [19] in a pioneering paper entitled 'The Chemical Basis of Morphogenesis'. Turing's paper stands alone in the biological literature and is independent of the mainstream of stability analysis in the physical sciences. It has been one of the factors motivating our work, which connects his ideas to the main stream, enlarging on them and extending them in ways he unfortunately did not live to pursue.

Besides a continuous closed loop, Turing analysed the origin of pattern in a ring of cells, that is, a compartmentalized loop. Now a logical question to raise is what effect has pre-existing nonuniformity on the appearance of new patterns in general. In our view the simplest way to attack this question is to analyse a regularly compartmentalized system—identical uniform regions separated by permeable barriers of negligible thickness, like Turing's cells. Incidentally, such barriers could be conceived of as arising at nodal lines of a regular pattern that might have previously been established in an originally uniform system. Results of our analysis of regularly compartmentalized systems will be presented elsewhere.

Coupling to mechanical stress and movement
The chemical state of a material influences its mechanical state—first the state of stress, and then the state of strain. Temperature-induced deformations commonly occur in everyday materials and are the subject of the well-developed theory of thermoelasticity (Nowacki [12]). Quite analogously, composition changes can induce elongation, swelling, shearing, bending, and twisting; the responses of wood products to humidity changes are familiar examples. Moving chemical patterns can in general produce moving patterns of mechanical stress which bring about movement. We focus here on such mechano-chemical coupling, although the chemical state of a material also influences its electromagnetic state and may do so sufficiently to activate electro-kinetic effects.

Moving stress patterns produce net forces and accelerations and thereby moving patterns of strain-displacement, which is simply mechanical motion. Thus, systems of coupled chemical reaction and diffusion of the sort we have been considering can give rise to pulsations, rippling, streaming, and so on, by what we call 'active-stress' mechanisms. An *active* stress is one depending on variables other than strain in a way that permits chemical energy and other forms of energy to be transformed directly to mechanical energy.

There are no reasons known to us to doubt that active stresses can arise in every category of mechanical stress. In line-like structures the only meaningful category

200

is tensile stress, which of course includes contractile stresses; in rods and tubes: tensile, bending, and torsional (twisting) stresses; in surfaces and membranes: isotropic tangential stress (often called membrane or surface tension), pure shear tangential stress, and internal torsional stress. In films, sheets and plates: the pertinent categories are these plus various types of bending and torsional stresses. In continuous and internally structured bulk phases there are the three basic types: isotropic stress, or pressure; pure shear stress, occasionally referred to as 'shifting' or 'shearing' forces; and antisymmetric stress, or internal torsional stress. Of these, only so-called contractile forces and two or three others seem to have been considered in the literature on mechano-chemistry.

By invoking various of these possibilities we can envisage chemical patterns producing the mechanical deformations shown on the right of figure 11. The efficacy of isotropic stress, or pressure, is firmly established by findings in thermo-elasticity (Nowacki [12]; Bupara [4]). In all cases it is easiest to deal first with situations of 'weak' stress-coupling, wherein the motions produced by active stress mechanisms do not affect appreciably the underlying chemical patterns and rhythms. The motions can have rather interesting consequences, however, from the points of view of hydrodynamics and, perhaps, cellular and organelle motility.

Fibres can displace themselves along their length through viscous surroundings by worming and squirming, including bending in planar and helical waves, as shown by G. I. Taylor [17] and Hancock [10]. Translation of a cylinder through liquid and transport of liquid contents of a tube by peristaltic waves are implied by Taylor's analysis of self-propulsion of a sheet by means of progressive bending waves [16]. That self-propelled rotation of a cylinder and a sphere can be produced by progressive waves of surface deformation has been established (Bupara [4]). So also has self-propelled translation of a sphere by means of axisymmetric zonal waves of the sort suggested in the seventh row of figure 11 (Lighthill [11]; Bupara [4]).

The theoretical analyses, prior to Bupara's recent one, simply postulated surface-deformation waves without going into their origin and control. He demonstrated that such waves can be caused by concentration-dependent pressure that is weakly coupled to moving chemical patterns. Analysis of systems with strong stress coupling is feasible though considerably more complicated. But the mechanical motions permit convective transport which may influence appreciably the other transport and transformation processes, and so should be included from the outset. Mathematically, the convective transport terms should be returned to the equations of change, from which they were omitted here [Eqs. (1)–(6)],

201

and an equation of momentum and stress must be brought in. Convection along and to a fibre or membrane gives rise to additional means of controlling the emergence and properties of pattern and rhythm.

SUMMARY

Elucidation of principles governing the spontaneous appearance of pattern and rhythm is an important problem in contemporary physical and engineering science, and one that surely relates to many levels of biological science. The key factors evidently are transformation processes and transport processes—one of each, at least—coupled together. There are many purely physical samples, but we have chosen to study a class of physicochemical systems of coupled chemical reactions and diffusion. They are open systems in that every point on a surface or a line is adjacent to the surroundings; in geometric configuration they may be either open or closed.

We find that a uniform steady state of almost any such system, beginning with the simplest, is potentially able to propagate chemical signals or to give birth to progressive chemical waves or standing chemical patterns. Possibilities in the simpler systems are drastically limited, but as the number of participating chemical species goes up there is an ever-increasing richness of possibilities for accomplishing any of these things, and for doing so within narrower design specifications, so to speak.

Analysing the possibilities involves sets of dependent concentration variables and a multiplicity of reaction rate, exchange, and diffusion coefficients. The mathematics of ordered sets known as matrices proves to be a powerful tool; the concepts and language are useful in thinking about still more complex systems.

Composition affects the local state of stress in a material, and thereby can develop net forces, acceleration, and motion. In this way travelling concentration patterns supply a physicochemical basis for spontaneous movements. Convective transport, which accompanies movement, raises additional possibilities of 'feedback' and regulation of spontaneously appearing pattern and rhythm.

ACKNOWLEDGMENTS

This work was supported by Air Force Office of Scientific Research Grant No. AF AFOS R-219-63. A travel grant by the American Institute of Biological Sciences made possible the participation of L. E. Scriven in this symposium.

John I. Gmitro and L. E. Scriven

References

1. Adler, R. B., Chu, L. J. and Fano, R. M. 'Electromagnetic Energy Transmission and Radiation.' (Wiley, New York, 1960.)

2. Bell, E. T. 'Mathematics—Queen and Servant of Science.' (McGraw-Hill, New York, 1951.)

3. Bellman, R. 'Introduction to Matrix Analysis.' (McGraw-Hill, New York, 1960.)

4. Bupara, S. S. 'Spontaneous movements of small round bodies in viscous fluids.' Ph.D. Thesis in Chemical Engineering, Univ. Minnesota, 1964.

5. Carson, J. R. 'Electric Circuit Theory and the Operational Calculus.' (McGraw-Hill, New York, 1926.)

6. Chalmers, B. 'Principles of Solidification.' (Wiley, New York, 1964.)

7. Chandrasekhar, S. 'Hydrodynamic and Hydromagnetic Stability.' (Oxford Univ. Press, London, 1961.)

8. Frazer, R. A., Duncan, W. J. and Collar, A. R. 'Elementary Matrices.' (Macmillan, New York, 1946.)

9. Gantmacher, F. R. 'Matrix Theory.' 2 vols. (Chelsea, New York, 1959.)

10. Hancock, G. J. *'Proc. Roy. Soc. (London)* Ser. A217*, 96 (1953).

11. Lighthill, M. J. *Commun. Pure and Applied Math., 5*, 109 (1952).

12. Nowacki, W. 'Thermoelasticity.' (Addison-Wesley, Reading, Massachusetts, 1962.)

13. Sani, R. L. and Scriven, L. E. 'Convective Instability.' Unpublished Report on Grant AF AFOSR-219-63 (1964).

14. Scriven, L. E. and Sternling, C. V. *J. Fluid. Mech., 19*, 321 (1964).

15. Sokolnikoff, I. S. and Redheffer, R. M. 'Mathematics of Physics and Modern Engineering.' (McGraw-Hill, New York, 1958.)

16. Taylor, G. I. *Proc. Roy. Soc. (London)* Ser. A209*, 447 (1951).

17. Taylor, G. I. *Proc. Roy. Soc. (London)* Ser. A211*, 225 (1952).

18. Tolstov, G. P. 'Fourier Series.' (Prentice-Hall, Englewood Cliffs, New Jersey, 1962.)

19. Turing, A. M. *Phil. Trans. Roy Soc. (London)* Ser. B237*, 37 (1952).

Self-reproducing Automata–some implications for Theoretical Biology

Michael A. Arbib
Stanford University

Abstract In Section 1 we review various models of self-reproducing automata which share with life the property of repetitive production of ordered heterogeneity. In Section 2 we note that these systems also share with living systems a dependence on a separate stable description of the reproducing organism, and we discuss a number of reasons why this condition may be necessary in more complex organisms. In Part 3 we face up to the fact that our automata are extremely prone to large malfunction as a result of small damage, and make a first step towards rectifying this deficiency by exhibiting an automaton model of a worm which, upon being damaged, will readjust its composition so that the front third is head, the next third is body, and the final third is tail. Part 4 concludes the paper with a brief discussion of how our model may be extended to deal with problems of evolution.

▶ *Introduction.* Life has been defined as *the repetitive production of ordered heterogeneity* [1]. Elsasser [2] has noted the importance of the term 'heterogeneity', since repetitive production of order *per se* could describe the operation of physical laws in a lifeless universe. Since many biologists would agree with Hotchkiss [1], it is of no small interest that the self-reproducing automata we present in Section 1 satisfy his definition. This suggests that automata theory may help refine theoretical biology by allowing us to completely describe systems which share more and more of the properties we ascribe to living systems, so that we may avoid oversimplifications which are almost inevitable at the level of verbal discourse. Note that our strategy may be a fruitful one whether or not one holds the reductionist view that all vital processes can be reduced to physicochemical terms, i.e. whether one believes that one is converging to an understanding of the nature of life, or sharpening an inevitable distinction between living things and man-made machines. I personally believe reduction is possible [3], but cannot share the confidence expressed by Crick in its imminence [4], since I believe that we are at the very beginning, and require immense breakthroughs in the mathematical theory of automata and other complex systems, before we can hope to really understand the hierarchy of processes that extends from molecule to man. It

204

seems to me that the notion of $DNA \rightarrow RNA \rightarrow$ enzyme transduction is of as vital importance to understanding life as the conversion of decimal numbers to binary notation is to understanding digital computers. However, just as computers may be built to operate in non-binary mode, so may there be life without DNA; and we are no more entitled to say we understand embryology when we understand DNA than we are entitled to say we understand computation when we understand radix two. Perhaps our automaton models may break us of too parochial a view of the goals of theoretical biology.

▶ *1. Summary of Basic Results on Self-reproducing Automata.* It should be emphasized that the theory of automata is usually concerned with devices which transform information from an input string to an output string, changes in the automaton being regarded as incidental. It was von Neumann in his Hixon symposium lecture [9] who shifted emphasis to the way in which initial information serves to regulate the growth and change in structure of an automaton. He noted that we associate with machines used for construction a certain degenerating tendency—we expect an automaton to build an automaton of less complexity. However, when organisms reproduce we expect their offspring to be of complexity at least equal to that of the parent. In fact, due to long-term processes of evolution, we even expect to see increases in complexity during reproduction. In view of this apparent conflict, von Neumann felt it worth while to see what could be formulated rigorously in the way of construction theorems for automata.

To appreciate von Neumann's ideas we should briefly recall the basic work of Turing [10]. A. M. Turing was one of the people who, in the thirties, were worried about evolving a precise notion of effective computation. We are all familiar with the idea of an algorithm; one has a recipe such that if one follows it one always gets the right answer—if there is one. One may run on indefinitely if there isn't an answer. Turing produced a formalization as follows: Let us consider a box which has finitely many states, say q_1 up to q_n, and which operates on a tape divided lengthwise into squares, each of which can bear any one of a finite number of symbols, x_1 up to x_m. We start with a finite tape on which are printed the initial data, the machine is started in state q_1, and operates synchronously. At each time ($t = 0, 1, 2, 3, \ldots$) the symbol it scans and the state it is in determine that the machine stops *or* determine three things: the new symbol (perhaps the same old one, or it may be a blank) it prints on the scanned square, the moving of the tape at most one square, and the change of state. Then it is ready to repeat the cycle, obeying the instruction keyed by the new state-symbol pair. If it ever comes to an end of the tape, then it will add on a new square.

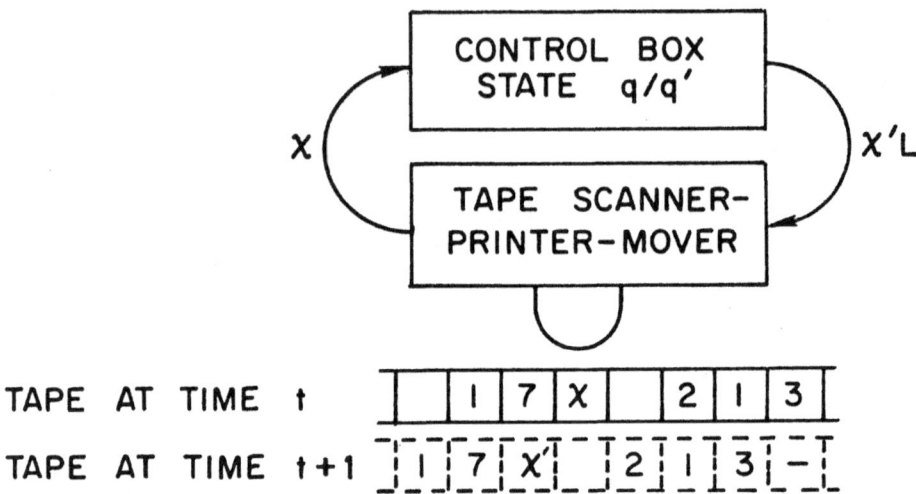

FIGURE 1
A Turing machine, whose program contains the quintuple qxx 'Lq'.

Now suppose we have an algorithm for effectively producing one integer from another. In the past, such a rule has always been transcribable into a set of instructions for a Turing machine, and, of course, it is quite clear that a Turing machine computation is in itself a recipe, an effective process. At the present time, the usual thing is to identify the intuitive notion of an effective computation with the formal notion of computation which can be carried out by a Turing machine.

For our present purposes, the important thing that Turing discovered in his paper of 1936 was that you could build a 'universal' Turing machine. How does it work? Usually, given a Turing machine, Z, we use it to compute a function, f, by placing the number x coded on the tape and letting the machine run. If and when Z stops the result on the tape is decoded to give us $f_Z(x)$. Every such machine Z is given by a finite list of instructions, one for each state-symbol pair. Thus a Turing machine Z can be represented by a string of symbols e(Z) which encodes its program. Turing gave a program for a Turing machine U which was universal in that if you wrote on U's tape the ordered pair e(Z) and x, then, at the end of U's computation, it would produce $f_Z(x)$ – precisely what Z would have computed with input x. If Z wouldn't have stopped, of course U won't stop.

Turing's result that there exists a universal computing machine suggested to von Neumann that there might be a universal construction machine, that is an automaton A which, when furnished with the description I_N of any other automaton N in terms of appropriate functions, will construct a copy of N. In what follows, all automata for whose construction the facility A will be used are going to share with A the property that their description will include the specification of a place where an instruction I can be inserted. We may thus talk of 'inserting a given instruction I into a given automaton'. The reader may at this stage think of the automaton A with the description I_A inserted into it. This entity will proceed to construct a copy of A. But note that this does not make A self-reproducing, for A with appended description produces A without an appended description – it is as if a cell had split in two with only one of the daughter cells containing the genetic message. Such a consideration suggested to von Neumann that the correct strategy might involve 'duplication of the genetic material'. He thus introduced an automaton B which can make a copy of any instruction I that is furnished to it – I being an aggregate of elementary parts and B just being a 'copier'.

Von Neumann then combined the automata A and B with each other and with a control mechanism C which does the following. Let A be furnished with an instruction I. Then C will first cause A to construct the automaton which is described by this instruction I. Next C will cause B to copy the instruction I referred to above and insert the copy into the automaton referred to above which has been constructed by A. Finally, C will separate this construction from the system $A + B + C$ and 'turn it loose' as an independent entity.

Let us then denote the total aggregate $A + B + C$ by D. In order to function, the aggregate D must have an instruction I inserted into A. Let I_D be the description of D, and let E be D with I_D inserted into A.

E *is* self-reproductive. Note that no vicious circle is involved. The decisive step occurs in E when the instruction I_D, describing D, is constructed and attached to D. When the copying of I_D is called for, D exists already, and it is in no wise modified by the construction of I_D. I_D is simply added to form E. Thus there is a definite chronological and logical order in which D and I_D have to be formed, and the process is legitimate and proper according to the rules of logic.

We thus see that once we can prove the existence of a universal constructor for automata constructed of a given set of components, the logic required to proceed to a self-reproducing automaton is very simple. [Though there is something somewhat whimsical in the idea of a universal constructor, as if a mother could have offspring of any species, depending only on the father. While this may be appropriate to Greek myths, it does not seem appropriate to biological modelling. We shall come back to this matter in later sections.] Our concern now is to examine the difficulties involved in actually providing a universal constructor. Von Neumann did not do this in his original paper, and the task involves hundreds of pages of his book on *The Theory of Self-Reproducing Automata* [11] which was published in 1966 on the basis of a manuscript left at von Neumann's death in 1956. The problem is essentially this. A Turing machine is only required to carry out logical manipulations on its tape, sensing symbols, moving the tape, printing symbols, and carrying out elementary logical operations. A universal computer only has to carry out these operations. But a universal constructor must also be able to recognize components, move them around, manipulate them, join them together. Thus, presumably, constructors of Turing machines require more components than do Turing machines themselves. We are immediately confronted with the possibility of an infinite regress. Given a set of components C_1, to construct machines which build all the automata made of components from C_1, you may need a bigger set of components C_2. To build all machines constructed of components C_2, you may need machines put together from a bigger set of components C_3. The question is: 'Is there a fixed point?' Can we find a set of components C such that all automata built from components of C can be constructed by automata built from the same set C? I have called this [12] *the fixed point problem for components.* This is the fundamental problem in the theory of self-reproducing automata. Once we have found a set of components C in which for each automaton A there can be found an automaton c(A) which constructs A, it turns out to be a fairly routine matter to prove the existence of a universal constructor. We then know from von Neumann that it is a simple matter to prove *the construction fixed point theorem,* namely that there exists a self-reproducing machine U which can construct a copy of U. There have been several procedures following on von Neumann's to exhibit a set of components which satisfy the component fixed point theorem. Von Neumann [11] used 29-state components and gave an elaborate construction taking about 200 pages. James Thatcher [13] used the same components but gave a much more elegant construction taking less than 100 pages. E. F. Codd [14], with remarkable ingenuity, showed that a construction similar to von Neumann's could go through using components with only eight states. I showed [12, 15] that the construction

208

could be done with great simplicity in a matter of eight pages, if one allowed the use of much more complicated components. My rationalization for this use of complex components is that if one wishes to understand complex organisms one should adopt a hierarchical approach, seeing how the organism is built up from cells rather than from macromolecules. We might add that Myhill [16] has given an axiomatic theory of self-reproduction, but that the axioms are formulated in a way which does not allow them to be directly applied to different sets of components but only allows one to generate theorems about self-reproduction when one already has theorems about universal constructors. However, Myhill's paper shows that results in recursive function theory can lead to rather startling conclusions about finite programs containing the possibility of infinite improvements in successive generations of offspring without requiring any randomness in the mutations.

Rather than go into any details of my construction I shall just briefly present three pictures which give some idea of the basic notions involved. Figure 2 shows a CT-machine which under the control of a program in its logic box can read and write on a one-dimensional tape in just the way a Turing machine does, and which can write but not read on a two-dimensional tape. The idea is that the two-dimensional tape is to be thought of as a construction area, and the writing of a symbol is to be thought of as equivalent to the placing of a component. Our task is to find a set of components from which we can build tape, logic box, and construction area. Such a component as shown in Figure 3 is a finite-state module which can contain up to 22 instructions from a rather limited instruction set. We are to think of a two-dimensional plane in which these cells are repeated in a sort of Cartesian array *ad infinitum*. An automaton is then represented as an activated configuration of these cells. I might mention that the little boxes marked W are weld registers which serve to 'weld' squares together so that a number of squares may be 'welded' into a one-dimensional tape in such a way that when any one square of that tape is instructed to move, all cells will move in the indicated direction. The assumption of such a weld operation greatly simplifies our programming. In Figure 4 we see an overall plan of an embedded CT-machine. We see that the logic box has been broken into two pieces, a one-dimensional tape which contains the program and two cells which form a control head. The idea is that on activation by the control head squares of the program tape may either be used to guide the control head in manipulating the computation tape in a Turing machine fashion or else may be used to place selected components in the constructing area and move welded blocks of components around. We

P

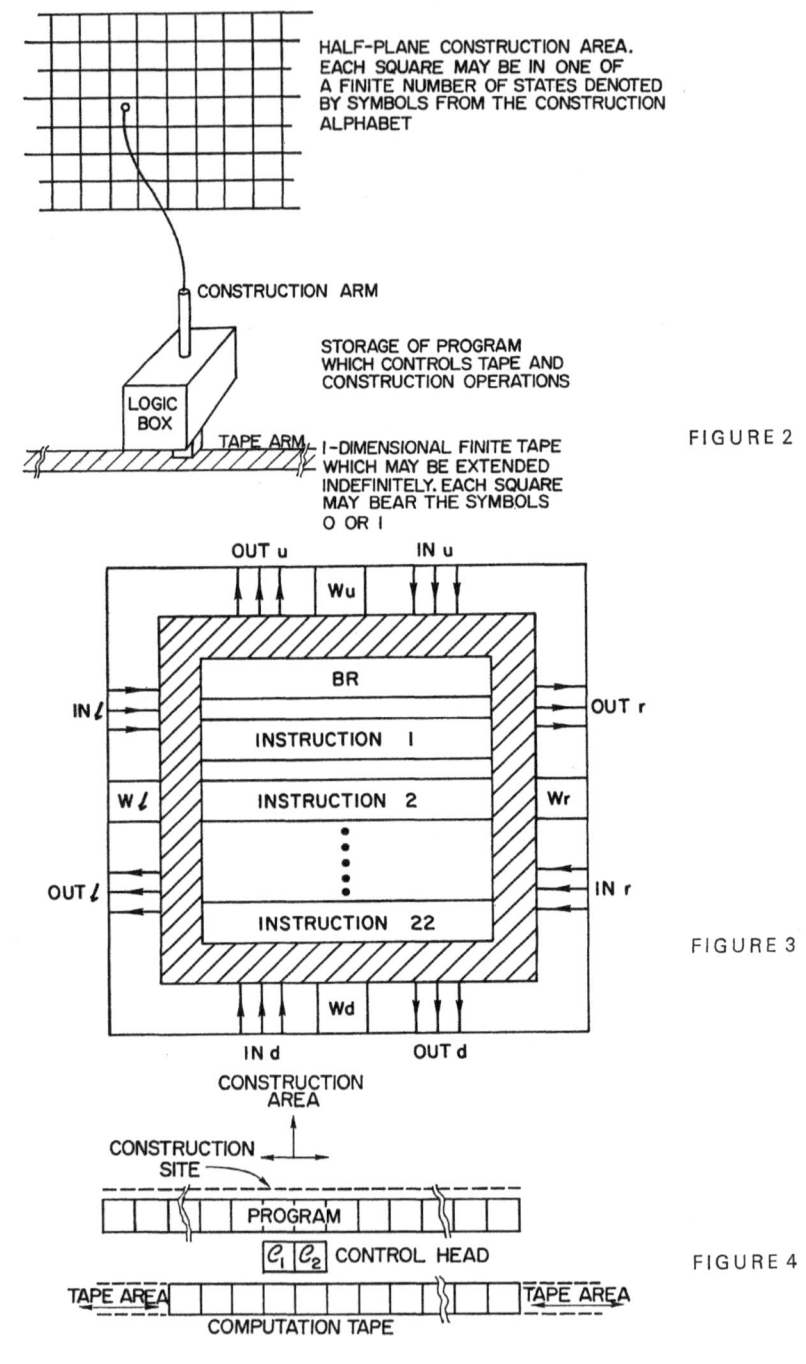

HALF-PLANE CONSTRUCTION AREA.
EACH SQUARE MAY BE IN ONE OF
A FINITE NUMBER OF STATES DENOTED
BY SYMBOLS FROM THE CONSTRUCTION
ALPHABET

CONSTRUCTION ARM

STORAGE OF PROGRAM
WHICH CONTROLS TAPE AND
CONSTRUCTION OPERATIONS

LOGIC
BOX

TAPE ARM I-DIMENSIONAL FINITE TAPE
WHICH MAY BE EXTENDED
INDEFINITELY. EACH SQUARE
MAY BEAR THE SYMBOLS
0 OR I

FIGURE 2

OUT u   IN u

Wu

BR

INSTRUCTION   I

INSTRUCTION   2

INSTRUCTION   22

IN ℓ   OUT r

W ℓ   Wr

OUT ℓ   IN r

Wd

IN d   OUT d

FIGURE 3

CONSTRUCTION
AREA

CONSTRUCTION
SITE

PROGRAM

C_1 C_2 CONTROL HEAD

TAPE AREA   TAPE AREA

COMPUTATION TAPE

FIGURE 4

show [12, 15] that in fact an instruction code can be specified for our basic modules so that it is not only possible to embed arbitrary Turing machines in the array of those components in the indicated fashion, but in fact to program these machines so that they can construct other such machines in the construction area and to go on from there to show that there exists a universal constructor made of these components. It is then a standard procedure, following von Neumann's argument, to present an actual self-reproducing machine.

Having thus presented the basic logic of the theory of self-reproducing automata, our aim in the rest of this paper will not be to present new results in the theory— save for an amusing example of a regenerating automaton in Section 3—but rather to try and list some of the problems posed for the theoretical biologists by the attempt to compare the mode of reproduction employed in biological systems with that employed in our self-reproducing automata.

▶ *2. What is the role of descriptions ?* It is a striking fact that in both real biological systems and in our self-reproducing automata we see a distinction between a description of the system and the active portion of the system. This goes back to the Weissman assumption underlying the current form of the theory of natural selection—namely that one needs to distinguish the genotype (a set of instructions) from the phenotype (their functioning embodiment) and that, whereas different environments produce different phenotypes from a given genotype, this change does not itself produce a change in the genotype. However, changes in genotype do cause marked changes in phenotype, and it is the genotype changes that propagate, whereas phenotype changes do not. Thus selection is of genotypes, but it acts on phenotypes. In Section 4 we shall briefly consider what steps might be taken to fit our automaton-theoretic considerations for a study of evolution, but in the present section we wish to consider the question (whose importance was emphasized to me by Howard Pattee) : 'Is a *description* of the object a necessary complication to obtain an interesting self-replicating object ?' In other words, must an interesting self-replicating object A made out of our elementary units contain within itself an object e(A), a description of A also made out of these units ?

Von Neumann [11] distinguished two methods of self-reproduction. In the passive method the self-reproducing automaton contains within itself a passive description of itself, and reads this description in such a way that the description cannot interfere with the automaton's operations. This is the method used in our self-reproducing automata, and if we identify DNA with an encoding, it appears to be the method used in living organisms. We may contrast this with the active method, in which the self-reproducing automaton examines itself and thereby constructs a description of itself. Von Neumann has suggested that this method would probably lead to a logical paradox. However, I am not at all convinced of this at the time of writing. It would seem to me that DNA does indeed replicate itself by this active method and we must ask not is there a logical paradox inherent in the active method, but rather is there some well-defined cut-off point at which the active method is no longer applicable ?

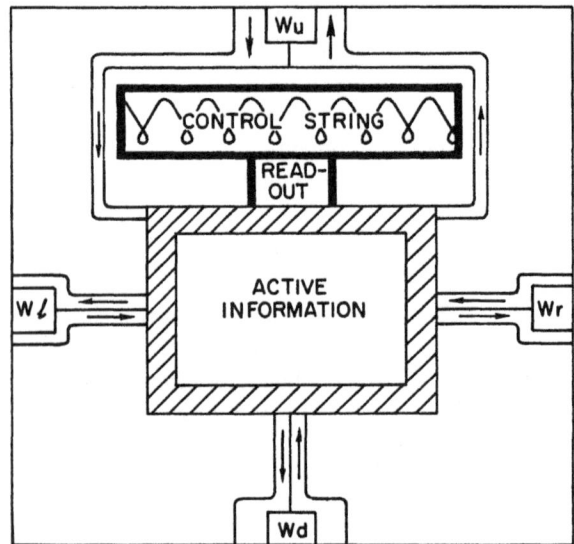

FIGURE 5

Pattee [8] sees the central biological aspect of hereditary evolution as the fact that 'the process of natural selection acts on the actual traits or phenotypes and not on the particular description of this phenotype in the memory storage which is called the gene'. He sees this as essential biologically because 'it allows the internal description or memory to exist as a kind of virtual state which is isolated for a finite lifetime, usually at least the generation time, from the direct interaction which the phenotype must continuously face'.

Goodwin [17] has pointed out that 'in a study of the dynamic properties of a certain class of biological phenomena ... it is necessary to extract a manageable number of variables from the very large array which occurs. ... Variables which are major to the phenomenon being investigated become the quantities that define the systems which one intends to study, while [minor] variables become

212

either parameters of the system, thus defining its environment, or . . . noise.'

'The *relaxation time* of a system is, roughly speaking, the time required for the variables to reach steady state after a "small" disturbance. The significance of this concept is the fact that if two systems have very different relaxation times, then relative to the time required for significant changes to occur in the "slower" system, the variables of the "faster" one can be regarded as always being in a steady state. . . . On the other hand, the variables of the "slow" system will enter into the equations of motion of the "fast" one as parameters, not as variables.'

The description, then, is presumably a portion of the system whose relaxation time is so great relative to the dynamics of the system that it may serve as a permanent record insulated from the vagaries of the existence of the organism. In terms of this notion of insulation from the vagaries to which the organism is subjected, we might note that Michie and Longuet-Higgins [18] compare the *germ plasm* (DNA specifications) of a cell to the *program* of a user of a large computer system, and they compare the *soma* (cellular machinery of implementation) to the *monitor* of the computer system (the program which controls input/output devices, assigns priorities to different users, translates programs using various compilers, etc.). They suggest that the genotype-phenotype distinction may be compared to the segregation of user's software and system's software. If a user writes a program which could modify the monitor during execution, difficulties could be created for users of other programs using the same monitor. Perhaps similar difficulties could arise if germ plasm and soma were not segregated.

Perhaps there is even something in the horribly naive thought that if an object is genuinely three-dimensional it cannot be copied in space without completely cutting it apart. Thus we copy it in the four dimensions of space-time instead, by using a genotype to *grow* a replica.

We end this somewhat inconclusive discussion of the nature of descriptions by noting that Thatcher [19] and Lee [20] have shown that there are Turing machines which can print their descriptions on their tapes when appropriately triggered although their description is not explicitly stored within their program. Perhaps this means that an automaton need not contain within itself an explicit separate description at all times, but need only retain the ability to generate such a description when it is necessary. In other words, we may visualize an organism in which the genome is *potential* rather than *actual*. The big question here, then, is one of reliability. If we do not explicitly segregate the genotype from the phenotype, can we guarantee that damage to the phenotype will not yield dangerous alterations to the potential genotype, as suggested by Michie and Longuet-Higgins ?

While our self-reproducing automata share with biological systems the use of a completely separate description, we should note that they are abiological in that the automaton which is being reproduced is completely passive until it is completed. It might be worth analysing why complete passivity is used and what might be done to avoid it.

▶ *3. Morphogenesis.* We examined [12] some of the ways in which our model diverged from biology and indicated the way in which one of these flaws could be corrected, namely we showed how the model could be modified so that the biological program became a string stored in *each* cell, rather than a program embedded in a *string* of cells. This Mark II module is shown in Figure 5. We have kept the tessellation structure, side-stepping the morphogenesis of individual cells. The control string is segmented in words which correspond to possible internal instructions of the original module. The whole control string corresponds to the overall program in our original model. Only a small portion of the control string can be read by an individual cell. *Every cell in an organism has the same control string.* Individual cells differ only in the portion of the control string which can currently be read out. *The change in activation of portions of the control string is our analogue of differentiation. The increase in the number of cells in a co-moving set is our analogue of growth.* Rather than rehearse here that discussion we shall turn to another criticism of the automaton-theoretic approach, namely that it gives us no insight at all into the reaction of embryos to damage. The program of the self-reproducing automaton is completely explicit, specifying the automaton in every detail. Any damage to the program will yield corresponding damage in the automaton. Our purpose here is to indicate a way in which such an array model of an automaton can exhibit some of the embryonic properties of resistance to damage. The problem is similar to the counting problem raised by Maynard Smith [21] and the French Flag problem raised by Wolpert [22], and we may phrase it here in the following form: 'How may we design a cell for our array so that a string of any length whatsoever of these cells, all identical, will so organize itself that the first third of these cells are in one state, we may say they form a head, the second third of the cells are in another state, we may say they form a body, and the final third are in yet another state, we may say they form a tail ?' Furthermore we demand that if this array is cut in any way, each portion of the array will rearrange itself so that the portion will after a certain period of time have a recognizable head, body, and tail. In other words,

214

how can an array count up to three irrespective of the length to which it grows
or the length to which it is reduced by damage ?

We shall indicate how arrays of one such type of cell may do the job in
Figures 6–9 (see pages 217 and 218). We shall see that they have one great
drawback from a biological modelling point of view, and then find to our pleasant
surprise that the rectification of this deficiency yields a cellular behaviour
which is more biological than that of the original.

We are to imagine an array which can conduct five types of pulse, an A pulse,
a B pulse, a head pulse, a tail pulse, and a body pulse. An A pulse will propagate
with unit velocity in a given direction until it hits the end of the string of cells,
whereupon it is reflected and moves with unit velocity in the opposite direction.
We decree that if there are ever two A pulses travelling in the array, then upon
their meeting the pulse travelling to the left is annihilated whereas the pulse
travelling to the right is propagated as if it had not encountered the other pulse.
B pulses are propagated with half the speed of A pulses and are created at a
boundary each time an A pulse is reflected there (see t_0 in Figure 6). If an A pulse
overtakes a B pulse going in the same direction, then the two pulses do not
interact but continue on their way at their specified velocities. If, however, the
A pulse and the B pulse are going in opposite directions, then a fairly complex
response is triggered.

If the A pulse is moving right and the B pulse is moving left, then the B pulse
is annihilated whereas the A pulse will continue on its way. In addition to this
effect, however, a tail pulse will be propagated with unit velocity towards the
right-hand end of the string where it will be annihilated, and as it reaches each
cell it will turn that cell into the tail state. At the same time a body pulse is
triggered moving left at a velocity of 1/3 (which leads to some problems in
discrete systems which I shall overlook, using a continuum in all my diagrams).
In general, this pulse will not propagate all the way to the left-hand end of the
string but will be annihilated on meeting an A pulse.

If an A pulse moving left meets a B pulse moving right, then the B pulse will
again be annihilated, and this time a head pulse will be propagated left at unit
velocity, turning each cell it encounters to the head state until it is annihilated
upon reaching the left-hand end of the string.

To appreciate how these rules work look at Figure 6, in which the array of cells
is graphed in space from left to right while its evolution in time is graphed with
time progressing as we move down the page. We see that at time 0 an A pulse
and a B pulse are both initiated at the left-hand end and they propagate until

215

they meet at time t_1, whereupon the B pulse is annihilated, but we see that the tail pulses and the body pulses work to cause cells to differentiate into the appropriate states. Note, however, that at time t_0 the A pulse was reflected from a boundary, at which time a new B pulse was generated, and so we see that at time t_2 the new B pulse meets our A pulse, at which stage the cells in the front third of the array are told to convert to the head state. You will notice that the body pulse was annihilated by meeting an A pulse at just the same point in space and time at which the head pulses were triggered. It was to assure this coincidence that the strange propagation speed of 1/3 was chosen for the body pulses. We now see that as time goes by the A pulses and B pulses ricochet back and forth, interacting as they do so, but the messages they send out do not change the configuration and so in fact at time t_3 the stable state with properly differentiated head, body, and tail is attained.

Let us now see what happens to our array when pieces are cut out of it. We have marked on Figure 6 three sections which are to be removed. The evolution of Section I is indicated in Figure 7, that of Section II in Figure 8, and that of Section III in Figure 9. We add one more condition. When a portion is cut from an array, an A pulse is triggered at the point or points of damage. We shall see how this works by tracing through the fate of Section I in Figure 7 (the reader is invited to carry a similar step-by-step analysis of Section II and Section III). We see that at time t_0 not only is there an A pulse and a B pulse travelling in the section but also there is a new A pulse initiated as a result of injury, but this A pulse only lasts until time t_1, when it is annihilated on encountering the first A pulse. The B pulse and the first A pulse meet at time t_2 just as they would have in the uninjured specimen, and so the head pulses propagate forward from time t_2 so that by time t_3 we do have an array in which the first portion is head, the second portion is body, and the third portion is tail. However, the proportions of the three components are not equal and so a stable state has not been attained. As we follow the evolution we see that at time t_4 the adjustments are made which ensures that exactly one-third of the array is in the tail state, but it is not until time t_6 that another encounter between an A pulse and a B pulse sets up the final equilibrium, in which head, body, and tail each occupy one-third of the array. Note that the propagation forward of a body pulse at time t_4 has no apparent effect until time t_5, when it first emerges from the front end of the string of cells that were already in the body state at t_4.

The reader will note one unpleasant feature of this evolution. Just after time t_4, a piece of body is caught between two pieces of tail. Similarly, in Figure 8, we

Michael A. Arbib

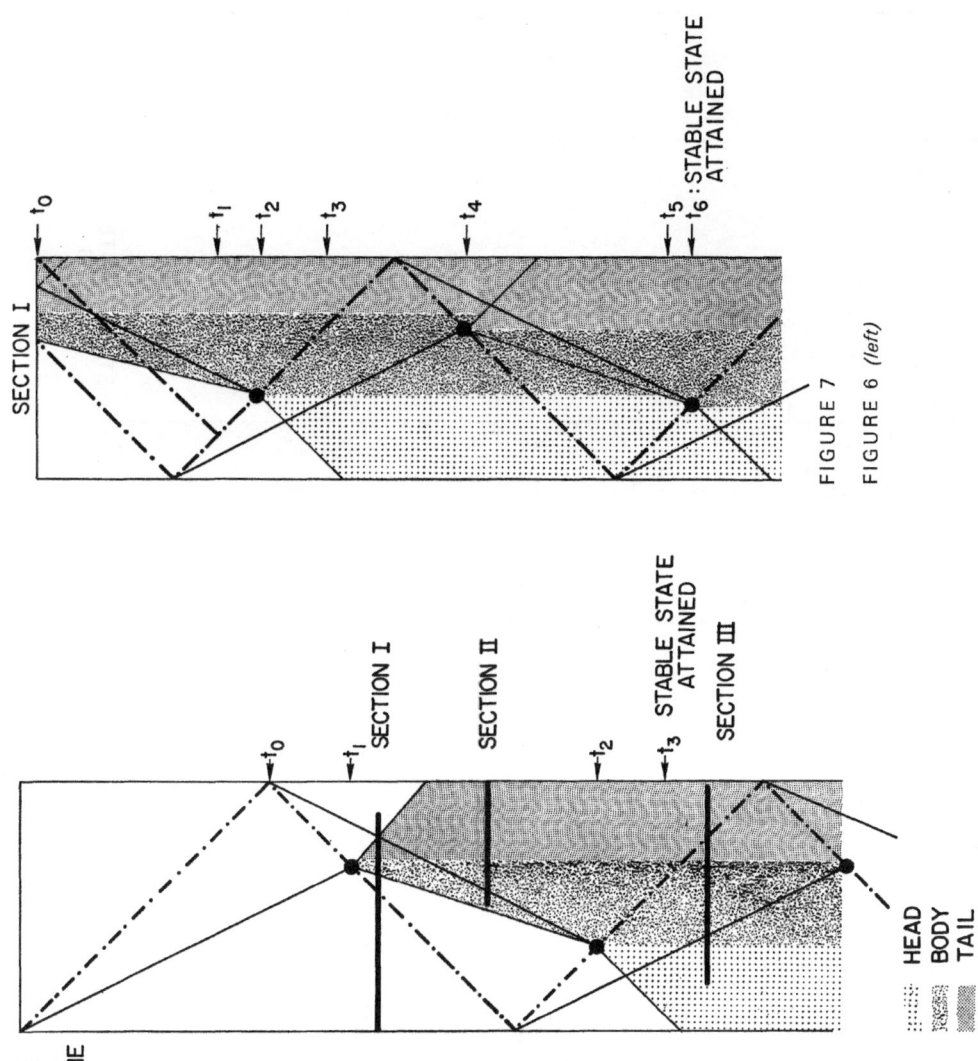

SECTION I

t_0

t_1
t_2
t_3
t_4
t_5
t_6 : STABLE STATE ATTAINED

FIGURE 7

FIGURE 6 *(left)*

t_0
t_1
SECTION I

SECTION II

t_2
t_3 STABLE STATE ATTAINED

SECTION III

HEAD
BODY
TAIL

TIME

217

Self-reproducing automata

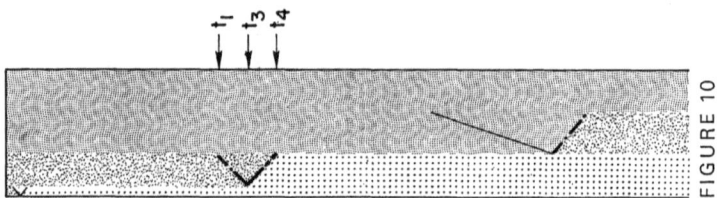

SECTION II

SECTION III

STABLE STATE
ATTAINED

FIGURE 10

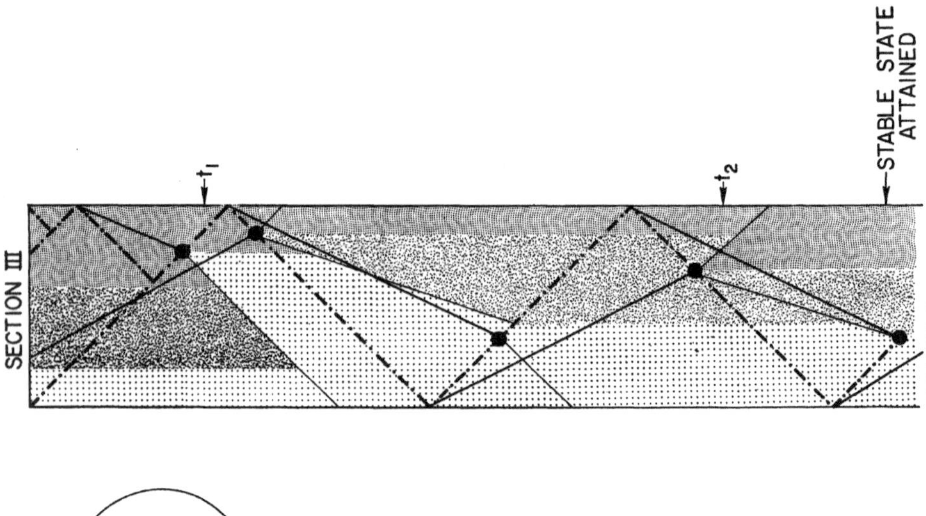

FIGURE 9

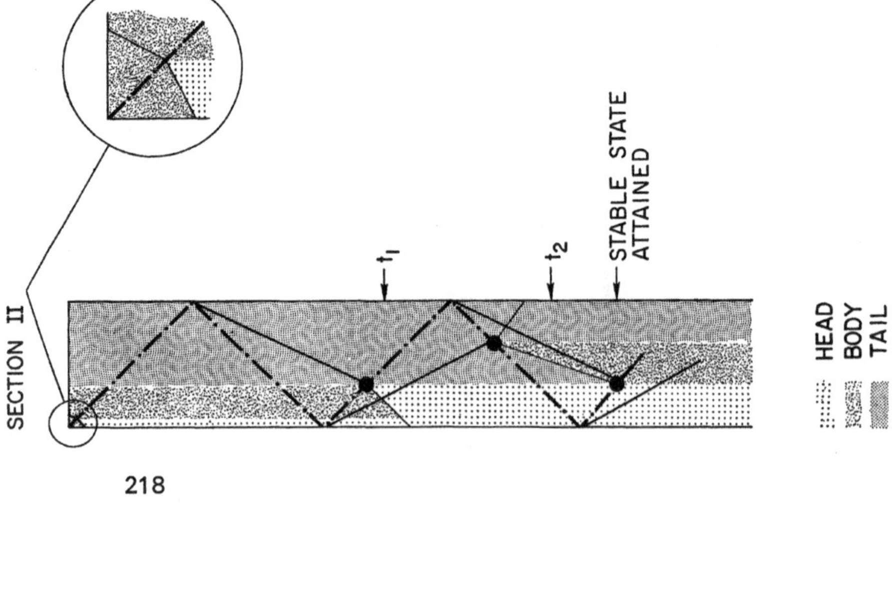

t_1

t_2

t_1

t_2

STABLE STATE
ATTAINED

t_1 t_3 t_4

HEAD
BODY
TAIL

FIGURE 8

218

Michael A. Arbib

see head-body-head-tail at t_1 and head-tail-body-tail at t_2; while in Figure 9 we see head-body-tail-head-tail at t_1!

I do not know whether embryologists have seen such effects, but it seemed to me worth while to look for a simple modification in my model which would avoid them.* We simply postulate that the states shown in my diagrams are not *actual* states, but *potential* states.

Let x be the location of a cell, $x-1$ the location of the cell to the left of it, $x+1$ the location of the cell to the right of it. Let A_{xt} be the actual state of cell x at time t, P_{xt} the potential state of cell x at time t.

We then have the following rules:

If $P_{xt}=$ head, then $A_{x,t+1}=$ head unless $A_{x-1,t}=$ body or tail.

If $P_{xt}=$ body, then $A_{x,t-1}=$ body unless $A_{x-1,t}=$ tail or $A_{x+1,t}=$ head.

If $P_{xt}=$ tail, then $A_{x,t+1}=$ tail unless $A_{x+1,t}=$ head or body.

In the other cases $A_{x,t+1}=A_{x,t}$.

Thus the cells are polarized, and a tail cell can repress the cell behind it from turning into a head cell or a body cell, etc.

In Figure 10 we have shown how the potential states of Section II, shown in Figure 8, give rise to actual states without any 'islands' when we make the repression assumption. For example, at time t_1 a pulse is sent forward changing cells to a *potential* head state, but until time t_3 each cell so changed finds a cell in front of it still in the actual body state. It is only at time t_3 that a cell can first change its actual state, and when it does so, the cell behind it can change one moment later, until at time t_4 all cells in *potential* body state have changed to *actual* body state.

I shall not explore this model further here. My purpose has simply been to show that the regenerative properties of the embryo are not incompatible with the automaton approach to development. It becomes, of course, a vital question to study how we may construct arrays which combine the properties of those in the first section with the regenerative properties we have presented here.

▶ *4. The bridge to evolution.* Among the many other questions which our discussion of self-reproducing automata raises are 'Whence come the components out of which our automata are made?' and 'Given that such automata exist, how might one imagine them to evolve?' It is not our purpose in this section to answer these questions—would that we could—but rather to suggest some interesting avenues towards their solution.

---

* Note, too, how in Section II, there is a period of time when no body cells are visible. Does such resorption occur in real preparations?

Self-reproducing automata

Pattee [8] has suggested that the crucial problem of the origin of life may be reduced to the problem of finding a molecule which is both stable enough to survive but yet not so stable as not to evolve. In our automaton-theoretic approach we tend to neglect these crucial problems of reliability. However, we may note in passing that Cowan and Winograd have erected a theory of reliability in neural nets which proceeds by the judicious application of redundancy, and it seems to me quite conceivable that a similar redundancy scheme might be used to replace the function of one cell in our iterative array by the function of a number of cells in the new array, it further being the case that each cell in the new array shares the properties of several cells of the old array. However, the question still remains of what is the simplest component with which one may hope to build up interesting self-reproducing automata. It has been pointed out that if the only system one wishes to reproduce is a falling domino, then the problem of self-reproduction becomes very trivial. If, however, one's criterion of interesting self-reproduction is that the organism involved can have computational power akin to that of a universal Turing machine, then the problem becomes non-trivial. It is known that if one embeds one's automaton in a tessellation in which the state of each cell is affected by only the states of its four immediate neighbours, then eight states will suffice to yield computation universality [14]. Codd has further shown that two states are insufficient, but no minimal number of states for such an array has yet been shown. We should note that even an eight-state cell is rather complex in that each transition, depending as it does on the state of the cell and its four neighbours, requires the specification of actions for each of 32,768 possible circumstances. Unfortunately, it is not at all clear how this sort of discussion can give us any insight into the minimal complexity required of the macromolecule which can enter into a process of hereditary evolution. This is especially so since we cannot be sure that the present DNA system occurred at any stage at all near the origin of life. Just as a virus, which appears to be intermediate in complexity between simple molecular systems and the cell, could not have evolved until after cells had evolved to provide an environment for their reproduction, so it may well be possible that DNA could only evolve after various other subsystems, such as those now available in the cytoplasm, were available to help it in its replication. At the present stage it still requires cytoplasm or Kornberg for the replication of active DNA! While on this topic of possible precursors for present replicating systems we would draw the reader's attention to the papers of Pattee [8] and Cairns-Smith [23].

Turning now to the broader questions of evolution, once we have already obtained a suitable supply of existent automata from which to evolve, we see that a radically different point of view must be imposed on that of our basic automaton-theoretic approach. We have already mentioned, in Section 3, the transition to the Mark II module model in which the whole program of the organism is contained in each cell, differentiation serving to activate different portions of the program. To study evolution, one must proceed to the Mark III model, in which these internal programs are themselves modifiable (mutable) programs. It will be a matter of some interest to see how necessary stochastic factors are in modelling this situation. We must then study not only what is the simplest program for a self-reproducing organism but also how the program may be complicated to yield 'fitter' programs. Let me just briefly sketch the framework one might employ for such a study (the only attempt in the literature of which I know at erecting such a framework is in the work of John Holland; see, for example [24]). In such a study we move to a higher level at which we neglect the details of the dynamics of cell division and work instead, perhaps with some measures of mitosis intervals, and stable life-times of cells. In fact, it is not at all clear to me whether study at this level will depend upon its automaton-theoretic base or will in fact become indistinguishable from Fisher-type statistical studies of natural selection. Having set up such basic parameters we might then study free movement in the tessellation and consider for various aggregates what is their reproduction time and stable life-time. One would hope to prove that the initial aggregates arise by a transition from symbiosis to functional reproducing organism (compare [23]). One might expect that aggregates would be better suited to utilize material. One should also be able to prove theorems that parallel operation would be favoured, as would information and material transmission channels, to increase the use of materials encountered by the cell (Crick [4] has observed that if one removed all the DNA from a single human being and stretched it end to end, it would span the whole solar system—surely a very strong argument for parallel processing in any organism!) We might then ask for theorems which contrast the short generation time of individual cells with the long generation time of aggregates. Simon [25] presented a rather relevant parable of two watchmakers, one of whom had his watches arranged in a hierarchical organization so that whenever his work was interrupted the only damage done was that his work was put back to its component subsystems, whereas another watchmaker was in the unfortunate position that whenever he put down a watch it dismantled right down to the very smallest components. Clearly, the first watchmaker could get much, much more done than the second

221

watchmaker. With such parables in mind we may hope to proceed to a really rigorous theory of evolution in which we can show the selective advantage of systems which are not only highly parallel but also hierarchical in structure.

Let us turn now to the relation between the origin and evolution of life on the one hand and the laws of thermodynamics on the other, and see wherein a statistical mechanics may prove useful.* Attempts so far to deduce thermodynamics from mechanics require an *additional hypothesis,* namely one of random phase, and I suspect that living systems are ones for which random phase hypotheses do not apply. There may be subsets of phase space, albeit of very small measure, where thermodynamics does not hold true, and such regions may contain the trajectories of populations of living systems. More strongly, we may even question whether thermodynamic parameters are well defined in such regions. Thus it is only as a sop to a limited statistical mechanics that in what follows I shall follow the standard line of reconciliation between evolution (increasing order) and the second law of thermodynamics (decreasing order) by saying that 'in the *total* system, the entropy must increase'.

I hope that developments in statistical mechanics, such as discussed by Prigogine [26], will eventually show how this simplistic statement may be refined.

Thermodynamic approaches to problems of organized complexity, integration, adaptation, or evolution fail to expose the mechanisms underlying these phenomena, and at best deal only with the direction in which a situation will develop and not with any quantitative aspects. In the same vein, we viewed evolution and progressive adaptation as directed processes. But, fine conceptually though this may be, it cannot tell us what will evolve, or what properties of the organism are going to emerge. A self-organizing system is characterized by change, whether it be growing or adapting. Thus we need to find a mechanism for the description of such a system which is essentially dynamic in nature. It must be further noted that the bulk of present thermodynamics and statistical mechanics is designed to deal with reversible systems which are essentially homogeneous. Self-organizing systems, on the contrary, are highly structured, and have a definite 'positive arrow in time', and so great care must be exercised in applying thermodynamic notions.

Be that as it may, let us now consider in more detail what we want our dynamic mechanism to be. It seems clear that whether we wish to analyse biological systems or synthesize adaptive machines, we are going to need a theoretical

* The rest of this section owes much to discussions with Warren McCulloch in May of 1962.

study of automata. We may think of automata as messages, and we may do so in two interconnected ways:

a. An automaton is a structure—as such it is essentially characterized by its form, and so contains information. In this sense of form (not of signal) we may consider it to be a message.

b. We may interpret as a message the regularity of the event the automaton realizes, i.e. its information-handling behaviour.

These messages are self-perpetuating and propagating in two senses:

i. They endure. (Endurance has three aspects: growth, learning, and homeostasis. Homeostasis must repair what would otherwise be permanent damage. The correction of purely temporary errors is a problem of information handling rather than self-perpetuation.)

ii. They multiply.

Although, in the first three sections, we gained some insight from the study of deterministic systems, it is clear that the eventual form of our theory of biological automata must be statistical, since they are continually subject to noise, both environmental and internal. The noise affects both aspects of the message:

a. It interferes with enduring and multiplying, and if uncompensated would cause the automaton to disintegrate rapidly and would prevent it from begetting its like.

b. It interferes with the information handling, and if uncompensated would cause drastic computational errors.

Clearly these aspects are not independent. We have already mentioned the work of Cowan and Winograd on combating computation errors in automata by applying information and coding theory to the problem. A major task for our theory is to learn how to apply coding theory to the problem of enduring and multiplying in the presence of noise.

Thermodynamically, the probability of making a living creature is very, very small, given a box of 'components'. But this is not the way it came about. The great trick of the living system is that it slows down the entropic degradation. The huge structural complexity of a human being is coded into the chromosomes selected for the job the organism is to do. The genetic instructions interact with the environment to produce the organism. (N. B. The environment of a cell includes adjacent cells, e.g. a zygote usually gives rise to an organism, but if the two halves are separated, two organisms will often be formed.) This organism is, as a result of the interaction, adaptive both in structure and in function. When something new hits the organism, be it a virus or an idea, it has to make new

223

protein structures fitted to the job. There is a great difference between animals that learn much and those that learn little. A human baby is relatively immature when he leaves the womb, and a lot of his formation comes from subsequent learning and adaptation. This means that he can learn more, and adapt to a wider range of environments.

We may define *adaptation* as the property of that which has been kicked once which enables it to change in such a way that when kicked again it undergoes less internal change. Nature uses this process for adjustment to the environment. Note that this adaptation can only occur if the organism has the requisite variety from which to choose—this is the important point that Ashby has made time and again. Whether we build a machine or investigate a biological system, we must have components capable of entering into sufficient relationships to each other to gain flexibility and reliability. (We note that bisexual reproduction also serves to increase the variety!) Greater diversity in function might also handle these problems. But variety is useless unless it can proceed from a sufficiently rich structure. It is this combination of natural structure and requisite variety which yields adaptability.

The necessity of the basic structure can be illustrated by an example [McCulloch, personal communication] from neurosurgery. If certain parts of the brain are removed, the patient becomes 'stimulus-bound'—he only acts in the present. He has a poverty of associational structure—his speech becomes less ordered in the large, his sentences shorter. He lacks adaptability. Now, his brain is still very complex and abounds in variety—but a vital part of its structure is missing. The warning is clear: in building a self-organizing system we must ensure that we transfer enough information in forming the machine to allow it the complexity of structure requisite for the adaptive tasks we set it.

We should also reiterate at this point the importance of hierarchical structuring. The unity of an organism overrides the unities which make up the lesser parts. In large groups of organisms, the regularities are greater than for the individual (this is reminiscent of the behaviour of fluctuations in the statistical mechanics of gases). One would like a theory that is adequate to treat, and distinguish between, on the one hand, 'genesis, growth, adaptation, and learning', all of which are in terms of a single device, and, on the other hand, effects due to the interaction of organisms, which thus develop 'language' in the sense of Occam's conventional terms and Pavlov's second signalling system. A human derives energy and information from food, perception, and verbal communication. (It is interesting to note that the method of gaining information by verbal inputs

224

Michael A. Arbib

is itself a product of learning—we have a regenerative loop of adaptation.)
The food is composed of useful proteins, sugars, etc., and so contains much
necessary energy and structure; perception and verbal inputs provide him with
information. He grows and learns. His output is informed when he communicates
and reproduces, but is otherwise degraded. Thus, his internal entropy is generally
decreased, (despite environmental and internal noise), but, again, the net entropy
of the universe may increase. In any case, we are reminded of the importance of
a richly structured environment in securing the development of an organism.

What we recognize as order is something describable in our language in a
small number of words. We strongly suspect that we have had the wrong way
of looking at living systems. We are beginning to see proper order. If we exclude
the case of a crystal, we find that, when we evolve something, there will be an
apparent disorder simply because we have not figured out how to look at it. Just
as the very concept of evolution itself brought a great deal of order to biology, so
our further studies should reveal to us even further order in the resultant systems.

When we analyse biological systems, we are trying to uncover a rich order. In
the optimum code for a noise structure each message may be encoded in many
ways. All possible words will be used as often as possible—and the result will
look like noise. Thus, *apparent* disorder may well be due to adaptation!

Our goal is to develop an automata theory which will be helpful wherever, as
in embryology, we have that integration of diverse processes which yields the
progressive emergence of increasing functional stability in complex systems.

Notes and References

This paper owes much to the fruitful discussion
at the Symposium on Theoretical Biology held
at the Villa Serbelloni at Bellagio on Lake
Como in August of 1967, and I should like to
record my gratitude to my fellow symposiasts
for the stimulation of their ideas.

Travel to the conference was made possible
by Martin Garstens under grant Nonr 225(87)
from the Office of Naval Research, while
research was supported in part by the U.S. Air
Force Office of Scientific Research, Information Sciences Directorate, under grant no.
AF-AFOSR-1198-67.

1. Hotchkiss, R. D. in (Gerard, R. W., Ed.):
Concepts of Biology, Behavioral Science 3,
No. 2 (April, 1958) 129.
2. Elsasser, W. M., *Atom and Organism: A
New Approach to Theoretical Biology* (Princeton University Press, Princeton, New Jersey
1966) 61.
3. The scheme we give for self-reproduction is
clearly realizable with electronic circuitry
which is subject to the laws of quantum
mechanics (questions of reliability will be
taken up in Section 4). It is thus disconcerting
to find Wigner [5] claiming to show that

self-reproduction is virtually impossible in a quantum-mechanical system (as part of a desire to show that the explication of much of biology, as well as of consciousness, requires 'biotonic' laws supplementary to those of physics, cf. Elsasser [6]. Section 6 of Arbib [7] shows wherein we believe Wigner's argument fails. However, Pattee [8] has pointed out that Wigner's argument is well taken in that the problems of transition from genotype to phenotype are inextricably bound up with the yet unresolved problems of the quantum theory of measurement. However, this does not counter the claims of the reductionist—it merely shows that reduction will not solve all our problems!

4. Crick, F., *Of Molecules and Men* (University of Washington Press, Seattle, 1966).

5. Wigner, E. P., in *The Logic of Personal Knowledge: Essays Presented to Michael Polanyi* (The Free Press, Glencoe, Illinois, 1961) 231–8.

6. Elsasser, W. M., *The Physical Foundation of Biology* (Pergamon Press, New York and London, 1958).

7. Arbib, M. A. in (Hart and Takasu, Eds.) *Computer and Systems Science* (Univ. of Toronto Press, 1968).

8. Pattee, H. H., in (C. H. Waddington, Ed.) *Towards a Theoretical Biology: I, Prolegomena,* (Edinburgh University Press, 1968).

9. Von Neumann, J., in (L. A. Jeffress, Ed.) Cerebral Mechanisms in Behavior, *Proc. of the Hixon Symp.* (Wiley, 1951) 1–31.

10. Turing, A. M., *Proc. London Math. Soc.,* Ser. 2, *42* (1936) 230–65; *43,* 544–6.

11. Von Neumann, J. (A. W. Burks, Ed.) *The Theory of Self-Reproducing Automata* (University of Illinois Press, Urbana and London, 1966).

12. Arbib, M. A., *J. Theoret. Biol. 14* (1967) 131–56.

13. Thatcher, J. W., *Universality in the von Neumann Cellular Model,* to appear in a book edited by A. W. Burks on Cellular Automata.

14. Codd, E. F., *Propagation, Computation and Construction in 2-dimensional Cellular Spaces,* Technical Report (Univ. of Michigan, March 1965).

15. Arbib, M. A. *Information and Control, 9* (1966) 177–89.

16. Myhill, J., in (M. D. Mesarovic, Ed.) *Views on General Systems Theory* (Wiley 1964) 106–18.

17. Goodwin, B. C., *Temporal Organisation in Cells: A Dynamic Theory of Cellular Control Processes* (Academic Press, London and New York, 1963).

18. Michie, D. and Longuet-Higgins, C., *Nature 212* (1966) 10–12. [Reprinted in (C. H. Waddington, Ed.) *Towards a Theoretical Biology: I, Prolegomena* (Edinburgh University Press, 1968), with a comment by H. H. Pattee.]

19. Thatcher, J. W., in *Proc. Symp. Math. Theory of Automata* (Polytechnic Press, Brooklyn, 1963) (Vol. XII of the Microwave Research Institute Symposia Series) 165–71.

20. Lee, C. Y., in *Proc. Symp. Math. Theory of Automata* (Polytechnic Press, Brooklyn 1963) (Vol. XII of the Microwave Research Institute Symposia Series) 155–64.

21. Maynard Smith, J., in (C. H. Waddington, Ed.) *Towards a Theoretical Biology: I, Prolegomena* (Edinburgh University Press, 1968).

22. Wolpert, L., in (C. H. Waddington, Ed.) *Towards a Theoretical Biology: I, Prolegomena,* (Edinburgh University Press, 1968).

23. Cairns-Smith, A. G. in (C. H. Waddington, Ed.) *Towards a Theoretical Biology: I, Prolegomena* (Edinburgh University Press, 1968).

24. Holland, J. H., *J. Assoc. Comp. Mach., 9* (1962) 297–314.

25. Simon, H. A., *Proc. Amer. Phil. Soc., 106,* 6 (Dec. 1962) 467–82.

26. Prigogine, I., *Mededelingen van de Klasse der Wetenschapper, Koninklijke Academie van België* (*Bulletin de Classe des Sciences, Académie royale de Belgique*), Series 5, *53,* 4 (1967) 273–87.

What Biology is About

Christopher Longuet-Higgins
University of Edinburgh

At the Villa Serbelloni we have been trying to make up our minds whether there ought to be a subject of Theoretical Biology, and if so what it should be like. Assuming that biology needs a general theory, should the theory be a mathematical one—a kind of dynamic topology in Thomist or Waddingtonian vein ? Should it, perhaps, be more like theoretical physics, or theoretical chemistry ? Or a sort of amalgam of evolutionary theory and population genetics, properly brought up to date under the title of neo-Darwinism ? Perhaps it should be an outgrowth of chemical engineering (Scriven), or of thermodynamics (Kornacker), or of metaphysics (Bohm), or of Fourier analysis (Iberall) or of classical dynamics (Lieber) ? To judge from our conversation, Theoretical Biology would seem to be all things to all men, no part of science being less relevant to the theory of life than any other. If this were so, I submit, we should have been wasting our time in this idyllic spot. But I do not believe this ; I think that we must do some ruthless pruning on the periphery if we are going to reach a clear appreciation of what lies at the centre of this new subject.

One trouble with regarding theoretical biology as a branch of theoretical physics or theoretical chemistry is that biology is at the same time a narrower and a wider subject than physics or chemistry. Narrower, in that organisms are, by common consent, physical systems and hence presumably subject to the laws of quantum mechanics and statistical mechanics. (Bohm has stressed, here and elsewhere, the metaphysical weaknesses of present-day mechanics, but I am sure he would agree that these weaknesses matter no less for electrons than for snails, which are at least fairly large systems.) Wider, because physics and chemistry lack concepts adequate for distinguishing life from non-life. As a sign of the times, the idea that organic chemistry is specially helpful to a biologist has become indefensible ; the name 'organic' applied to carbon chemistry rings very hollow nowadays, when we are beginning to realize that the interest of an organism lies, not in what it is made of, but in how it works. On a personal note, that is the reason why my consuming interest in living things has led me away from the physical sciences, to which concepts such as organization and function are entirely foreign. Kornacker might say at this point that biological organization is just a manifestation of thermodynamic coupling—and I would agree that this

227

What biology is about

idea can be most illuminating when applied to specific phenomena such as membrane excitability—but surely there is more to biology than just that. An astronomer might quite correctly assert that quasars are just a manifestation of the conversion of mass into energy; but such an assertion would do little to satisfy our curiosity about quasars. I rather incline to follow Pattee's strategy of trying to see what must be added to physics—that is, to its language—before we can describe inheritance and evolution in physical terms. But I cannot think, with all respect to Lieber, that we shall make much progress with the construction of Theoretical Biology if we feel it necessary to go right back to the universal constants of nature for our primary concepts. Variability is a much more con-spicuous feature of the living world than constancy, as Elsasser has emphasized; and anyway, Theoretical Biology ought presumably to begin at the point where living things start to differ in important respects from non-living matter. A theoretical biologist cannot escape the necessity of defining his field so as to exclude some phenomena as firmly as it includes others.

All this may seem rather negative, but one gets nowhere with difficult problems unless differences of opinion are as clearly stated as points of agreement. Con-tinuing, then, with our catalogue, let us consider Fourier analysis, topology, chemical engineering and evolutionary genetics. Which of these subjects is of most importance to Theoretical Biology? Set as a question in an intelligence test this would probably evoke the answer: evolutionary genetics. Plainly there *is* an honest discipline of evolutionary genetics (I wish Maynard Smith could have held forth at greater length about it) and it deals quietly and unpretentiously with—not metaphysical, astrophysical, or astrological problems—but genuine biological ones. All right, so evolutionary genetics must take an honoured place in the Theoretical Biology of the future. Also, as most of us seemed to agree, must chemical engineering. Not perhaps the old mix-it-and-hope variety, but the science which enables the engineer to design and operate complex processes with optimal efficiency, whether under steady-state or limit-cycle conditions. Nature is a chemical engineer of vast experience, and to understand the internal economy of the cell we may find it helpful to make models ourselves, either concrete or abstract. What about Fourier analysis? Of course we must pay due attention to the oscillatory character of some biological processes. But a good idea becomes a positive menace if it is allowed to dominate one's whole thinking, and I had a feeling that Iberall's emphasis on frequencies and relaxation times was in danger of taking this course. One swallow does not make a summer.

How about dynamic topology—if Waddington and Thom will allow this name

228

for their related ideas about the development of forms ? Here is an original and sophisticated approach to a wide class of problems, both evolutionary and morphogenetic. But is its descriptive power equalled by its explanatory power ? Most bacteria cannot live in a broth which is rich in sulphathiazole, *but some can* (and more are learning to do so, regrettably). One may say, with a wave of the hand, that the chreods of most bacteria peter out when they run into high concentrations of sulphathiazole, whereas the chreods of the sophisticated bacterium do not; but this is hardly an explanation. Actually, of course, *we know* in general terms why some bacteria are drug-resistant and others are not: it is because the resistant bacteria have special genes for coping with the drug, and the others do not or cannot use them. We must not make the mistake which might be made by a statistician studying the movement of cars through Central London who tried to account for everything in terms of a hydrodynamic sort of theory. He might do quite well until suddenly one day there was a complete absence of buses. Unless his theory of London traffic included the concept of a bus strike, he would be entirely unable to account for such an extraordinary fluctuation. Like the traffic in London, the macroscopically observable contortions of a cell are in the last analysis under the control of programs which are not directly apparent, and in a sense this is the most significant single fact about organisms as opposed to pretty, inanimate systems like Liesegang's rings or the patterns in a convecting fluid.

If they have read thus far, Arbib and Gregory will begin to understand why I have not so far referred to their recommendations about how we should think of constructing a coherent theory of biology. It is because basically I agree with them that the most fruitful way of thinking about biological problems is in terms of design, construction and function, which are the concrete problems of the engineer and the abstract logical problems of the automata theorist and the computer scientist. Indeed, it was on this basis that Gregory, Michie, and I decided to set up shop together in Edinburgh. But what justification can one offer for this view ?

To make a sweeping statement—but one which seems to be much more than half the truth—it seems to me that the problems of biology are all to do with *programs*. A program is a list of things to be done, with due regard to circum-stances. (This is not a formal definition, but it conveys the spirit of the notion.) It may be a recipe in a cookery book, an algorithm for calculating the zeros of the zeta function, a set of rules for writing a fugue on a given subject, or a set of genetic instructions for making a mouse. At the first Serbelloni Conference

229

What biology is about

some people were exercised about the question whether a man contains more *information* than the strands of DNA which his parents put together. Waddington felt that the problem was like the question whether a textbook of Euclidean geometry contains more information than the axioms of Euclid—a nice parallel—but nobody was prepared to say how the concept of information might be applied precisely to either problem, so that we got no further. In the end we felt ourselves moving towards the idea of a set of *instructions*, and Michie and I tried to demonstrate the relevance of this idea in our Party Game Model of Biological Replication, from which we ventured to draw some tentative biological conclusions. At the second Serbelloni Conference Arbib put forward some thoughts on the logic of automatic self-replication, and by that time everyone seemed to feel at home with the idea that the secret of an organism is the program which its life expresses. In the rest of this note I want to explore a few of the ramifications of this idea, expanding the remarks which I made at the conference itself.

Here, to begin with, is an irreverent little parable *a propos* of Kornberg's synthesis, *in vitro*, of an infective virus. Once upon a time there was an applicant for a post in computer programming. He was asked: 'Have you ever done any programming ?' 'Oh yes: here's one of my tapes.' 'How long did it take you to write ?' 'Only about ten minutes ; I copied it from a tape I found in a drawer.' 'You *copied* it ?' 'Yes ; I used the tape copier in the computer room.' The applicant was not appointed.

A thoroughly unfair gibe at a fine scientific achievement. But it may help to put our present biological knowledge into perspective. What we would dearly love to know is how existing biological programs originated (the origin of life, that is to say), how they have developed (the course and mechanism of evolution) and how they are implemented (the principles of morphogenesis). Considering the extreme youth of modern biology, a promising start has been made. We know the alphabet (adenine, thymine, guanine, cytosine) and the vocabulary (the set of triplet codons) and even something about the compilation (protein synthesis) of Nature's programming language. But we are a long, long way from being able to write our own DNA programs without cribbing from Nature. Not that this is particularly shameful, of course. The development of human computing would be an impossibly slow process unless computer scientists frequently and shamelessly lifted subroutines from one another's programs, and genetic evolution certainly has involved—and still does—a great deal of cribbing, particularly of new tricks such as how to survive the onslaught of man-made antibiotics. Nevertheless there is all the difference in the world between writing a program oneself, or understanding someone else's by reading it through, and discovering the meaning of a program by running it through a computer and seeing what comes out. Not until we can interpret the DNA of a new species without actually growing an individual from it will we be able to claim a full understanding of epigenesis.

Now perhaps we can see a little more clearly how the various components of theoretical biology ought to be fitted together. Take for example the problems connected with the origin of life. Pattee's work, and the abstract models being studied by Arbib, are concerned with complementary aspects of the matter: the physical limitations and the logical demands which must be satisfied by hereditary systems. To understand the origin of life it is not enough to know what chemical substances (proteins, nucleic acids, sugars) can be formed under pre-biological conditions, interesting as such information is. Indeed, Katchalsky has argued that our ability to produce such materials by sparking gas mixtures means that if we find them in objects arriving from extra-terrestrial regions we *cannot* conclude that they are of biological origin ! Not until we find clear traces of functional programs shall we be able to infer the existence of life on other planets.

Or again, take the problems of cellular metabolism. Here the chemical engineer's role is to provide a set of principles which will enable us to see how far a complicated chemical system can be made to regulate itself without continual intervention from outside, and how far, in order to achieve a specified pattern of behaviour, it must be directly controlled by an operator or a *program* of operating instructions. At certain stages in the life of a cell—probably during the process of cell division, for example—it may be that things will not go according to plan without direct reference to the DNA; whereas at other times the cell may be able to carry on quite automatically using the enzymes which it has already synthesized.

Turning to morphogenesis, we may note a striking parallel between the concept of a repressor or de-repressor in the biological case and the idea of a 'conditional jump' in computer programming. In a high-level computing language there is a facility for labelling chosen points in the program and later returning to such points—or not—according to the outcome of other operations of the program. 'If such-and-such conditions are satisfied, then return to (or advance to) L' is a typical conditional jump. 'If such-and-such an enzyme or substrate is present, then start making this other one' would be the logically equivalent instruction for a cell. The biological case is, of course, extremely complicated in that control can be exercised at various levels—by reference back to the genetic material or in the cytoplasm; but I feel that the analogy with the man-made program is bound to be illuminating if one can explore it more thoroughly.

At the evolutionary level, in the Darwinian spirit, we can discern two kinds of problem: how variation occurs, and how selection operates upon it. We now know that the point mutation is by no means the only mechanism of genetic variation; phenomena such as lysogeny—a close analogue of 'subroutine borrowing'—and gene duplication—to which there is no obvious analogue in human invention (or is there?)—are beginning to resolve some of the most formidable problems about the rate of evolutionary change. When it comes to selection we are faced with issues which are already the subject of much useful work in classical evolutionary theory, and which are beginning to be tackled by the powerful statistical methods which Kerner outlined for us.

Finally, on a rather different level, one may try to see whether the theory of programs can be fruitfully applied to one of the most lively areas of present research—the functioning of the central nervous systems of animals. (The word 'function', incidentally, has a message for us: to the mathematician a function is a rule for generating one set of entities from another, and there is a very real sense in which biological function means, basically, the same thing. To a programmer a function is nothing more or less than a subroutine in his program; to the biologist the function of an organ is nothing more or less than the operational contribution which it makes to the life of the organism. The concept of biological function need not, therefore, be a loose and undisciplined concept, if the concept is employed always *within* a wider context. It may not mean much to ask what is the function of an elephant, but it does make sense to discuss the function of its trunk.) The difference between a genetic program and the program which an individual implements by his behaviour is that the former is constant for the individual over his life span whereas the latter is being modified all the time in response to his experience. Evolution is the development of genetic programs; learning (a faculty of much less general occurrence) is the development of an individual's own behavioural program. The task of the psychologist is to define the relation between the two. In both cases the program is modified by the data on which it operates—using the word 'data' to include not only sensations but also environmental pressures. And in both cases the overall result is that the programs 'rewrite themselves', a phenomenon which it would be very nice to be able to reproduce in the world of man-made computation.

To sum up this line of thought: the organic thing about organisms is that they

232

organize themselves and their environment in relation to themselves. Organization (in this active sense) is the following out of a program. Just as a program consists in the evaluation of a series of mathematical functions, so an organism lives by the performance of its various biological functions. If—as I believe— physics and chemistry are conceptually inadequate as a theoretical framework for biology, it is because they lack the concept of function, and hence that of organization. In a sense the ideas of structure and of function are complementary in Bohr's sense of the word ; a clock is not just a rather curious distribution of matter, it is a device for telling the time. This conceptual deficiency of physics and chemistry is not, however, shared by the engineering sciences ; perhaps, therefore, we should give the engineers, and in particular the computer scientists, more of a say in the formulation of Theoretical Biology.

Comments by C.H.Waddington

I should like to make three comments—

1. *'The problems of biology are all to do with program.'* So far as I can see I agree a hundred per cent, but is there perhaps something more concealed in this statement than was included in my precirculated memorandum (Prolegomena, p. 8) 'that in the transition from the zygote to the adult the "information" is not merely being transcribed and translated, but is operating as instructions—if you want to put it in fancy jargon, as "algorithms".' The Waddington-Thomist concept of a chreod is a structure in a vector field—but a field under the control of programs.

2. *'But is the descriptive power (of dynamic topology) equalled by its explanatory powers ?'* The examples Christopher gives, to turn this question into a 'nasty' one, seem to me unfair. Of course if 'thinkers in chreods', like myself and René, come across situations in which some strains of bacteria can live in higher concentrations of sulphathiazole than others we can't at first say anything more than 'its chreod has terminated'—but nor could conventional biochemists say anything more than that 'the concentration of a deleterious drug has risen above what its resistance can deal with', which amounts to the same thing. The real point about the explanatory power of the notion of chreods is to ask 'Has this idea suggested a more plausible explanation for any category of phenomena than we had before ?' I should argue that it has done so by suggesting the (experimentally confirmed) hypothesis of genetic assimilation—namely that selection of a population for the capacity for the modification of a chreod in relation to environmental stresses will lead to the establishment of a chreod

233

which leads to the selected-for end result more or less independently of the environment. This provides a convincing explanation of all those cases of adaptations which look as though they could have been produced by physiological responses to environmental circumstances but which turn out to be 'genetically assimilated' to a degree which renders them almost independent of the impinging environment. The Mendelian-Darwinist theory of evolution can then stretch out its hand to take in the whole body of phenomena in which Lamarckism has found its main support. And the process is—apart from some pre-Mendelian and today almost uninterpretable speculations attributed nationalistically in America to Baldwin and in Britain to Lloyd Morgan—a radically new one. This is, so far as I know, the only new evolutionary process which has been discovered since the cytologists learned about allo-polyploidy and other mechanisms of chromosomal evolution in the twenties and thirties. What more can you demand of a piece of theoretical science ?

3. *'The overall result is that the programs "rewrite themselves".'* This seems to me to be the pay-off line in Christopher's contribution—and see the remarks made by Arbib in his synopsis of the second meeting (p. 327). But here Christopher throws it away as though it were a mere aside. The point is, is it a characteristic of biological systems that they rewrite their own programs ? The chreod approach emphasizes only the point that they rewrite them to the extent that they improve their error-correcting ability. Longuet-Higgins is, I think, suggesting something much more radical and far reaching. To build up an error-correcting chreod is after all no more than to preserve the *status quo.* The system performs a function akin to memory. Is not Christopher perhaps hinting at the possibility that biological systems may be able to rewrite their programs, not merely to incorporate the past, but in such a way as to be not inappropriate for a future which is speculatively, but not irrationally, forecastable ? And, if this is his meaning, would this be so far from my suggestion (*Prolegomena*, p. 21) that a long-term future in evolution goes to those biological systems which have both accumulated for themselves a good Bridge hand and developed strategies which tell them when to 'pass'—drawing back into hibernation or some other mere subsistence ecological niche—or aim to keep up their sleeves some other defensive trump card, such as an efficient antibody-production mechanism ? The point of being able to rewrite your programme is not to keep on doing it all the time in relation to every changing breeze—a Lamarckian exercise—but to do it only as you graduate from the amateur to the competition, and finally to the world champion, class of games players.

234

Reply by C. Longuet-Higgins

Reply by C. Longuet-Higgins

1. It was good to see Wad, in the *Prolegomena,* promoting the idea that the right way of thinking about morphogenesis was in terms, not so much of information as of *algorithms* (a fancy word, perhaps, but a very precise one !). But in my essay I said more : '. . . the problems of biology are *all* to do with programs.' Biogenesis, evolution, morphogenesis, cerebral function—the lot.*

2. No, I was trying hard to be fair, but at the same time to wave a red flag. Where Hinshelwood went off the rails (in the opinion of many of my biological friends) was that he thought of the bacterial cell *just* as a delicately balanced homeostatic physicochemical system, ignoring the fact that it is under the control of a mutable set of genetic instructions. The trouble with Hinshelwood's account of bacterial adaptation was—so it seems to me—that he tried to make do with chreods *alone,* without reference to the fact that what determines their observable forms in real biological systems is the controlling genotype.†

3. I'm glad Wad has put his finger on that sentence, because I certainly wouldn't like it to be regarded as an 'aside'. Plainly *some* biological programs are not only adaptive but *predictive;* certain types of learning undoubtedly involve inductive generalization, and one must not overlook the possibility that other sorts of biological development also involve a predictive element. The only problem is one of mechanism ; how could such a notion be fitted in with orthodox evolutionary theory ? Or was Darwin wrong in supposing that the variations on which natural selection operates are random—with no particular rhyme or reason ?‡

Footnotes by C.H.W.

* *Touché*—I agree.

† The concept of a chreod is capable of very wide application, as René Thom has shown. But I introduced it (*Strategy of the Genes*) in the context of the epigenetic working out of a set of genetic programs.

‡ Is it perhaps just because the raw materials offered to natural selection are produced by random mutations that it is possible to select adaptive mechanisms, which are 'predictive' because they do more than is immediately necessary ? (e.g. the vertebrate immunological system).

On how so little Information controls so much Behaviour

R. L. Gregory
University of Edinburgh

Perhaps the most fundamental question in the whole field of experimental psychology is: How far is behaviour controlled by currently available sensory information and how far by information already stored in the central nervous system ? Considering the origin of neurally stored information, we believe that this has only two origins: (1) ancestral disasters, changing neural structure according to the principles and processes of other phylogenetic changes occurring by natural selection; (2) previous sensory experience of the individual, stored as 'memory'. We may call these two ways of gaining stored information phylogenetic and ontogenetic learning respectively.

It is important to distinguish two quite different kinds of stored information. We learn *skills* and *events*. Some skills (e.g. walking, swimming, fighting) may be inherited (though often showing as behaviour only after sufficient maturation) and so are examples of gaining information phylogenetically; while learning or storing *particular* events is always ontogenetic. For examples of inherited skills, babies walk without special training at about fifteen months, and as Coghill [1] showed, salamanders kept from all movement by anaesthesia will nevertheless swim normally as soon as allowed, once the neural connections of the spinal cord (visible in the living animal) are complete. For examples of learned skills we may take games such as tennis, piano playing and chess. We may be able to recall the odd particular games or concerts, but as skills it is not individual past events which are stored, but rather appropriate behaviour and strategies which give more or less complete success in later similar situations. Evidently crucial generalized features of the original situation are stored and used when appropriate. But sometimes stored features are used when inappropriate: then we have an example of 'negative transfer of training'—for example playing table tennis with the straight arm movements appropriate to tennis. This serves as a handicap.

It is an open question just how far individual events are stored as such, and how far they have to be 'constructed' for recall (cf. [2]). What is certain is that information gained phylogenetically is always of the general 'skill' kind. We are not able to recall individual events experienced by our ancestors.

R. L. Gregory

We know quite a lot about the stages by which skills are learned by individuals. I would like to suggest that this can provide clues to the nature of how behaviour is controlled by sensory information. It suggests that control is not direct, except in the special cases of reflexes, but is via internal neural models of reality. These internal models are essential for skills—including perception of the external world.
▶ *The Learning of skills.* It has been clear ever since the experiments of Blodgett [3] that 'latent' learning occurs—that is, some information storage which does not at once show itself in behaviour nevertheless occurs during the early stages of developing a skill. We find two features of learning curves characteristic of ontogenetic skill learning : first, in learning discriminations—which seem vital to 'map the ground' in the first stages of learning—learning curves are positively accelerated ; there being at first no progress, then later progress appears at an increasing rate. Experiments have shown that the animal (generally a rat) is responding to other, and it turns out irrelevant, features of the situation. Secondly, learning curves of skill show marked 'plateaux', during which no progress is observed but each plateau is followed by a sudden jump in performance, associated with a different strategy. In learning Morse code, typing, or the piano, increase in speed of performance occurs in steps as the input is handled in larger and larger units. Thus in typing, while each letter remains a unit, speed is limited to about two letters per second ; but later, letter groups up to whole words and finally groups of words become the neural units. Speed is then far greater than is possible with the maximum decision rate of about 0·5 sec per decision possible for the human neural system. Lashley [4] has described the process for piano playing :
'The finger strokes of a musician may reach sixteen per second in passages which call for a definite and changing order of successive finger movements. The succession of movements is too quick even for visual reaction time. In rapid sight reading it is impossible to read the individual notes of an arpeggio. The notes must be seen in groups, and it is actually easier to read chords simultaneously and to translate them into temporal sequence than to read successive notes in an arpeggio as usually written.'

This grouping of what is at first discrete inputs is however done at the cost of complete flexibility. Unusual combinations of inputs may be missed, or accepted as though they were in a more usual order, with consequent errors. Random music is very difficult to play and random letters very difficult to type.

A system which makes use of the redundancy, in space and time, of the real world has the following advantages :

237

On how information controls behaviour

▶ *Advantages for a system utilizing input redundancy*
1 It can achieve high performance with limited information transmission rate.
(It is estimated that human transmission rate is only about 12 bits/second.)
The gain results because perception of objects—which are always redundant—
requires identification of only certain key features of each object. Some kind of
search strategy for these features would save a great deal of processing time
for object recognition. (This is open to experimental investigation and has
implications to pattern recognition, which is *not* the same as object recognition,
which is perhaps an artificial concept.)
2 It is essentially predictive. In suitable circumstances it can cut reaction time to
zero. (Experimental situations for demonstrating reaction time are somewhat
artificial, seldom occurring during actual skills, such as driving, typing, piano
playing, etc.)
3 It can continue to function in the temporary absence of any input, e.g. turning
the music page, blinking, or sneezing while driving.

Loss of input is very different from loss of output control (e.g. the steering
wheel coming off), and this difference seems important for investigating these
internal selected groupings, or as we call them 'models', of reality.
4 It can continue to function when the input changes in kind. Thus in maze
learning, rats can continue to run a maze once learned though each sensory
input in turn is denied it—vision, smell, kineasthetics, etc.

(There is an important implication here for interpretations of brain ablation
experiments, for so-called 'mass action' might appear though each sensory and
corresponding learning system were precisely located in the brain, for the other
specific systems might take over after destruction. The fact that rats can swim a
flooded maze after learning to run it dry is particularly striking, for evidently it is
not primarily patterns of motor movements which are learned. This is important
evidence for cognitive learning at the level of the rat, and we believe that it gets
even more important higher up the phylogenetic scale.)
5 It can extract signals from 'noise'. If the internal models are highly redundant,
they can be called up with minimal sensory information. This means that the
models can enormously improve the effective signal/noise ratio of sensory
systems.
6 Provided a particular situation is similar to the situations for which a 'model'
was developed, behaviour will generally be appropriate. This, in the language of
experimental psychology, is 'positive transfer of training'.

We come now, however, to disadvantages of conceivable systems (including

238

robots) in which behaviour is based on internal models.

▶ *Disadvantages of internal model systems*

1 When the current situation is sufficiently similar to past situations which have been selected and combined to give an internal model, but the current situation differs in crucial respects, then the system will *be systematically misled by its model.* This is 'negative transfer'.

2 Internal model systems will be essentially conservative—showing inertial drag to change—for internal models must reflect the past rather than the present. (This implies that rapid change of environment or social groups is biologically dangerous, and of course it favours young members of such groups.)

▶ *Further implications of internal models.* Since no model can be complete, and few if any are entirely accurate in what they represent, biological or computer systems employing internal models can always be fooled. They are fooled when characteristics which they accept for selecting a model occur in atypical situations. It is always possible that a wildly wrong model may be selected when this happens. It will happen most often when only a few selection characteristics are demanded or are available to the system. We know from many learning and perceptual experiments that there are great individual differences in what kinds of features are demanded. (In general 'brighter' individual animals, such as rats, demand where possible non-visual features while the dimmer brethren are largely content with visual features. This is curious in the case of the rat, which is generally regarded as rather a 'non-visual' animal.)

A model may be selected on purely visual data, but once selected it is generally used for non-visual predictions. Thus, in driving a car, the road surface is 'read off' the retinal image: what matters is whether the road is slippery. Slipperiness, though not a property of images, can be read from the retinal image.

In general, the eye's images are only biologically important in so far as non-optical features can be read from the internal models they select. Images are merely patches of light—which cannot be eaten or be dangerous—but they serve as symbols for selecting internal models which include non-visual features vital to survival. It is this reading of object characteristics from images that *is* visual perception.

Gross errors may occur when a wrong model is selected. Errors of scale can also occur; and these, I believe, are the familiar perceptual distortion illusions. These illusions are interesting because they can tell us something of how internal models are made to fit the precise state of affairs in the outside world (cf. [5]).

We cannot suppose that there are as many internal models as there are perceptible objects *of all sizes, distances, and positions in space.* But it is important for the models to represent the current sizes, distances, and positions of external objects if they are to mediate appropriate behaviour. To solve this problem we may suppose that the models are flexible. They can be adjusted to fit reality. They are adjusted by 'size scaling' visual features, such as perspective convergence of lines—though not always appropriately.

In the absence of any available scale-setting data, perception is determined by average sizes and distances. These are modified by 'scale-setting' sensory information when available. When scale-setting information is inappropriate to the prevailing reality, then perception is systematically distorted. On this view we can use distortion illusions as quite basic research tools. In the Muller-Lyer, Hering, or Orbison visual illusions, typical perspective depth features are presented on a flat plane. Features which would be distant if these figures were truly three-dimensional are expanded in the flat illusion figures. Thus expansion is normally appropriate—since it is object size and not retinal image size which is biologically important—but here the system is misled by the scaling information and systematic distortions occur. By studying these distortions we can discover experimentally just how flexible the internal models are; what sorts of information are used to give object scale, and also something of how internal models are built by perceptual learning.

Biologically important features of the world must be read from available sensory information. To be useful, visual features must be related to the weight, hardness, and chemical properties of objects which have to be handled or eaten. Now it is well known that a small object of the same weight as a larger object feels up to fifty per cent heavier. This is the 'size-weight' illusion. Vision selects a model calling up appropriate muscle power for lifting the weight, but when the internal model is inappropriate the power called up is inappropriate—and we suffer an illusion corresponding to the error.

The weight setting adopted by the nervous system in the absence of information of the size of the weight corresponds to a density of one—about the average density of common objects.

It is interesting that scale distortion illusions are (a) similar in different individuals from the same culture, but differ somewhat in different cultures when the available characteristic features are different, and (b) are very slow to change in adults. (On the other hand, systematic changing of *all* inputs, with e.g. distortion glasses, does produce rapid appropriate adaptation in adult humans.)

In a case of adult recovery from infant blindness we found [6] that the newly available inputs were only accepted when they could be directly related to previous touch experience. In our present terms, vision was only possible after the corneal grafts when visual data could select *already available* internal models, based on earlier touch experience. Building new models was very slow, taking a year or more. The use of vision for size-scaling occurred within a few months, the initial distortions being very great in situations where touch or other information had not previously been brought to bear—as when looking at the ground from a high window, when the ground appeared almost within touch range though actually forty feet below. The normal systematic distortion illusions did not occur: I suppose that there was no 'negative transfer' of perceptual learning where there had been no opportunity for learning of the normal size-scaling features, such as perspective.

▶ *Sensory discrimination and the appropriateness of models.* I have distinguished between (a) selecting models according to sensory information, and (b) size-scaling models to fit the orientation, size, and distance of external objects.

Now let us consider an experimental situation which may tell us something about the 'engineering' nature of the models in the brain. The experimental question is: What happens to sensory *discrimination* when there is a scale distortion?

Consider the following paradigm experiment. We have two sets of weights, such as tins filled with lead shot. Each set consists of say seven tins all of a certain size, while the other set has seven tins each of which is, say, twice the volume of the first set. Each set has a tin of weight, in grams, 85, 90, 95, 100, 105, 110, 115. The 100 gram weight in each set is the standard, and the task is to compare the other weights in the same set with this standard and try to distinguish them as heavier or lighter. The tins are fitted with the same size handles for lifting to keep the touch inputs constant except for weight. Is the discrimination the same for the set of *apparently* heavier weights but which are in fact the same weights? The answer is that discrimination is *worse* for weights either apparently *heavier* or *lighter* than weights having a specific gravity of about one [7]. Why should this be so?

Suppose that sensory data are compared with the current internal model—as they must be to be useful. Now if it is not only *compared* with it, but *balanced against it,* then we derive further advantages of employing internal models. We then have systems like Wheatstone bridges, and these have useful properties. Bridge circuits are especially good (a) over a very large input intensity range and (b) with components subject to drift. Now it is striking how large an intensity range sensory systems cover ($1:10^5$ or even $1:10^6$), and the biological components are subject to far more drift than would be tolerated by engineers in our technology confronted with similar problems. So balanced bridge circuits seem a good engineering choice in the biological situation.

Consider a Wheatstone bridge in which the input signals provide one arm, and the prevailing internal model the opposed arm against which the input is balanced. Now the internal arm is part of the model—and will be set wrongly in a scale distortion illusion. In the size/weight illusion, visual information has set the weight arm wrongly. This means that the bridge will not balance. The illusion is the misbalance of the bridge. Now an engineer's bridge which is not balanced suffers in its ability to discriminate changes in its input, for it is no longer a null system but relies on scale readings of the galvanometer or other misbalance detector. Thus the supposed biological system gives just what a practical engineer's bridge would give—loss of intensity discrimination associated with an error in balancing the bridge. This is some evidence that internal models form arms of bridge circuits in the brain.

▶ *Speculations on mental events—normal and abnormal.* On this general view, perception is not directly of sensory information but rather of the internal models selected by sensory information. Indeed, the current perception *is* the prevailing set of models.

There are well-known situations in which the sensory information calls up two or more incompatible internal models with equal probability. The best-known example is the spontaneously reversing Necker cube. The available information is insufficient to decide between rival internal models, and each comes to the fore in turn. It is interesting that in this case the addition of tactile information— provided by holding in the hand a luminous cube viewed in darkness—does not serve to abolish visual reversals, though it does reduce their rate of occurrence [8]. Evidently the visual internal model system is largely autonomous, though it is partly under the control of other senses. Visual size and distance can be set by other senses, especially touch. It is also worth noting that size scaling follows not only currently available sensory information, but also changes in the internal model. Thus, a luminous cube appears as a cube when seen correctly—though the further face is smaller at the retina—but as a truncated pyramid when depth- reversed. Here there is no change at all in the sensory input, only in the internal model, so the scale changes *with the model*, though the sensory information remains constant.

Generally, the internal model is reasonably complete and appropriate, but a wrong model may always be selected, and even if appropriate it may be wrongly scaled. We know from perceptual experiments in situations where only minimal information is available that both selection and scaling can be quite wrong. So it is a small step to say that in the absence of any sensory information entirely wild models might be called up. This could be the case in dreaming, and in drug or fatigue-induced hallucinations. Hallucinogenic drugs might call up internal models either by increasing cortical noise or by reducing the threshold criteria for acceptance of the stored models.

Abnormal conditions such as schizophrenia might be caused by inappropriate models being built in the first place, or by wrong selection criteria being employed. Greater knowledge of the processes and conditions for perceptual learning might have implications for psychiatry. If the models *are* our internal world, we should find out more about them.

▶ *Implications for the design of robots*. Devices which respond to sources of information are commonplace. There is no difficulty in arranging for a door to open itself when someone breaks a beam of light to a photocell. But such devices do not 'see' or 'perceive' in the sense that we do. Similarly, our reflex blink to a sudden bright light is not 'seeing', 'perceiving', or 'observing'.

Theories of perception (especially the Gestalt theory) lay far too much stress on sensory characteristics, giving insufficient weight to the vital point about perception: perception is geared to *objects*, for it is objects which are biologically important. Objects are dangerous or useful, food or disaster; but retinal images, and vibrations of the tympanum, are of no importance except to indicate the identity of external objects. The patterns of sensory activity are but symbols from which reality may be read. This involves far more than the recognition of patterns. Pattern recognition is only an early stage of perception, for objects are more than patterns, and it is objects that matter. Objects have all manner of vitally important properties which are seldom sensed, so current sensory information cannot be adequate for dealing with objects.

On this theory, perception allows behaviour to be appropriate to the hidden properties of objects, when the internal models sufficiently reflect their properties. This is very like the notion of a medical syndrome—a few spots may indicate the past, present, and future course of a disease, together with an appropriate strategy for dealing with it. Once recognized, the syndrome—or perceived objects— may be accepted for guiding the most complex behaviour with but little current information.

The special feature of perception is that it does not mediate behaviour directly

On how information controls behaviour

from current sensory information, but always via internal models of reality—which themselves reflect the redundancy in space and in time of the external world. This is where perception differs from devices such as photocells actuating doors, or biological reflexes, for these give control directly from the inputs. They do not use the current information to call up appropriate models, giving information drawn from the past of the hidden features of the present situation. The past is usually a reliable guide, and our memory contains vastly more information than can be transmitted in reasonable time by the sensory channels even when the relevant information is available—which is rarely the case.

One might be tempted to think that objects, as perceived, are no more than statistical groupings of sensed events—syndromes of sensation. But to say this is to miss a vital point. Sensed events are categorized also in terms of the use made of them. A book, for example, is seen as a single object. This is because we handle the collection of pages as one object. Sensory inputs are grouped according to the repertoire of behavioural skills of the owner of the perceptual system.

One man's object may be another's pattern—or be nothing but randomness.

This brings out the kind of difficulty we have in imagining the perceptual world of animals, or even people whose interests are very different from our own. It also has implications for designers of robots—machines to see and act on what they see. If they are to respond to objects via internal models—and all the biological advantages will apply to the machine—then its models must be appropriate to *its* sensory inputs and to *its* repertoire of actions. These will differ greatly from ours. But could we communicate with a robot having internal models very different from our own ? We should expect the same extreme difficulty that we have in trying to communicate with other animals or with schizophrenics. Even though we design and build our own robot, and know exactly how its circuits function communication could be impossible when its internal models are not ours.

▶ *The status of perceptual brain models.* We suppose that perceptual models are aggregates of data about objects, and about how objects behave and interact in various circumstances. Perceptual models bear a resemblance to hypotheses in science. We may think of sensory data suggesting, testing, and sometimes modifying perceptual models in much the same way that scientific data suggest, test, and modify theory and hypothesis in science. A precise comparison of perceptual processes with the logic and method of scientific inquiry could be highly rewarding.*

* This project was planned by Norwood Russell Hanson with the present writer, but tragically Russ Hanson was killed in his private plane.

R. L. Gregory

We are concerned here with not only the logical but also the biological and the engineering status of brain models. Whatever they are, one thing is quite clear—they are not isomorphic pictures of external shapes. The Gestalt theory misses the point here, for all sorts of information about objects must be stored but pictures can only represent specific shapes and colours. Shape and colour have only indirect significance: what matters is whether the object is useful, a threat, or food. It is non-optical properties that are important. When we look at a picture, we can read all kinds of significance beyond mere shape and colour. The picture serves to evoke our internal models, which have been developed by handling objects, so that non-optical features have become associated. Similarly the pictures in the eye, the retinal images, only have significance when related to non-optical properties of objects. Without such correlations all pictures, including retinal images, would be meaningless—mere patterns. The artist by presenting selected visual features plays games with our internal brain models, and may quite drastically change them by evoking new associations. It is clear that the brain models cannot be logically at all like pictures, for though pictures can evoke models, their appropriateness is in terms of objects, not pictures, which in themselves are utterly trivial.

The computer engineer will ask: are these supposed brain models digital or analogue? This distinction is important to the engineer, because analogue and digital systems have very different design features and advantages and disadvantages for various purposes. Indeed, it is possible to make an informed guess as to which system is adopted by the brain in terms of speed of operation, types of errors, and other characteristics typical of analogue or digital engineering systems cf., [9]. The engineering distinction arises from the fact that in practice analogue systems work continuously but digital systems work in precisely defined discrete steps. This difference is immensely important to the kinds of circuits or mechanical systems used, and vital practical implications follow. Discontinuous systems can have much higher reliability in the presence of 'noise' disturbance. Analogue devices can have much faster data transmission rates, but their precision is limited to around 0·1–1·0%. There is no limit in principle to the number of significant figures obtainable from a digital computer if it has space enough and time.

Because of the clear engineering distinction between continuous and discontinuous systems, there is a temptation to define analogue in terms of continuous, and digital in terms of discontinuous. But this will not do. We can imagine click stops fitted to a slide rule: this would make it discontinuous, but it would still be an analogue device. We must seek some deeper distinction.

245

On how information controls behaviour

The point, surely, is that analogue and digital systems both represent things, and so in both cases their internal states represent something else. The essential difference between them is not in their engineering, but rather that they represent logically different kinds of things. The distinction is between *actual events in the world,* which occur continuously, and *symbolic representations of events,* which are always discontinuous. (Even the continuous functions of differential calculus have to be handled as though they were discrete steps.)

A continuous computing device can work without going through the steps of an analytical or mathematical procedure. A digital device, on the other hand, has to work through the steps of an appropriate mathematical or logical system. This means that continuous computers functioning directly from input variables necessarily lack power of analysis, but they can work as fast as the changes in their inputs—and so are ideal for real-time computing systems provided high accuracy is not required. The perceptual brain must work in real time, and it does not need the accuracy or the analytical power of a digital system following the symbolic steps of a mathematical treatment of the situation. Perceptual motor performance only has an accuracy of around one per cent. It seems that a continuous analogue system is appropriate for perceptual data processing. This holds both for actual brains and future robots.

Perceptual learning involves not the development of software programmes for programming a digital system according to mathematical analyses of the behaviour of objects, but rather by developing quite crude continuous analogues of the organism's input-output functions, in the presence of recognized objects. From the point of view of the perceptual computer, objects represent transfer functions between the organism's input and output in various situations. Behaviour is given by selecting the appropriate transfer functions, which are stored in the perceptual model elicited by the recognized object.

To build a seeing machine, we must provide more than an 'eye' and a computer. It must have limbs, or the equivalent, to discover non-optical properties of objects for its eyes' images to take on significance in terms of objects and not merely patterns. The computer must work in real time. It need not work according to analytical symbolic descriptions of the physical world—all it requires are quite crude analogues of input-output functions selected by distinguishing features of objects. These collections of transfer functions give appropriate behaviour through predictions, made possible by the redundancy of the world of objects. Ultimately the perceptual brain models reflect the redundancy of the external world—when they do so correctly we see aspects of reality without illusion.

Comments by C. H. Waddington

References

1. Coghill, G. E. *Anatomy and the Problem of Behaviour* (C.U.P., 1929).

2. Bartlett, F. C. *Remembering* (C.U.P., 1932).

3. Blodgett, H. C. *Univ. Calif. Publ. Psychol. 4* (1929), 113–34.

4. Lashley, K. S., in (Jeffress, L. A. ed.), *Cerebral Mechanisms in Behaviour* (Wiley, N.Y., 1951).

5. Gregory, R. L. *Nature, 199* (1963), 678–80.

6. Gregory, R. L. and Wallace, Jean G. *Recovery from Early Blindness: A Case Study* (Heffers, Cambridge, 1963).

7. Gregory, R. L. and Ross, Helen E. *Percept. and Motor Skills, 24* (1967) 1127–30.

8. Shopland, C. and Gregory, R. L. *Quart. J. Exp. Psychol., 26* (1964), 66–70.

9. Gregory, R. L. *Brit. J. Phil. Science,4* (1953), 15, 192–7.

Comments by C. H. Waddington

Richard Gregory asks the question—a very good one—how so little information controls so much behaviour, and answers it by arguing that the information triggers off pre-existing models of objects in the perceivable world which have already been formed within the brain. This is an attractive thesis, but I find the word 'model', with its suggestion of miniature clockwork railway locomotives and other object-like hardware, rather inhibiting to an imaginative grasp of what is involved. I should prefer to think of these 'models' as chreods. If X denotes a set of (informationally inadequate) external stimuli, these are mapped into the brain-states Y by some function $f : X \to Y$ dependent on the neural connections between brain and sense organ. Now we have to suppose that the space of brain-states is divided into a number of domains, each characterized by a vector field which, at any given time, is dominated by a particular attractor (cf. René Thom's discussion of words as chreods in his forthcoming *Stabilité Structurelle et Biologie*). The information contained in the brain-state y which corresponds to the external stimulus x then has added to it the information embodied in the chreod (vector field plus attractor) into which it falls. In such a picture it is easy to realize the provisional character of the 'internal information', i.e. the boundaries of the various domains, the characteristics of the vector fields and attractors, and to appreciate that this must be subject to continual change and adaptation through processes of learning.

247

Cognitive Processes in Physics and Physiology

Karl Kornacker

Massachusetts Institute of Technology

It at first seems trivial to note that cognitive processes are involved whenever we perform experiments. Perhaps the clearest non-trivial example from physics is the measurement of heat, where the observed process stands in direct contradiction to the universal work-energy theorem of mechanics, thus forcing the formulation of statistical mechanical theories of macroscopic observation. The recording of a single quantum mechanical event is another example from physics where the experimenter's cognitive process seems to be explicitly involved. By mentioning these two examples I do not mean to imply that either is understood.

It appears to me that a striking aspect of physiological systems is that cognitive processes occur inside the system. The recognition of patterns by neurons in the central nervous system is an obvious case in point. Less obvious cases are enzyme-substrate specificity and the active transport of ions across membranes. In this paper I will outline a general theory of the cognitive process and show how it applies to heat, active transport, and neuronal pattern recognition.

In brief I argue that cognition, or the recognition of form, always comes down to the calculation of correlations. The usual form of a correlation calculation for the variables A and B is $(\langle AB \rangle - \langle A \rangle \langle B \rangle)$, where $\langle \, \rangle$ denotes an averaging operation

$$\langle A \rangle = \int A(x)p(x)dx \qquad \qquad \dots (1)$$

$$\langle B \rangle = \int B(y)p(y)dy \qquad \qquad \dots (2)$$

$$\langle AB \rangle = \int\int A(x)B(y)p(x,y)dxdy \qquad \dots (3)$$

and p denotes a density function. Randomness is not required. Note that a non-zero value of $(\langle AB \rangle - \langle A \rangle \langle B \rangle)$ implies that $(p(x,y) - p(x)p(y))$ is non-zero for some x and y, but that the converse may fail, allowing some correlations to pass undetected.

The quantity $(\langle AB \rangle - \langle A \rangle \langle B \rangle)$ is inherently macroscopic, since it vanishes if A, B, and AB are measured directly. Furthermore the calculation of this quantity requires non-linear interactions (multiplication) within the cognitive device. Let us now consider the relevance of these considerations to heat; active transport, and neuronal pattern recognition.

Despite the tradition in statistical mechanics which follows the ensemble theory of Maxwell and Gibbs, a tradition which emphasizes the non-reproducibility of molecular states in the repetition of thermodynamic experiments, the fact remains that spatio-temporal averaging completely defines the thermodynamic (macroscopic) point of view. For example, pressure is the finite spatio-temporal average of mechanical forces exerted by molecules on the transducer surface of a pressure gauge. The need for spatio-temporal averaging is basically the need to increase the signal-to-noise ratio of the measurement. Spatial and temporal averaging are complementary in the sense that, for a given acceptable noise level, the smaller the transducer area the longer must be its response time. If for example one wishes to use local equilibrium assumptions in a general theory of irreversible processes, then clearly the implied local measuring devices must have a long response time.

Now suppose that the time-averaging operation performed by a local thermodynamic measuring device is invariant under translations in time, that is

$$\langle A \rangle = \frac{\int_{-\infty}^{t} f(t'-t)A(t')dt'}{\int_{-\infty}^{t} f(t'-t)dt'} \qquad \dots (4)$$

For example, in viscous mechanical damping, or capacitative electrical smoothing,

$$f = e^{(t'-t)/\tau} \qquad \dots (5)$$

where τ is the time constant (response time). For any such invariant averaging operator which in addition has 'finite memory', meaning that $f(-\infty)A(-\infty)$ is zero, it is easy to show by direct calculation that

$$\frac{d\langle A \rangle}{dt} = \left\langle \frac{dA}{dt'} \right\rangle \qquad \dots (6)$$

Now consider the effects of time-averaging on the work-energy relation. The relation itself is

$$\frac{dE}{dt} = F \cdot v \qquad \dots (7)$$

where the force vector F and the velocity vector v are considered in the system phase space, so that the product F·v includes summation over all coordinates of all particles. For invariant finite memory time averaging we have, combining equations 6 and 7:

$$\frac{d\langle E \rangle}{dt} = \langle F \cdot v \rangle \qquad \dots (8)$$

On the other hand, we calculate the macroscopic rate of work done on the system as

$$\frac{dW}{dt} = \langle F \rangle \cdot \langle v \rangle \qquad \dots (9)$$

249

so that we recognize the dynamic cross correlation $(\langle F \cdot v \rangle - \langle F \rangle \cdot \langle v \rangle)$ as the rate of heat flow dQ/dt into the system, based on the first law definition

$$\frac{dQ}{dt} = \frac{d\langle E \rangle}{dt} - \frac{dW}{dt} \qquad \dots (10)$$

In this case the time-averaging is performed by the measuring devices and the multiplication is performed by the observer.

Active transport, like heat, has no microscopic counterpart. By definition active transport is a net transmembrane flow which is not caused by any macroscopic energy gradient. The energy for active transport comes from chemical reactions which have no macroscopic spatial direction, and the coupling of this energy to the macroscopically directed flow takes place in an anisotropic membrane structure. I will now show that such coupling requires a dynamic cross correlation calculation in the membrane.

The electrical conductance of a passive membrane, with respect to the γ charged species, is defined as

$$g_\gamma = \frac{I_\gamma}{\mu_\gamma} \qquad \dots (11)$$

where I_γ is the transmembrane current carried by the species and μ_γ is the electrochemical potential difference for the species. If there is active transport of the γ species, then, treating active transport as if it were a current generator, the modified conductance equation is

$$g_\gamma = \frac{I_\gamma - J_\gamma}{\mu_\gamma} \qquad \dots (12)$$

where J_γ is the active transport current.

Let us now consider explicitly the electrical time-averaging performed by biological membranes. The averaging is due to the membrane capacitance (about one microfarad/cm^2) which lies across the passive membrane resistance (about one thousand ohms/cm^2), giving a time constant of about one millisecond. Introducing the fluctuation potential

$$e_\gamma = \mu_\gamma - \langle \mu_\gamma \rangle \qquad \dots (13)$$

Equation 11 may be rewritten as

$$I_\gamma = g_\gamma e_\gamma + g_\gamma \langle \mu_\gamma \rangle \qquad \dots (14)$$

Averaging equation 14 gives

$$\langle I_\gamma \rangle = \langle g_\gamma e_\gamma \rangle + \langle g_\gamma \rangle \langle \mu_\gamma \rangle \qquad \dots (15)$$

which, rearranged, becomes

$$\langle g_\gamma \rangle = \frac{\langle I_\gamma \rangle - \langle g_\gamma e_\gamma \rangle}{\langle \mu_\gamma \rangle} \qquad \dots (16)$$

Comparing equation 16 with equation 12 allows us to identify

$$J_\gamma = \langle e_\gamma g_\gamma \rangle = \langle \mu_\gamma g_\gamma \rangle - \langle \mu_\gamma \rangle \langle g_\gamma \rangle \qquad \dots (17)$$

so that the active transport current is a calculated cross correlation between the fluctuations of conductance and electrochemical potential. In this case the calculation is done by the membrane. The molecular mechanisms which could generate the required correlated fluctuations are considered in detail elsewhere [1].

Let us finally consider how it would be possible for neurons to calculate cross correlations between various synaptic inputs. From the physiology of synapses we suspect that excitatory actions generally add with each other, inhibitory actions generally add with each other, and inhibition generally divides excitation. The nervous system must therefore find a special way to obtain subtraction and multiplication if it is to calculate any correlations. The following method is not unique, but its physiological implication is so striking that further close examination seems warranted.

We note first that an exponentially weighted averager, as in equation 5, has the property that

$$\left\langle \frac{\partial A}{\partial s} \right\rangle \quad \frac{A - \langle A \rangle}{\sigma} \qquad \qquad \dots (18)$$

If s is time, then σ is the temporal averaging time constant γ; if s is space, then σ is the spatial averaging space constant λ. Equation 18 shows that the required subtraction can be obtained if the gradients or velocities of certain physiologically important quantities, rather than the quantities themselves, are fed into the neuronal correlation calculator. A correlation calculation of the form $\langle (A - \langle A \rangle) B \rangle$ could then be performed without explicit subtraction.

As regards the required neuronal generation of multiplication, two successive divisive interactions would accomplish this. In other words, disinhibition is a multiplicative form of excitation. I propose therefore that the central nervous system could be organized as an analogue computer for calculating spatio-temporal correlations between various neuronal activities. The above arguments might then point to the fundamental functional significance of disinhibition and mathematical differentiation in the nervous system. The cerebellum, for example, seems to function as an organ of disinhibition and might therefore be expected to connect to major correlation calculators (cognitive centres).

References

1. Kornacker, K. in (Douben, R. M., ed.)
Biological Membranes (Little,
Brown Co, in press.)

A General Property of Hierarchies

Ted Bastin

Cambridge Language Research Unit

At the 1967 meeting at Bellagio the idea of a hierarchy frequently entered the discussions. This was to be expected. If one thinks at all about biological systems, it is natural to simplify the problem of getting a comprehensible picture of what is going on in them, by drawing boundaries around blocks of activities which have a fairly obvious coherence and autonomy; and having one such block exercise control over others, while being controlled by yet another or others. When we have done this, we have already imposed the concept of a hierarchy. Sometimes it is clear that the imposition of hierarchical structure is natural, because the 'blocks' are very apparent in their action. Other times it is extremely hard to find them at all, yet the hierarchical idea has such a grip on us that we still find biological systems of this second kind being thought of in terms of these blocks. This latter situation is particularly common in theories of brain physiology.

My task in this note is to think of hierarchical structure as abstractly as possible, in order to get a simple picture of the logical and mathematical relations that must exist between the 'blocks' if they are to constitute a hierarchy. The picture that I shall put forward will be remote from detailed biological experimentation, but this requires no apology, for, if it is really the case that biologists are in the habit of using hierarchical ideas as a conceptual framework for biological systems, then it is important to be as clear as we can what consequences this habit really imposes on us.

To pursue this task I shall use mathematical ideas from a hierarchy theory that was developed by Amson, Bastin, Parker-Rhodes, and Kilmister [1] for application to foundational questions in quantum theory, and I go back to the root from which the hierarchy idea sprang—namely the priestly and then political hierarchy— to provide a simple picture to which these mathematical ideas apply. I do not of course claim that this simple picture of a hierarchy formalized in this way is the only one that can be formalized, or even that the formalization in question is the only possible formalization of that picture. It is, however [2], usual in science, if one has a model or a theory—particularly one with some degree of mathematical articulation—which has been applied to the explanation of phenomena in some field, to try to apply it in other fields. Partly, doubtless, this is because it is natural to expect to be able to generalize; partly, again, the discovery of a piece of mathematics which fits the world in a new way is a rare event (a fact that is emphasized in the paradigm philosophy of Kuhn and that was disastrously overlooked in the old hypothetico-deductive view of the use of mathematics in science).

I shall call my picture the 'similar units model' (the 'units' being the discernibly separate blocks to which I have already referred). In operating with these units there is one question which hits us at the outset. We may be able to discern coherent entities, or 'blocks', or units, which control each other in a hierarchy, when we look at some biological—or other—system; but what is it, experimentally speaking, that we discern? Bohm, at the Bellagio conference, answered that when we strip our observations to bare essentials what we discern is 'order', and he has written at length on the importance of this concept. He also maintains that 'order' is the colligating concept to which we should first look when we wish to assimilate ideas from biology to those from the foundations of physics. I think his position would be that order (defined within finite sets of recognizably similar entities) is a concept that really comes into its own as soon as we think in terms of hierarchical control, so that 'order' and 'hierarchy' are mutually supporting concepts. I am sure that this association is a correct one, and that it would prove a rewarding study to justify it from the general standpoint of the logical foundations of mathematics, but I cannot attempt this investigation here, and for the present I shall assume the validity of the association.*

A brief account of the stages which led to formalization of the similar unit model, though not of the mathematics, is given in appendix I. For my present purpose I need only the central idea which underlay the work and which became more definite as the formalization process progressed, which is this: If units in a hierarchy of control are similar in their intrinsic properties and differ only in their function, then the only possible distinction that can serve to classify them into levels (and hence delineate the hierarchy) arise through their behaving with different characteristic reaction times. Hence *levels of control must be distinguished by different time constants.*

There were indications of a similar idea in the remarks made by several speakers at Bellagio (notably Goodwin and Iberall), which fact possibly supports my belief that this characteristic may be a general property of control hierarchies. In

---

*At the Bellagio Conference I illustrated the concept of order in the context of Bohm's remarks by describing one particular *'ordering relation'*—namely that which we employ when we ascribe 3-dimensionality to the space continuum. This illustration is given as appendix II.

A general property of hierarchies

order to illustrate it I shall go back to the primitive case of a hierarchy from which the idea comes and to which we refer when we use the concept—namely to a political (originally more specifically a priestly) hierarchy of control or authority. ▶ The feudal system of medieval England provides an extremely simple idea of a hierarchy of command with its different levels of authority each consisting of a set of similarly privileged and similarly powered individuals (king, barons, knights, squires, yeomen, villeins) in descending order of authority. It is unlikely that this simple picture would ever have been accurate: some barons would matter more than other barons, and some would have more rapid or easy access to the ear of the king, and so on, so that the simple type of order specified by the hierarchy with the set of individuals at a given level having equivalent places in the structure (and hence unordered in respect of command) would be true, at best, only as a first approximation. Nevertheless the mere fact that ranks were devised and allotted to individuals demonstrates that the ideally simple pattern of control presented by the simple hierarchy was necessary, at any rate to enable those individuals who themselves formed the hierarchy to have a basic picture of its function, even if that picture had then to be modified in detailed ways.

This then is the picture that is to be generalized into a hierarchy of orders, with higher levels consisting of abstractions from the structure at lower levels, in order that the picture may serve as a useful theoretical model of aspects of nature. For such a model to give an essentially new way of thinking, moreover, which is not reducible again to a simple set of mechanical relations or orders at one level* the hierarchical structure must be a dynamical one, with changes at each level being dictated from other levels, and themselves dictating changes at other levels. This dynamical character is quite evident of course in the case of the feudal structure where there is constant conflict, or at any rate interaction, between the ways the individuals at a given level would act if left to themselves and the dictates of higher levels of command. (This conflict of interaction is not

---

*Bohm emphasized the 'non-mechanical' character of mechanically interdependent ordering relations, which follows indeed from the fact that we certainly have no mathematics at present that is capable of specifying such a scheme in its general form. (My similar units hierarchy model is a special case which will—I hope—give clues to treatments of greater generality.) Bohm argues that in so far as a mathematical specification can be found, explanation is to that extent reduced to one level. This view would mean that the phenomena of life always elude exact specification. This I find too extreme. More likely there is a class of mathematical constructions, a case of which is familiar to us as the mathematics of classical dynamics, which could not afford explanation at more than one level. However, I do not think that all mathematics has to be of this type, and in particular I think recursion theory is not.

necessarily one way, but I do not want at present to discuss the need—or lack of need—for a one-way restriction.) What is not evident—is indeed strange—is the idea that an individual at a given level—say a baron—could be an abstraction from the structure existing at lower levels. However, when we look into it this idea, too, is reasonable. The baron, as an individual, is no different essentially from individuals at all other levels. As a baron—a member of the level of barons; on the other hand—he has his particular status in recognition of the fact that he acts on behalf of, and with authority over, the whole segment of the total hierarchy that is under his command. Thus the concept 'baron' itself is an abstraction, it refers to the power of acting on behalf of a sub-hierarchy.

The situation can be clarified further if we imagine a being from another planet—a Martian, say—to visit a feudally structured society. We shall suppose that he is so different himself from the individuals composing the hierarchy that he sees no essential differences between them that accord either with the status they have in their own eyes or in those of the other individuals that compose the hierarchy. If he notices differences at all between the individuals (in circumstances where all seem to him completely remote from himself) he will be as likely as not to pick up differences that have nothing remotely to do with the hierarchy structure. Thus he may notice the height of the individuals, or their weight, or whether they wear blue or red clothes. Let us therefore idealize the position in which the Martian finds himself, by considering all individuals to be identical. This Martian will then have to tell them apart by their *function*. We will suppose that the Martian is familiar with the logic of control, so that he is all set to discover a hierarchical structure, with individuals at different levels. Also, we will suppose that he can impose limited constraints in the behaviour of any given individual to observe the effects of the constraint on the other individuals.

This latter power does not enable him directly to infer control. For example, in a tightly knit society, a constraint on one member at a given level will have repercussions on some or all of the other individuals at that level, as well as on the individual that it controls or that controls it.

So where does the Martian look for his *first evidence* on which to classify individuals? I argue that he can find it at all, only in cases where there is a great difference in the *time constant* (or relaxation time) associated with individuals at different levels. This argument seems to accord with our everyday experiences, for we know that if we are members of a hierarchy of command, then we cannot operate at all if commands come to us with a frequency comparable with the rate at which we can implement them in detail. Since we are similar individuals to the individual who sends (or the individuals who send) us commands, we must perform each detail at approximately the same speed as that at which the overall command—implying many details—was delivered. We need far more time.

The Martian, accordingly, will be able to make a consistent start in discerning the levels of the hierarchy if and only if he can classify individuals by the rate at which they interact with their own kind. Or—to put it another way—he will be able to discover which are the same kind as a given individual by their all having a distinctive time-constant for their mutual interactions.

I conjecture that the intuitive argument that I have sketched is in fact capable of general application to cybernetic systems where control (or command, as I have hitherto called it) is hierarchical in character. If this is so then we have a non-formalized (but probably formalizable) theorem underlying hierarchical control which asserts that two levels are specifiable if and only if the ratio of the time constants characterizing the two levels is large (numerically). The *ratio* rather than either of the separate time constants is the important quantity here, for it is clearly only the ratio which has any experimental meaning in the absence of an independent time reference.

I have given a simple illustration of the thinking underlying a model of hierarchical structure by abstracting certain essential features from the hierarchy of the political type. It was such thinking, using these abstractions, that permitted mathematical development of this very difficult concept (the hierarchy), and this development proved to be possible only in the case of one simplifying assumption (namely that the time constants characteristic of different levels were very dis-parate numerically). There is a *prima facie* case therefore for trying to use the concept of a hierarchy with levels differentiated by their possessing different time constants, in biology, and I shall now look more closely at this possibility.

▶ A biological system usually shows some clear differences from our idealized feudal hierarchy. Firstly, the complex biological structure is not obviously an association of essentially similar individuals whose differentiation into levels is a matter of function rather than of design, as it is in the case of the feudal men. The latter we think of as men first, and as elements of a feudal hierarchy second. Thus there is a contrast; true, we immediately find ourselves looking for reasons to establish similarity between the two cases. A feudal man, we may say, who is trained as a baron is really very different from a villein, and, so we can argue, the real entity—a man—is not the biologically separable mechanical entity, but is that mechanical entity together with all its learned functions in the society of other individual men. However, that we find ourselves arguing in ways like this

256

Ted Bastin

really only re-emphasizes our dependence on the hierarchy concept, for in macroscopic biology there is little which is conspicuously set before us to emphasize the build-up of diverse structure out of assemblages of essentially similar elements.

When we get to the cell level the situation is different; there is direct evidence for hierarchical organization in the very existence of cells, and the cells are the feudal men in our pattern of control. To make this comparison would be pathetically simpliste (indeed, that is the reaction we immediately feel) if it were not that current ideas about patterns of control among biologists at the cell level and below often seem to be even more lacking in subtlety.

Indeed, I think there is a danger that biologists may be inhibited from formulating their underlying assumptions about control relationships just because these ideas would seem too crude if brought to light. Certainly a great deal of detailed knowledge of control relationships—particularly of the combinatorial sort that is provided by modern molecular biology—does exist, but little of it is systematized in such a way as to provide general guides to thinking. In the absence of such systematization it seems that what the scientist must do is to work on hunches to guide him in his estimates of the plausibility of hypotheses, and presumably these hunches incorporate implicit judgments about control. Nevertheless, however crude these judgments would turn out to be if formalized, it would surely be much better to have them out in the open. For example, if the analogy with feudal villeins has something to do with our feeling that there ought to be a basic, undifferentiated unit of control in the shape of a cell, and thus the existence of such a thing is something that one would expect, whereas anything to the contrary would need explanation (and I guess that this is indeed what people mostly think) then I should very much like to have this strand in the biologists' thinking thoroughly out in the open and explicit.

The mere existence of biological structures built out of large numbers of roughly similar cells hardly constitutes a call for the 'similar unit hierarchy' model. The structure is usually simply too evident, and there is no need for elegant mathematical methods to be applied to answer questions about the functions of cells in—say—a bird's wing. We already know the answers. That is not to deny that an enormous mystery remains, but the mystery of control has been pushed back into the microscopic and even into the molecular conditions of development of the cells. By the time we can look at a functioning bird's wing there remains very little of the flexibility of function to individual cells that would constitute good ground for application of the 'similar units model'.

We might find what we were looking for in something like the slime fungus

257

where the cells can wander about in conditions of slow, highly viscous, mass flow and yet the organism retains some sort of primitive structure and identity.

However, the case we are most likely to find worth looking at is the brain. There is obviously a great deal of structure of some sort to the brain, as is obvious from how our memories can work. But this is not structure in the same sense as we apply the term to the bird's wing. It is indeed more like the structure of a feudal society as our Martian sees it, where the hierarchical status of an individual has to be inferred from his function. A good deal of brain-physiological research does take the form of investigation at the cell level into the effects of stimulation of individual cells by different means, either on the organism as a whole or upon other cell complexes in the brain. This research is rather like the efforts of my Martian to map out the control patterns by putting spanners into the works at particular points to stimulate or inhibit particular individuals, and the general conclusion to be drawn from the mathematics of the similar unit model would be that, in conducting such research, one should look first for any evidence of a division of individuals into classes according to their reaction rates.

In making these suggestions I have spoken rather loosely of 'cells' and of 'units' without definitely asserting that cells are to be regarded as the units that compose the hierarchy. In fact I want to keep open the possibility that the correct identifi-cation for the hierarchy individuals are small complexes of cells. There are a variety of cells known to exist in the cerebral cortex, so that some structure is visually evident which shows that it would certainly be inappropriate to treat members of different varieties as similar individuals. On the other hand, I think it is widely assumed that some repeated structure of cell complexes must exist at some level in such a large mass of cells as constitutes the cerebral cortex. The alternative –that no repeated structure exists–is logically possible but hardly conceivable. And wherever repetition sets in, we find the units for our hierarchy.

I have suggested how models like my 'similar unit hierarchy' may provide concepts which will assist in the understanding of control relationships in cerebral tissue. A critic, however, may argue that a model is not likely to be of much use if its applicability depends upon our *lack* of knowledge of structure. This model– it seems–is going always to be in retreat from the progressive discovery of what the detailed nerve net shape actually is. I think this criticism contains a serious error. Obviously one should be prepared to trace control chains as far back as one can. On the other hand, if one supposes that in doing so a stage will come at which one suddenly meets a kind of uncaused cause, or beginning of the control chain, one is obviously in a muddle. One may even find oneself asking the nonsense question 'at what point does consciousness enter the control process ?' Concepts appropriate to hierarchical organization and to the centralization of control are needed for their own sake. I think moreover that it is a fact strongly in support of my position that biological tissue gives every appearance of becoming more homogeneous and less mechanically differentiated as we approach the centre of control.

This article is at best only the prolegomenon to a piece of work. It is not the work: that would be on a larger scale. It would go on the assumption that there was scope for a real application of the similar unit model in biology (for example to the cerebral cortex) and assume a sorting of units into levels by time constants. So far we have already discussed the problem. But then the work would go on to deduce from the mathematical properties any similar unit hierarchy must have, what form detailed structures within the hierarchy must have. This problem would be important as soon—for example—as one wished to ask how particular memories are stored. Clearly in such an enquiry there is a conflict between the requirement for similarity of the units and the requirement for diversification of structure. This conflict has to be resolved by considering the dynamical characteristics that are allowed by the hierarchy mathematics within the similar unit model, and using these to create diversity of structure.

Appendix I

Note on the stages in the development of the 'similar unit model' for hierarchies.

Stage I   An attempt to replace the intuitively defined continua of space and time by patterns of ordering or *ordering relations* [3].

Stage II   Realization that the ordering relations could never be mapped on to the physical continua. The latter had to be constructed in terms of operationally defined interpolation of points starting from finite (discrete) sets. See also [4].

Stage III   Hierarchical construction seemed to be the only way to extend the sets of points

Stage IV   In absence of any rigorous way of understanding such hierarchical relationships, and with help from Gordon Pask, a three-level continuously variable feedback system with binary switches acting at thresholds, was constructed [5, 6].

Stage V   These models suggested to Parker-Rhodes [1, 6] a mathematical method which made it possible to represent the relations between levels in a hierarchy by using matrix algebra over the binary field with symbols 0, 1 as elements of matrices and vectors. An information preserving mapping of $p \times p$ matrices on to $p^2 \times p^2$ matrices was the key step [1, 6].

Stage VI Amson, Bastin, Kilmister [1] showed that this calculus required for its basic logical operation a process whereby one entity could discriminate the other entities in a set by elementary decisions of the type, like/unlike. Its history of such decisions then defined its place in the hierarchy. Intrinsically it had no place. In this way the concept of a similar unit hierarchy was given a definite shape.

Stage VII It was then found that for a hierarchy of a given depth of level (number of levels starting from the simplest) there existed an upper bound to the number of discriminable entities. A way was found to relate this upper bound to the quantum limit of measurement in the physical world, with numerical success.

Stage VIII It is now possible, with the conclusion of Stage VII, to look back to the control model of the hierarchy and fit into it the insights supplied by the formalization. We then see that the different reaction times which characterize the levels as a matter of mechanical expediency in a hardware hierarchy of the Bastin-Pask type, must exist as a matter of logical necessity if we apply the constraint that units must be intrinsically similar. This gives us the main argument that we used in the body of this paper.

Note. It is of course very difficult to construct a similar unit hierarchy in practice and one has to use imagination and guesswork in interpreting its behaviour. A part of the motivation of this whole line of work, indeed, so far as the present writer was concerned, was an idea that quantum uncertainty must be due (if one were assuming a hierarchical construction of the physical continuum) to the fact that one could never with complete certainty settle the allocation of hierarchy units into levels because complete certainty could only be attained at the expense of infinite ratio of time constants of the levels. This reasoning was based upon a sort of engineering guesswork.

Appendix II

An example of an ordering relation in a hierarchy

Consider the *quadratic group S* which consists of the elements *a, b, c, e,* and in which all the operations in the group can be read off from the following 'multiplication table' : thus the result of the operation $i \cdot j$, where *i, j* are elements and where the dot between them represents the group operation, can be found as the content of the square in the table where the ith row and the jth column intersect.

	a	b	c	e
a	e	c	b	a
b	c	e	a	b
c	b	a	e	c
e	a	b	c	e

The table gives a set of relations, therefore, between all possible triples of group elements, which together define the group. This set of triples is usually presented as equations : $i \cdot j = k$, or as transformations : $i \cdot j \rightarrow k$, but for my purpose it is more convenient to write them in the form $R(ijk)$ which we can read as a triadic relation, defined by an ordered set of 3 elements. Looking at the table with these relations, $Rijk$, in mind, we see immediately that one of the elements—namely e — differs from the other three in its behaviour. In fact it is, in group theoretic terms, the unit element. However, we can find no distinction of this sort or any other among the other elements. Their relations seem to be completely symmetrical. In fact we should like to say that the sub-set a, b, c of the elements has the particular *ordering relation* that no order can be defined among a, b, c by the group operation. This fact can be restated by saying that all the relations $Rijk$ which can be formed by giving i, j, k values taken from the set a, b, c, and taking each once only, are true. We call this the S-ordering relation (S for simultaneous).

How much importance, if any, we may give to this statement about the symmetry of a, b, c is unclear. In a paper [3] by Kilmister and the writer an attempt was made to relate the 3-dimensionality of physical space to the S-ordering relation of a, b, c in this algebraic structure. I think that no difficulty really arises from the fact that the S- ordering relation specifies the *absence* of any *preferred* order rather than anything positive ; the statement of the S-ordering relation can be put in positive form by saying that everything that is true under any one ordering is true under every other. This last way of putting the matter, indeed, leads straight to the intuitively strongest case for relating the algebraic structure of the quadratic group to the dimensionality of space. We do know that any equation from physics has to be precisely equally valid in every way if the coordinates are interchanged—a fact which strongly suggests the S-ordering relation.

However, none of the remarks I have so far made about the significance of the S-ordering relation are precise, and a brief account of the efforts made to get them so (p. 253) and how these efforts forced us to look for a new mathematical tool—namely ordering in hierarchies—may be useful here. These efforts were the first attempt to formalize the order concept in anything like the sense that was being used at Bellagio, though, particularly in relation to space-time, they owed a lot to Whitehead's theory [7] of the logical construction of space-time from ordered events.

Let us discuss some criticisms that were made of the position that the ordering relation (S) of a, b, c in S underlie the 3-dimensionability of space, to show the stages by which the attempt to state an ordering relation led to a hierarchical point of view.

Criticism (*1*). The specification given in the S-ordering relation to take elements of the group S other than the unit element must be a remark in some metalanguage. (It cannot be part of the group theoretic calculus. If it could, then the deduction of the ordering relation S might indeed be a deduction about groups instead of merely about the number 3.) The importance of this criticism appears when one analyses the intuitive argument that was used to relate the ordering relation S to physical space—namely that physical equations must be invariant under interchange of coordinates. This argument gets its appeal from the (unstated) assumption that the S group will be mapped on to a number field to give a 3-dimensional Cartesian space with a, b, c as unit vectors. In fact, however, the group itself does not provide the ordering relation S, and so no such development could be undertaken.

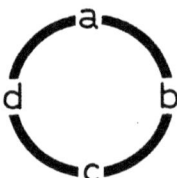

Criticism (*2*). The alleged ordering relation S is simply a property of the number 3 wrapped up in a suggestive way, for *Rijk* cannot be given values from a, b, c in just *any* way. Each must be chosen only once. (For example, *Raab* is not true, since *Raae*.) Moreover, in view of criticism (*1*) this restriction is arbitrary. Hence, all that is being said is that if three things are arranged on a ring then nothing we can say about the relative positions of the things will be rendered untrue by interchanging them. This—so the criticism admits—would not be true of four things arranged *abcd* on a ring, for a (for example) has a different relationship to b and d (which are neighbours) from what it has to c (which is not). The algebraic equivalent of this demonstration is simple: if *Rabc* and *Rabd*, then $c = d$, and there are not four things but three, and if it is not the case that both *Rabc* and

Rabd, then the rules distinguish *c* and *d*, contrary to hypothesis. Hence the *S*-ordering relation cannot be extended to four, or therefore to higher numbers of things. It is just a property of the set of 3 things. But it is no more.

Criticism (*3*). Physical reality has to be attributed first to space-time and only subsequently to 3-space. This could never be the sequence of development within the proposed basis for the space-time structure.

Criticism (*4*). The proposed basis for the 3-dimensionality of physical space is most implausible. The 3-dimensionality really specifies the number of *independent* assertions that can be made about the position of an object. The nature of the symmetry or otherwise of these assertions can therefore not determine their number which is just a fact about the world.

I include criticisms (3) and (4) because physicists will usually make them, but discussion of (4) is not relevant to my treatment of *S* as an ordering relation (though it opens up fundamental questions of physical interest). Of (3) it will have to suffice to say that if we examine with care the processes we actually might use to ascertain the number of spatial dimensions by means of actual operations, we see that we in fact reject many that are actually necessary to give complete specification of the kinematic state of a test-particle at a point (such as the electric potential) and it is actually plausible to suggest that we do so just because these are quantities of diverse sorts.

The substantive criticisms are therefore the related ones (1) and (2), and the *point d'appui* in finding a way around the difficulty that these criticisms raise — namely that what we wish intuitively to say seems to be forbidden mathematically — is likely to be the distinction upon which our imaginary critic has made these criticisms completely hinge : namely that between formal language and metalanguage. Clearly, simply to insist that statements from the metalanguage be allowed in the formal language will do no good, since we should then have no criterion of what to allow in and what not. We may not take that course even if we are doubtful of the validity of the distinction as an ontological necessity.

To avoid the impasse in which we seem to be, let us look at the mathematician as he manipulates the symbols of the structure that we have been discussing. How does he know that he is meant to try *every* operation that the calculus allows him, and not, for example, to stop short at those operations that for our purposes we should like to forbid ? There is no answer to this, logically, except that he has been trained as a mathematician, and that is what the other mathematicians do. So we

A general property of hierarchies

seem to have demonstrated that two can play at the metalanguage game, for we have caught the mathematician allowing himself to be guided by a rule that can only come from the metalanguage.

What sort of assumption would the mathematician have to make in order to avoid appealing to the way in which he happens to have been trained ?* The mathematician might well argue that he could have been replaced by a machine which was programmed or otherwise constrained to obey the rules of the calculus but by nothing else, and in which the starting positions are subject to random variations. Then he could be sure that the machine would, in the course of sufficient time, run through all the cases without there being any question of appeal either to the mathematician's insight or to his upbringing.

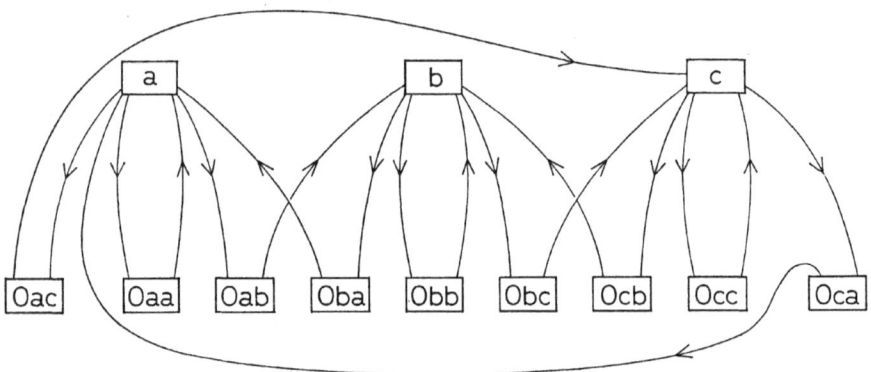

Now, however, an interesting change has come over the total picture. Because we have introduced a machine to supplement the mathematician we have raised a quite new sort of question. What status now have the rules of the calculus which before seemed in no need of investigation ? The mathematician was forced to justify his conduct by an appeal which, like the magician's apprentice, he has no power to call a halt to. In particular, the choice out of the original rules of operation which we desired to make in order to represent principles of ordering turns out to be more natural for a wide range of mechanical systems than does the conventional set of group operations, 'complete'. Let us therefore be guided by the logical

---

* And the validity of my argument depends upon the mathematician's being compelled to make some such appeal to a mechanism as the one I have outlined.

Comments by C. H. Waddington

implications of this train of thought and try to think of the structure of that machine which will embody our ordering ideas with the fewest *ad hoc* assumptions. It will incorporate the metalanguage spectre quite explicitly as a level of operators upon the original ordered entities—one for each pair of the latter, as shown in the figure.

To get a self-functioning system, of course, there has to be a second level of operators that operate upon the first level of operators, and so the total system rapidly goes off into indefinite complexity which we are unable to follow and have to represent by a random input at some level not too far distant.

The fact that we draw a diagram to illustrate certain features of a hierarchy of orders by no means implies that we understand in detail how it could work (and in those circumstances it is doubtful whether we should describe the construction as a machine at all). Nevertheless our immediate task is accomplished if we have mapped the stages in the argument which lead from a certain simple mathematical structure to a crude realization of the hierarchical organization of orders of which, we claim, it is an abstraction. (Even if the immediate reaction of the mathematician, confronted with the horrors thus exposed, is to set the whole argument in reverse as fast as he can.)

References

1. Amson, J. C., Bastin, E. W., Parker-Rhodes, A. F., Kilmister, C. W. Work summarized in Bastin, E. W., *Studia Phil. Gandensia, 77* (1966) 4.
2. Kuhn, T. S. *The Structure of Scientific Revolutions*. (Chicago, 1962).
3. Bastin, E. W. and Kilmister, C. W. *Proc. Camb. Phil. Soc., 50* (1953) 2.
4. D. Bohm. *Proc. Int. Conf. Elementary Particles*. (Kyoto, 1965) 252.
5. 'Self-Organization and the Notion of Level'. U.S.A.F. Office of Scientific Research Technical Note. (Cambridge Language Research Unit, Information Structures Project 1960.)
6. *U.S.A.F. O.S.R. Annual Reports* 1961—5. Information Structures Project.
7. Whitehead, A. N. *Phil. Trans. Roy. Soc.,* 1905.

Comments by C. H. Waddington

I think it is worth drawing attention to the way in which several contributors to these papers have, from different points of view, approached one another in attributing great significance to the association within biological systems of a number of time-scales differing markedly in magnitude. From a purely biological standpoint, I have myself been remarking on this for some years. For instance, in 'The Strategy of the Genes', 1957, pp. 5, 6: 'Perhaps the main respect in which the biological picture is more complex than the physical one, is the way in which time is involved in it . . . to provide anything like an adequate picture of a living thing, one has to consider it as affected by at least three different types of temporal change, all going on simultaneously and continuously. These three time-elements in the biological picture differ in scale: . . . evolution . . . life history . . . turnover of energy or chemical change.' And the same point was made in my summary of the results of the first meeting (*Prolegomena*, p. 218). These statements were made as descriptions of what we in fact come across in biology, where we are confronted by systems affected by controls at the levels of evolution, development, and functioning, with very different time scales.

It is more significant to find the same, or at least a very similar, argument being advanced by several of the physicists on grounds of logical necessity. Bastin, in the article above, argues that it is possible to analyse a complex control system into a hierarchy of subsystems only if 'the time constants of the different levels are very disparate numerically'. At the first symposium Pattee (*Prolegomena*, pp. 69–93) came to the conclusion that such a fundamental biological process as heredity could only operate with adequate reliability if it involved non-holonomic constraints, i.e. 'metastable configurations with relatively long relaxation times compared to our time of observation'—an example being a DNA polymerase enzyme molecule which persists for a time long in comparison with the time required to catalyse the bonding of one pair of bases. In the same vein, surely, but at a more abstract level, would seem to be David Bohm's insistence that an order of orders ultimately demands a reference to a 'timeless order'—which perhaps need be timeless only in the sense of having a time scale much greater than that of the lower orders, since David explicitly contemplates the possibility of its changing (p. 57). Finally, there would seem to me to be at least an analogy, and perhaps something more, between non-holonomic constraints and Christopher Longuet-Higgins' programs, which have relaxation times much greater than the time required for performing each item in the program.

It is true that there have been other contributions, dealing with time, which have not laid such stress on the importance of discontinuities between time scales. This was, for instance, in the main true of the discussions of limit cycles by Kerner, Iberall, Goodwin, and Cowan. But one may point out that these authors were mainly concerned with some one level of biological operation, e.g. Iberall found his continuous spectrum of limit cycle frequencies all within the realm of metabolism.

It is surely significant when so many of those who approach the problems of general theoretical biology from the standpoint of professional physicists advance

266

Comments by C. H. Waddington

such closely allied arguments attributing major importance to changes of time scale. It seems probable that we may here be approaching a major theorem of theoretical biology, such as it was hoped these symposia would produce.

Physical Problems of Heredity and Evolution

H. H. Pattee
Stanford University

My purpose in this paper is to present questions about the nature of life and evolution which can still generate deep curiosity among physicists in spite of the assurances of molecular biologists that the physical basis of life is now clearly understood. I have learned that this is not a popular purpose, to say the least; for these days to admit uncertainty about the basic nature of heredity, enzymes, or evolution is often regarded only as evidence that you have lost contact with modern biological knowledge.

It was not always this way. In fact, credit for some of the origins of molecular biology must go to physicists who were puzzled by the peculiarities of living matter [1]. The attitude of physicists twenty years ago towards biology was recently characterized in the following words [2]: 'Thus it was the romantic idea that "other laws of physics" (Schrödinger) might be discovered by studying the gene that really fascinated the physicists. This search for the physical paradox, this quixotic hope that genetics would prove incomprehensible within the framework of conventional physical knowledge, remained an important element in the psychological infrastructure of the creators of molecular biology.'

One example of this type of so-called 'romantic idea' was the speculation of Delbrück [3]: 'It may turn out that certain features of the living cell, including perhaps even replication, stand in a mutually exclusive relationship to the strict application of quantum mechanics, and that a new conceptual language has to be developed to embrace this situation.'

But today, twenty years later, these questions do not often arise. It is now commonly asserted, not only that the secret of heredity is understood in terms of physics, but that 'No paradoxes had been encountered, no "other laws of physics" had turned up' [4], and that '...up to the present time conventional, normal laws of physics and chemistry have been sufficient....' [5].

Now while these evaluations may be true in one sense, I believe it is correct to say that no 'strict application of quantum mechanics' to the living cell, including replication and specific catalysis, has even been attempted by molecular biologists. I am not even aware of any 'loose' treatments of the physics of heredity by molecular biologists. Therefore I do not see any evidence that the physical basis of heredity has been worked out.

268

Of course there is no question that the molecular biological revolution has uncovered facts of life in incredible detail. There is no doubt, either, that all this knowledge will be increasingly useful to the medical sciences and helpful in unifying the many areas of biology which have not had, until now, a common foundation. My discussion, on the other hand, has to do with whether these new complex facts of life really have, as yet, any relationship to the discipline of physics.

What is it then about the fundamental facts of living matter which are still not easily understood in terms of physical theory ? To answer this question we must have some idea of what it means to a physicist to understand his observations in terms of a theory. Biology and physics have long had different traditions with respect to the relation between observations and theories, so that it is not unexpected that this difference in style often causes misunderstanding. Biologists have always had to struggle with an enormous variety of facts and with great complexity and lability in the simplest living units. Consequently molecular biology appears from this perspective as a set of relatively simple unifying obser- vations about all life, and the ability to reproduce many of the basic functions of cell components in the test tube is often interpreted by the biologist to be a 'reduction of life to physics'. Indeed, this certainly is one type of significant reduction in the complexity of even the simplest living cell.

But the physicist sees a different problem. He has already learned a set of unifying foundations or theories which quite accurately describe an enormous range of his experience. The physicist also collects facts, but it is the overwhelming concern of the physicist to find out whether or not these facts can be predicted by or 'reduced' to theory. Moreover, his rules for 'reduction' require relatively high experimental and formal precision compared to that expected in biology [6]. Historical experience has also taught physicists that elementary concepts are clarified not by applying them to more and more complicated systems, but rather by asking deeper questions about the elementary concepts themselves. For example, the laws of classical dynamics would not have evolved into the more profound relativistic or quantum theories simply by applying these classical laws to systems with more and more degrees of freedom. For this reason it does not seem reasonable to expect the current interpretation of heredity in terms of the Watson-Crick template model and the formal description of the coding and synthesis process to develop into a more profound theory of life simply by applying it to more and more complex organisms or by exhaustive description of, say, one strain of bacterium.

269

Physical problems of heredity and evolution

▶ *What is the central question ?* If living matter differs from non-living matter because of its 'hereditary' property, which allows for evolution by natural selection, then the central question for the physicist is to explain what 'heredity' means in the language of physics, and to explain how hereditary structures would arise and persistently propagate and evolve in a relatively disorderly environment. In other words, if the nature of heredity is 'the secret of life' then surely we must know something of the physical significance of this concept if we are to claim we understand this secret.

Let me make clear at this point that I shall assume that the current laws of physics are the most universal laws we have for describing the behaviour of all matter, dead or alive. I am not searching for 'other laws of physics'; but neither am I claiming that life is now 'explained in terms of physics' just because we know the average structure of D N A or can make enzymes work in a test tube. I am not questioning whether matter in the living state obeys the laws of physics, but rather how the laws of physics can actually explain the *difference* between living matter and non-living matter. Karl Pearson [7] put the question long ago, 'How, therefore, we must ask, is it possible for us to distinguish the living from the lifeless if we can describe both conceptually by the motion of inorganic corpuscles ?'

We are all familiar with the traditional biological generalizations about life, e.g. living matter depends on enzyme-controlled reactions, living matter self-re-produces, living matter is cellular, living matter evolves, etc. We can also find more general distinctions, e.g. life is distinguished by adaptability, order, design, purpose, etc. These distinctions between the living and the lifeless forms of matter not only suffer from ambiguity at the most elementary levels of living processes, but they are not in the language of physics. Thus it is difficult to say exactly where protein enzymes and simple catalysts differ, or where self-reproduction and crystal growth differ in terms of physical observables [8].

▶ *Two approaches to the central question.* Molecular biology has evaded this general question by using only explicit descriptions of life; and this has satisfied many biologists. No longer do textbooks begin with discussion of the general 'characteristics of life' but rather with the specific 'building blocks' of life. Textbooks emphasize that it has been demonstrated that life is built out of normal atoms and molecules, and since most cellular reactions can now be demonstrated *in vitro* there is the implication that life is just a certain collection of D N A, R N A, enzymes, and other molecules in the proper spatial configuration. So if there can be said that there is any difference between living and non-living matter, then according to the molecular point of view, this difference lies only in how these molecular parts are organized.

This is a deceptively simple and convincing attitude, for we do understand the essential behaviour of many types of molecules in terms of physical laws. What is deceptive is the idea of 'how these molecules are organized' because the laws of physics do not account for special types of structure or organizations of matter called 'living' as distinct from the organization of non-living matter. So our question of how we distinguish life and non-life has simply been displaced by the question of what is the physical difference between organizations which live and organizations which do not.

In the language of physics, matter is 'organized' only by the 'energy of interaction' or equivalently by the 'forces' which act on it [9]. For example, the structure or organization of the solar system is largely the result of gravitational interaction or forces, the structure of molecules is partly the result of electric forces, and the structure of nuclei depends on nuclear forces. A few other types of forces are known, and all types may be combined in a great variety of ways, but as yet there is no clearly recognized special force or special combination of interactions or forces which can be uniquely associated with living organizations.

We find, then, that there are two basic attitudes toward the nature of life. I will simplify these attitudes somewhat to contrast them more clearly. By far the predominant attitude today is that of molecular biologists who do not worry about the obvious physical distinction between the behaviour of living and lifeless matter. Their point is simply that they have looked at the parts of cells down to their essential molecules without being able to perceive any special physical distinctions. Since all the parts of living systems work in a test tube, it is asserted that life has been reduced to physics [10].

The physicists, along with many older biologists, focus on the obvious differences between the behaviour of living and non-living matter, principally the evolutionary fact that living matter becomes more and more orderly, and non-living matter, less and less. Since they presently see no basis in physical law for this difference, it is asserted that some principle is missing from physics, or if no such principle exists, that life is irreducible [11].

Both of these assertions may have some truth in them, but I do not believe that either of them provides a satisfactory answer to the question which was asked. The first position says, in effect, we discern no basic physical difference in detail between living and lifeless matter, therefore life can be reduced to physics. The second position says, in effect, we discern a clear difference between living

271

and lifeless matter but there is no physical basis for this difference, therefore life cannot be reduced to physics. Stated so bluntly, there are obvious logical fallacies in both these statements which are often obscured or softened by more elaborate formulations. But I do not think the lack of logic is the main difficulty. The fallacies in both cases, I believe, are more in the nature of seeing too narrow a view. Strangely enough, the biologist often takes a narrow view of the conditions for life to exist, while the physicist often takes a narrow view of the conditions for physics to exist. In particular, molecular biologists tend to ignore the nature of evolution and underestimate the exceptional physical requirements for persistent hereditary processes, whereas the physicists tend to ignore the nature of observation and underestimate the exceptional hereditary requirements for measurement processes.

▶ *Summary of a third approach: Three problems.* My approach will be to assume that the potential for *hereditary evolution* is the primary characteristic of life which distinguishes it.from other collections of matter. The central question will then reduce to formulating *first* a physical description of an elementary hereditary process; *second,* the physical conditions which allow hereditary machinery to persist in the face of the inevitable disordering interactions with its environment; and, *third,* the nature of descriptive symbolism in hereditary systems which have the potential for evolving the enormous degree of 'self-determination' that we observe in living matter.

The first problem, the formulation of elementary hereditary events in terms of physics, will lead to the conclusion that hereditary transmission depends entirely on specific rate-controlling mechanisms. Macroscopic hereditary machines (automata) require special flexible constraints (non-holonomic constraints) which are not directly derivable from equations of motion, but can be interpreted by non-equilibrium statistical mechanical descriptions. Molecular hereditary machines must satisfy similar conditions, but the quantum mechanical analogue of non-holonomic constraint is a more difficult conceptual and formal problem that has not been adequately solved. The second problem, the formulation of conditions for persistence of hereditary machines in a disordered environment, will lead to the thesis that classical hereditary machines cannot have exceptional reliability, and that only quantum mechanical hereditary machines, without thermal dissipation, can operate more reliably. The third problem, the conditions for evolving increasing self-determination, will lead to the requirement for a separate *description* or *symbolic representation* of the hereditary machinery, along with the actual hereditary machinery itself which can read (decode) the descriptive

272

symbols. This is closely related to the condition for universality in abstract automata [12]. It is represented biologically by the genetic description, the code, and the synthesis machinery found in all cells we now observe and, at a higher level of symbolism, in the brain.

▶ *The first problem: Physical conditions for hereditary processes.* Hereditary processes require the existence of a set of relatively fixed objects or traits any one of which can be transferred in a recognizable form from parent to offspring in the course of time. The first essential condition for hereditary propagation is the possibility of a set of more than one trait. For example, even though the energy or velocity of an undisturbed mass or the number of degrees of freedom of a system will be 'propagated' unchanged in the course of time, we do not speak of these constants of motion as hereditary traits since there is no possible alternative allowed by the laws of motion. The biological idea of a hereditary 'trait' also implies the possibility of an alternative trait, which in turn implies a record for identifying which one of the more-than-one alternatives is to actually be propagated. This record for identifying traits may be called the description of the trait or the genome.

In order to see more clearly what the physics of heredity requires, consider a very simple hereditary tactic copolymerization in which two types of monomer are added to a chain in alternating sequence, *ababab* . . . The fact that such an alternating copolymer exists is, in itself, no assurance that a hereditary reaction has occurred, for it is quite possible that this alternating sequence is the *only* stable sequence. In other words, the homopolymers *aaaa* . . . and *bbbb* . . . , or even dimers *aa* and *bb,* might be energetically unstable so that no alternative sequence is energetically allowed. In this case the sequence is uniquely determined directly by the equilibrium configuration and hence no hereditary propagation can be said to occur.

Therefore one condition for heredity propagation in such a copolymer is that more than one sequence is more or less equivalent energetically. Let us say then that all bonds *aa, ab, ba, bb* have approximately the same ground state energy so that the probability of each distinguishable sequence is approximately the same at equilibrium. But now for heredity propagation we must have a second condition that the probability of the growth of one (or a subset) of the possible sequences is increased relative to the others because of the previous existence of one of these sequences—the genetic sequence. In other words, although the initial sequences were required to be initially equiprobable (degenerate) so as to have more than one alternative, we then require that the existence of these

273

sequences leads to unequal addition probabilities in the growing sequences. How can this be ? Since the bonds being formed are the same between all monomers, we must assume that no difference can occur in the equilibrium energies of different sequences and therefore only the relative *rates* of monomer addition can be changed. In other words, in such a copolymer, *hereditary propagation must depend on specific catalytic control of the rates of monomer addition* [13]. This also implies that to the extent that sequences are determined by the equilibrium energies, they are not properly hereditary. It is also important to realize that in so far as some other *inherent* physical property of the bonds joining the monomers controls relative rates of monomer addition, i.e. different activation energies or affinities, then the sequence has no alternatives and is not hereditary. In other words, only when bond formation is *correlated* to specific, rate-controlling elements *other than the bond itself* can we imagine the physical alternatives necessary to call it a hereditary process. The same distinction holds if the monomer addition step requires energy. If there is an energy coupling reaction that joins all monomers with equal facility then this is not a hereditary process. It is the specificity interaction correlated with the catalytic reaction which effects the reduction of alternatives necessary for the hereditary process.

A more general physical description of a hereditary process can be given in terms of the initial conditions and equations of motion. A physical system is defined in terms of a number of degrees of freedom which are represented as variables in the equations of motion. Once the initial conditions are specified for a given time, the equations of motion give a deterministic procedure for finding the state of the system at any other time. Since there is no room for alternatives in this description, there is apparently no room for hereditary processes. Alternatively, since the specification of the state of a system at one time serves to completely determine the course of the system at past and future times, there is no consistent way to introduce the additional concept of a record or memory of past events within this deterministic system. The only useful description of memory or heredity in a physical system requires introducing the possibility of alternative pathways or trajectories for the system, along with a genetic' mechanism for causing the system to follow one or another of these possible alternatives depending upon the state of the genetic mechanism. This implies that the genetic mechanism must be capable of describing or representing *all* of the alternative pathways even though only *one* pathway is actually followed in time. In other words *there must be more degrees of freedom available for the description of the total system than for following its actual motion in the course*

274

of time. This can be accomplished classically only by introducing dynamical constraints which are additional equations relating coordinates to the trajectories in a way which is not derivable from the ordinary equations of motion or initial conditions. Such constraints are called *non-holonomic* following Sommerfeld [14], although other definitions are found.

Almost all man-made machines are interesting and useful because they introduce non-holonomic or hereditary constraints. From the simplest clocks to the largest computers, the essential dynamical elements such as ratchets, relays, switches, and escapements, which make the machine function according to some human design, depend upon non-holonomic constraints. Of course, it is seldom questioned if these machines can be explained by the laws of physics. As a matter of fact, non-holonomic constraints are seldom discussed in physics texts except for very simple cases, because of the serious formal mathematical problems which arise and because of the difficulty in making generalizations. All that it is necessary to say here is that since a classical non-holonomic constraint has more static than dynamic degrees of freedom it cannot be described by formal systems with a fixed number of specified degrees of freedom. Therefore, in order to effectively reduce the number of dynamic degrees of freedom we must introduce the *statistical* idea of time-averages or time-dependent *correlations* between degrees of freedom. But then the dynamical system becomes essentially irreversible and dissipative. It is for this reason that any classical device which operates as a logical element by reducing the number of alternatives must dissipate energy (approximately kT per bit) for each reduction. This must be true for the simplest reset mechanism [15] as well as for a complex Maxwell demon [16], otherwise the Second Law of Thermodynamics could be violated. In general a statistical mechanical treatment of 'discrete' classical non-holonomic or hereditary constraints will require highly non-equilibrium and non-linear equations with at least two distinct time scales, so that it is always easier to design the logical behaviour of machines and treat the physical reliability as a separate problem.

The first essential point I want to emphasize here is the inherently statistical nature of non-holonomic constraints and hence the inherently noisy nature of hereditary or rate control processes. In the copolymer example recall that all sequences had to be energetically possible in the absence of the specific catalytic control in order to satisfy the hereditary requirement for physical alternatives. Therefore the overall accuracy with which a given sequence is propagated depends on the precision of the monomer recognition steps and the correlated catalytic rate difference in the addition step. For example, if the catalytic mechanism

275

increases the addition rate of a on to a terminal b, or vice versa, by a factor of 10^4 over the uncatalysed rate for all monomer additions, then there will necessarily be a mutation rate of 10^{-4}; i.e. there will be on the average a non-alternating pair per 10^4 additions. The second point, which is also fundamental, is that although the *reliability* of a hereditary mechanism can be very high or very low in a given situation, our formal mode of describing the situation is not a matter of degree. Either it has or it has not the hereditary property. That is, in the alternating copolymer example, the mutation rate could be much lower, as in DNA replication, or much higher, say one error for every two additions, on the average, but both reactions would be hereditary provided we are describing *a reduction in the set of possible alternatives in the course of time*. It is significant that this formal distinction between non-hereditary and hereditary descriptions is the same type of distinction as that between reversible and irreversible descriptions, or between strictly deterministic or probabilistic descriptions.

The biologist should be aware that these distinctions have appeared irreducible to many physicists [17] and that hereditary and non-hereditary behaviour is therefore a serious problem even in simple physical systems, quite apart from the question of whether the hereditary reliability of living systems is exceptional or not. For practical calculations it is sometimes possible to pass over the gulf by 'approximations' without any physical bridge, but it is important not to ignore the depth of this gulf. Biologists should also realize that once we have decided to pass this gulf and operate, isolated, without any physical bridge, on the hereditary side, then we are in the unreal world of deterministic automata in which we can largely design what amounts to our own error-free laws of motion (though we still will have unsolvable problems) [18]. Yet some matter has not only crossed this gulf, but also literally lives on the other side with exceptional persistence. To explain by physical laws the bridge over which matter passes into a hereditary world, I would call the reduction of life to physical laws. Nothing less could be called the secret of life.

▶ *The second problem: Physical conditions for hereditary reliability.* We have argued that in order to introduce the idea of hereditary behaviour into physics we must allow an internal (genetic) selection of alternative trajectories which amounts to a dynamical reduction in the number of degrees of freedom in the system, i.e. a non-holonomic constraint. But any process in which we lose track of degrees of freedom has passed from the dynamical world into the statistical world, and therefore in spite of the formal precision used to symbolize abstract hereditary rules, no real physical system can follow these rules without statistical error.

Let us look at the physical nature of error in more detail. What is the simplest classical system in which we would recognize error ? If we choose to describe our observations by the equations of motion of classical mechanics we know that there is no room for error, no matter how complicated the system. We do not consider that the motion of the stars or the galaxies could be in error. If the earth is struck by a meteor it might be called a catastrophe, but not an error.

In order to speak of a system making an error, it is necessary to have alternative trajectories possible within the system in the same sense that we find alternatives necessary to speak of hereditary systems. For a simple example of error, let us return to our copolymer, where we defined the monomer sequence as hereditary if from a set of energetically possible sequences there was a sequence-dependent, rate-controlling rule of growth of new sequences. This rule of hereditary propagation we saw could be executed with high or low reliability without jeopardizing the existence of the abstract hereditary rule itself. In this context I believe it is useful to speak of one type of error as simply the failure of a physical system to follow a formal hereditary rule. This type of error occurs in the dynamic motion of the hereditary machinery and therefore interferes with a time-dependent course of events. A second type of error is said to occur in time-dependent patterns or memory storage structures, such as D N A, in which an unpredicted alteration in the relatively static memory pattern occurs. Most error-correction schemes operate on this second type of error, since there exist general methods of coding or checking against error. However, the concept of memory storage or message has no meaning without specifying the code or the hereditary rules for reading-in and reading-out the memory or message patterns. Furthermore, any error-correcting code or checking scheme for messages must use the hereditary rules of coding to determine what is correct. Therefore I believe that the concept of error as *the failure of a dynamical rule of hereditary transmission* [19] is the most fundamental for this discussion.

Now we are ready to turn again to the fundamental question of whether classical non-holonomic machinery is reliable enough to assure the persistent hereditary evolution in a disordered environment, which is our basic criterion for life. As we said earlier, it is difficult to generalize about hereditary devices or non-holonomic constraints, because once you admit their simplest forms they can be concatenated to produce the most ingenious dynamical behaviour that man can conceive. On the other hand, man is by now reasonably familiar with the practical characteristics of classical machines, such as automobiles, clocks, and computers, and in most cases we know that high speed, low error rate and small size are incompatible, and that eventually they all wear out. The limitations on the speed and accuracy of man-made machines can of course be stated more elegantly [20], but I can foresee that even with a more formal treatment of error someone would raise the objection that although it is true that our man-made machines do not have the high speed, high precision, small size, and overall reliability of biological machines, this could still be only a matter of better design on the part of nature. It could be claimed that if we knew more about the error-correcting and self-repair properties of cells we could then match their behaviour with our macroscopic machines. Now this appears very unlikely to me, for the following reason: The elementary rate-controlling or logical machinery of living systems are not macroscopic organs but individual molecules — the enzymes. So the only relevant question is whether the reliability and speed (or, in chemical language, the specificity and catalytic power) of enzyme molecules can be predicted quantitatively by classical models.

To briefly summarize the argument, we began by agreeing that persistent hereditary evolution is the essential characteristic of life, but that the abstract hereditary process itself is clearly not the distinction between living and non-living matter, since many man-made machines exhibit elaborate hereditary behaviour. However, we find that no hereditary rules can be represented by physical systems without introducing statistical description and hence error in hereditary propagation. Furthermore, small size, high speed, and low energy dissipation is not consistent with low error rate in hereditary transmission in classical machines. On the other hand, the living cell executes all its hereditary rules with incredible speed and reliability using single molecules.

Therefore I conjecture that one fundamental physical distinction between living and lifeless matter is the exceptional *reliability* of hereditary transmission in living systems. Specifically, this implies that the dynamical behaviour of enzymes (and possibly other single, non-holonomic molecules) cannot be quantitatively described by classical models, that enzymes do not dissipate kT per bit of hereditary information transmitted, and that their hereditary logic is in some sense isolated from the thermal motions in the classical environment [21].

Let me make it clear that I have no doubt about the continued role of classical enzyme models both for pedagogical and technological advances. But that is not the fundamental question. We have asked if there is a clear physical distinction between living and lifeless matter. If there is none, and if the present classical descriptions of the hereditary rules of replication, transcription, coding, and synthesis tell the essential story, then the molecular biologists do indeed approach the truth in saying that we understand the secret of heredity and that life can now be explained in terms of physical principles.

But, on the other hand, if there is a crucial distinction between living and lifeless hereditary machinery, and if it cannot be explained classically, then the physicist will not only have a very strong point but also a profound problem. The point is that the secret of life is neither simple nor understandable by classical models; the problem is to express hereditary rules in the language of quantum mechanics, that is, to describe how an exceptionally reliable non-holonomic constraint can arise in a single molecule. This is not primarily a problem arising from the complexity of enzymes or our inability to calculate solutions to certain equations. It is of the same nature as the problem of interpreting the measurement process in quantum mechanics where we must use both reversible (deterministic) and irreversible (statistical) descriptions for a single physical situation. But in the case of molecular hereditary processes it is even less clear where to apply each type of description [22].

▶ *The third problem: Physical conditions for descriptions.* So far we have paid attention only to the most elementary physical conditions for hereditary machinery. Such machinery associates the specification of some action with the action itself, the description of the trait with the trait itself, or ultimately, the genome with the phenome. All hereditary machinery controls the rate or path of a system towards equilibrium, but in living matter this hereditary transmission is performed by specific catalytic molecules which associate the shape (or other interactions) of a molecule with a change of rate of reaction of a particular bond. We have conjectured that one fundamental physical distinction between living and artificial hereditary systems is in the exceptional reliability of the non-classical correlations between the specificity and catalytic power of enzymes. But clearly the course of biological evolution requires more than this. No hereditary machinery can operate entirely isolated, free from perturbations outside the system.

We are therefore led to my second condition for life, which is closely related to the traditional necessary condition of self-reproduction. I believe it is just as sensible to say that for persistent evolution of hereditary machines they must have an inherent capability for self-repair. But it is not primarily the question of whether self-reproduction or self-repair is the better concept which I think is fundamental. Rather it is the physical significance of the concept of 'self'. Now while it is difficult to find anyone who will dispute the necessity of self-replication

279

Physical problems of heredity and evolution

for biological evolution, it is almost as difficult to find anyone who will consider the nature of 'self' in this context as a serious fundamental problem. The problem has been made most apparent to me in thinking about the origin of life and the physics of heredity. When can we say, with some physical justification, that a certain collection of molecules has a 'self' to reproduce ? For example, does ordinary crystal growth or the occurrence of a repeating copolymer sequence demonstrate some aspect of self-replication ? In fact, one should even ask exactly what aspects of the Watson-Crick model of DNA template fitting may be usefully called self-replication.

To answer these questions I would like to return to the simple alternating copolymer example. There we noted that the mere existence of a chain with alternating monomers, *ababab* . . . , is not sufficient to establish a process of hereditary propagation. A necessary condition for a hereditary sequence is that a different sequence could exist, and that the particular sequence which occurs is the result of a specified rate control process. For a hereditary step there must be both a 'specification' and a 'catalysis' with the hereditary machinery correlating the two. The same type of condition must hold for self-replication if it is to be considered as a type of hereditary process. The idea of 'self' here must refer to the specifications of the catalytic reaction and not to the reaction itself. Therefore the idea of 'self' in self-repair and self-replication can only make sense as a subclass of hereditary descriptions in which *the hereditary machinery is a part of what is symbolically described.* This concept of self-replication is considerably more stringent than the simple idea of copying or repeating or growing the same thing again. The 'self' prefix is used here to mean not primarily the specification of the incidental traits being copied, but the description of the hereditary mechanism which executes the reading of the description itself (i.e. the coding mechanism). This concept of self-replication is very nearly equivalent to the concept of the condition that the automaton be capable of producing a more complicated automaton [23].

To return to the physical meaning of symbolic 'descriptive' function in molecules, let us assume that our first and second problems are solved and that we have a quantitative physical description of the specificity, reliability, and catalytic power of a molecular hereditary machine (e.g. an enzyme). This is difficult enough, but our third problem is to explain what a *molecular description* of this machine could mean in physical language.

Now most biologists would say that a particular base sequence in DNA is a specification of the amino acid sequence in the enzyme, and in this sense the

280

DNA molecule functions as a symbolic description of the enzyme. This is a useful logical statement as long as the concepts of 'description' and the 'object described' (the DNA and the enzyme) are kept logically and physically separated, that is, as long as the real physical dynamics is not followed too closely. The cell actually physically separates the descriptive, coding, and synthesis functions quite clearly, so that in a sense the cell helps obscure the physical laws which in simpler systems appear to determine all behaviour. But of course that is exactly what any automaton is designed to do. It is designed to 'follow' its own symbolic laws with as little physical interference as possible.

However, there is no escaping that the symbolic description molecules must be read-out by interacting with hereditary molecules, and in fact the entire genetic meaning of the description depends only on how the hereditary machines are constructed. Yet it is just this construction which is being described in a self-reproducing system. Von Neumann was the first to recognize and indicate one solution of this logical problem, but it is far from clear that his or subsequent formal discussions [24] have much to do with any real physical system, certainly not a quantum mechanical system [25].

Now we may state the third problem again: How do the laws of physics account for the assembly of molecules into a hereditary machine which includes a *description* of this machine so that it can be a self-reproducing hereditary system?

This is hardly more than the chicken-egg problem, and hardly less than the matter-symbol problem reduced to a primeval level. I would call the physical account of such a molecular system from reasonable initial conditions the fundamental problem of the origin of life.

CONCLUSION

Living matter is an exceptionally reliable form of molecular hereditary machinery. The most general and exact formal language for describing molecular behaviour is quantum mechanics, whereas the most highly developed formal language for describing symbolic hereditary behaviour is automata theory. One might expect then that a satisfactory account of the essential material and symbolic aspects of living matter could use both quantum mechanics and automata theory. But this would now require a dualistic theory of life, for in their present forms quantum mechanics and automata theory have very little in common with each other. Furthermore, where they do share a common language there arise suspicious inconsistencies.

281

Physical problems of heredity and evolution

For example, the intuitive foundation for the idea of an automaton is an idealized classical machine, as in Turing's machine [26]. Gödel has gone even further to define any 'formal' symbolic system of logic or mathematics as a 'mechanical procedure' for producing formulas called 'provable formulas'. He also equates the idea of 'finite' procedure with 'mechanical' procedure [27]. But as we have pointed out, any real mechanical device which is designed to execute logical or hereditary steps is non-holonomic and hence dissipative in some sense. In other words, the laws of physics say that logical or hereditary operations are necessarily statistical in nature and hence inherently subject to error. This physical situation should disturb logicians who rely so heavily on idealized models of macroscopic machines to guide their intuition. Modern molecular biologists should be even more acutely aware of the problem of hereditary reliability, since the hereditary machinery they observe in cells is truly microscopic and can be even less accurately idealized by large computing machinery.

It is also by no means clear that quantum mechanical description will easily solve the hereditary reliability problem; first, because of the completeness principle which says that there are no 'hidden variables' to statistically correlate in forming hereditary dynamics, and second, because of the indeterminacy principle which says that complete deterministic precision is impossible. However, there exists a suggestive complementary relationship between the quantum mechanical picture of the material world and the automaton picture of the formal symbolic world. In the quantum mechanical picture it is assumed that there exists a complete formal description of the state of a system with the consequence that attempts at exhaustive observations on the system are necessarily indeterminate. On the other hand, in the automaton picture it is assumed that all operations and observations are strictly deterministic with the consequence that attempts at exhaustive description are necessarily incomplete. These two pictures were developed in the first place to describe entirely different sets of observations, and yet in living systems there appear at a fundamental level both the material and symbolic aspects of each of these pictures.

Although the idealized models of living machinery will undoubtedly improve in detail and utility, any profound reduction of life to physical laws will require understanding the quantum mechanics of elementary hereditary processes. This must include not only an account of how molecular codes and descriptions can originate, but also how they can continue to operate so reliably in a disorderly environment. To the physicist this still appears as a deep enigma.

H. H. Pattee

This work was supported by the Physics Branch of the Office of Naval Research Contract Nonr 225 (90) and the Environmental Biology Division of the National Science Foundation Grant GB 6932.

Notes and References

1. One fruitful area of molecular biology is the study of the structure of macromolecules by X-ray diffraction. This area was pioneered by physicists, beginning with W. T. Astbury. Another well-known example is M. Delbrück (see J. Cairns, G. S. Stent, and J. D. Watson, eds., *Phage and the Origins of Molecular Biology*, Cold Springs Harbor Laboratory on Quantitative Biology, 1966).

2. G. S. Stent in *Phage and the Origins of Molecular Biology*, 4.

3. M. Delbrück *The Connecticut Acad. of Arts and Sciences 38* (1949) 173. (Reprinted in ref. of note 1, 20.)

4. G. S. Stent in *Phage and the Origins of Molecular Biology*, 6.

5. J. C. Kendrew *Scientific American 216* (1967) 142 (reviewing *Phage and the Origins of Molecular Biology*).

6. The physical explanation of the chemical bond is an appropriate example. A predictively useful concept of valence was developed in the 1850s (E. Frankland, A. Kekulé), and by the 1870s the idea of the chemical bond permitted the development of three-dimensional structural chemistry (Van't Hoff). But, however useful such models are, the concepts of valence, saturation, bond angles, isomerism, etc., were not explained or reduced to physical laws until Heitler and London (1927) showed how these concepts could be calculated from quantum theory.

7. K. Pearson *The Grammar of Science* (J. M. Dent and Sons: London 1937) 287 (first published 1892).

8. For some examples of earlier discussions of the physical nature of life see Lottka *Elements of Physical Biology* (Dover Publications: New York 1956) ch. 1 (first published 1924); A. E. Oparin *Life: Its Nature, Origin, and Development* (Academic Press: New York 1964) ch. 1.

9. We exclude as a scientifically verifiable hypothesis the fortuitous occurrence of initial conditions which are equivalent to a state of living matter simply as the result of rare fluctuations from a more probable configuration. This would amount to a 'special creation'.

10. Examples of some molecular biologists' attitudes may be found in J. D. Watson *The Molecular Biology of the Gene* (W. A. Benjamin: New York 1965), esp. p. 67; F. Crick *On Molecules and Men* (Univ. Washington Press 1966), esp. pp. 56–7.

11. Examples of some physicists' attitudes may be found in N. Bohr *Atomic Physics and Human Knowledge* (John Wiley and Sons: New York 1958) papers 1 and 2; W. M. Elsasser *Atom and Organism* (Princeton Univ. Press: 1966).

12. For an elementary discussion of *universal computing machine* see B. A. Trakhtenbrot *Algorithms and Automatic Computing Machines* (D. C. Heath: Boston 1963). Also M. Minsky *Computation: Finite and Infinite Machines* (Prentice-Hall: N. J. 1967) Part II.

13. This idea is expanded in H. Pattee (A. D.

Physical problems of heredity and evolution

Ketley, ed.) *The Stereochemistry of Macro-molecules* Vol. III (Marcel Dekker: New York 1968) 305.

14. A. Sommerfeld *Mechanics* (Academic Press: New York 1952) 80. Also E. T. Whittaker *A Treatise on the Analytical Dynamics of Particles and Rigid Bodies* 4th ed. (Dover Publications: New York 1944) ch. 8.

15. R. Landauer *I B M J. Res. and Development* 5 (1961) 183.

16. L. Szilard Z. *Physik 53* (1929) 840. This paper is discussed in L. Brillouin *Science and Information Theory* 2nd ed. (Academic Press: New York 1962) 176 *et seq.*

17. For example, M. Planck from his *Survey of Physical Theory* (Dover Publications: New York 1960) 64: 'For it is clear to everybody that there must be an unfathomable gulf between a probability, however small, and an absolute impossibility'. And on p. 66: 'Thus dynamics and statistics cannot be regarded as interrelated'.

Also H. Weyl in *Philosophy of Mathematics and Natural Science* (Princeton Univ. Press 1949) 203: '... we cannot help recognizing the statistical concepts, besides those apper-taining to strict laws, as truly original'.

And of course we should include von Neumann in his discussion of measurement in quantum mechanics: 'In other words, we admit: Probability logics cannot be reduced to strict logics, but constitute an esssentially wider system than the latter, and statements of the form $P(a,b) = \theta(0 < \theta < 1)$ are perfectly new and *sui generis* aspects of physical reality.' [By 'P(a,b) = θ' von N. means: If a measurement of a on a system has shown a to be true, then the probability of an immediate subsequent measurement of b showing b to be true is equal to θ.]

18. I am referring to Gödel type proofs of unsolvability and incompleteness theorems. See M. Davis, ed. *The Undecidable* (Rowen Press: New York 1965).

19. For more discussion on the physical basis of error in hereditary molecules see H. Pattee in (C. H. Waddington, ed.) *Towards a Theoretical Biology, I* (Edinburgh Univ. Press 1968) p. 86 *et seq.*

20. A collection of fundamental theorems on fluctuations and noise can be found in C. Kittel *Elementary Statistical Physics* (John Wiley and Sons: New York 1958) Part 2.

21. The idea of dissipationless change of shape conceivably playing some biological role was suggested by F. London *Superfluids* Vol. I (Dover Publications: New York) 8.

22. The relation of the quantum measurement process to molecular heredity is discussed further in H. Pattee *J. Theoret. Biol. 17* (1967) 410.

23. J. von Neumann in (A. W. Burks, ed.) *The Theory of Self-reproducing Automata* (Univ. of Illinois Press: Urbana 1966), esp. Lec. V.

24. M. Arbib *Information and Control 9* (1966) 177; C. V. Lee in *Mathematical Theory of Automata* (Polytechnic Press: New York) 155; J. W. Thatcher, ibid., 165; J. Myhill in (M. D. Mesarovic, ed.) *Views on General Systems Theory* (John Wiley and Sons: New York 1964) 106.

25. E. P. Wigner in *The Logic of Personal Knowledge* (Routledge and Kegan Paul: London 1961) 231, remarks on the '... "tailoring" of what substitutes for equations of motion ...' being only the result of hard, macroscopic systems with discrete variables. He also notes that the 'reliability' of classical models of self-replication needs to be evalua-ted and compared with experience.

26. See reference, note 12.

27. See reference, note 18, p. 72.

Statistical Mechanics and Theoretical Biology

Martin A. Garstens

University of Maryland

It is hardly an understatement to say that nothing in the history of biological research has appeared and been found acceptable as a formulation of theoretical biology. A variety of disciplines have evolved in recent times contending for this title. Included among these are cybernetics, information theory, theory of automata, systems analysis, mathematical models, analyses based on computer technology and adaptations of physical models. Current philosophic discussion on a possible formulation of theoretical biology is endless but with little consensus. A large school of thought works on the assumption that biology is in a state similar to that of the early days of physics and that, given time, a theory will develop naturally, as it did in physics. The analogy is not good, since physics has always had more substantial relative strength of theory to empirical content than biology ever had. There is a feeling among some that biology is inherently incapable of yielding any theory; that it is the field of the unique and unpredictable; that its uniformities exist only in its physical aspects.

Not only has no acceptable general working theory appeared, in spite of the many attempts to generate one, but even more frustrating, there have not been indications as to which theoretical or mathematical techniques show promise of leading to a theory acceptable to biologists. It is also not clear how the above-mentioned disciplines (i.e. cybernetics, automata theory, etc.) relate to each other or to the establishment of a theory for the field of biology. It is the object of these remarks to clarify the issue and suggest a possible direction for more hopeful coordination of current activities in theoretical biology.

It is the thesis of this approach to the problem that the field of statistical mechanics, properly developed and expanded, contains the clue for a theory of biology. Statistical mechanics can be a coordinating area for all the disciplines which have attempted a theory of biology and supplies a general outlook suitable to settle many philosophical controversies such as those which arise between the reductionists and the non-reductionists. In addition, statistical mechanics is believed to show promise of infiltrating operationally into professional biological research, allowing a deep probing into the nature of biological measurement, a necessary activity if biological theory is to emerge.

285

Statistical mechanics and theoretical biology

There are several obvious reasons why statistical mechanics should be considered as a clue to theory in biology. Statistical mechanics is the only area where a large degree of success has been attained in dealing with complex systems. The problems which remain to be solved in the field are reminiscent of those to be solved in complex biological systems. A most important reason for remaining 'close' to statistical mechanics is that it diminishes the tendency, particularly in philosophic discussions of biology, to wander too far afield from the realities of the physical world, without necessarily succumbing to a reductionist point of view.

The primary objective of statistical mechanics is to determine the connection, both in equilibrium and non-equilibrium, between the microscopic properties of ensembles of objects, such as atoms or molecules, and the macroscopic variables and laws observed in these ensembles. Such objectives are clearly manifested in thermodynamic studies where there also arises the important question as to the relationship of macroscopic measurement and microscopic events. How does a microscopic event result in a determinate macroscopic quantity during measurement ? (This problem is concerned with what is known as the reduction of the wave packet in quantum theory, a topic extensively discussed in connection with the theory of measurement in physics.) One of the main problems in irreversible thermodynamics is determination of the conditions for the existence of macroscopic quantities. It is also important to enumerate the complete set of macroscopic variables of a many-body system, given the Hamiltonian of that system.

The fact that some of the macroscopic quantities in statistical mechanics are impossible to observe, although they show their existence by the presence of a relaxation time, is of interest to biology. In some cases in statistical mechanics there are indications of the presence of several levels of macroscopic states manifesting themselves in different relaxation times. Both situations are suggestive of possibilities in biological phenomena where interactions between microscopic and macroscopic processes also give rise to such levels.

If the comparison of ensembles of biological systems with those occurring in statistical mechanics is valid, an expanded statistical mechanics would be expected to predict reliably those biological phenomena which are reproducible, in the sense of being common to all individuals. An expanded statistical mechanics should, in principle, be able to determine those aspects of behaviour which are common to systems of a given class. It is clear how this happens in inanimate systems. An examination of the foundations of statistical mechanics makes plausible the

286

extrapolation of its methods into biology, constituting the sought-for foundation for theoretical biology.

In physics we deal with collections of identical microscopic particles, each of which is described by the laws of quantum theory. These particles, however, manifest their existence in terms of macroscopic measurements. Thus a particle passing through a cloud chamber leaves a macroscopic trail of cloud behind it. To deal with the macroscopic aspects of matter, through which one learns all that is known of nature, including its atomic properties, one proceeds in two ways. First the macroscopic properties of the physical world around us are observed, including the relationships among them. This is the subject of such branches of classical physics as thermodynamics, hydrodynamics, mechanics, etc. For example, in thermodynamics one deals with the macroscopic properties of heat, temperature, entropy, and pressure, and one finds them related through the first and second laws of thermodynamics. These properties and laws, historically, were inductive and empirical findings. Secondly, the microscopic or atomic properties of matter are observed and the macroscopic properties are reinterpreted in terms of them.

The discovery of atomic phenomena meant that certain macroscopic events were indicative of, and could be used to measure, microscopic events. The theory of measurement which attempts to trace this connection between macroscopic and microscopic events is still unresolved in modern physics. When clarified, it should play a fundamental role in theoretical biology.

Statistical mechanics has the task of relating the atomic domain with the macroscopic, determining how macroscopic measurements can lead to microscopic information and how microscopic events are recorded by macroscopic ones. It must show the connection between observed macroscopic variables and averaged microscopic properties. In order to do so, two types of assumptions are made. One consists of all the assumptions underlying quantum theory. The other is a set of auxiliary assumptions required to ease the difficulty in dealing with complex systems. To suit the needs of physical systems, various standard auxiliary assumptions are made. However, for a long time there has been an ongoing activity, called ergodic theory, to reduce all auxiliary assumptions to quantum theory or mechanics. This is now known to be impossible [1]. It is now seen that there cannot be any rigorous mathematical derivation of macroscopic equations from microscopic ones. Additional postulates or principles must be assumed to make this transition. This is precisely what must be done to establish the theory of complex systems in biology. Among the postulates which have been in common use in physics are : (a) the postulate of random phases, (b) the ergodic hypothesis, (c) the master equation assumption, (d) the assumption of equal *a priori* probability, (e) the metric transitivity assumption, and (f) the hypothesis of maximum entropy.

The hypothesis of maximum entropy is found capable of replacing most of the above assumptions. This replacement, first suggested by Elsasser [2] and recently extensively amplified by Jaynes [3], is receiving a great deal of current attention.

It is important to note that the whole set of auxiliary assumptions required to go from quantum theory to statistical mechanics are no less basic than quantum theory itself and are not reducible to the latter. However, while quantum theory, as applied to atoms and molecules (excluding high-energy phenomena) seems very well substantiated, the same is not true of the auxiliary hypotheses needed in practice to deduce the macroscopic variables from microscopic foundations. In fact the 'art' of obtaining good statistical mechanical analyses consists in knowing how to choose suitable auxiliary hypotheses (not reducible to quantum theory). The same art, much extended, needs to be applied to biology.

Accepting this account of the relation between the macroscopic and the microscopic in non-living systems, how can we profit by it in attempting an account of the living ? There seems to be no reason to believe that quantum theory, the theory of the microscopic, does not carry over without change into biology. There has never been any evidence that individual particles violate Schrödinger's equation, or its equivalent, when applied to the atoms or molecules making up biological systems. However, in trying to describe such systems auxiliary assumptions are needed, as in the statistical mechanics of inorganic systems. An important question which arises is : Are unique auxiliary assumptions required to lead to observed biological macroscopic variables and laws ?

If unique assumptions are required, they cannot violate presently accepted physical principles, nor can they be reduced to them. Evidently no amount of searching among current concepts in physics will fill the bill if biology is indeed irreducible. Just as in present-day fundamental particle research there is widespread 'feeling' that new principles are needed to explain the newly observed particles, so there is reason to 'feel' that more general ideas are needed in biology.

One method of approach to statistical mechanics, which shows promise of generating such general ideas, is that of E.T. Jaynes. In a series of articles [4] he has been able to relate information theory to statistical mechanics. By injecting the methods of statistics more directly into statistical mechanics it appears to be possible to obtain the required auxiliary principles and macroscopic variables by inductive methods. Assuming only that the entropy

$$S = -\kappa \sum_i P_i \ln P_i$$

is maximized in any system, observed physical quantities are equated to the average values of the operators representing them (see Appendix). Subject to these constraints, the density matrix ρ of the system can be obtained. Knowing ρ, additional macroscopic observables characterizing the system can be calculated from the inductively obtained partition function and checked by experiment. This procedure can and has been applied to physical systems in a straightforward manner. For biological systems one is faced with difficulties of complexity, although the method seems sufficiently general to be applicable. One must expect of course that, at first, prediction must be confined to common aspects of species behaviour rather than to individual behaviour. But before a prediction can be made or a biological species characterized the knowledge of an excessive number of essential parameters seems to be required. So many as to make the problem seem hopelessly complex unless a means is devised for selecting only parameters which are most essential and thus greatly reducing their number.

There would appear to be some ways of accomplishing such reductions. One method is to eliminate from consideration those aspects of behavioural systems which do not seem essential to continued existence. It is known that certain parts of living systems can be removed without ensuing death. Secondly, by suitable restrictive hypotheses, gained from experience, one could diminish the generality of Jaynes' approach and make it more adaptable for biological considerations. Such restrictive hypotheses might involve topological constraints on the system rather than quantitative ones. These restrictive hypotheses combined with quantum theory could yield laws of developments describable in topological language. Hypotheses of this type would correspond to the auxiliary principles mentioned above, which are also required in physical systems in order to extract the familiar macroscopic variables.

G. Ludwig [5] has pointed to the need for specific additional hypotheses for distinguishing macroscopic variables in microscopic systems. The macroscopic variables in his analysis are characterized as continuous observables with respect to a certain defined macroscopic measure of discernibility $d(W_1, W_2)$, which is of the commensurable type and for which groups of continuous, weakly continuous, and strongly continuous operators coincide. Different measures of discernibility are then defined with respect to the ensembles W_1 and W_2.

Statistical mechanics and theoretical biology

Microscopic discernibility, for example, is defined as the distance:

$$d(W_1, W_2) \equiv \frac{1}{\sqrt{2}} \| \sqrt{W_1} - \sqrt{W_2} \| = \frac{1}{\sqrt{2}} (\sqrt{W_1} - \sqrt{W_2} \, ; \, \sqrt{W_1} - \sqrt{W_2})^{\frac{1}{2}}$$

This is followed by a similar but special definition of macroscopic discernibility. The axiomatic method is not the only approach to statistical mechanics nor even the most desirable. In practice one may use an empirical procedure as was done in thermodynamics and arrive at the variables inductively. In biology all methods will be needed and will play a rôle, since each presents a special picture and a unique point of view.

The importance of the above techniques for biological theory is that they maintain an intimate and necessary connection with the quantum theory of matter and still do not demand reducibility to the latter. This should produce great flexibility of method without separation from physics.

The auxiliary hypotheses under consideration should not be reducible to known principles, since they would then be redundant. They also should not be contrary to established principles. Finally, they must be capable of synthesis with quantum theory in the sense in which the above-mentioned principle of maximum entropy is used in conjunction with quantum theory. This means that they should not constitute an independent doctrine making no conceptual contact with quantum theory.

Special definitions and postulates, capable of being grafted on to basic physical foundations, are needed for biological theory. It is likely that all quantitative regularities in biology can be explained in terms of established physical principles. The required auxiliary hypotheses must therefore be qualitative in character, i.e. topological or statistical. Among the regularities in organisms which seem unique and demand explanation are their stability, unity, development, reproduction, memory, and evolution. All of these phenomena and many others have in common the incredible difficulty of explanation in terms of current physics.

There is a particular problem in trying to set up auxiliary postulates that could lead to or explain regularities of the above type. Whereas in physics it is generally easy to spot the macroscopic variables, usually small in number, this does not seem to be the case in biology. The concepts here are fuzzier, overlapping, and of far greater number. To postulate conditions giving rise to such variables and laws will therefore be difficult. Ultimately a large number will be necessary. At all times the questions of their interdependence will be a concern.

An example of the modelling of development or morphogenesis in biology can be obtained by setting up the conditions in irreversible thermodynamics under which a macroscopic variable grows, fades away, and then is transformed into another variable. It would be of interest to study the conditions under which several macroscopic variables interact so as to enhance some, weaken others, and eliminate the rest, or the conditions under which a given macroscopic variable is transformed into two, or the conditions under which two variables join to give rise to a third (reproduction?). Many variations of these situations resembling processes in living organisms can be imagined. The importance of these considerations are that they are not too divorced from accepted, successful, and fruitful procedures in physics and that they supply ready-made formalisms. The mode of analysis involved also indicates the directions to be taken in search of needed mathematical techniques.

Since statistical mechanics is the only well-established physical science of the complex, it should have bearing on biological systems. At the same time statistical mechanics does not rule out empirical or inductive findings which must ultimately be incorporated within it. An amplified statistical mechanics can allow for pluralistic (or not completely integrated) functioning of systems, if such exist. It avoids the narrowness of an approach like the ergodic, which attempts to reduce all phenomena to quantum mechanics.

Through the proper use of auxiliary hypotheses, statistical mechanics can become sufficiently flexible to act as the much-needed area of integration of the various new disciplines which attempt to set up theoretical foundations for biology. The need is not only to integrate such disciplines with each other but to show their connection with the basic discipline of physics. The approach through statistical mechanics shows much promise of accomplishing these aims.

Appendix

The method used is called maximum entropy inference, using the Shannon expression for entropy:

$$S_i = \kappa \Sigma P_i \ln P_i$$

or $\quad S_i = -\kappa T_R(\rho \log \rho)$

where T_R stands for trace and κ is an arbitrary constant. S_i gives a measure of the amount of uncertainty in the probability distribution P_i or the density matrix ρ. A density matrix or probability distribution is chosen which maximizes S_i subject to the constraints supplied by the available information. Thus the most uniform assignment of weights to the states in the ensemble representation consistent with the given information is obtained. Any other assignment of probability over states would be equivalent to assuming some arbitrary additional information not warranted by the observed macroscopic measurements. The constraining equations are:

$$\sum_i P_i = I$$

and $\quad \sum_i P_i \, g_r \, (X_i) = \langle g_r \rangle, \qquad r = 1, 2, 3, \ldots$

where the $\langle g_r \rangle$, the average or expectation values of g_r, are considered known and constitute the given data. The Lagrange method of undetermined multipliers maximizes S_i subject to the above constraints, yielding:

$$P_i = e^{-\lambda_o - \lambda_1 \, g_1 \, (x_i) - \lambda_2 \, g_2 \, (x_i)}$$

which is the solution, with the multipliers λ_i all determined, λ_o being the log of the partition function. In fact

$$\frac{\delta \lambda_i}{\delta \lambda_r} = -\sum_i P_i \, g_r \, (X_i) = \langle g_r \rangle$$

Thus, knowing the partition function, the macroscopic observables can be determined.

References

1. I. E. Farquer *Ergodic Theory in Statistical Mechanics* (1964).

2. W. M. Elsasser *Physical Review 52* (1937) 987.

3. E. T. Jaynes *Information Theory in Statistical Physics* (Benjamin: New York 1963).

4. E. T. Jaynes *Physical Review 106* (1957) 620; *Physical Review 108* (1957) 171.

5. G. Ludwig 'Axiomatic Quantum Statistics of Macroscopic Systems' in *Ergodic Theories* (Academic Press: New York 1961).

Aspects of Evolution and a Principle of Maximum Uniformity

Paul Lieber
University of California

General introduction and theoretical background. This essay is concerned with the identification of various aspects of evolution and their connection with a universal evolutionary process, which emanates from irreducible and universal processes identified here with the Dimensional Universal Constants of Nature. These Aspects of Evolution are specifically concerned with Constancy, Force, Process, Equilibrium, Stability, Adaptation, and the Nature of Information. This first paper is specifically concerned with the nature of force, equilibrium, non-uniformity, and stability, envisaged here as particular aspects of evolution, conceived as a universal process which reconciles everywhere in nature, constancy, and change.

In this paper an outline is given of ideas and reasoning which led to the conception of a proposition that may prove to be a general law of nature, fundamentally endowed with aspects of evolution. This proposition embraces the laws of classical mechanics, a general stability law, historical thrust and commitment, and information relevant to the formulation of a theory conditioning strongly non-equilibrium thermodynamical processes. The stability law so obtained bears the same kind of relation to stability, envisaged here as a general aspect of the performance of classical mechanical systems, as do the laws of classical mechanics to another equally general aspect of their performance, namely equilibrium.

The conception of the general proposition which embraces this stability law is inextricably linked with a conception of the nature of force. By this conception, force is the universal and fundamental global aspect of non-uniformity posited in nature to sense perception and sense awareness, and from which all sensation, experience, information, and consequently knowledge, ultimately emerge. In the particular case of classical mechanics, force is here conceived as the universal manifestation in sensation of global non-uniformity in nature, that is, as the resultant of all non-uniform connections that exist between an inertial body instantaneously situated at a particular location, and the universe in which it is contained. From these considerations it follows that the dynamical aspect of classical mechanics (more specifically the kinematical aspect) which is based on a conception and description of process ascribed to immutable bodies in motion, is significantly

293

more restricted and consequently less fundamental than is the aspect of nature symbolically designated by F in Newton's propositions. I use the word designated, rather than represented, in order to emphasize that this symbol, as it is used in classical mechanics, is not brought into correspondence with the anatomy and structure of nature's space-time manifold. Indeed, a critical examination of Newton's formulation and use of the known laws of classical mechanics does in fact suggest that he may have also tacitly conceived of force as an ultimate and global aspect of nature, and of his law of motion as a relationship between this ultimate aspect of nature and the motion of a body endowed with inertia. This point of view differs essentially from and denies the one taken by most of his followers, as well as from one expressed as a consensus among contemporary scientists, who choose to interpret his law of motion as a definition of force.

According to the ideas of this paper, force, as designated by the symbol F in Newton's propositions, in fact dominates the established laws of classical mechanics which are here understood to express only some of its fundamental aspects in nature. According to this paper, F in fact assumes a fundamental and dominant role in Newton's propositions. It dominates the dynamical term appearing in Newton's law of motion, which expresses only one of its particular manifestations within the domain of classical mechanics, and consequently does not define it. Instead, by this symbol Newton implicitly designated the resultant and thus total connection between a body endowed with inertia and the universe in which it exists. In so doing he implicitly assumed that this connection is independent of the frame of reference in which the motion of the body is described and calculated. This is tantamount to postulating, by implication, that the global aspect of nature symbolically designated by F, and the connection it represents between a body and the universe, is covariant under all coordinate transformations. By treating force in this way Newton evidently displayed humility and wisdom. Humility, because he instinctively realized that the nature of the global connection between a body and the universe in which it exists is the most fundamental and least understood aspect of mechanics; and wisdom, by treating force as primitive, and thereby not imposing arbitrary restrictions on what is not understood. The present interpretation given to force as it appears in the propositions of classical mechanics is not explicitly represented in Newton's writings, but rather is inferred here from its usage and the way the symbol F is formally treated in his propositions. From the above considerations it follows that the global connection designated by F, and called force, has the same stature in classical mechanics as does inertia, interpreted according to Mach's principle, according to which inertia is also a manifestation of a global connection between a body and the universe—a connection which, however, is characterizable by a scalar, and is therefore intrinsically endowed with high uniformity.

What is strictly local in Newton's propositions refers to their kinematical content on which their dynamical aspect is based. It is this dynamical and local aspect which restricts their covariance to inertial frames and consequently limits their generality. From this it follows also that it is naïve to interpret Newton's law of motion as a definition of force, as it is nonsense to define a fundamental aspect of nature which has unrestricted covariance in terms of an aspect whose covariance is limited to inertial frames. We see here again, from this point of view, that force does in fact dominate the laws of classical mechanics.

These considerations show that, in classical mechanics, the presence of a resultant force impressed by the universe on an inertial body, which is consequently not free, implies a non-symmetrical and thus non-uniform connection between the body and the universe. When the connection between an inertial body and the universe is symmetrical, and thus uniform, in the particular sense that individually impressed forces cancel vectorially—the body is then said to be free, according to the established laws of classical mechanics, and consequently moves according to Galileo's principle. We shall show in the section subtitled 'Hierarchies of Uniformity' that there in fact exists a hierarchy of free bodies, that is, bodies which can with meaning be distinguished as being more or less free, but all of which are equivalent and therefore not distinguishable by the established laws of classical mechanics.

These and other considerations concerning the nature of force, made within the edifice of classical mechanics, are sufficient to demonstrate that *all forces* in nature may be conceived as manifestations of the existence, in nature's space-time manifold, of non-uniform connections between inertial bodies and the universe. According to this thinking, forces that are revealed in the domain of classical mechanics emerge from the same ultimate and universal processes in nature as do all other forces. Force, thus conceived as the universal manifestation of non-uniformity in the space-time manifold, posited to sense-perception and to inertial bodies embedded in this manifold, brings into universal correspondence the various domains of physical theory which we have by convention learned to distinguish as classical and modern. These ideas and considerations are particularly designed to point out the fundamental connection between non-uniformity in nature and force, and to establish the thesis that force is the universal manifestation of these non-uniformities invoked in sensation.

295

Evolution and maximum uniformity

I shall now introduce the observations and ideas which led to the conception of the principle of maximum uniformity, in which force is there conceived as the fundamental physical aspect of non-uniformity. The identification of a natural law is not an exercise in formal logic; nor is what its propositions assert, provable. General propositions about nature can be tested only by experience— by what they predict and explain of it. In the present case the principle of maximum uniformity was discerned by fully generalizing explicit global information, which was obtained as a theorem for a class of dynamical systems, by suitably modifying and using Gauss's and Hertz's formulations of the principles of classical mechanics [1]. This information pertains to a global, positive, definite, scalar measure of the internal forces generated at each instant within such a system. The modifications of the Gauss-Hertz Variational Principles of Mechanics which render this general information explicit and without quadrature, consist of ascribing to force *the* dominant role in mechanics, and of identifying all forces in nature with an onto-logical-geometrical basis for the production of stringent geometrical constraints, which were [1–4] originally conceived to emerge from the impenetrability of matter understood as a property of position. This information, which bears directly on the fundamental problem of continuum mechanics, has not been made explicit, and as far as I see *cannot* be made explicit by Newtonian Mechanics, in which the only representation, given to force in its propositions, is vectorial. This means that in the significant sense of information rendering, the various formulations of the principles of mechanics are only conditionally equivalent. This development led me inexorably to the concept of 'Categories of Information', in terms of which questions concerning the equivalence and non-equivalence of various formula-tions of the principles of mechanics can be rationally examined and resolved. This led to identification of eleven distinct, yet related, Categories of Information by examples derived from familiar as well as more sophisticated aspects of experience. Once cited, these examples invoke consensus [5].

The global information, so explicitly obtained as a theorem on the distribution of internal forces, asserts that a positive, definite, *scalar* measure of all the internal forces is *instantaneously* less for the actual motion than it is for any other motion which satisfies the initial conditions and the geometrical constraints (as well as the external forces) which are instantaneously impressed upon the dynamical system. This theorem was established for a particular (non-trivial) class of dynamical systems. For this class, the scalar measure of the internal forces can be directly interpreted as a global measure of non-uniformity in momentum space.

The Principle of Maximum Uniformity, as it pertains to classical mechanics and classical continuum mechanics, was obtained by (a) interpreting the information obtained from the above theorem as a particular aspect of a general law which holds in all mechanical systems, and (b) introducing the concept of conditionally stringent geometrical constraints and relating these to material properties through which they are implemented in nature. This brings the Principle of Maximum Uniformity into correspondence with thermodynamical aspects of the equations of constitution of various materials and relates the idea of conditionally stringent geometrical constraints to uncertainties in the initial conditions from which historical commitment, causality, and a general stability principle naturally emerge.

The phenomenological description of the performance of classical mechanical systems reveals two general and mutually independent characteristics, namely equilibrium and stability. The known propositions of classical mechanics refer strictly to equilibrium, by invoking the condition that forces be instantaneously in equilibrium everywhere and for all time in the system. This is their information content. They report nothing of stability, which is an equally general and funda-mental aspect of the behaviour of classical mechanical systems. The laws of mechanics give but limited expression to the Principle of Maximum Uniformity by asserting that the forces acting everywhere in a system sum vectorially to zero in all directions. This restriction allows a multiplicity of directional and spatial distributions in the magnitude of the forces impressed upon a body, without therefore exercising a condition on preferred distributions which the stability principle presented here in fact does.

1.1. Concerning the nature of evolutionary adaptation
The considerations noted above help demonstrate that force, equilibrium, and stability are particular manifestations of an overriding tendency in nature to increase a global measure of uniformity identified with the global structure of the space-time manifold. This process is envisaged here as universal and conditioned by the principle of maximum uniformity, namely (a) that force is the instrument for increasing uniformity in nature, or what is equivalent, the instrument for effecting reduction of global non-uniformity existing in the space-time manifold, (b) that all forces in nature emerge from these global non-uniformities, and constantly act to reduce them, and (c) that forces are the universal manifestations of non-uniformities in nature as they are directly posited to sensation.

Evolutionary adaptation is envisaged here as a universal aspect of all process in nature, an aspect which reconciles constancy and change in all of their ramifica-tions in natural phenomena. The thrust of evolutionary adaptation, so conceived, derives from the ultimate processes, embedded in the space-time manifold, which

297

drive and construct the manifold by irreversible connections which must necessarily exist between these ultimate processes and the manifold. The irreversible connections are implied by the immutability of these ultimate processes, called here the universals, as they are reflected in, and revealed by, the Dimensional Universal Constants of Nature with which they are here identified.

The universal adaptive process described above has been conceptually identified with, and emerged from, a conceptual model of nature's space-time manifold which is endowed with certain essentially ontological features inferred from the Dimensional Universal Constants [6]. These (essentially) ontological characteristics were independently discerned in concurrent studies that initiated in 1947, which is based on C. F. Gauss's [7] and Heinrich Hertz's [8] formulations, of the principles of classical mechanics which in each case was motivated by a quest to grasp the nature of force by attempting to establish force on a strictly geometrical foundation. This endeavour was initiated by Gauss in 1829 and culminated at the turn of the century by Hertz's last and monumental work entitled 'Principles of Mechanics'. In this profound and beautiful work Hertz formally constructs a $6N$ dimensional Euclidean Manifold in which the motion and state of a classical mechanical system consisting of N bodies free of prescribed forces are described and represented. Hertz restricts the admissible motions and states of the mechanical system by formally subjecting the coordinates of its bodies to constraints which in the most general case are considered as non-integrable and therefore non-holonomic. As the application of these geometrical constraints to a body restricts its freedom geometrically, these restrictions must emerge in the Newtonian scheme as forces.

The study based on the Gauss-Hertz formulations* has produced two results which bear on the conception of the Principle of Maximum Uniformity and on the identification of a physical, that is, of an ontological-geometrical basis for the production of actual stringent holonomic as well as non-holonomic constraints in nature's space-time manifold. This ontological-geometrical basis gives physical support and justification for the existence in nature of the geometrical restrictions which Hertz used to effect a formal reduction of force to geometry, and serves to identify the formal representations he gave to non-holonomic constraints, with experience and thus with nature.

The same study revealed that Hertz's edifice, in which he formulated a general law governing the motion of forceless mechanical systems subjected to non-holonomic geometrical constraints, and which renders obsolete all previously

* Some results of this study are presented in [2].

known formulations of the laws of classical mechanics, also accommodates the formulation of a new and general stability law cited above. This stability law, which bears the same kind of relation to stability as do the established laws of classical mechanics to equilibrium, is found to be independent of the known laws of mechanics and to embrace fundamental and general information not included in these laws. This information bears on historical thrust and commitment and derives from an adaptive-evolutionary process ascribed directly to the geometrical restrictions which impress non-uniformities on the space-time manifold, and from which all forces are understood here to emerge. This entails the identification and classification of holonomic and non-holonomic ontological-geometrical constraints into the following types : (1) Active Stringent Constraints, (2) Passive Stringent Constraints, (3) Conditionally Stringent Passive Constraints. This classification led naturally to the idea that the annihilation of conditionally stringent passive constraints which are ascribed here to universal congruence restrictions impressed on the space-time manifold by the irreducible universals identified by the Dimensional Universal Constants constitutes a fundamental and general instrument of adaptation in the space-time manifold. It is this crucial instrument which allows one to posit a general stability law for classical mechanical systems and which affords according to the Principle of Maximum Uniformity the mechanism which is essential for physically producing *the required many to all mappings evident* in biological systems.

The annihilation of conditionally stringent constraints is accompanied by consequent modifications of the forces emanating from the non-uniformities induced by them in nature's space-time structure. According to the observations and reasoning of this paper, the annihilation of conditionally stringent constraints is envisaged as an essential feature and instrument of adaptation, conceived here as a general and universal aspect of all process in nature seated in the space-time manifold. This universal process of adaptation in nature, and consequently the process of annihilation of conditionally stringent constraints upon which it incisively depends follow by the thesis of this paper the Principle of Maximum Uniformity.

1.2. Concerning aspects of uniformity and non-uniformity

Some significant aspects of uniformity and non-uniformity revealed in human experience are cited in this section. This is done in order to point up their universal role in natural phenomena, and as a consequence the strong implications of the principle of maximum uniformity noted here and which will be further examined in depth in subsequent papers under the present title. These aspects of uniformity include :

Evolution and maximum uniformity

1. Symmetry
2. Equilibrium : local, global, spatial, and temporal
3. Stability : local, global, spatial, and temporal
4. Isotropy : a local aspect
5. Homogeneity : a global aspect
6. Constancy
7. Invariance
8. Covariance
9. Law
10. Correspondence
11. Element
12. Order
13. Reproducibility : the ultimate criterion and requirements of scientific investigation

The Corresponding Aspects of Non-Uniformity include :
1. Force : the most fundamental and universal aspect of non-uniformity posited to sense perception and sense awareness
2. Asymmetry
3. Information
4. Curvature
5. Symbol
6. Language
7. Anisotropy : a local aspect
8. Inhomogeneity : a global aspect
9. Gradient
10. Structure
11. Shear
12. Constraints
13. Uncertainty
14. Fluctuations
15. Disorder

To each set of conditionally stringent constraints there corresponds a positive definite scalar measure of non-uniformity as manifested in experience by the internal forces. The relaxation of such constraints increases uniformity, and the selection among a possible set of conditionally stringent constraints is made to maximize global uniformity in adherence with the Principle of Maximum Uniformity amplified in the following section 2. The universal constants embrace constancy and process and thus both uniformity and non-uniformity. This is the synthesis in the elementary processes which they reveal.

The process of reducing the non-uniformities in nature's space-time manifold is here envisaged to be the ultimate aspect of all adaptive phenomena in nature. Evolution becomes then a word labelling this universal adaptive process. An aspect of evolution that is both essential and universal is *force,* and its nature we evidently do not grasp more in physics than in biology.

1.3. Concerning the nature of non-holonomic constraints

Non-holonomic geometrical constraints are constraints that emerge physically from contact. These constraints are therefore directly tied to the printing process, and consequently, necessarily to local impenetrability, conceived here as the ontological-geometrical basis and support for all printing and sensation in nature. Indeed, non-holonomic constraints are significant aspects of local impenetrability. The question concerning how non-holonomic constraints are made in nature is necessarily linked to the fundamental question which emerged from the study of Hertz's Mechanics, cited above, namely : What is the ontological-geometrical basis in nature of the stringent geometrical constraints formally invoked by Hertz to geometrize force ? These considerations, which bear on Pattee's [9] interesting discussion of the role of non-holonomic constraints in relating the genotypes and phenotypes of organisms, will be developed in a separate essay.

The process of randomization in gas flows has been elucidated in terms of a dissipation mechanism [3]. This mechanism is based on the propositions of classical mechanics and unilateral non-holonomic constraints, which come directly from the ontological-geometrical property of matter described here as local impenetrability. This dissipation mechanism bridges conceptually unilateral non-holonomic constraints, oblique collisions, asymmetries in momentum-configuration space and randomization, with a process of intrinsic irreversibility which emerges from them and the laws of classical mechanics. The evolution of shock waves in gas flows and their associated zones of high temperature and dissipation have been explained by this dissipation mechanism. As randomization is under certain conditions an aspect of uniformization, we are able to establish a correspondence between the Principle of Maximum Uniformity and this mechanism of randomization.

2. Concerning the nature of force, equilibrium and stability

Among all the symbols which appear in the formal statements of the propositions

301

of classical mechanics, the symbol F, used to abbreviate the *name* Force, desginates what is recognized here as the universal and the most fundamental aspect of nature posited to experience within the domain of classical mechanics. Although, as previously noted, Newton did not explicitly associate with this symbol the universal aspects of nature which are revealed in physics by the Dimensional Universal Constants, his instinct for its dominant role in the propositions of classical mechanics is nevertheless suggested by the constancy he implicitly ascribed to it, under transformations connecting non-inertial as well as inertial coordinate systems. Since the symbol F is simply a designator, that is, a name, and not a representation of a well-defined aspect of nature relating to the space-time manifold, it is in fact inconceivable how Newton could have, in principle, formally prescribed to it geometrical conditions which refer to coordinate systems and to their transformation. To do so he would first have to describe geometrically the architecture in the space-time manifold, from which *all* forces derive. I emphasize *all* forces because this is the meaning that we must with logical necessity ascribe to the symbol F, since it must necessarily designate the sum of all connections which exist between a non-free body and the universe in which it is contained. This way of thinking is amplified below, and is sufficient to draw the inference, on strictly logical grounds, that the aspect of nature designated by the symbol F in the propositions of classical mechanics must necessarily be endowed with information content that refers to the Universals of Nature and which is therefore directly related to Dimensional Universal Constants of Nature. Although this logic, which by itself is sufficient to infer that the symbol F necessarily labels universal aspects of nature, is clearly not included in the familiar body of knowledge reported under the aegis of classical mechanics, it nevertheless evolved from a critique of the concepts which led to the emergence of this symbol in the history of classical mechanics.

In this regard it is important to emphasize that the unrestricted covariance Newton tacitly ascribed to what the symbol F designates in the space-time manifold allows the interpretation I have given to it here and therefore the universal information content which is here ascribed to it. In other words, Newton's intuition led him to a formulation which allows us to assign to the symbol F appearing in his propositions, the role of designating the connections between the most fundamental and universal aspects of nature and all experience manifested within the domain of classical mechanics. This inference, which reports that in classical mechanics the symbol F necessarily designates universal aspects of nature, was derived by implication from a critical examination of the concepts which led to its emergence in the development of the known laws of classical mechanics.

Historically, according to Aristotle, the motion of a body calls for the constant application of an externally applied impetus to sustain it. This deeply entrenched view was subsequently denied by Galileo on experimental grounds which led him to the conception of a free body in motion, that is, a body which can sustain a uniform velocity indefinitely, free of an externally applied agent. Newton, as did Galileo before him, conceived of inertia as an innate property of a body, and consequently as a local property of its location in space. Mach, however, conceived of the inertia of a free body as the manifestation of its global connection with the whole universe in which it is embedded. According to Mach's Principle, a so-called free body is nevertheless joined to the universe, but in a very special way, as revealed by a scalar connection manifested by its inertia. The connection between a free body and the universe is not necessarily strictly isotropic, that is, completely and thus maximumly uniform in all directions. The particular and indeed very special case of a maximumly uniform connection may be envisaged as equivalent to a strictly isotropic force field. Although for this very special case the resultant of all forces connecting a free body with the universe vanishes, this condition may be satisfied under much less stringent conditions of uniformity ; all of which equivalently satisfy the definition of a free body as presented by the established laws of classical mechanics. This observation led me naturally to the concept of 'Hierarchies of Freedom' which correspond to differences in the degree of uniformity or order of the connections existing as forces between bodies and the universe, all of which equivalently meet the condition of a free body as required by the laws of mechanics.

By Newton's view of inertia, a free body must necessarily be understood as disjoined, that is, as insulated from the universe in which it exists. This on reflection violates intuition, unless the body is envisaged as existing in a strict vacuum, that is, in a strictly empty space which insulates the body from all matter and radiation in the universe. With this view in mind, Newton extended Galileo's principle by in effect posing the question : what is the case when the body is not free ; that is, not strictly insulated ? In so doing he vectorially *designated* the sum of all the individual connections that exist between a body and the whole universe by the symbol $\vec{F}$, and evidently understood these individual connections as emanating from sites strictly external to the body-sites which are individually manifested by individual agents called forces, externally applied and which follow the law of linear superposition.

303

Evolution and maximum uniformity

2.1. The principle of universal correspondence

The symbol F is strictly a name, that is, a label which is devoid of representational content of the space-time architecture and anatomy from which the connections named forces emerge in the space-time manifold. In classical mechanics we must therefore necessarily interpret F as a dummy label designating the resultant of all connections existing between a body and the whole universe. Accordingly we must associate with the symbol F a universal and completely global aspect of nature, endowed with the highest information content* manifested within the domain of classical mechanical experience—a content which it however does not explicate in the sense of a representation.

We shall see that the interpretation given here to the symbol F, which is indeed compatible with its usage in Newton's formally stated propositions of classical mechanics, allows the accommodation and satisfaction within the domain of classical mechanics, of a fundamental principle which was conceived in a study based upon the Universal Constants of Nature [6]. This principle stems directly from a philosophical axiom which, if denied, leads to a profound philosophical catastrophe, that is, to a conclusion which is conceptually absurd, and which denies common sense as well as philosophical intuition. This axiom asserts that the universal and elementary aspects of nature are prominent and indeed imminent in *every* natural phenomenon, every process, and every experience, and in particular therefore in experiences which transpire within the domain of classical mechanics. In reference [1] the universal and irreducible aspects of nature are identified with the universal constants of nature which are revealed in the domain of experience, conventionally called physics, by the Dimensional Universal Constants. The Dimensional Universal Constants of Physics are accordingly conceived as being manifestations in the realm of what we by convention call physical reality, of the truly elementary processes in nature which universally and immutably transpire in the space-time manifold. This realization is, I believe, tantamount to a partial fulfilment of man's search and sustained quest for the identification of the elements of nature, which of course derives from his faith and conviction that they indeed exist. The axiom cited may be paraphrased to read that what is universal and fundamental in nature can in no sense be incidental, either phenomenologically or in the laws that condition the phenomena incurred in various realms of scientific experience. To claim the contrary is considered here absurd and to deny common sense and philosophical intuition.

---

*The notion of information content is further considered in the section 3, in which 'Hierarchies of Information' was conceived as aspects of 'Hierarchies of Uniformity'.

Paul Lieber

The extension of experience immediately accessible and manageable by the human hand to domains that are remote from everyday experience and that require particular technological skills to engage has revealed natural phenomena which cannot be accommodated by the laws discovered earlier in physical phenomena that were and remain immediate in everyday experience. The theoretical accommodation of these newer phenomena demanded the formulation of new laws in which certain Dimensional Universal Constants first emerged in physical theory, as for example in the case of the Special Theory of Relativity and Quantum Mechanics.

To avert a philosophical catastrophe in the progressive development of physical theory, Bohr invoked 'The Correspondence Principle'. This principle demands that the formal statement of the laws conceived for the neo-classical and modern domains of physical experience correspond asymptotically to the laws conceived earlier for the classical domain in such cases where the phenomena being experimentally examined belong strictly to the domain of everyday experience. Bohr's Correspondence Principle thus leads to the result that the terms which in the neo-classical and modern physical theories include the Dimensional Universal Constants become asymptotically small, and thus negligible, in comparison with the terms which are retained in going over from the neo-classical to the classical domains of experience. Although Bohr's Correspondence Principle averts in this way a theoretical catastrophe, it does not however address itself to what might be a much more serious theoretical catastrophe, namely the violation of what is here called the 'Principle of Universal Correspondence'.

This principle, which stems directly from the above cited axiom and which asserts that the universal and irreducible aspects of nature are *imminent* in *all* natural phenomena, demands that in this specific sense *all* natural phenomena and their laws bear to each other universal correspondence. By this principle every aspect of nature is endowed with the universals, reflected in physics by the Dimensional Universal Constants, and are understood to be controlled by a universal law of evolution pertaining to the global action of these universals upon the space-time manifold in which they are conceived to be embedded, and which they internally drive, structure, and organize by a universal process of evolution. Every aspect of nature, including therefore such phenomena which by convention we refer to as classical, bears in this sense universal and complete correspondence to every other phenomenological aspect of nature, be it classical or non-classical according to the usual definitions.

By this thinking every natural phenomenon, classical as well as non-classical, is in effect a micro-cosmos of the universe, in the sense that they are all

305

Evolution and maximum uniformity

pre-eminently endowed with the universals which irreversibly drive, structure, and organize them in the space-time manifold, according to a universal evolutionary law, which operates in every phenomenological domain. Accordingly, all natural phenomena bear to each other, and to the universe as a whole, complete phenomenological correspondence, as do their laws which have been progressively forged out of experience in both the classical and non-classical domains.

The Principle of Universal Correspondence immediately presents a dilemma when it is invoked conceptually and formally in the domain of classical mechanics. From the above considerations, the symbols which label the Dimensional Universal Constants in the laws of physics are understood as designators, that is, as simply names with physical dimensions, that reflect the presence in the space-time manifold of immutable, and consequently ultimate processes called here universals. These symbols are accordingly the formal designators of these universals, within the formal statements of known physical laws.

According to Bohr's Correspondence Principle, which for obvious reasons I shall refer to as the principle of asymptotic correspondence, the terms in the formal statements of physical laws which include the symbols labelling the Dimensional Universal Constants are asymptotically eliminated when these laws are put into correspondence with the domain of classical mechanics. In Relativistic Mechanics, for example, the terms including the Universal Constant C, the speed of light, modify only the dynamical part of Newton's classical laws and do not and indeed cannot, for reasons given earlier, modify the unrestricted virtual covariance *a priori* ascribed to the symbol F for all coordinate transformations.

The present dilemma concerns reconciling the Principle of Universal Correspondence with the elimination within the domain of classical mechanics, of the symbols which label the universals in the physical laws pertaining to the non-classical domains. The obvious question to answer is: wherein are the universals manifested in the domain of classical mechanics, as demanded by the Principle of Universal Correspondence? The answer is Force, that is, in the aspects of nature which are formally designated by the symbol F in the propositions of classical mechanics. This is the symbol to which we earlier and inexorably ascribed on broad but incisive conceptual grounds an information content which dominates the information content designated by all the other symbols appearing in the formal proposition of classical mechanics. Now, however, we see from the Principle of Universal Correspondence that the universal aspects of the designator F and its consequent dominant position in the formal statement of the laws of classical mechanics can and must be attributed to its rôle as a designator within

306

the domain of classical mechanics, of the universals, as they are directly posited to experience within this domain. By this thinking, all forces in nature are endowed with the universals, and are manifestations in direct experience of a universal law of evolution which refers to the action of the universals on the space-time manifold to which I ascribe everywhere the ontological-geometrical-temporal property I call local impenetrability. Moreover, it follows from these considerations that a long-standing and prevalent view among mechanicians and scientists generally, which holds that Newton's laws define force, is untenable and indeed absurd. What these laws do is simply state certain connections between force and parameters that describe the behaviour of classical mechanical systems according to some preconceived ideas which are now in fact challenged by more recent developments in modern physics.

The known laws of classical mechanics do not define force, but instead give only limited expression to its information content, much of which in fact remains untapped by them. As a consequence, forces as they are directly posited to experience in the domain of classical mechanics accommodate a statement of a new fundamental and general law governing the performance of classical mechanical systems, which reports fundamental and general information which is not contained in the long-established laws of classical mechanics. The statement of this new law is made possible by the inherently high information content of force which derives from the universal aspect of all forces in nature, and from the fact that this information is only partially explicated by the known laws of classical mechanics.

3. Concerning aspects of hierarchies of uniformity
As shown earlier in this manuscript, we can interpret the resultant force posited to a non-free body as the vector sum of all non-uniform connections which exist between the body and universe. Each force individually contributing to this sum posits to the body a non-uniform aspect of the universe. In cases when the vector sum of these individually applied forces vanishes, we previously considered the body as free but not disjoined from the universe. In these cases the individual forces may be envisaged as existing in mirror symmetric pairs, the forces in each pair being consequently equal in magnitude. However, according to the usual laws of classical mechanics, the definition of a free body does not demand that the magnitudes of the individually applied forces be uniform for all pairs.

From these considerations we learn that there exist hierarchies of free bodies, all of which are equivalent according to the known laws of classical mechanics and

Evolution and maximum uniformity

which are therefore not discernible or identifiable by these laws. The hierarchies
of free bodies may be identified and designated by either the degree of uniformity
or non-uniformity of the magnitudes of the individual forces that are the immediate
manifestations immanent in experience of particular aspects of non-uniformity
existing between a body and the universe. Since all free bodies which belong to
these various hierarchies (of freedom) are equivalent according to the presently
established laws of classical mechanics, these laws cannot in principle render
conditions which select among the many actual-possibilities these hierarchies
afford at each instant a particular one that belongs to a particular hierarchy of
freedom. The concept 'Hierarchies of Freedom' is a particular aspect of the concept
'Hierarchies of Uniformity'.

It is helpful to point out some other equivalent aspects of this concept, as it
assumes a crucial role in the statement of a general principle of evolution which is
in accord with the Principle of Universal Correspondence, and which is consequently understood to operate universally in all natural phenomena, including
those which belong to the domain of classical mechanics. Some equivalent and
related aspects of the concept 'Hierarchies of Uniformity' include 'Hierarchies of
Symmetry', 'Hierarchies of Certainty', 'Hierarchies of Order', 'Hierarchies of
Information', Hierarchies of Compatibility', 'Hierarchies of Harmony', 'Hierarchies
of Forces', and 'Hierarchies of Consistency'. Moreover, in all of these cases it is
important to distinguish between what in each case corresponds to the local
aspects of uniformity and to its global spatial-temporal aspects. It is clear that the
established propositions of classical mechanics do not and cannot make such a
distinction, because the restrictions they impose on mechanical systems apply
instantaneously and locally, everywhere as well as for all time. As the conditions
they invoke, namely that forces be instantaneously in equilibrium everywhere and
always, are uniform in space and time, they do not implicitly describe or define,
nor do they condition the existence and the spatial-temporal evolution of local
and global non-uniformity in their various hierarchies. This is the reason they are
inherently devoid of historical thrust, causality, and evolutionary process.

It is the universality of all forces in nature, and therefore in particular of those
forces which in the classical domain are designated by the symbol F, that facilitates
invoking and applying the principle of evolution cited above, in the domain of
classical mechanics. The established laws of classical mechanics, in all of their
equivalent formulations, express a particular and restricted aspect of the Principle
of Maximum Uniformity, an aspect which, as explained earlier, is independent of
location and independent of time. These laws consequently express universal

propositions, that is, truths which are necessary in the strict logical sense, and are therefore not contingent upon space and time. These laws are in the sense of Liebniz logically universal, that is, necessary and analytic, as they are not contingent upon conditions that refer to particular places and to particular times. It is important to emphasize in this regard that these laws refer to a particular and restricted aspect of uniformity which is characterized and defined by the equilibrium of forces, and they assert that this particular aspect of Hierarchy of Uniformity is constantly maintained at all locations and is therefore not contingent upon space or time. In other words, the laws of classical mechanics as well as the particular Hierarchy of Uniformity to which they refer, namely the hierarchy characterized by the Equilibrium of Forces, and which as laws they report to be a general aspect of nature, are *both constant* in space and time, and are thus both free of contingency. If we follow this way of thinking, the usual laws of classical mechanics may be conceived as developing in two steps. The first consists of a definition of equilibrium, in which force is the aspect of nature to which the word equilibrium in the definition refers. The second uses this definition to express the universal law which asserts that equilibrium so defined is constantly maintained in nature, that is, everywhere and at all times. In other words, the particular Hierarchy of Uniformity which is characterized by the equilibrium of forces, as well as the laws of classical mechanics which ascribe this aspect of uniformity to all nature, are both constant with space and time, and thus free of contingency.

The existence in nature of Hierarchies of Uniformity, which as in the particular case of equilibrium are all directly revealed in experience by forces, leads here naturally to the identification of a universal law, which although free of contingencies in assertion, nevertheless conditions aspects of nature which are in fact contingent upon the evolution in space and time of distinct Hierarchies of Uniformity. The law does not in this case constantly refer to a particular hierarchy, but reports a universal proposition that governs a process of evolution which is contingent upon the emergence in space and time of the various Hierarchies of Uniformity. The usual laws of mechanics which are indeed embraced by this general law are a very special case of it, in so far as the particular Hierarchy of Uniformity in terms of which they are expressed is always *fixed* and therefore not contingent. This is precisely the reason why the established laws of mechanics are inherently and completely devoid of contingency in all aspects and consequently of historical thrust, causality, stability criteria, and evolution. This is, of course, also true for all of the so-called equivalent formulations of the laws of classical mechanics, and in particular therefore for their formulation in terms of the principle of least action. The reason I refer here particularly to the principle of least action is its power, and unifying role in physical theory. The power of this principle in the formulation given to it by Hamilton is seen by the fact that not only classical mechanics of particles and rigid bodies, but also elasticity and hydrodynamics, electromagnetism, and all modern field theories connected with ultimate particles (electron, proton, neutron) can be formulated with its help. All of the theories which are formulated with its help, therefore, share with Newton's formulation of the laws of classical mechanics the important feature of being devoid of historical commitment, causality, and inherent stability criteria. In other words, all of these theories are free of historical content, and consequently essentially devoid of an evolutionary principle.

4. Concerning the principle of maximum uniformity and a general stability law
We have shown earlier that the formulation of the laws of classical mechanics may be conceived in two essentially distinct steps, the first being a definition of equilibrium, whereas in the second the proposition is made that equilibrium as defined by the first step holds constantly everywhere and for all time. The notions of stability and equilibrium both developed by observing and critically examining the phenomenological behaviour of classical mechanical systems. As explained in the case of equilibrium, a general operational definition based on forces was established on the basis of experience, and then used in the formulation of the known laws of mechanics, which inherently report nothing about stability for reasons already gone into. Whereas the notion of stability has been described by many definitions, these have led to various stability criteria which are statements of convention rather than a general law—to a law that refers to stability as do the laws of mechanics to equilibrium. I shall now endeavour to formulate a statement of such a law, that is, of a general stability law, which refers to all of the Hierarchies of Uniformity and bears to them the same kind of relation as do the known laws of classical mechanics, to the particular hierarchy of uniformity characterized by the equilibrium of forces. For this purpose it is first necessary to identify and descriptively define the Hierarchies of Uniformity in terms of forces, which as explained above are interpreted here as the most fundamental, universal, and direct manifestation in experience of the non-uniform connections existing between the universe and bodies contained in it.

   We start by considering in some detail the very special and fundamental Hierarchy of Uniformity to which the known laws of classical mechanics pertain. This special hierarchy is defined by the characteristics that the vector addition of all the non-uniform connections existing between a body and the universe which are posited in experience, and which we designate by the name force, sums to zero. It is clear that there can exist a conceivably infinite number of distinct configurations of forces impressed on a material point which individually designate the individual non-uniform connections between it and the universe, all of which equally belong to the very special hierarchy of force equilibrium. It is the *differences* between these distinct but otherwise equivalent force configurations which I define as the 'Hierarchies of Uniformity'. The figure below illustrates pictorially how we can conceive of an infinite number of distinct force configurations, all of which belong to the hierarchy of uniformity defined by the equilibrium of forces and which by their differences here define the 'Hierarchies of Uniformity'.

The 'Hierarchies of Uniformity', so descriptively defined in terms of force fields, are now used to formulate a Principle of Maximum Uniformity which virtually includes the known laws of classical mechanics as well as a general stability law. This stability law bears the same relation to stability which will be defined here in a particular way, as do the established laws of classical mechanics to the equilibrium of forces as it is defined in classical mechanics.

The Principle of Maximum Uniformity asserts that among all the force configurations which can be collectively and instantaneously accommodated in a finitely extended material domain, non-uniformly connected to the universe by maintained forces, and which individually belong to the special hierarchy of uniformity characterized by force equilibrium; the particular set of force configurations which actually evolve and which satisfy the instantaneous and stringently exercised geometrical constraints, instantaneously maximizes a global positive definite scalar measure of uniformity obtained by summing local measures of uniformity that depend on the local force configurations over the entire domain.

This statement of the Principle of Maximum Uniformity differs essentially from the statements of the established laws of classical mechanics. Whereas, as explained above, the laws of classical mechanics are essentially atemporal,

311

acausal, and consequently devoid of historical commitment and evolutionary process ; the Principle of Maximum Uniformity, though conceived here as a universal proposition, nevertheless refers to essentially contingent aspects of nature expressed in terms of 'Hierarchies of Uniformity' which in general evolve non-uniformly in space-time. It is precisely because the universal and established laws of classical mechanics constantly refer to one, and only one, hierarchy of uniformity that they are free of contingency in all respects, and are consequently in principle amenable to mathematical formulation. This follows, as all mathematically stateable propositions are essentially free of contingencies which refer to space-time and therefore in principle devoid of historical content.

The Principle of Maximum Uniformity is indeed a *procedure* rather than a formally stateable proposition—it is the description of a process, of a universal process, that is, a process which is understood to operate universally. In this process the existence and operation in the space-time manifold of contingently stringent geometrical constraints, as well as absolutely stringent passive constraints, are among its essential features. Because (1) time, conceived as duration rather than the times of events ordered as points on the real time line, (2) the ontological-geometrical ground for passive-stringent geometrical constraints, is ascribed here to local impenetrability of matter, (3) force is the essential instrument in nature for effecting compatibility and excluding contradiction in nature by reconciling its universal and contingent aspects, and (4) the temporal and spatial contingencies expressed by the spatial-temporal evolution of various and distinct hierarchies of uniformity ; are all involved in the operation in nature of the Principle of Maximum Uniformity, it follows that its description and statement cannot in principle be completely mathematically formulated.

This conclusion has direct bearing on the questions that concern the nature of biological theory and the kind of laws we can expect it to produce. It of course also bears on the nature of physical theory and the fundamental implications inherent in the formal statements of its laws. It is precisely because of the fact that they can be given mathematical expression that they are in principle devoid of all contingency, and consequently of historical content and thrust, inherent stability criteria, causality, and evolutionary process. Conversely, it is because the laws of physics are essentially ahistorical and acausal that they can be given mathematical formulation.

The second law of thermodynamics is indeed unique among the laws of physics. Whereas the other laws of physics do not know about ageing, and therefore about history, the second law does consider and compare earlier and later states of

312

systems, but not how they evolve from the earlier to the later states.

We can sum up by saying that the physical laws as they are known are space-time invariant and thus not contingent, and that the aspects of nature to which they refer are devoid of ageing process. Laws of nature may however be space-time invariant and still refer to fundamental aspects of nature which are nevertheless contingent, and which therefore essentially include historical and evolutionary aspects. The Principle of Maximum Uniformity appears to be such a law, and laws which we may expect to emerge in biological theory will be essentially of this character. The Principle of Maximum Uniformity will be considered in a larger context and in much more detail from the biological side in a volume concerned with the constants of nature and biological theory, categories of information, and aspects of evolution, and in which it will assume a unifying role.

Stability according to the present definition is a characteristic of the instantaneous state of a system, just as is equilibrium ; moreover, the stability so defined has both local and global aspects, which again correspond to the case of equilibrium. The instantaneously stable state is defined as the force configurations belonging to the highest Hierarchy of Uniformity which instantaneously satisfies all the conditions cited above in the statement of the Principle of Maximum Uniformity. According to this definition, instantaneous *global* stability is defined as the collection of instantaneous locally stable force configurations. The definitions given here to 'Hierarchies of Uniformity' and to stability are descriptive, pictorial, and conceptual, and not analytic or quantitative in a mathematical sense. For this purpose it is natural to consider continuously extended material domains, in which the forces joining an element to the universe are characterized by a stress tensor. The Principle of Maximum Uniformity and the general stability law that derives from it will be in part formulated in more analytical (terminology) language in another volume, in which it is planned to treat this subject in a more comprehensive manner, and particularly its biological ramifications.

The Principle of Maximum Uniformity is manifested in the domain of classical mechanics, as required by the Principle of Universal Correspondence, by the evolution in time at different locations of various and distinct *force* configurations. Each of these force configurations belongs to the Hierarchies of Uniformity, and have in common a particular member of the hierarchy, which is defined here by the equilibrium of forces. The progressive evolution in time of the Hierarchies of Uniformity is revealed in all experience and therefore in the classical domain, in particular, by the progressive evolution of different force configurations, each of which may also be interpreted as a hierarchy or order. As noted earlier, all forces

Evolution and maximum uniformity

are understood here to give direct expression in experience to the universals, which are reflected by the Dimensional Universal Constants and consequently to what is referred to in reference [6] as the domain of the universals. By this way of thinking, the operation in nature of the 'Principle of Maximum Uniformity', and the conception of its operation, demands the existence, and the consideration of the relation between, and interaction of, the domain of the universals and what I call in reference [6] the domain of the observables. This of course, applies equally to the operation in nature of the universal stability law manifested in every domain of experience, and which derives, as do the conventional laws of mechanics, from the Principle of Maximum Uniformity.

The Principle of Maximum Uniformity and the Universal Stability Law attendant upon it have been made operational, within the realm of classical mechanics, that is, exercised computationally in this realm by the development of an algorithm, by modelling certain aspects of the domain of universals by a potential theory. This model allows the formal description of the interaction between viscous flow fields which belong to the domain of the observables, and an ideal domain characterized by the potential theory from which, according to the algorithm, they emerge by what is analogous to a process of evolution. This has produced mathematical representations of viscous flow fields that evidently satisfy the fundamental partial differential equations of classical hydrodynamics and realistic boundary conditions.

The interaction between the domain of the universals and the observable domain brings necessarily under consideration multiple scales and the realization that they assume an essential role, especially their interrelationship, in the interaction between these domains. From the standpoint of classical mechanics, for example, such scales may be identified with temperature fluctuations in a heat bath which are related to the Universal Boltzmann Constant, and the production of inelastic deformations in a solid subjected to forces impressed by the universe from the outside. These points and considerations, as well as the relationships between the Principle of Maximum Uniformity, the stability law, the role of the Constants of Nature as the foundation of natural law and the development of biological theory, and the connection between these, and the existence in nature of Categories and Hierarchies of Information, all will be comprehensively examined and in concert in a volume more specifically directed at their ultimate biological aspects.

The considerations outlined above point to an ordering and designing principle in the universe, a universal evolutionary principle which acts constantly and

immediately in accordance with the universe and the universals as they are immediately and everywhere presented in experience and in specific events. I now see
the questions that have led in some cases to a search for teleological explanation
as essentially clarified and possibly resolved. It is resolved by seeing in context
that purpose and design in nature as well as history, and its thrust may be comprehended by considering the immediate relations between the universals and their
constant operation in and upon the universe at each instant, and the fact that this
is sufficient to give purpose, design, and direction toward a constant goal which
may indeed conceptually be projected into the future. All action in the present
includes goals directed at the future, in so far as they all inherently and necessarily
include thrust toward the future.

5. Concluding remarks

A feature that strikingly and significantly distinguishes physical theory from
biological theory is that its laws which have emerged from a description and
critical examination of modes of experience which we by convention call physical
are formulated in a very special language known as Mathematics. This language,
its rules, its operations, and the primitive notions and abstract objects to which
they refer are all essentially free of contingency and are consequently in principle
atemporal, acausal, and ahistorical. It is a language which therefore cannot in
principle describe or report on history, on evolution as a process, on selection, on
identification, on force, on sensation, and on the nature of information, all conceived here as fundamental manifestations in experience of this process. All the
laws of physics which have been given complete expression in this particular
language are in fact devoid of historical content, causality, and evolution. There
is nothing they report which distinguishes the past from the future or which refer
to or condition the universal phenomena of birth, ageing, and death. Conversely,
from the fact that the language called Mathematics is essentially atemporal, it
follows immediately that a law formulated as a proposition in this language cannot
in principle refer to or condition historical development and evolution.

A prevalent attitude among scientists and laymen as well holds that the rank of a
scientific law, that is, its importance, depends upon whether or not it can be
formulated in the language of Mathematics. The above considerations should
dispel this misconception by pointing out some of the essential limitations of this
language from the standpoint of describing, embracing, and conditioning certain
fundamental aspects of human experience and scientific experience in particular.
In my view, the most fundamental scientific law is a universal law which pertains

to and conditions a universal evolutionary process and which therefore cannot be fully embraced by and be totally expressed in the mathematical language. This paper is in part committed to the identification of this law.

An equally significant and striking fact which concerns biological theory is that the ideas and laws which have emerged from observations and thinking based on the critical examination of biological materials have hardly been expressed, if at all, in the language which we call Mathematics. A careful examination of the ideas and laws which make up biological theory as we know it today reveals a profundity, an imagination, and a creative power no less penetrating than what has emerged in physical theory. This means, I believe, that what emerges as characteristic and prominent in a critical examination of biological materials and phenomena cannot be appropriately described, expressed, and communicated in the mathematical language. The reason is now becoming clear: it is simply that what is characteristic and eminent in biological phenomena is essentially the developmental and evolutionary aspects of nature, aspects which I have endeavoured to show here are in principle extra-mathematical, and that consequently the laws pertaining to them cannot in principle be given full expression in the language called Mathematics.

It is planned to amplify and extend the ideas and conclusions outlined here in subsequent essays to be presented under the same title in a separate volume. Direct contact will then be made with descriptions, experiments, and theories which have come from work with biological materials. Particular attention will be given to Professor C. H. Waddington's pioneering and experimentally revealing work concerned with the genetic assimilation of acquired characteristics [10–14] and the penetrating question he raised, namely: 'How does evolutionary adaptation *work*?' [15]. By so doing it is hoped to be able to present the design of a crucial experiment using biological materials, by which hypotheses and questions derived from the ideas presented here may be incisively tested.

References

1. Lieber, Paul. 'A Principle of Maximum Uniformity obtained as a theorem on the Distribution of Internal Forces.' Institute of Engineering Research. University of California, Berkeley, Nonr-222 (87), No. MD-63-8, April 1963.

2. Lieber, Paul and Farmer, Arthur. 'Studies on Wave Propagation in Granular Media.' *Trans. American Geophysical Union*, Vol. 39, No. 2, April 1958.

Paul Lieber

3. Lieber, Paul. 'The Mechanical Evolution of Clusters of Binary Elastic Collisions and Conception of a Crucial Experiment on Turbulence.' Volume of the *Proceedings of the Symposium on Second-Order Effects in Elasticity, Plasticity, and Fluid Dynamics,* sponsored by the International Union of Theoretical and Applied Mechanics, April 1962.

4. Lieber, Paul and Wan, K. 'A Minimum Dissipation Principle for Real Fluids', *Proc. IX Int. Congress of Mechanics,* 1957.

5. Lieber, Paul. 'Categories of Information.' Office of Research Services, University of California, Berkeley. Nonr-222 (87) AM-65-13, October 1965.

6. Lieber, Paul. 'Constants of Nature ; Biological Theory and Natural Law', in (C. H. Waddington, ed.) *Towards a Theoretical Biology, I* (University of Edinburgh Press, 1968).

7. Gauss, Carl F. 'On a New General Fundamental Principle of Mechanics', *Creele's Journal f. Math, 4* (1829) 232. Also appears in *Werke, 5,* 23.

8. Hertz, Heinrich. 'Collected Works, Vol. II', 'Principles of Mechanics' (Macmillan, New York, 1896).

9. Pattee, H. H. 'The Physical Basis of Coding and Reliability in Biological Evolution', in (C. H. Waddington, ed.) *Towards a Theoretical Biology, I* (University of Edinburgh Press, 1968).

10. Waddington, C. H. *Nature, 150* (1942) 563.

11. Waddington, C. H. *Evolution, 7* (1953) 118.

12. Waddington, C. H. *Evolution, 10* (1956)1.

13. Waddington, C. H. *J. Genet., 55* (1957) 241.

14. Waddington, C. H. *Nature, 183* (1959) 1654.

15. Waddington, C. H. 'Evolutionary Adaptation', *Evolution after Darwin,* Vol. I (University of Chicago Press, 1960).

Two poems by Mary Reynolds

Conferenza di Bellagio

('I'm nobody. Who are you ? Are you nobody too ?') — Hello !
At Berkeley, yes. Last summer, in a sem-
inar on nucleotide sequences. Not published yet.
— You did ? I know the Michigan department,
Stronger than Harvard, and — Campari, grazie.
— Stochastic independence ; I heard at last year's meetings,
Rigorous to a fault, an elegant presentation.
He has a Guggenheim. — Sorry, I didn't catch
The name ? No, Sussex, since July.
— Behavioural approach ; the larger context,
The influence of varying kinds of crisis. Did you see
My latest book ? — Yes, yes, a lovely place,
So quiet. Well, a background paper only. The deter-
minants of allocative processes. — I think
Not tenure. Oh, he's sound enough, I know,
But there are other factors. — And I ques-
tion the equation too, he said, and is it cen-
tral to the argument would you say. — Hello, yes not
Until this morning. Plane delayed. — The food
Is great, their cellar must be something
Out of this world. — The Basle Club agreed.
Convertibility ? Perhaps a hundred million.
— But not statistical. — Hello ! I heard you, Friday,
The BBC, that panel ; who arranged it ? — Not compute
the coefficient of friction. — Earth is finite.
Standards require universal constants. The velo-
city of light is constant because I choose
To say it is. — The wind is rising. — Yes, you're right,
You're right, a lake trip might be even wiser,
Give them a chance to talk outside the sessions. If (— Hello !
'I'm nobody, who are you ? Are you nobody too ?
Then there's a pair of us. Don't tell,
They'd banish us, you know.')

August 1967

318

Mary Reynolds

Sestina

(Conference on Highly Theoretical Biology and Machine Intelligence)

This ancient dwelling underneath the hill
Graced by the cypress and the singing bird,
Blessed for a thousand years by turning stars,
Has made us welcome, last upon this shore
Of all the long processional of man
Who comes and dwells and has his little day.

Now comes the breakfast bearer, as the day
Dawns with Italian voices down the hill;
Fiats will bring the staff (for every man
Cannot afford a Lancia) ; and the bird
Is on the wing above Pescallo shore
Searching abandoned garbage. Under the stars

Last night, the boat-borne songsters praised those stars,
The moon, the wine, their memories of the day;
And cruised, inebriate, closer to the shore
Than pleased the tired sleepers on the hill.
And Tom Cat, hunting, hunting for a bird
Invoked contempt for all the rule of man.

Noon. On the terrace, strident voice of man,
Who knows he soon will reach and pass the stars,
Considers how the essence of a bird,
Computerized and jet-propelled, one day
Will soar above the highest human hill
And come to rest on planetary shore.

Then will he find, upon that alien shore,
A simulated counterfeit of man?
Clanking to harsh command, on lunar hill,
Directed by equations based on stars,
And disregardful both of night and day?
Was it for this he dreamed himself a bird?

319

Two poems

Will the computer, whirring like a bird,
Then make a dry martini on this shore
As now Vincenzo closes down our day ;
All unaware that pre-computer man
Once knew high revelry beneath the stars,
Ten thousand years ago, upon this hill ?

Then man-constructed bird, computer-man,
Will range Pescallo shore, under the stars,
And have their day, upon this ancient hill.

August 1967

Appendix: notes on the second symposium
by Michael A. Arbib

(Arbib was one of the very few who kept notes throughout most of the discussions. They contain a rather thorough survey of the discussions that went on between the longer presentations, and they are printed here for the stimuli they offer for further reflection on some of the points raised. Like most lecture notes, they are written in very condensed form; in fact, they can be taken as an informal examination — if you can understand them all, you have done your home-work properly in reading the rest of the book! — Editor).

1. OVERVIEW

Is there a theoretical biology? Is there a *special* mathematics for biology?
Waddington (i) General Processes in Biology:
A. Evolution
B. Epigenesis from phenotypes to ecosystem
(Note importance of selection for adaptive phenotypes)
C. Molecular biology—the study of mutable heritable algorithms
D. Origin of Life—is D N A : R N A : Protein the only system in which life could be produced?
E. General Structure of the Universe. General Physics. Metaphysics
(ii) The central nervous system is special both because it is an organ of thought, and because it affects the way we perceive, and thus the way we build theories.

Kornacker's Scheme

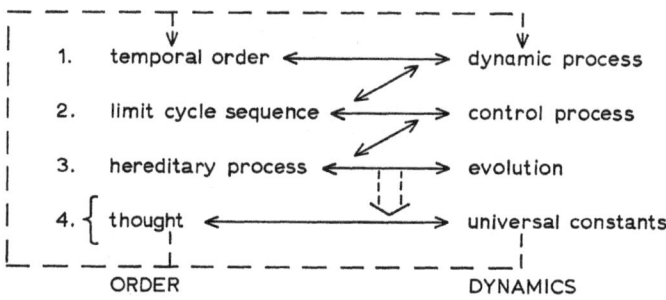

Kornacker Diagonals are obtained by calculation of correlations over the left-hand side of quantities on the right-hand side to get the next order of the hierarchy. In

321

Appendix: notes on the second symposium

Ted Bastin's model, matrix transformation cycles at one level are elements at the next level.

Arbib In what way do we have: temporal order :dynamic process ::thought: universal constants? I don't understand:

Correlation of dynamic processes over time yields limit cycle sequences?

Correlation of control processes over a limit cycle sequence yields hereditary processes?

Correlation of evolution (?) over hereditary processes yields thought?

Correlation of universal constants (what?) over thought (how?) yields time?

Kornacker Information transmission is the attempt to impose correlations between sender or receiver.

The states in the temporal order are moments of time.

The states in the limit cycle sequence are limit cycles. (*Arbib* So, in automata theory, we ignore transients between states — these must be reinserted if we desire a physical description.)

The states in the hereditary process are generations. (*Arbib* Thus the order, unlike the temporal order, is asynchronous — a fact we often choose to ignore in our theorizing.)

Are theories the states of the thought order?

Pattee In studying Brownian motion, the transition from a law of motion to noise marks a transition in your *design* or choice of *function*.

Coding and measurement take us from a dynamic process to a control process. Reading and writing then take us to the genome or program of the hereditary process.

Kornacker At the 2 to 3 level we obtain function for the first time, and with it:

code		communication
memory		learning
experience	and	maturing
perception		insight
thought		

and with these, thought can study dynamics and time to close the loop.

2. EVOLUTION

The phenotype is the genotype's model of the environment.

Maynard Smith Neo-Darwinism seeks to explain the facts of adaptation, and the

322

ascribability of functions to parts of organisms. Evolution is assumed to be the result of multiplication, variation, and heredity.

Heredity : says that there are at least 2 types of entity, A and B, such that A tends to beget A and B tends to beget B.

Variation : says that the above doesn't work completely reliably.

The theory requires that there exist differences in rate of multiplication of different entities. 'Fitness' of a type is a measure of this rate of multiplication — and is thus an ensemble concept, rather than a property of the individual. Unfortunately the theory cannot evaluate for you the fitness of a given type in a given environment.

One needs to distinguish the genotype (a set of instructions) and the phenotype (their functioning embodiment).

The only really strong assumption of the theory (the Weissman assumption) is that, whereas different environments produce different phenotypes from a given genotype, this change does not itself produce a change in the phenotype.

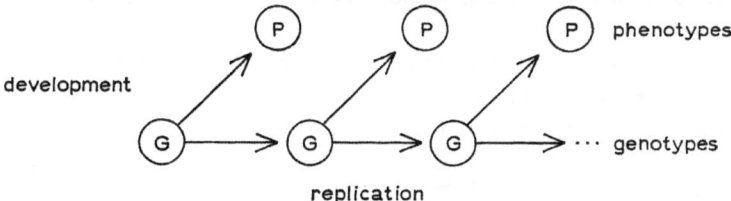

Genotype changes propagate, but phenotype changes do not. (Lamarckianism is the contrary of this.)

Of course, a change in the environment, such as a focused beam of radiation, can change the genotype !

It does *not* seem necessary to posit randomness of variation, but one does assume that *most* changes of genotype lower fitness — or you wouldn't need natural selection !

In the beginning was the word   W O R D
                                        W O R E
                                        G O R E
                                        G O N E
and by the mutations came the gene   G E N E

One believes that the laws of physics and the earth environment permitted the sequence

$$G_1 \rightarrow G_2 \rightarrow \ldots \rightarrow G_N$$

such that G_1 could arise 'spontaneously', G_N specifies a human, each step increased fitness, and no step needs to be regarded as a highly programmed change.

Kornacker To study this one needs to know how large is the range of phenotypes reachable from a genotype via one mutation. The theory of evolution would not be very interesting if we could take $N = 1$ or 2.

Waddington Selection is of genotypes, but it acts on phenotypes. To specify the situation we need at least 2 genotypes and 2 environments. A genotype might develop in one environment but be selected in the other. We might then get selection for the most frequently occurring environment. We often get selection for ability to learn a skill rather than for the skill itself.

Each organism introduces new organismic relations – and then a new organism can evolve to exploit the new ecological niche provided by this relationship.

Evolution is not driven – there just exist metastable situations.

Arbib Consider the Volterra equations. They show that selection is not for simple maximization of species population, but involves time averages in an environment of multiple species.

What can we say about neutral changes of genotype which can accumulate until a combination may be reached which is advantageous or disadvantageous? Cf. a random walk with absorbing barrier.

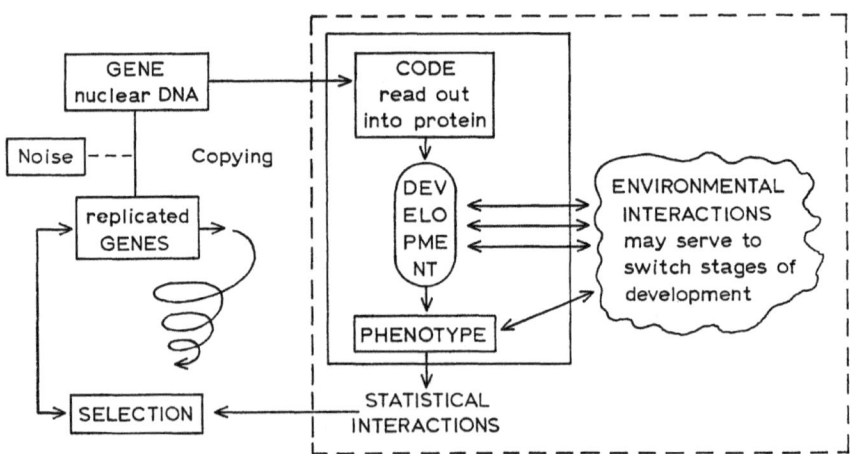

Pattee The essence of neo-Darwinism is that ⌐ ⌐ is a black box; Waddington would have the theory only consider ☐ as a black box.

Maynard Smith wants a theorem that species formation is more common than species extinction. It takes 10 generations to establish a gene change. (*Waddington*. Take care—this estimate comes from a theory with many unobservable parameters. A new species of fish was formed in Africa in 400 years.)
Lysogeny speeds up evolution.
Symbiosis acts as a higher-level store of genetic information :
> lichen = fungus + moss ; and each is the environment of the other.
What was the evolution of differentiation ?

Note that evolution is *not* synonymous with increase of complexity, since the viruses are presumably simpler than their ancestors—but they could only evolve when there existed cells in the environment to support them.
Pattee Some bacteria have survived for billions of years without apparent dissatisfaction, yet man evolved too.

So perhaps one requires a theory of competitive and cooperative games which can throw light on coalition formation. (Cf. Waddington's comment on multiplication of niches.)
Maynard Smith We have ecotheories for non-evolving populations, and evolutionary theories for 'stable' ecosystems, but no stability criteria for general evolution. Is evolution a stable process ?

If you evolve into an ecological niche, it may not still be there—you might be so efficient, that you would eat everybody else or die of starvation. (Cf. remarks on Volterra equations above.)

3. MORPHOGENESIS
Waddington What is a form ? There seems to be some notion of regularity.
General categories :
1. Monotypic : e.g. crystal forms coming from one type of component ;
vs. Polytypic : many biological forms come from many types of components.
> *Drosophila* wing form can be altered by changes in 40 to 50 genes.
2. Synchronic : a form which is complete as soon as it appears ;
vs. Diachronic : a form which is slowly elaborated and built up.
3. Element elaborated : like a jigsaw ;
vs. Whole controlled : with feedback from the developing whole controlling each
> subsequent step.
In the biological case, the chreod usually corresponds to a polytypic or diachronic form which is whole controlled, corresponding to a morphogenetic field.
Various modes of form generation :

Appendix: Notes on the second symposium

A. Unit generated: (a) Particles; (b) Fibres; (c) Sheets. How far up the scale
can you go with chemistry? Perhaps (cf. July 1967, *Scientific American*) as far
as virus head construction. Perhaps cells use specific attachment points (desmo-
somes as at synapses) to fix on to one another.
B. Instruction generated, of which a special case is
C. Template generated: contrast of simple and complex coding
D. Condition generated, which may involve stochastic conditions (cf. Turing's
morphogenesis paper)
In intersusceptive template action, growth takes place all through the whole organ,
not just at the edges as in apical growth.

Scriven Certain aspects of morphogenesis are to be studied in terms of temporal
or spatial rhythms. Scriven has a treatment, developed from Turing's paper on
morphogenesis, based on transport processes to move things from one place to
another. (Robin Grands has a Turing manuscript for the non-linear case.)

Dynamic forms known to the chemical engineer:
   1. Eddies on oil burners
   Blobs form to hold the flame under certain
   dynamic conditions.
   2. Rotating the inner of two cylinders between
   which is liquid.
   Counter-rotating toroidal eddies
   form Taylor vortices.
   Increasing speed yields a switch to eddies —
   standing waves? As speed increases
   these start to progress.

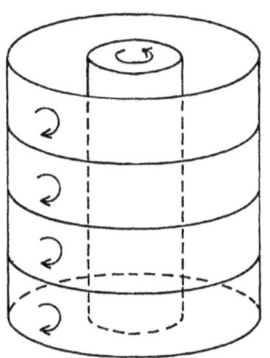

3. Bunsen burner      **vs**   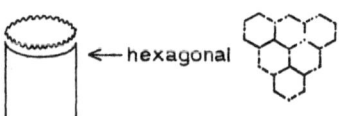 ←—hexagonal

Michael A. Arbib

Can also get this structure in crystallization of binary melts, under certain conditions. Arctic tundra polygons from alternate thawing and freezing.

4. If a dishpan with a shallow (to avoid turbulence) layer of liquid is uniformly heated from below, a hexagonal pattern will form after convection is set up. In a circular dish, pattern formation starts from the perimeter inwards, with cylindrical units. In a small dish you never get hexagons, whereas in a large dish you eventually do. Perhaps the determination of the actual pattern depends on nonlinear effects, with 'noise' getting things started.

4. AUTOMATA THEORY AND THE ORIGIN OF LIFE

Longuet-Higgins A great unifying concept: mutable hereditary algorithms.
The 'secret of life' is the ability of living creatures to improve their programs.

1. Morphogenesis involves differentiation—change of cellular programs.

2. Evolution—change of organism's programs.

3. The brain may be regarded as a program for processing inputs, and is modifiable by learning.

4. Biogenesis—Where does the program come from? How is it modified?

To be appropriate automata should have an unbounded set of states, and be capable of rewriting their programs. Finite-state automata do not suffice for grammar—so why should they suffice for theoretical biology?

Arbib Rewriting programs is quicker (and safer!) and more compact than consulting a list of ancestral actions (or repeating ancestral mistakes!). How do we mix this up with evolutionary theory?

Note that my self-producing automata grow in tessellations – here there is no limit on the size of the automaton, though at any time a finite-state space will describe the automaton, in that temporal neighbourhood. Perhaps in formulating a proper theory of chreods we should consider them as moving through an expanding sequence of finite-dimensional subspaces of a 'Hilbert-Waddington biospace'!

Longuet-Higgins Gödel proved his incompleteness theorem via a self-referential 'I'm not provable' statement, which is still unprovable if you change the code. What (if anything) has this got to do with genetic coding?

Pattee What can be said about playing off the two principles of impotency: undecidability and uncertainty?

Goodwin We've been doing differential equations for years—then along came automata, which are irreversible. So they look good for biological irreversible processes—but don't let's get carried away.

327

Appendix: Notes on the second symposium

Arbib Note that my 'Common Framework' paper indicates ways in which differential equations and automata can be combined.

Kornacker Groups $\xrightarrow{\text{damping}}$ Semigroups? Do correlation terms destroy inverses but preserve the associative law?

Arbib As soon as you partition a set, a group which acts on it loses its inverses:

$$\begin{pmatrix} 1 \\ 2 \\ 3 \\ 4 \end{pmatrix} \rightarrow \begin{pmatrix} 4 \\ 1 \\ 3 \\ 2 \end{pmatrix} \text{ has inverse } \begin{pmatrix} 1 \\ 2 \\ 3 \\ 4 \end{pmatrix} \rightarrow \begin{pmatrix} 2 \\ 4 \\ 3 \\ 1 \end{pmatrix}, \text{ but if we identify 3 and 4,}$$

labelling it S, we get problems! Not only do we lose inverses, but we also lose determinism. We might approximate with $\begin{pmatrix} 1 \\ 2 \\ S \end{pmatrix} \rightarrow \begin{pmatrix} S \\ 1 \\ S \end{pmatrix}$ which has no inverse, or with

$$\begin{pmatrix} 1 \\ 2 \\ S \end{pmatrix} \rightarrow \begin{pmatrix} S \\ 1 \\ 2 \end{pmatrix} \text{ which } \textit{does} \text{ have an inverse.}$$

A large system is deterministically (Newtonian) describable in time and space — if you ask the right questions about it. A small system must be described at the wave function level. Can we talk of medium-scale systems (hopefully including RNA, etc.) which, with reference to the questions of interest to us, are describable as probabilistic automata? i.e. they have enough stability, with reference to our questions, for the phase shifts of the wave functions to be of 'no' relevance.

Questions for the automata theory approach to tackle:
Waddington Bone growth
Pattee Think about virus construction (cf. July 1967 *Sci. Amer.*). Suggests one program this as a test. It's not a 'hydrodynamic instability' type of form generation.

Pattee The logico-mathematical approach to theory is the only practical one for complex systems — 'don't use the Hamiltonian to build a watch' — although the physical approach helps us with components, general concepts, etc.

[?] The origin of life problem seems the appropriate place to make an interface between these two approaches in our attempt to link up the three great mysterious peaks, of development, evolution, and thought. (The components of self-reproducing automata are too complex.)

328

Michael A. Arbib

We must conform to the fact that part of the cell consists of tactic copolymeri-
sations, with 20 subunits for amino acids, 4 for nucleotides, and several for the
saccharides. With the closed matter cycle on the surface of the earth, fed by energy
from the sun, how did we reach metastable 3D configurations with hereditary
properties? We also needed degradation mechanisms – to prevent the earth filling
up with self-reproducing automata! Building up and degrading down gave an
ecocycle driven by the sun – the first cycle must (? cf. Marcus Goodall's idea on
cell boundaries) have been non-living and outside the cell.

Fox has a preparation in which within 20 minutes of starting with amino acids
and hot water, proteinoid micropheres arise which bud spontaneously. Pattee
isn't happy with this as an origin of life contender – it's not necessarily hereditary.
(*Maynard Smith* To be hereditary, more than one property must be propagatable.
Thus, the flame is a poor model of reproduction because the offspring depends not
on the 'parent' but on the substrate.)

So we're really interested in 'hereditary propagation which is modifiable by
natural selection'.

We may think of death as an error-correction mechanism – which must be slow
enough to allow living organisms to accumulate the information derived from
natural selection.

What, then, is the simplest system which will operate at the origin-of-life level,
with enough error-correction to have the possibility of evolution? We're interested,
then, in the occurrence of a specific reliable catalyst – it's not the permanence of
the structure, it's the reliability of the catalyst that's important – the greater the
capacity of replication, the less the demand for reliability.

The question of which chemical automata survive is not just a question of logic
alone, but of how reliably a quantum-mechanical system can approximate the
logic. (Cf. Swanson & Landauer on reliability of components, in the I B M Journal.)
Monro A single template can lead to the production of 10^{12} enzymes (in a cell, or
the whole organism?). Thus error-correction of templates is crucial – since $10^{12} \times a$
small error can be a whopper. Anyway, these figures show that D N A can control
macrostates. A single error in the egg cell can cause death in the organism – it's
incredible that anything survives!

There are two types of errors in the templates:

A letter gives a non-letter: there are enzymes to correct this.

A letter gives another letter: there's nothing you can do.

Errors in later stages of division – somatic mutations – are not as crucial, e.g. 1/10%
errors in a haemoglobin molecule might not matter.

329

Appendix: notes on the second symposium

Evolution of protein structure is related to evolution of organism—proteins are both structural and functional, cf. enzymes in active transport.

Typical metabolic patterns—protein synthesis, Krebs cycle, carbohydrate metabolism—recur in most species.

Another aspect of enzymes is the control of their synthesis and function. Control at point of enzyme synthesis through controller genes. The repressor is a protein. Presumably the repressor controls whether or not the operator can get going (in what sense?)

Waddington DNA amplification (in toad egg, not in bacteria): Ribosomal cistrons —for synthesizing ribosomes—are represented about 2,000 times in the toad. An interesting situation would occur if you could modify the amplification, e.g. to synthesize more actin and myosin in muscle cells (but it's not known whether or not this occurs).

In toad egg, 5,000 rings of DNA are thrown off to float around; this release is under the control of the cell as a whole. This presents the possibility of a rather flexible control beyond that we can discover in bacteria.

Longuet-Higgins In computer programs we can bring in subroutines, or skip them, on the basis of the computation state—surely a paradigm for repressors or derepressors.

Waddington In Arbib's self-reproducing automaton we may consider

   V: copies the DNA.

   U: gets the DNA to produce the new cytoplasm.

   W: resets.

Question: Why are the genotype and phenotype separate?

Longuet-Higgins For 'sanitary' reasons to prevent overwriting of a 'master routine'.

Arbib If the alternative is that the objects are to copy themselves, we seem bound to get into problems of destructive readout with a 3D object in 3D space.

Pattee For reliability, one would want to make the genotype very stable. But then the dead time is too great for it to do anything. Solution: use ultra-reliable DNA to store the message, but build up a phenotype of other components to obtain a flexible organism. (We need a fast and accurate read-out mechanism.)

Monro There exist bacterial cells with different processors (sRNA) which instead of terminating will put in a new amino acid.

Arbib But even stronger, one could imagine a different cytoplasmic machine which would consistently assemble the DNA-decodings differently.

Maynard Smith A cell can 'accept' into its genes a useful 'subroutine' from an

330

invading virus, thus speeding up evolution. Perhaps mitochondria were originally cell invaders which took up permanent residence, their division cycles became entrained, and we obtained an extra-nuclear genetic mechanism.

Kornacker For automata which model human thought, we need pattern recognition and generation with

input : spatio-temporal patterns in many sensory modalities.
output : patterns of muscular coordination.

Kerner thinks of biology as comprising *in vivo* biochemistry, genetics, and ecology and these are all amenable to theoretical treatment. The automata theory is 'overprinted'.

'What is the connection of software to hardware ?'

Arbib In time-shared systems or parallel computations we have a hierarchy of cycles : bit times, word times, and job times. In choosing between synchronization (every operation takes the time of the slowest) and queuing (do a job as soon as the prerequisites are available) we may play off simplicity of operation and speed of operation.

An evolutionary theory will probably have to face up to a preponderance of breadth over depth in that fundamental principles will tend to rest on the accumulation of evolutionary 'accidents' or the importance of arbitrary decisions.

Note the different formations required to model different aspects of information processing.

decomposition of nets vs. semigroups vs. events.
X for deterministic vs. $X \times Y$ for stochastic automata.
Hierarchies of computational complexity.

Contrast the questions :

Bellagio '66 : Is random search necessary for evolution ?
von Neumann : Is a minimum complexity needed for self-reproduction ?
I suspect the answer to both is no—I can imagine a machine which computes according to some valuation function to see which of its features should best be modified. However, I might expect that there is a minimum complexity for a machine which can achieve this 'self-conscious evolution'.

331

Appendix: notes on the second symposium

5. NONLINEAR OSCILLATIONS

Iberall started teasing out a broad spectrum of nonlinear oscillators – macro-spectroscopy– from heart beat and below to circadian rhythms and beyond, and then realized that the oscillators *were* the system, with DC changes in the milieu changing operational points of the oscillators. This emphasized homeokinesis–the dynamic nature of regulation.

'Any compact system containing a complex of sustaining nonlinear oscillators and a series of algorithms to let it operate in a wide variety of ambient conditions ...' is a living system, which may thus involve many types of successful mechanism.

We have dynamic regulating chains, be they stable, unstable, or marginally stable; self-activated motor activity; and, when time is adequate, entrainment of the oscillations.

Iberall would extend this scheme from involuntary to psychological systems: adolescence is an instability preparatory to reproduction; self-activation seems more important than pre-programmed activity.

Kerner Do the cycles matter, or are they just details of other things–e.g. in a factory, consider changes in the waste-paper-basket-emptying cycle.
Goodwin The periodic environment we live in imposes periodicities on the system, and once we've got a few it's easier to put the whole lot in.
Elsasser Poincaré's theory says that any stable system with one parameter changed will start to oscillate. All nonlinear systems oscillate – not just living systems.

Goodwin wants to describe, and then explain, the sequence of events within the cycle of cell division.

The naive view is that everything goes on in parallel, and when all cell concentrations are doubled, the cell splits. This is not so. Nor is there a causal chain, with completion of stage A triggering the start of stage B, etc. Rather, Brian found, we have phase-locked processes which interlock so that appropriate concentrations exist at various times.

There are then stability problems for temporal sequence of events (of macromolecular synthesis) during cell division. Neoplastic growth (cancer) is an instability–one wants a theory which can also describe this instability.

He assumes the relevant variables are enzyme concentrations. The functional units are then taken to be control circuits based on a functionally linked set of genes whose end-product metabolites can feed back to inhibit the genes.

There are 200–300 such functional units, roughly, per bacterial cell—two units are judged distinct if you can turn one off without affecting the other. The units are *not* localized—messages can diffuse from one place to another, and one doesn't want to only consider operons controlled by a single gene.

One assumes variables are exclusive to one of the units (?). Of course, there are non-specific interactions, due to sharing a common pool of ribosomes. Glutamate appears in two cycles and yields strong interactions, while pH gives weak interactions.

Studies of 10 of these circuits showed them operating in an oscillatory mode, with the oscillations giving the basis for sequential production of enzymes—the systems being phase locked with one another, yielding different points for maximal concentration of various enzymes. This presumably serves to minimize the time required for cell division.

Scriven Chemical reactors are non-isothermal and thus highly non-linear. There are *many* steady states, both stable and unstable, and some of the latter can be stabilized by appropriate control to give higher yields than the stable steady states. There exist limit cycles more efficient than steady states.

They are attempting to study this at the molecular level, and to join statistical mechanics and continuum mechanics, as well as to use optimization theory. They are also concerned with improving the design of interlinkages in chemical plants.

One needs constitutive relations to define the system, as well as boundary conditions in time and space to define the rest of the world for the system,

biology ⇌ chemical engineering,

e.g. catalysis and optimization of complex chemical systems.

Relation of this work on chemical process optimization to Goodwin's work.

1 Complex reaction kinetic schemes.

Nature of steady-state hypothesis for low concentrations of steady-state intermediates.

2 Multiple chemical reactions of greater than first-order nonlinear, consider relations between phases, e.g. in polymerisation.

Characterizing a chemical plant: (i) what's it for?; (ii) inventory; (iii) flow diagram; (iv) material balances—check they're closed.

Maynard Smith Surely the strange thing about the cell is that it has so few steady states.

333

Appendix: notes on the second symposium

Arbib Goodwin only distinguishes the growing state and the dividing state–but surely these are too few?

6. STATISTICAL MECHANICS

Kerner Gibbs ensemble method is not restricted to use in mechanics. Biological systems are so complex that you may need an analysis which gives an overview.

We know more about 10^{23} molecules in a gas than we do about the moon–earth–sun system. Temperature is a useful measure for the former, not the latter.

Gibbs gave a statistical theory of differential equations. D.E.s are useful in biology, e.g. in ecology (the dirtiest form of biology, but with a long mathematical history going back to Malthus & Benjamin Franklin). An ecosystem may be defined by 10^6, say, variables, one for each species number.

The Gibbs strategy says you must know

(i) one conserved quantity of the observed system

(e.g. energy in classical physics, something similar in Volterra-Lotka dynamics, mass in a closed chemical kinetic system).

(ii) a good way to introduce statistics

e.g. a probability distribution on the ergodic surface (i.e. a surface of constant value for the conserved quantity).

(iii) nothing else–so all you know is which constant surface you are on.

What probability distribution should be placed on the surface? The uniform distribution is O.K. if you have a Lionville theorem to assure that it's preserved by the dynamics–and you pray that the ergodic hypothesis is valid. You have to find the appropriate time scale, and check that there are no isolated 'pockets' in the motion.

You don't need a Hamiltonian for a Gibbsian analysis, just a conserved quantity. Let N_r be the number of animals in the rth species. Volterra set up the equations

$$N_r = e_r N_r + \frac{1}{\beta_r} \sum_s a_{sr} N_s N_r \quad \text{with} \quad a_{sr} = -a_{rs}$$

With $V_r = \log \dfrac{N_r}{q_{r\cdot}}$ (q_r being a stationary value) you get equations for which Lionville's theorem holds. The conserved quantity is then of the form

$$G = \Sigma \tau_r (e^{V_r} - V_r) \equiv \text{const.}$$

What are the observables of the system? Suppose we could think classically and observe the motion of one of the 10^{23} particles in a gas (cf. Cowan's one neuron and the EEG). A Brownian particle is a good thermometer. We want to know

334

Michael A. Arbib

averages telling us amplitudes of fluctuations, mean levels of crossings, etc. All these are expressed in terms of temperature. Using an ensemble $e^{-G/\theta}$, where θ, the modulus of the distribution, is the temperature, we get a thermostatics. Here is the idea of a Gibbs ensemble—a description of complex phenomena with many averages expressed in terms of a single parameter.

Kornacker To get from mechanics $\left(\dfrac{dE}{dt} = F_e \cdot V\right)$ to *heat* is to introduce averaging over all values consistent with our partial state of knowledge.

If $\bar{E} = \int Epdx$, then $\dfrac{d\bar{E}}{dt} = \int\left(E\dfrac{dp}{dt}\right)dx + \int\left(p\dfrac{dE}{dt}\right)dx.$

The central dogma of statistical mechanics: you gotta be ignorant. Who you, the observer, are is crucial. In classical mechanics you can ignore the nature of the measuring devices. Measuring devices perform averaging—a pressure measure is an integration of molecular forces. A time average looks like

$$\frac{1}{\tau}\int_{t-\tau}^{t} g(x(t))dt = \frac{1}{\tau}\int_{t-\tau}^{t}\left[\int g(x)\delta(x-x(t))dx\right]dt = \int g(x)\overline{(x-x(t))}\,dx$$

and this is where the ensemble comes from. This requires the time constant in the measuring device to be long enough so that ... (*Arbib*, what?)

By this criterion, one molecule could be macroscopic for suitable time intervals. The less the number of particles, the lower the frequency, and so the longer the time interval required—in general.

A working hypothesis could be that $\delta(x-x(t,x_0))$ is the microcanonical distribution.

                               ↑
                          initial condition

Of course, with probability 0 (or greater?) the average could be degenerate, e.g. if all particles are initially moving in parallel.

Let $\langle\ \rangle$ denote a time average on a single system over a time which is several relaxation times for the measuring device, but is short compared with the relaxation time of the system measured.

$\dfrac{d\langle E\rangle}{dt} = \langle F_e \cdot V\rangle$, whereas work is given by $\dfrac{dW}{dt} = \langle F_e\rangle \cdot \langle V\rangle$, and the difference between work and energy is

$\dfrac{dQ}{dt} = \dfrac{d\langle E\rangle}{dt} - \dfrac{dw}{dt} = \langle F_e \cdot V\rangle - \langle F_e\rangle \cdot \langle V\rangle$

and so heat is a macroscopic measure of the degree of *correlation* of force and displacement.

There are many ways of measuring the strength of correlation:

linear $\qquad \overline{xy} - \overline{x}\,\overline{y}$

probabilistic $\quad$ Does $p(x,y) = p(x)\,p(y)$?

$\qquad\qquad$ If so, they are independent, which implies uncorrelated, but the converse is not true: $\overline{\sin\,.\,\cos} = \overline{\sin}\,.\,\overline{\cos}$

? A more powerful way of approaching correlation is via an entropy measure:

$$S(p_x) = \int p_x \ln p_x \text{ and ask, does } S(p_x) + S(p_y) = S(p_{xy})\,?$$

Heat is only sensitive to linear correlations, and Kornacker claims the trouble with entropy is that it picks up too many correlations, and this causes many problems.

Heat is intrinsically indirectly observable.

The second law of thermodynamics will be set up to within fluctuations, but Kornacker claims that it is irrelevant to biology.

Temperature depends on momentum fluctuations $\langle (p - \langle p \rangle)^2 \rangle$.

7. METAPHYSICS

Elsasser believes biology is not the kind of subject you can unify, but that there exist unifying ideas like those of Darwin, or Watson and Crick.

Longuet-Higgins thinks biological systems are physico-chemical, but that the questions we ask about cells are not those for which we have operators in quantum mechanics. There is an energy operator but not a kidney operator!

It's more expedient to build a theory in terms of the concepts we're really interested in – since it is inexpedient to do everything in terms of quantum mechanics, need we worry whether it's adequate?

Elsasser Starting from quantum mechanics, you'd like to make inferences. But, as Bohr said, you can't measure something as complex as an organism without destroying it. So there are enough loopholes to give it a living character or *autonomy*: a class is autonomous if, as a matter of principle, you can't deduce its regularities (the essence of scientific predictability) from quantum mechanics.

Arbib Quantum measurements which suffice to convince one an earth thing is a cat might not suffice to determine an extraterrestial creature? How does this relate to Elsasser's terminology?

Bohm Nobody has reduced statistical mechanics to quantum mechanics, and they may be inconsistent. Perhaps there is a contradiction between biology and quantum mechanics – and this would demand the creation of a new system embracing both.

A wave function does not describe an object in time or space.

Bohm (*Rev. Mod. Phys.*, 1966) showed that von Neumann had built his con-clusions on observable classes into his axioms – if you take Euclidean geometry as a fact, then non-Euclidean geometry yields a contradiction!

Gregory notes that an engineer succeeds by, e.g., dividing a radio into com-ponents – functional units. This should be our lead to overcoming the quantum mechanical problem of immense numbers.

How much should the biologist be an engineer, how much a physicist?

Goodwin The discovery of appropriate variables for biology is itself an act of creation.

Bohm Compare Piaget's study of the evolution of object perception in children.

Kerner The only question of creation he allows is: 'Where does the next instant of time come from?'

A model is a dynamic map – what is a dynamic model?

What is organization?

Kornacker would reserve the term for functional organization (crystals having order rather than organization) – parts related to a whole geometrically, perhaps, but with the emphasis on action.

Physics says irreversible processes, of which one is scalar and one is vector, cannot be coupled isotropically – an anisotropic membrane which couples two such processes is organized, but one which is poisoned is not.

Waddington Surely the organization is in the whole system of which the membrane is part.

Iberall Kornacker would accept a laser as an organized system of which a crystal is *part*.

Maynard Smith At least in biology, an organization is something of which you can ask 'What is it for; to what does it contribute?'

Longuet-Higgins There's a hierarchy. Subsystems have function in relation to systems having functions in relation to . . .

Bohm Distinguish the organization and that which organizes it.

Arbib We are interested in systems whose dynamics can be described using far

337

fewer parameters than are required to describe the dynamics of its individual parts (non-holonomic constraints ?). This implies maintenance of a certain structure which permits these few parameters to remain an adequate description. When they cease to be adequate, the organization has changed—cf. ageing and death. It all depends on the level of observation and questioning we apply.

Grene This notion doesn't exclude gases. For biology we concentrate on organizations with properties of not just survival, but hereditary and developmental properties. A simpler characterization of biological organization may follow.

Waddington Not just homeostasis but homeorhesis—stabilized paths, not stabilized parts.

Pattee The problem still remains of reconciliation with physics. An electron has no function. What is the simplest system with a function ? What is the property (is there such a property ?) of a thing that marks it as alive, or as an artefact of a living system ?

Elsasser Incredible—a scientific conference where people can talk metaphysics without being shot down !

List of Participants

(Second Symposium. 3–12 August 1967)

Michael A. **Arbib**   Automata Theorist. Stanford University.

E. W. ('Ted') **Bastin**   Physicist, Computer Scientist. Language Research Unit, Cambridge.

David **Bohm**   Theoretical Physicist. Birkbeck College, University of London.

Jack **Cowan**   Neuroscientist. Imperial College, London (now Dept. of Mathematical Biology, Chicago University).

W. M. **Elsasser**   Physicist. Princeton University (now University of Maryland).

Martin A. **Garstens**   Physicist. Office of Naval Research, Washington (now University of Maryland).

Brian **Goodwin**   Theoretical Biologist. University of Sussex, Brighton.

Richard L. **Gregory**   Neuroscientist. University of Edinburgh.

Marjorie **Grene**   Philosopher. University of California (now University of Texas).

A. S. ('Art') **Iberall**   Systems Analyst. General Technical Services Inc., Pennsylvania.

Edward H. **Kerner**   Physicist. University of Delaware.

Karl **Kornacker**   Neuroscientist. Massachusetts Institute of Technology.

Paul **Lieber**   Physicist. University of California, Berkeley.

Christopher **Longuet-Higgins**   Theoretical Chemist and Computer Scientist. University of Edinburgh.

John **Maynard Smith**   Geneticist. University of Sussex, Brighton.

Robin E. **Monro**   Molecular Biologist. M R C Unit for Molecular Biology, Cambridge, England.

Howard H. **Pattee**   Physicist. Stanford University.

L. E. ('Skip') **Scriven** Chemical Engineer. University of Minnesota, Minneapolis.

C. H. ('Wad') **Waddington** Biologist. University of Edinburgh.

Secretary
  Miss D. **Manning** University of Edinburgh (now Internat. Inst. Genetics and Biophysics, Naples).

Author Index

References to extended treatments are given in italics

341

Author Index

Author Index

Author Index

344

Subject Index

Subject Index

Subject Index

347

Subject Index

laws
  biogenetic, 78
  biotonic, 226
  of conservation, 132, 133, 138
  of motion, 25, 57, 254
  Newtonian, 25, 293-316
  physical, 35, 36
  physical, new, 268, 270
learning, 223, 224, 225, 235, 236, 237, 324
  cognitive, 238
  maze, 238
  perceptual, 240
life, definitions, 27, 64, 122, 169, 204, 229,
      233, 235 270, 271, 272, 276, 332
limb buds, 180
limit cycles, 6, 12, 17, 81, 266
  efficient, 333

Mach's principle, 303
machines, 68
Malthusian parameter, 108, 129
mathematics, 23, 24, 254, 315, 316, 321
maximization of populations, 90, 93, 94, 95,
      100, 101, 104, 324
maximized variable in evolution, 114, 125,
      126, 128
measurement, 18, 248
  quantum mechanical, 226, 248, 279, 284,
      286, 287, 336
  and statistical mechanics, 335
mechanism, 27, 28, 29, 41, 65, 67, 70, 81,
      227
mechano-chemical systems, 196, 200, 201
membranes, 14, 250, 337
memory, 21, 234, 236
  store, 115, 118, 277
Mendelism and evolution, 82, 106, 124, 127
menstrual cycle, 172
metabolism, 4, 5
metalanguage, 263, 264
metaphysics, 2, 8, 9, 10, 41, 44, 51, 72, 79,
      80, 81, 227, 336-8
  of process, 42, 49, 52, 61, 75, 96
  tacit, 41, 46, 47, 90, 96, 99
mice, 121
mind-matter, 62, 63
mitochondria, 110, 331

models
  as chreods, 247
  flexible, 247
  and hypotheses, 244
  neural, 237-47
molecular biology and quantum theory, 35,
      268-82
molecular evolution, 116, 117
molecular hybridization, 124
morphogenesis, 6, 162, 181, 214, 230, 231,
      290, 325, 326, 327
motility, 201
mRNA, 145
muscles, 169, 170, 176, 177
music, 25, 26, 38
musical themes, 173, 180
mutation rate, 276
mutationism, 106, 117
mutations
  new, 107, 116, 117
  random, 7, 84, 109, 116, 118, 122, 235
  systemic, 123, 124

natural necessity, 45, 70
natural selection, fundamental theorem, 108,
      125
nature, artist or engineer, 49, 68, 76, 104
neo-Darwinism, 6, 60, 65, 79, 82-128, 227,
      228, 322, 323, 324
  definitions, 82-5, 127
  and metaphysics, 91, 92
  refutations, 85, 86, 102, 103, 105, 110,
      111, 124, 125, 126, 127, 128
nervous system, central 13, 14, 55, 174, 175,
      180, 220, 232, 244, 248, 251, 252, 258,
      273, 321
neurosurgery, 224
noise, 28, 38, 68, 133, 134, 184, 225, 238,
      322
non-chromosomal genes, 110
non-holonomic constraints, 266, 272, 275,
      276, 279, 282, 298, 301, 338

object
  perception, 337
  recognition, 238-46
objectivism, 62

Subject Index

349

Subject Index

talandic
 temperature, 134, 135, 136, 153, 154, 156, 158, 159, 162
 variables, 5, 153
tautology, 66, 82, 83, 85, 87, 98, 101, 102, 110, 114, 125, 126
teleology, 66, 67, 85
temporal templates, 179
theoretical biology, 10, 18, 122, 227, 285, 321
thermodynamic coupling, 227
things, 42, 73, 75, 76
time, 1
 averaging, 249, 250, 275, 324, 335
 constants, 266
 in quantum mechanics, 57
 scales, 146, 265, 266, 267, 325
timeless order, 11, 54, 56, 57, 63, 68, 69, 70
topology, dynamic, 227, 228, 229, 233, 234
traffic, 229
training, transfer of, 236, 238, 239, 241
transcendental, 69, 70
transfunctional, 45, 46, 49, 66, 67, 69, 70, 85
transport, cellular, 14, 184, 190
Turing machine, 205, 206, 207, 208, 209, 211, 213, 220, 282

uncertainty, 2, 5, 7, 32, 260, 282, 327
undecidability, 327
uniformity, maximum, 293-316
universal constants, 228, 302
universal construction machine, 207, 208
universal correspondence, 304

valence, 283
value, 63
variational principles, 296
vibrating plates, 182
vicious circles, 44, 69, 70
virus, 220, 326, 328
vitalism, 1, 81

watches, 221, 328
water snails, 119
wave - particle duality, 32, 33, 56
Weismannism, 82, 84, 94, 124, 126, 127, 323
wombat, 117
words as chreods, 247
work, 136, 158, 335
work - energy relation, 249
world egg, 73

351

Foreword

I am honored to write this forward to Sanni Paljakka's and Tom Carlson's book, *So You Want to Do Narrative Therapy? Letters to an Aspiring Narrative Therapist*. While I have written this story elsewhere, I find it particularly apropos as a way to introduce the importance of this book as it relates to the reimagining of narrative therapy. In late 2007, after spending several years completing our two seminal projects, *Michael's Maps of Narrative Practice* and my *Biting the Hand that Starves You*, Michael and I were determined, as we put it, "to start all over again!" In our conversation, Michael reassured me that he had much he also wanted to review and revise and "go further" than narrative therapy had gone. Of particular interest to both Michael and I was to finally settle two questions of great importance to us:

♦ What do we mean by rich story development?
♦ How can narrative therapy better achieve its stated aim of being a therapy of literary merit?

We kept telling each another we couldn't wait. We finally agreed on a date in the late spring of 2008. However, Michael's untimely death in April of 2008 meant that "starting all over again" never took place in the way I had imagined and anticipated. Ever since that time I have being trying to fulfill the promise I had made to Michael to reimagine narrative therapy and to revise our work and go further than it had gone before. Along the way, I sought out and mentored narrative therapists who were engaging in their own revisioning of narrative therapy practice. Sanni and Tom were two such therapists who had concerned themselves with refining and expanding the idea of rich story development in narrative therapy.

I wish Michael were alive to consider the following by Rebecca Solnit (2014) who elaborates on both the centrality and power of stories:

> What's your story about? It's all in the telling. Stories are compasses and architecture; we navigate by them, we build our sanctuaries and our prisons out of them, and to be without a story is to be lost in the vastness of a world that spreads in all directions like arctic tundra or sea ice . . . We tell ourselves stories that save us and stories that are the quicksand in which we thrash and the well in which we drown . . . Not a few stories are sinking ships, and many of us go down with these ships even when the lifeboats are bobbing all around us . . . We think we tell stories, but stories often tell us, tell us to love or to hate, to see or to be blind. Often, too often, stories straddle us, ride us, whip us onward, tell us what to do, and we do it without questioning. The task of learning to be free requires learning to hear them, to question them, to pause and hear silence, to name them and then to become the storyteller.
>
> (pp. 3–4)

This book goes a long way in returning story and rich story development to its rightful place as the center of narrative practice. *So You Want to Do Narrative Therapy?* is both intellectual and practical in its reach. Although written in the kindly and generous pedagogical style of Letters to a young poet (Rilke) and Letters to a young writer (Colum McCann), whereby an elder seeks to pass over their wisdom to an enthusiastic newcomer, this book goes far beyond doing merely that. How it does so is by immersing the reader in the author's practice by way of their commentaries on the edited transcripts that are provided. I can think of nothing more apposite for the purpose of having a sense of being there – on the inside of practice rather than the outside looking in. The very heart and soul of it.

Of particular importance to me, this book provides a thoroughgoing revamping of Michael White's enthusiasm for what he referred to as "rich story development" and what I have posed elsewhere – "What is a good story?" This text has us return to and reconsider the question raised by Michael in 1989 at the very outset of the development of narrative therapy. Can narrative therapy be considered a "therapy of literary merit"? This book proposes this to be the case and shows a myriad of ways of doing so. The authors are to be congratulated for taking on such an overdue and arduous task and providing these letters to a young narrative therapist for the next generation to come.

David Epston, co-founder of narrative therapy

Prologue

Dear student,

Hello my dear. Your tears on Tuesday were not for naught. They are the reason I am writing to you. You said you felt alone and weren't sure of your belonging in our field of therapy and these expressions are causing echoes in my mind and transforming into a resolve in my soul.[1]

What I would like to do in response to your longing is to write you a series of letters about narrative feminist therapy during this fall as the leaves are turning and as you are preparing for your work to begin in the dead of winter. If you will have me. Yes, that is a real question with an absentee question mark.

You see, this is a selfish endeavor because I have so much I want to say, and yet I come up against all manner of firewalls in saying anything at all outside of a real relationship with a person. When I write, not for a phony show, but for a relationship that is real tears and a question to me, I can barely help myself: the writing spins and spools and curls out of me with an annoying urgency that I cannot manufacture in any kind of spool factory – although my grandmother worked in one by a lake in Finland where I was born. In the evenings, her and her girlfriends amused themselves by secretly climbing up on the logging boats in the harbor that brought the wood for the spools and by daring each other to jump into the freezing lake. I hope to make you an offer that carries the spirit of those young women in the middle of nowhere, wasting time on laughter, shrieking, and challenge. What say you?

Wait. Before you say anything, I probably should expand on my hopes, so that you can understand them better and have more of a chance to suss out how to take my hopes and what you want to say back to them. I want to write you a series of letters about the most useful discoveries during my years of practicing in a narrative feminist agency, both as a therapist and in recent years, as a supervisor. I have pages upon pages of writing hoarded on my computer about work with clients and work with students already, but it is all lifeless chatter to me unless I can press it into the responsibilities and privileges of a real relationship that matters to me. I want to make the disjointed pieces of writing come together to a whole to serve your person at your beginning, one letter a time. Letters addressed to you will force me to sort birdsong from pretty feathers as well as to order them for human usefulness. In short, letters to you, dear student, will make the writing real and the writer honest.

Inside these letters to you, I want to share with you some of my most significant lived experiences with clients and with students up against a backdrop of anxiety and failure – in the form of case stories, transcripts, therapeutic documents, and some conceptual explication. In fact, I aim to combine "story work" with unique persons and "idea work" with general tendencies in a way it's not clear where one ends and the other begins. See, an idea is either significantly applicable in my conversation with a person next Saturday or I don't much care for it; similarly, my person on Saturday is either urgently forwarding an idea in their stories or I am not hearing them well enough yet.

And either the adventures respond back to the chatter of anxiety and failure in that odd feeling of satisfying focus (in my mind, it sounds like "fuck it, here goes nothing") or they shall be rightfully forgotten.

May I tell you a sketch of a story about me to embody my voice further?

There have been two times over the course of my career when I almost resigned. The first was right at the beginning, as a new therapist. On that day, I was watching a big-screen image of my face on an old-fashioned TV in a windowless office. I knew the session had not gone well, but I forced myself to watch it, and what I saw was the birth of a stutter. On the video, I set up to speak sentences, again and again, only to trail off, in mid-word. It was a silent debate with myself that sounded like "I . . .", (pause) "could . . .", (pause), "um . . .", (pause) "is . . ." (pause) and so on. I could not have said this on that day, but in hindsight, I know I was caught between what I thought I should say and could not bring myself say, and what I thought I wanted to say, and had no way of saying yet. Terrifying, hey? If nothing else, I aim to write you these letters to help you fail better than that disaster. Stuttering confusion is not what I aim for.

The other time I wanted to resign was as a beginning supervisor. The therapist I was working with brought a session of hers to me for review that had not gone well because she had felt entirely disconnected from the responses of the client. In speaking about what had happened for the therapist, she said to me: "I did what I thought you told me to do." And does this now terrify you? It does me. More so than a stutter, much, much more. Compliance is not what I aim for.

Both times I spent weeks thinking about resigning, in fits of excessive idealism run amok that I can never quite temper, that stupefying beast that rides me. I envisioned other work for me, and it was clear that it could only be work that would not center me speaking to other people very much – I wondered if could shelve books or arrange flowers somewhere.

When I finally decided to get a grip on those fantastical beasts of mine and delete my name out of the Guinness Book of Excuses once more for the time being, I resolved to do something else about these two moments of failure that had real effects on the clients in question. This series of letters is one visible expression of that something other.

So, my pedagogy[2] is a series of promises to you:

1. In these letters, I will show you real transcript excerpts of my own work. Let's face it, arm-chair talk about client work without looking at the client work is nice, but judging the one in the arena by her real words is much more fun, I swear.
2. I will sweat over[3] choosing those transcript excerpts that constituted real hinge moments in my work that helped me fail noticeably better at the task of becoming a narrative feminist therapist. This means, I'll choose the ones that I liked the best for the purposes of the point I'm trying to make, and I'll abbreviate extraneous stuff that would make the transcripts 30 pages long, because no one would read that. You're welcome.

3. I will really break a sweat in finding clear and concise words for the theoretical discoveries behind these hinge moments. Conciseness is not a virtue of mine, but I'll try to hack it for you.
4. And finally, my hope is that you would understand me. This last one is both a promise and a request: please don't leave me at "read" but work with me.

See, if I can labor to be understood and you can make yourself available to trying to understand me, then it means you can imagine my decisions, and if you can imagine them, it means that you can imagine them alongside decisions you have made and might make in the future.[4] And that means that you and I can be encouraged in a venture of reciprocal "thinking as well as we can" about the thing we are currently thinking about. And that is a space of siblinghood between us that can counter confusion with curiosity and compliance with consideration that is the closest space to moments of an actual paradise I have sometimes known.[5]

I know you're worried about whether you must have some training and education in narrative therapy to read or understand these letters. Short answer is no. I wrote these letters while disciplining myself into the assumption that you did not. Sometimes, when students grow in courage to tell me things, they whisper to me of an urban legend that reading Michael White and David Epston would give them migraines for being hard to understand. I'm all too happy to dispel this fear here and encourage you to give those lovely freaks the time of day! If you already have, wonderful, then you'll understand my sweaty endeavors ever more richly. But for the purposes of these letters, let me reassure you that I won't assume any afore-knowledge about narrative therapy or feminism. Most of our students come to our agency having heard the phrase "the person is not the problem, the problem is the problem"[6] and they might also have stumbled over the word "externalizing" once before, but that's pretty much the gist of it and that's where these letters begin.

Finally, dear student, I'm relieved that you said that my freedom of language use is not a hindrance to you, but that instead, it drew you to seek me out. (I may have muttered "Mary, Joseph and all the carpenters" or some such.) But seriously, to expand on the matter and make it even a slight bit more alarming – these letters come to you from the midst of client work with adults at a barrier-reduced therapy agency. I need to tell you that the topics of conversation in the transcripts pertain to moral hinge moments of interesting adult protagonists wrestling with sex, fat, suicide, drinking, yelling, hitting people in the face, being locked in rooms for three-hour yelling matches, and the difference between wanting to jump out of windows and sneak out of windows for ruin or triumph. In addition, swearing and satire are not unwelcome visitors but desired helpers to me and my people in endeavoring to find our second – best words for any moral matter and laugh while we're at it. If you're wont to be queasy about the word "fuck" or if you don't want to read me calling venerable people "freaks" or myself a "dumbass", then please be forewarned: forgive me, and I did it on purpose, and with tenderness, as you'll hopefully discover.[7]

See, a young woman once wrote before she died: "remember me when the cats are being themselves." I think of her sometimes, like now, and I smile. Quite simply, I want to write you letters to accompany you in your endeavor to be a therapist because that's what you told me you want more than anything, and one day we'll die and then it'll be too late. But before that day, there might come a day when I, if I'm very lucky, catch a glimpse of you defiantly and joyfully in the midst of your work, and as a "storyteller at the center of your world."[8] Then I'll surely have reason to think of that young woman again and smile. I aim for that day.

What say you?

Love,
Sanni

Notes

1   "Social justice has brought an 'ethic of belonging' to all of my work, paid and unpaid." Reynold, V. (2016). "Hate kills: A social justice

response to suicide." In J. White, I. Marsh, M. J. Kral, & J. Morris (Eds.), *Critical Suicidology: Transforming Suicide Research and Prevention for the 21st Century.* University of Chicago Press, pp. 169–187, 176.

2   "Is there such a thing as a pedagogy of the joyous, and if so, where do I find it?" Cheng Thom, K. (2019a) *I Hope We Choose Love: A Trans Girl's Notes from the End of the World.* Arsenal Pulp Press, p. 35.

3   "By using the idea of 'sweaty concepts' for this kind of descriptive work I was trying to say at least two things. Firstly, I was implying that too often conceptual work is understood as distinct from describing a situation: and I am thinking here of a situation as something that comes to demand a response, a situation is often announced as what we have ('we have a situation here') as well as what we are in. Concepts in my view tend to be reified as what scholars somehow come up with (the concept as rather like an apple that hits you on the head, sparking revelation from a position of exteriority) as something we use to explain by *bringing it in.* For me, concepts are ways of understanding worlds that are in the worlds we are in." Ahmed, S. (2014). "Sweaty Concepts," *Feminist Killjoys,* https://feministkilljoys.com/2014/02/22/sweaty-concepts/

4   "Reading a good story, we begin living it; the words disappear and we find ourselves thinking not about the word choice but about the decisions the characters are making and decisions we have made and might make some day, in our actual lives." Saunders, G. (2021). *A Swim in a Pond in the Rain.* Random House, p. 223.

5   "The strongest lesson I can teach my son is the same lesson I teach my daughter: how to be who he wishes to be for himself. And the best way I can do this is to be who I am and hope that he will learn from this not how to be me, which is not possible, but how to be himself." Lorde, A (1984). *Sister Outsider: Essays and Speeches.* Crossing Press, p. 77.

6   "Within the context of practices associated with externalizing of problems neither the person nor the relationship between persons is the problem. Rather, the problem becomes the problem and then the person's relationship with the problem becomes the problem." White, M. & Epston, D. (1990). *Narrative Means to Therapeutic Ends.* W.W. Norton, p. 40.

7 "I could say, 'Dismantle the patriarchy.' Or, 'smash the patriarchy.' Or use any number of verbs that signal urgency, but I don't. I am a writer, and I understand how language works. I understand how audiences – and readers – react to the language I use. I know exactly what I am doing. And I say, 'Fuck the patriarchy,' because I am a woman, a woman of colour, a Muslim woman and I am not supposed to say 'fuck.' In my experience, almost nothing can match the power of profanity delivered by a woman at a podium unapologetically." Eltahawy, M. (2019). *The Seven Necessary Sins for Women and Girls*. Beacon Press, p. 56.

8 "A people who doesn't live at the center of the world, as defined and described by its poetics and storytellers, is in a bad way. The center of the world is where you live fully, where you know how things are done, how things are done rightly, done well." Le Guin, U. (2019). *Words Are My Matter: Writing of Life and Books*. Mariner Books, p. 4.

1

What's So Narrative About Narrative Therapy?

A Letter About Stories and Story-Work

Dear student,

YES, you say. Bloody marvelous.[1]

Shall we get to it then?

On behalf of your beginning, I want to use the closest thing I have, which is my beginning. I have chosen the first page of transcript from my beginning as a narrative therapist for us to look at in this letter in the hope that it will serve your beginning well. I am purposefully inviting you into a brief transcript study of an actual session of mine because it will anchor the ideas in this letter in their practical application.[2] And then I will talk you through the main conceptual reach of this letter, which is the difference between an unstory and a story.

But before we read the transcript, humor me if you will a bit with some context for the transcript.

Last we wrote, I left you with an image of me stuck in a windowless room watching a therapy session gone to the birds for my inability to spit out actual sentences.

DOI: 10.4324/9781003478478-1

Now why would I sketch out that story? In fact, if you do not have a conscious space for this yet, I encourage you to start making room for a colorful storage space in your mind entitled "Why Would You Tell This Story?" – in short, WWYTTS.[3] This will be the one and only acronym appearing in these letters, so use it wisely. Thinking about why we tell the stories we tell is the beginning of questioning power relations. I'll explain: therapy, at its core, unlike some other professions, is a venture of bandying words. Narrative therapy, at its core, is a venture of bandying stories, with the hope of shaping healing stories into being, as in "narrative means to therapeutic ends" – see? One of the most difficult and tacit center weights of narrative therapy is a keen interest in power relations between the people engaging in this kind of bandying of words.

Words have aims, dear student. Stories have aims. Aims and purposes and intentions as diverse and technicolored as the reasons why humans get out of bed and do anything at all.[4] It would fill the rest of my life to endeavor to write out all the possible aims of stories and every new encounter I have would only add to the task, so suffice it to say, stories have aims to cause both the storyteller and its recipient(s) to think, feel, or do something. Behold, a great power. In order to help make this power more visible to us, we might engage in a practice of a few questions in the wake of our own, and others' storytelling:

- ◆ Why would you tell this story?
- ◆ What does this story incite me and you to feel, think, or do in its wake?
- ◆ If I were to take up the incitements of this story, where would they take me and you?
- ◆ And do I agree with these incitements?[5]

So why did I begin to sketch out that story of the day of the stutter?

At first, I experimented with two other stories to tell at the outset, but I deleted them and began again, because I needed a story that would better fulfill my aims. My aim was to speak into the heart of your fear. I wish to start lessening the anaphylactic shock reaction to stories of failure.

See, students in their moments of honesty whisper to me about walking a horrible tightrope of feeling very bad, like fake little impostors while, concurrently, nursing a winged wish to fly by the seat of their pants and be awesome in their sessions already. At the risk of sounding like a school-marm for a second, forget that bullshit.[6] Take your heart in your hands because here it is: you're going to eat many suck-it-sandwiches in your sessions, because achieving a healing conversation is difficult. But here's the other thing: the more you're willing to bear the cringey feeling of sucking and to return to look at sessions notwithstanding, the more you're also going to suck a little bit better next time.[7] In my experience, the willingness to suck and the ambition to suck a little bit better are your friends in this particular future endeavor. They will help you more than anyone in causing that electrically-steep learning curve toward the flexibility and innovation of practice that I know you so long for in your heart.[8]

And here's more good news: I can at least prevent a stutter for you! My past students are proof of that. If nothing else, your engagement with these letters can serve you at your beginning to already suck a little less than I did at mine and speak. Yay. So read on.

Okay, so you probably couldn't help but notice that my story as a therapist did not end that day in that room and in a stutter. The next scene in that story is the transcript excerpt in this letter. This transcript excerpt was recorded on my first day of my first job with my first client in my first session as a narrative therapist. To say "I was nervous" is an understatement: to me, that day was of great consequence! I had longed and studied to become a therapist for much of my life, failed once, and unsure if I could talk myself out of another ruinous day. To prevent a repetition, I had prepared as best I could. For me, preparation meant practicing narrative questions for months, out loud, and even recording myself speaking them, and then listening to these recordings during car rides and my kids' naps.

But more importantly, preparation meant giving myself permission for a particular kind of poise that would refuse some of the usual ways and ideologies of speaking that had incited my idealism to default on speaking altogether and cornered me into a stutter in the first place.

I'm not sure if that makes much felt sense to you, so perhaps a detail would bring it to life. For example, I was prepared to ask "forgive me, but what is that?" in response to a client's "mental health diagnosis." It was a question that I could likely spit out despite a stutter for its brevity, but more importantly, I had magical hopes that it would transport my client and me into the heart of narrative therapy from the get-go.

Last summer, I tried to explain this point to my parents, who are not therapists. I explained to them that narrative therapy concerns itself with stories of lived experience as the center weight of therapy, much like solution-talk is the center weight of Solution-Focused Therapy or thought-talk is the center weight of Cognitive Behavior Therapy. This was very boring to them. I tried again and told them that most clients don't come to therapy with a story, but with something like "I have depression." The aim of the game for the narrative therapist is to catapult themselves and the client out of the circular reasoning of "I have depression because I feel bad, and I feel bad because I have depression" and into the midst of a story of living that would help me understand something about why they feel "some kinda bad" in their lives right now. I had their interest now: "but how would you do that?"[9] they asked. Exactly.

On this day of the below transcript excerpt, I was prepared to ask short and punchy and stutter-resistant "What is that?" and "Where was that?" and "When was that?" questions to aid me to seek stories of what my first client had been up against in life.[10] But behind the scenes my questions aimed to seek a response to the question *"why* are *you* here, my dear?" on the slant.[11]

I will never forget what happened next, and now you have given me the reason to return to the excerpt of this session with you.

I will take you through this transcript excerpt and discuss it step by step to raise the thinking about my doing and what my doing might be thinking. Please note that the printing of this excerpt also denotes an important request to you to consider recording your first session, if at all possible. Even if you decide to hide it and deny it ever happened for the next decade, it might prove unexpectedly useful to you at some time in your future to consider what in the world you did think narrative therapy was about and what sparks of wit and charm you may have relied on in the pinch of a first session prior to the colonizing of you into somebody else's rote conversational manners.

Sanni:	Alicia, I have come to think that no one makes the decision to come to therapy *lightly*. I am thinking that something must really *matter* a whole lot to you in your life right now, as we speak, for you to consider taking time out of your day and your life to come to speak with me about it. So my question is, what is it that matters so much to you right now – it might be a dilemma, or a question, or an experience, or something really important that won't let you go – that you would walk yourself here to meet me tonight?
Alicia (Low):	I have . . . depression. My doctor said that it would be good if I went to counseling.
Sanni:	And do you agree with that recommendation? *(Alicia looks up, a bit surprised)* You know, I hear many strange, and good and foolish things doctors say when the day is long. *(Alicia smiles)* Do you think your doctor is one of the good ones?
Alicia:	She's alright.
Sanni:	Did she . . . do you agree with her . . . did she cause an interest, and are you interested now in talking about your life?
Alicia:	Yeah. I want to get a handle on this depression.
Sanni:	Okay, I wonder Alicia, forgive me . . . this will make me sound like an idiot, which you know, maybe I am, but: what exactly is depression? *(Alicia laughs a bit)* . . . Like . . . what is that for you, "depression," right now at this moment in your life?
Alicia:	It's . . . I have a lack of motivation. Like I'm stuck.
Sanni:	Huh. What do you mean, a stuckness – when in your life does such a thing happen to you?
Alicia:	Like a lot. I have been crying, and just not knowing what to do.

Sanni:	Can you tell me . . . Of those days, or moments, or afternoons, or evenings, whatever they are . . . when this, how did you say it? . . . tearful stuckness comes to you?
Alicia (Laughs a little):	Well the weather isn't helping. All the rain. *(Sanni: yeah?)* . . . It's like this, it's like I'm standing there, like in front of a river, and I can't move.
Sanni:	A river, what kind of a river? What's it look like, can you help me imagine . . .
Alicia:	It's muddy and overflowing. And it's raining.
Sanni:	And you look upon it, in the rain, and the tears come, and all movement is stilled . . .?
Alicia:	Yeah, but I need to move and cross the river . . . *(Low)* I just don't know how.
Sanni:	But wait. Why? Why do you wish to cross it, why not just stay over here?
Alicia:	*(Shrugs, considering)*
Sanni:	What's on the other side, what do you see?
Alicia (Suddenly tearful):	My sister. My grandmother.
Sanni (Very softly):	And why is it so important to you to get to them?
Alicia:	Because I need to tell my grandmother something important before she dies. I'm afraid she won't understand me though. My sister knows, but . . . that's another issue. She's drinking now. And I don't know if I should tell my grandma. She is very Catholic, and I'm afraid of how she'll respond . . .

I'll stop it here, because for a little while from here on out, Alicia and I got distracted with the metaphor of the river. Perhaps you paid attention to it too, dear student? But a metaphor is not the point. A metaphor can be very helpful to a point, but the point of narrative therapy is a story. That's all I want to write about for the rest of the letter, because I have a hunch that I am not alone in getting distracted by techniques of storytelling, in this case, metaphor work, and forgetting that the techniques either serve as narrative means to story or they do not. I believe that we have collectively forgotten what a story is and what it is not. If I had been asked at age 6, having been raised in Finland on the formidable stories of Northern storytellers, I would have immediately and urgently known what a story is. "A story is dangerous!" 6-year-old me would have confidently known. In a story, people fight for something important even though they are so afraid. The antagonists in the story are fierce and smart challengers and almost succeed in capturing the protagonist's soul. The companions to the protagonist are helpful and sweet and also flaky in their own unique way. People are given gifts like a wondrous horse or sword that come to their aid when the night is darkest. The story takes place in a particular land with its own logic in which dragons fly overhead, mountains and stars and wolves speak, and rulers rule and people respond accordingly. The protagonist is a fallible hero whose purposes for fighting I can understand all the way down to my bones. I root for them and fear for them on every page to know whether they go home and give up on that which is beautiful . . .

I don't think I'm alone in this appreciation of what a story is. Our instruction as narrative therapists sometimes confuses this knowing, or at least does not serve to focus our attention on it – unless I was a fresh new fool who just didn't understand my teachers.

Do you know what Alicia told me over the course of this session and the next, once the metaphor of the river was downgraded to its proper place as only one means to help a person step into a narrator position of the most urgent story of what she was currently up against in life? She told me that over the past two years, she had grown estranged from her grandmother who had raised her. Her grandmother was the one person in Alicia's life whose scent and gestures and traditions had always spelled "home" and "love" to her in a setting in which most other caregivers were too drunk to notice her. This estrangement had begun with Alicia's "discovery" that she was gay and her subsequent desperate attempts to "straighten up" because she feared that her queerness would unbelong her from her grandmother's home who had embraced a strict Catholic faith. Not only did Alicia tell me in tears of the things she had done in an attempt to not be gay, but also of her relationship with her first girlfriend that was currently imploding under the strain of heated arguments marked by drunken-ness and jealousy. Alicia had not breathed a word of her girlfriend to her grandmother, and, because of her own moral code of sincerity, could therefore barely stand to be in the presence of either of these important women in her life. Alicia's grandmother had noted Alicia's distance with mounting hurt and worry. In this time of dramatic fights with her girl-friend, Alicia longed for her grandmother's place as a home to go and spend the night away from fighting as she had done when she was little, and to find solace and new strength in con-versation with her. In the two weeks prior to speaking with me, she had indeed gone for a visit on a particularly difficult night, but had been silently sitting on the couch, choking back tears, torn between her grandmother's concerned questions over the teacups, and her girlfriend's impassioned text messages on her cellphone. Her grandmother implored Alicia that night to see a doctor, and Alicia agreed to it to soothe her worry. Alicia's doctor referred her to me with what she had termed "depression."

The above, dear student, is a story. It's a dramatic story that answers the questions of "why are you here in therapy, my dear?" and "what's that got to do with your lived experience, my dear?" and "what's most urgently keeping you up at night, my dear?" The wild thing is that I didn't know that at the time. In fact, it is worse: I encouraged this story along with Alicia guiltily and secretly and proceeded to hide this transcript from view because I thought it was a distraction from what I was supposed to be doing as a narrative therapist. The idea I had gathered was that I was supposed to be doing more of some sort of "river work," complete with finding out when and how Alicia had influence over the river, and how she might reclaim her life back from the river, or how she remembered sweet moments about what her grandmother appreciated about her and so on. While in session with Alicia we kept telling the above story, and behind the scenes, I kept telling myself that we would "get" to the "real work" of narrative therapy soon enough. Only the "soon enough" never came. And then other clients joined Alicia in telling equally unique and memorable and deeply moving stories of the dilemmas they were currently up against in life, and I sat there with my mouth open, and writing notes, and wondering what exactly this "narrative work" was that I was supposed to be doing in response to such stories. I stalked it all up to my inexperience, and after that excuse grew too wobbly, I blamed my clients at our agency for being singularly interesting people who just didn't resemble other clients I had read about in transcripts and therapeutic documents.

And here is where you find me still to this day, dear student. The only difference is that now, I am able to claim what I was only secretly doing back then as a narrative therapy practice.

What is a story, the center piece of narrative therapy?

A story is a narrative that has:

- a unique setting,
- memorable characters,
- a plotline in which grave and substantial matters of every day living are at stake to all involved,
- a political and cultural context,
- and a protagonist who is puzzling over it all in a unique voice and full of feeling and sensations, and thoughts and actions.

A therapy story in your beginning sessions is a story with all of the details of the above that has narrative causality, meaning, and it answers the query of "Why are you in therapy?"

If the beginning story doesn't relate to this question, Alicia and I could go on and on about the one time she went to camp when she was little, or this other time a clown came to her birthday. These might be interesting anecdotes, but a beginning therapy story vividly and significantly responds to the query why someone has come to therapy.

In the above transcript excerpt, I sought such a story from the beginning and with my first question:

> Alicia, I have come to think that no one makes the decision to come to therapy *lightly*. I am thinking that something must really *matter* a whole lot to you in your life right now, as we speak, for you to consider taking time out of your day and your life to come to speak with me about it. So my question is, what is it that matters so much to you right now – it might be a dilemma, or a question, or an experience, or something really important that won't let you go – that you would walk yourself here to meet me tonight?

I invite you, dear student, to critique whether this is the most eloquent way of inviting an interesting story of why this person has come to therapy and translate this into your own mouth and vernacular for your beginning.

The idea of this first question for me is to welcome my person as an agentive protagonist with an interesting dilemma that is of substantial consequence to their living right now. I aim to tell my person that I mean to be interested in that thing that they are currently most interested in, that I think of human dilemmas as interesting, and that I wish to put all of myself to use to story this dilemma into being as well as we can. I did not come up with these aims for my clients alone.

See, narrative therapy has perhaps most famously of all told us whom people are *not*: if you read the prologue of this book, then you have already heard of this idea, that the "person is not the problem, but that the problem is the problem."[12] This is a wonderful clue and upon first reading such an idea, it was as if the mountains and waters of my own landscapes rumbled and moved, and nothing was ever the same again for me. I only came to discover later that this idea has also been translated into some vague yet totalizing humanist assumptions about the inherent goodness of persons, and the inherent badness of problems such that very few conversations with actual living persons can possibly lead to any interesting dilemmas or their subsequent revolutions after all.

But a very close read of narrative therapy theory can tell us something rather interesting about who persons are which shaped my conversations with clients, including the above with Alicia, in useful ways: persons are "agentive,"[13] "active mediators and negotiators of life's meanings and predicaments," "originators of many of the preferred developments of their lives,"[14] both "narrators"[15] as well as "protagonists"[16] of the stories of their lives.

What this spawned for me was the following imagination:

Imagine. What if every person you met had walked a long, dusty, and confusing road to come to this conversation with you. What if no one had made the decision to come to therapy lightly. What if they had waited a long time for this chance to speak with you. What if every person you met was full of vivid questions about living, had had countless arguments with themselves about what they were about to say, had lived sleepless nights and tears full of wondering, and was full of jumbled or clear or fierce words. What if every person came with longing, desire, and will – however low such longings had been driven underground by years and years of ideological belittling in the form of Master Narratives that determine who in our world has legitimate claims on dreams at all.

Would you treat them then as rambling philosophers of life, as sages and seers of powers unknown, as poets of the ordinary Wednesdays, as magical thinkers and feelers of things you have never even imagined let alone glimpsed or felt, as siblings to you in the long night where none of us hold the keys to paradise – and those who pretend to should be laughed out of the room – as protagonists (and not side characters like the "maid" or "love interest" to someone else more interesting), and simultaneously, as authors of their own stories, as purposeful, willful, original agents in their own most fascinating endeavors, as supervisors to you in your learning about life ("in our pain, we are teachers") and joy that you would never ever forget, as long as you shall live? If they each had ancestors, long lineages, legacies, waters, and rocks they came from, words of their own, yes, whole dictionaries of their own, unique ways of loving and hating, companions to their journeys who are not 100% trustworthy but still beloved, antagonists who challenged and irked and angered them but were not 100% garbage either and spare and strange reasons for getting out of bed in the morning – how would you speak to them then?

If each and every one of them was full of the thing called "lived experience" and it was up to us to invent the means to center and story this lived experience – what would you ask them then, as your first question?

Your clients will not tell you any of this. Even if you try to shape such an imagination into your first question, only some clients turn into storytellers and philosophers right there and then. Most times, they do not. Note Alicia's response: "I have . . . depression."

Alicia's response is a good example of an unstory.[17] Think about it: if a "story" has memorable characters, a setting, a plotline in which grave and substantial matters are at stake to all involved, a political context, and a protagonist who is puzzling over it all in a unique voice and full of feeling and questions – an unstory is the antithesis to a story. An unstory is a label, an identity conclusion, a disembodied description that is stripped of the context, the history, the relationships, and the personal every day lived experiences of a living human being.

Perhaps some examples of unstories that are common at our agency will make this even clearer: "I have depression." "I have trauma." "I have complex PTSD." "I wear these masks." "I struggle with jealousy." "I have anxiety." "I'm addicted to Adderall." "I have panic." "I freak out for no reason." "I have burnout." "I have trauma." "I have this pattern of unhealthy relationships." "I care too much." "I am an empath." "I have ADHD." "I need to be more resilient." "I am so broken." "I have borderline." "I need to feel more present." "I have trauma." "I don't have self-love." "I have codependency." "I have been abused." "I have an eating disorder." "I have no motivation." "I am not my best self." "I have self-sabotage." "I have trauma."

Perhaps you noticed that I snuck in the word "trauma" more than once into that list and that's not an accident. The "having" of "trauma" is among the most popular unstories of our current time, followed shortly by a re-interest in adult ADHD, especially among women and queer people.

I recently heard a new one, that was helpfully supplied to my client by her friend on the phone and translated to our conversation like this: "I have cognitive distortion disorder and need medication for it."

An unstory is:

1. characterless, plotless rhetoric
2. that purposefully burns away all political and relational context of lived experiences
3. by constructing a hollow label to the question of "why are you in therapy?"

In short, unstories do the work of unstorying people. They actively work to suspend, steal, and shush persons' capacities to account, to respond, and shape their lives forward as interesting agents in the midst of interesting social, political, and relational dilemmas.

The majority of my clients over the past years have responded to my first question with an unstory drawn from the medical lexicon, like Alicia did. Why might that be?

Since the inception of narrative therapy as one of the anti-therapies to the medical model, one of the master colonizers – the vernacular of the unstories of the DSM and its popular cousin, the self-help literature – has had unprecedented success in capturing people's minds and vocabularies. It has been on a march to rip up our diverse languages for our own experiences, to diminish our bodies and our thoughts and those of our ancestors, and to sound-proof our visions of our own worlds to sound generic, disposable, and replaceable. But my work in the midst of a non-profit agency, among the halting, bursting, shy, and fierce stories of people who were not born into privilege, of women, queer folk, poor women, and women of color, of chronically ill women, and of the stories of what happened to them in their lives, in their families, communities, workplaces, bedrooms, drug houses, backs of cars, and hospitals, has convinced me that unstories are proliferated particularly vehemently among some people. In fact, I have come to think that the speech acts of this convincing rhetoric and its particular pathologizing vernacular combined with the sexism, racism, transphobia, homophobia, and ableism of our world have particularly attempted to steal marginalized persons' rights to positions of authorship and roles as protagonists of their own life stories. To be an author, and simultaneously a protagonist of one's own stories, are claims of moral substance, belonging, authority, interesting-ness, and shaping rights and shaping responsibilities in one's own right. Put another way, those who have been relegated to the sidelines of history by dominant, privileged, and powerful others also have a history of being relegated to the roles of minor characters of more privileged protagonists' stories. In my meetings with clients, it has become my mission to return their own stories into their hands from the prejudice and subjugation of the unstorying attempts on their lives.

"But what if clients themselves cling to their labels?" my students sometimes ask me. Good question. The construction of lived experiences of struggle into unstories is multi-motivated and also has a multitude of effects on persons' lives. Sometimes, unstories and labels are exchanged as an expression of a person's hopes for care, for understanding, for grace, for respect, for solidarity, for belonging, for conspiratorial tips and tricks, for gentleness. Are these hopes realized in a particular client's life? Start noticing. Of the multitude of effects of unstories on a client's life, perhaps the most difficult for me to behold is the stifling of curiosity about one's own living. What in particular is difficult for me in my day-to-day life? What am I pointing towards when I say "I have depression?" I have never heard two identical descriptions of any struggle on the ground of one's life between many persons sharing the same unstory. "I don't know what depression is" – this throw-away line from my first conversation with Alicia is now truer for me than ever after a decade of practice.

Unstories, if you will, provide easy reductive answers to complex questions of living – often before the question is even over. Perhaps there is a sweet, gratifying certainty or relief to this reduction. But such reductions are an antithesis to curiosity as an attitude of mental and emotional engagement in the world, a continuous act of the frustration and pleasure of "figuring things out," a project of "becoming" rather than "being."

Now I'm not asking you, dear student, to set out to argue with clients' unstories. I am only asking you to think of unstories as a portal at the beginning to invite clients to translate them into living stories with you in your sessions. Think of it this way, if you will: unstories provide "universal, global, and stable" explanations and metaphors for persons' misery. "This always happens to me, because I am this way" – end of story. Your work of the translation of unstories into stories insists that matters of living become "particular, unique, and currently-on-the-move" instead. What can you or I or Alicia or anyone do with a global and stable explanation of "I have depression." And what can a person do with the particular unique life experience of "I'm fighting with my first girlfriend because I think she doesn't like me, not really, and I stand to lose the one person in the world who likes me?" That difference is what makes the difference between the pressure to dispense manualized advice to universal client problems, and the venture of storytelling in therapy. And the best part is: you do not have to step into knowing what any person ought to do. Your concentrated labor in translations to stories, and clients' own understandings of "oh so this is what I'm struggling with here this spring" – will prompt them to originate a unique response, driven by their desire in the world, that you couldn't have thought of if you tried.

Returning to the transcript of the conversation with Alicia, let's take a look at some ways to do so:

Alicia (Low):	I have . . . depression. My doctor said that it would be good if I went to counseling.
Sanni:	And do you agree with that recommendation? *(Alicia looks up, a bit surprised)* You know I hear many strange, and good and foolish things doctors say when the day is long. *(Alicia smiles)* Do you think your doctor is one of the good ones?
Alicia:	She's alright.
Sanni:	I wonder, did she cause an interest, and are you interested now, in how we could get up to talking about your life and what that could possibly do?
Alicia:	Yeah. I want to get a handle on this depression.

What does the above interlude about Alicia's doctor have to do with resisting unstories in our clients' lives? To me, it represents an important haggle to establish the authorship of a story. Alicia might well wish and have been trained to propose her doctor as someone with the rights to summarize her lived experience and to shape our work forward. I hoped that my query of Alicia's agreement with her doctor's recommendation and the invitation to evaluate her doctor's aims for her would raise a possibility of inviting Alicia into her own agentive authorial position in relation to me. Let me put it this way: had I accepted her doctor's input the way Alicia offered it to me - who would have been answering the query of "why are you in therapy, my dear?"

It is of utmost significance to me that the reason for therapy is negotiated between me and an active agent and a storyteller with authorial rights – even if we sometimes decide to agree to take another person's scribbles in the books of a person's life into account in these negotiations.

So: at your beginning, question unstories, question authorship. Translate unstories back into stories of lived experience, told by an interesting author right in front of you. Check, check, so far, so good? All this you may have already expected from narrative therapy. I'll raise the stakes one more time for you, dear student, before we conclude. Note that Alicia and I went from "depression" to "stuckness" to a "muddy, overflowing river" in her responses to my queries. And it is in these skips and hops over different words to express lived experiences that even narrative therapy loses its original intention to shape stories forward. Many narrative therapists would end all story efforts already at the word "stuckness," and, at the latest at the mention of a "muddy, overflowing river," just as I felt pressed to do. What commonly happens is that in our excitement to discover expressions outside of the vernacular of the DSM, we keenly embrace these words, call them a "story," and proceed to work with Alicia to reauthor her relationship to "the river." But the muddy, overflowing river is not a story. It is simply a more poetic way to say "depression." Remember what a story is?

A story for us as narrative therapists is a narrative that has:

- a unique setting,
- memorable characters,
- a plotline in which grave and substantial dilemmas of everyday living and relating are at stake to all involved,

- a political context,
- and a protagonist who is puzzling over it all in a unique voice and full of feeling and sensations, and thoughts and actions,
- and which answers the query of "why are you in therapy?"

Forgive me for this repetition, if you were already there. I tell you, sometimes I have felt like a madwoman wanting to yell "that is not a story!" in my best imitation of a menace-to-the-party voice that sometimes gripes out of me in my sessions when I am trying to impersonate so-called problem voices. The image of the "muddy, overflowing river" was only a means by which it became easier for Alicia to tell me the story of what she was currently up against in life. I call these initial therapy stories "Up Against Stories" to delineate them from structurally different "Counterstories" later on in the therapy sessions – but that is matter for some future letters, if you still wish to hear from me after this elaboration.

For now, know that the most radical decolonizing question you can ask in your sessions is "What happened?"

This often leads to other very difficult questions such as: "And then what happened? Where were you? What did you do then? What did she say? And what did you say then? And then what happened? How come? What made you so angry? And how did you two part then?" and so on.[18] This series of simple story-prompting questions is the catapult that my parents wondered about, and they are also behind my coming to know Alicia's story behind the metaphor of the river that I summarized above. Do not be fooled: despite their simplicity, such questions are rarer to be found in therapists' transcripts than needles in a haystack. It appears that we have accepted invisible marching orders to skip and hop straight into metaphor land or meaning-making land without any knowledge of clients' actual life experiences, on the ground of the magic of their ordinary moral moments among other human beings.

Alicia became my first experience for the notion that stories can tell people what to do. Alicia never showed any interest in returning to either the word "depression" or its poetic equivalent, the "muddy, overflowing river" in any of our future conversations. You are probably wondering what she was interested in instead. To the best of my ability, what I can say is that it was her own life. I can barely take credit for this as I know I was entirely out of my depth in all that I ever managed to ask of Alicia in our follow-up sessions whenever I even managed to remember to speak when my mouth was not wide open in wonder. Honestly, while swearing that I would get back to the doing of the "narrative therapy," I questioned Alicia as to why she drank when she would rather speak to her girlfriend, what happened in their yelling dialogue with each other, how their nights ended, what made Alicia so lonely in it all, whether she wished to break up with her girlfriend, how come she was feeling so jealous, what was said in her conversations with her grand-mother, what Alicia's objections to her grandmother's beliefs were, and why she needed to confess anything whatsoever to her grandmother about her girlfriend or her intimate life, and why she could not go on loving her girlfriend and lying boldly to her grandmother about it and so on.

And, to my astonishment, Alicia and I heard something that was far scarier than the rush of a muddy river inside these conversations: the snarl of a queer-loathing dragon that would have the heart of her and me. More about that snarl in the next letter.

Somehow, Alicia managed to understand the crudely formed queries out of my mouth as sincere curiosity on behalf of her life and somehow her voice grew steady over the land and the dragon. She left her girlfriend on purpose, because she decided that love had given way to angry obligations, and then she promptly stopped drinking due to wanting to keep up with a group of women on the sports field, and then she fell in love again on said sports field. Alicia and I met one final time right before Christmas later that year, and she proudly proclaimed to me: "This is it, Sanni! I am taking her to meet my grand-mother and see the pickles hanging on grandma's Christmas tree!" I laughed and we haggled out a plan for said meeting, that would include a pre-party, eye-level heart to heart with grandma over teacups. That was the last I ever heard of Alicia. Well, not exactly. Someone tells me that she got married to her girlfriend last year. So, was my aim in all this to tell you a pretty "They Lived Happily Ever After Story?" Maybe. But I am pretty sure that places us in the genre of fairy tales, and we are working in life, after all. And perhaps I'll venture a guess that both you and I know that fairy tales don't ever start with weddings . . .

So let the plot thicken, and welcome to the dilemmas rising all over again. In my most bursting moments of good hope, I swear I can almost hear the intermingled voices of three important women, Alicia, her wife, and her grandmother, as they are finding the words to tell each other the stories of their dilemmas that life is serving to them, all set to the sound of teacups finding their queer places on mismatched saucers.

Love,
Sanni

Notes

1   A British aid worker and my friend Frans Barnard described his freedom as "bloody marvellous." Barnard, F. (2010). "It's marvellous to be free." *The Independent.* www.independent.co.uk/news/world/africa/it-s-marvellous-to-be-free-says-abducted-aid-worker-frans-barnard-2111508.html

2   "Don't just tell me, *show* me." Epston, D. (2018). Personal communication.

3   "What is the heart of you, dear story? (Or, channeling Dr. Seuss, 'Why are you bothering telling me this?'" Saunders, G. (2021). *A Swim in a Pond in the Rain.* Random House, p. 84.

4   "We tell stories that make us seem adventurous, or funny, or strong. We tell stories that make our lives seem interesting. And we tell these stories not only to others, but also to ourselves. The audience for these stories, of course, affect the stories we tell. If we're trying to impress a date, we might tell a story that makes us seem interesting or witty or caring, whereas if we're trying to justify a dubious act to someone who is judging us (or perhaps ourselves), we might tell a story that makes us out to be without other recourse in the situation. In the latter case, what we are doing is dissociating

ourselves from a value we might be associated with and thus implicitly associated ourselves with a different one. Not all our stories about ourselves express values like these. However, many – perhaps most – of them do. This is so even where a story might seem to express a disvalue. Think, for instance, of people whose stories about themselves are often about things not working out for them. Whatever they try, they fail; the world conspires against them. These stories express values as well, values that often stem from resentment or even despair." May, T. (2017). "The stories we tell ourselves." *New York Times*.

5   "Okay. To encourage speakers to situate their opinions in the context of their purposes, we should ask questions like: So you have a strong opinion about what I should do. Tell me, in voicing your opinion in this way, what effect do you hope this might have on what I do?"White, M. (1995). *Re-authoring Lives: Interviews and Essays*. Dulwich Centre Publications, p. 129.

6   "But it is preposterous to imagine that we ourselves are determinate, and hence susceptible both to correct and to incorrect descriptions, while supposing that the ascription of determinacy to anything else has been exposed as a mistake. As conscious beings, we exist only in response to other things, and we cannot know ourselves at all without knowing them. Moreover, there is nothing in theory, and certainly nothing in experience, to support the extraordinary judgment that it is the truth about himself that is the easiest for a person to know. Facts about ourselves are not peculiarly solid and resistant to skeptical dissolution. Our natures are, indeed, elusively insubstantial – notoriously less stable and less inherent than the natures of other things. And insofar as this is the case, sincerity itself is bullshit." Frankfurt, H.G. (2005). *On Bullshit*. Princeton University Press, p. 68.

7   "Try again. Fail again. Fail better." Beckett, S. (1989). *Nohow On; Company, Ill Seen Ill Said, Worstward Ho!*, Grove Press, p. 7.

8   "The result [of practicing something] is a speed and directness of response comparable to that of mere habit, but unlike it in that the lessons learned have informed it and rendered it flexible and innovative."Annas, J. (2013). *Intelligent Virtue*. Oxford University Press, pp. 28–29.

9   "How can we assist people to name their experience? How do we ask questions in such a way that words come alive for people? How can we ask questions in such a way that people make such vocabularies of experience their own? How do we allow people to decide what words resonate for them? What words are capturing of experience and, in particular, that experience that has not had words before? Experience that has not been rendered in to an event before? What do we do in therapy talk that generates the new rather than merely reiterating the old? Shouldn't we take an interest in words that are alive with association? Shouldn't we think about the poetics of language and concern ourselves how words feel to people?" David Epston as quoted in Paljakka, S. (2021). "Christina and the Robin: A decidedly narrative response to rape." *Journal of Contemporary Narrative Therapy*, September release, pp. 6–31, 10.

10   "I'm really interested in people's accounts of their experience. I really want to understand what life has been like for them. So, I guess the first part of my work is to try to get some appreciation of what persons have been going through. I think it's important that I achieve a degree of understanding about this, and I think that it's important that persons are aware that I have achieved at least a degree of this understanding." White, M. (1995). *Re-authoring Lives: Interviews and Essays*. Dulwich Centre Publications, p. 21.

11   "For many years I was trying to figure out: What would a good narrative therapist do here? Why are my clients not following the logic of my questions? How can I improve my externalizing practices? How should I follow this conversational map? These are questions about 'technique.' However, my attempts to answer them left me feeling stifled rather than liberated as a practitioner, and I began to doubt my own competence. So, I changed tack, and began to tackle a very different kind of question: not 'What should I do?' but, 'who are *you?* What have you been made into? How would you like to live?' This lead me to other questions: 'Who is the person in narrative therapy? What sort of world does she or he occupy, and what impact does this have on him or her?' I began to de-privilege the question of how to do good narrative therapy, and became more curious about how I should think about the person sitting in front of me" Guilfoyle, M. (2014). *The Person in Narrative Therapy: A Post-Structural Foucauldian Account*. Palgrave Macmillan, p. 3.

12 "Within the context of practices associated with externalizing of problems neither the person nor the relationship between persons is the problem. Rather, the problem becomes the problem and then the person's relationship with the problem becomes the problem." White, M. & Epston, D. (1990). *Narrative Means to Therapeutic Ends.* W.W. Norton, p. 40.

13 "This is an account of personal agency, an account that emphasizes what could be called the person's 'agentive self.' It includes details about what the person has been up against in the performance of this personal agency, and, against this background, emphasizes the significance of any more recent steps that the person has been taking toward having more to say about how their life goes." White, M. (1995). *Re-authoring Lives: Interviews and Essays.* Dulwich Centre Publications, p. 143.

14 "Intentional states of identity are distinguished by the notion of personal agency. This notion casts people as active mediators and negotiators of life's meanings and predicaments. It also casts people as the originators of many of the preferred developments of their own lives: people are living out their lives according to intentions that they embrace in the pursuit of what they give value to in life." White, M. (2007). *Maps of Narrative Practice.* W.W. Norton, p. 103.

15 "Rather, this work encourages people to take up an observer or self-reflexive position in relation to their own lives, a position in which they become the narrator of events . . ." White, M. (1995). *Re-authoring Lives: Interviews and Essays.* Dulwich Centre Publications, p. 134.

16 "The narrative mode locates a person as a protagonist in his or her world." White, M. & Epston, D. (1990). *Narrative Means to Therapeutic Ends.* W.W. Norton, p. 82.

17 "As LGBTQ2+ individuals we very often come into the world with a story of what we are not: straight, whole, beautiful, enough. This story is the soul of colonization and homophobia: it drains us of the will to struggle, of the confidence to name ourselves and our ancestors, the vision to see each other and act in solidarity. Homophobia is an anti-story, an unstory; it erases and diminishes our bodies and rips away our stories." Cheng Thom, K. (2019b) "Storytelling and poetry workshops." https://moosejawpride.ca/event/kai-cheng-thom-storytelling-poetry-workshops/

18 "Psychologists should also familiarize themselves with culturally appropriate treatment modalities. Additionally, traditional healing stories in Indigenous culture involve transformation, rather than removal or erasure. Psychologists should understand the value of these stories and the power of the concept of transformation. Appropriate conceptualization should focus on finding solutions or bringing help and relief rather than labelling, diagnosing, or judging. For example, providing descriptive terminology of the behaviour is more helpful than a clinical name for a pattern of behaviour or thinking. Furthermore, conceptualization should be contextualized in community. Approaches should be collaborative and represent the more collectivistic social structure favoured by many Indigenous clients." Canadian Psychological Association & The Psychology Foundation of Canada. (2018). *Psychology's Response to the Truth and Reconciliation Commission of Canada's Report.*

For Alicia
My First Try

In the bathroom
– Why do turning points happen in the bathroom? –
Staring at the pregnancy test

I thought
How is it that a person who believes in
Old fashioned loyalty
Managed to
Fail
My best friend
My girlfriend
And my best grandma
All at once?

Maybe
If my loyalty
Lies
With this first best try
At this queer life
Of mine.

2

What's Riding Us?

A Letter About the Master's Narratives

Dear student,

The last letter denoted a beginning attempt to define stories by way of contrasting them with unstories. It was an effort to think alongside David Epston's question of "how come we call ourselves narrative therapists and know so little about what a story is?" (Carlson, 2020). I also claimed that the narrative invitation to side-step unstories in favor of bidding stories of lived experience is an anti-oppressive re-orientation of practice.

Today's letter is a thought piece about a concept called Master Narratives,[1] that can obstruct both the telling and hearing of such living stories.

(Disclaimer: I am writing to you from a therapy room with a specific location – in Canada – and in a specific time – in the year 2023. The particular examples of ideas are specific to this location and time, and I won't pretend to speak to dominant cultural narratives of other countries and times. In fact, it is worse: I am an immigrant to Canada, so I inhabit a peculiar insider-outsider position to cultural norms in the community I currently live in. I'll do my best to explicate my thoughts and ask you to take the particular peculiarity of my vantage point into account as you weigh my thoughts. For example, you may have already

DOI: 10.4324/9781003478478-2

gathered that I was never properly taught that voicing direct disagreement as clearly as you could as a woman was anything but
an offer of friendship and respect. Sheesh. I found out as my first
lesson in Canadianism that yelping "that's interesting! I totally
disagree with you!" and proceeding to explicate my reasons for
disagreement over the heads of other university students would
not lead to an offer of going to coffee afterwards to continue an
enlivening discussion as the first portal to life-long friendship. In
the 20 years I've lived and worked here, I've learned to not say
that. Just kidding. And before I get started on the obligations and
entitlements of Canadian small talk – end disclaimer.)

I want to begin this thought piece with the promise I made in
the last letter to elaborate on the "queer-loathing dragons" that
Alicia and I met as an embodied example of working with Master
Narratives. After a look at this example, I'll define the concept for
you and challenge you to think about the Master Narratives at
work in all that we have been taught about what it means to be
a "good therapist."

Inside my early conversations with Alicia, we stumbled upon
Master Narratives while I was encouraging her to tell me of her
lived experiences. In Alicia's descriptions of what pained her in
her life and her relationships on any given ordinary Saturday
nights, I could not help but wonder out loud about the seemingly inescapable misery of stuckness and the binds on her
freedom of movement that kept her feet glued and her imagination stale. Why could she not speak to her grandmother about
her girlfriend?

In Alicia's case, the question that opened a portal to
questioning the easy reign of Master Narratives was "and why
do you have to tell your grandmother about being queer at all?"
This was a hypothetical proposal that accidentally outed one of
the Master Narratives that was creeping on Alicia: that women
and queer people and polyamorous folks and trans people and
people with disabilities, and all other people who do not fall
in line with cisgender-heteronormative imaginations of love
owe confessions and veritable "comings out" of their intimate
relations to powerful others. "Lying is the work of those who
have been taught that their truths have no value," Amber Dawn

writes.[2] In one lively evening, it caused great hilarity for Alicia and me to consider lines of follow-up questions:

- ◆ "And why is it that you are charged with 'lying' only when it comes to confessions about your sexuality?"
- ◆ "What all else do you routinely 'lie' about, by way of omission, to your grandmother?"
- ◆ "For example, do you also owe her tellings of your greatest triumphs and sorrows at work in relation to projects and colleagues that really matter to you? Do you feel like you owe her a description of what made your heart so full of life today when you walked the dog and ran your hand through the first wet lilac blossoms of the summer and how it made your hand smell? Or perhaps you should make sure you tell her the brand of tooth-paste you use as well as why so as to ensure you are not a liar in relation to her?"

Alicia's laughter about these considerations emboldened me.[3] Now earnestly, I asked:

- "Do you wish to give your grandmother the authority to decide about your queerness?"
- "Does her approval, in the end, decide whom you love?"
- "Then how about this, do you give authority to the anti-gay bastion of the Catholic church to judge and guide whom you love?"
- "No," Alicia answered, slowly, to each of these questions, after serious consideration.
- "Who then do you give the authority of opinion to judge whom you love?"
- She didn't bother to answer this question in words but smiled at me.
- "Okay, I see." I smiled back. "Then what if it were possible to imagine a chat with your grandma that didn't seek her approval – so not an approval-seeking chat – but more of a . . . like an "opinion-sharing" or "experience-sharing" chat with her in which you can tell her what you want but she

doesn't become the "decider" of your life . . . do you know what I mean? What would that conversation sound like – are you interested to think with me about that?"

Alicia was interested, and we spent some time imagining a conversation in which she didn't owe a confession (countering the insidious Master Narrative that something is "wrong" with queerness), and in which Alicia's grandmother wasn't cast as an arbiter of a good life for Alicia (countering the Master Narrative that marginalized persons are obligated to performances of proper personhood to regain their worth under the gaze of representative authorities).

These queries of the errands of Master Narratives promised Alicia and me cracks of freedom in our conversations: once Alicia's supposed obligation to confess and be judged by norms she did not agree with was made visible – she responded with a new measure of freedom to consider possibilities of her own agentive response that had hitherto been held hostage. In other words, Alicia experienced a measure of freedom, defined as "more options and more say" in relation to her imagination of how she wished to respond to the events of her life, in this case, negotiate her queerness with important others. The wonderful epiphany to me was this: once Master Narratives are held to account for their prejudicial shaping of our lives and our moral worth, persons can experience some measure of agentive freedom to engage with their own conscience in envisioning a response to important matters in their life. It makes sense, does it not: once the impressiveness of the power of Master Narratives to degrade a person's worth is outed, persons are regraded to equal moral worth and can be invited to consider the bids of their own conscience in responding to life events again. In Alicia's case, when we removed her obligation to confess details of her intimate life to her grandmother and seek her approval that was based on ideologies that Alicia did not agree with, we were able to consider the possibilities of Alicia's moral code of honesty that Alicia did agree with![4] We'll return to these ideas in a future letter entitled Moral Reading Prompts, so no worries.

For now: welcome to the world of Master Narratives, those slippery creeps. Master Narratives like to resist inquiry, and the best way I can point to their incitements is inside particular client stories like the above which I will do in these letters. Master Narratives are powerful judgments as well as prescriptions for action that are riding all of us to attend to and interact with the world and ourselves in particular ways. We have all been formidably trained in Master Narratives to inform both our assumptions and our templates for relating to this world. Master Narratives don't helpfully announce their presence or their origin, like DSM labels do. Master Narratives nebulously resist their seeing, naming, and speaking, and yet they carry with them the whips to our practical obedience in very real-world conversations with real-world people around real-world teacups.

Master Narratives are sets of beliefs that we are not born with, but instructed in, that settle on our bodies, minds, and hearts as unquestioned codes of conduct for maintaining our survival, our safety, and our belonging in this world.

Master Narratives are:

- guides to living,
- drawn from master ideologies that define the moral worth of persons,
- that prescribe obligations and entitlements to different persons unequally,
- and that are powerfully enforced in everyday life in the form of social mores.

I learned the term "Master Narratives" from Hilde Lindemann-Nelson. She taught me that Master Narratives are dominant cultural ideologies[5] that play a powerful part in the construction and tightening of identity conclusions around a person. If you don't like any of the above attempts at definition above, think of Master Narratives as "stock plots" or "stock characters" in your stories. Think for example of princesses that are not white, pure, straight, cisgender, thin, and don't live their lives minding their "p's and q's" in wholesome pursuits in a tidy neighborhood. Think of the "other lives" that are being lived – and waiting to be storied – while our collective dreams of wholesomeness are being held hostage.

"The most subversive thing a woman can do is talk about her life as if it really matters," Mona Eltahawy writes.[6] For our purposes, we may substitute "queer person," a "trans person," a "person of color," a "person living in poverty," or any marginalized person in a community for "woman" in this quote. I forward that it is our responsibility and privilege as narrative therapists to concern ourselves with the encouraging of stories that matter, especially the stories of those folks whose life experiences have been routinely and systematically shunned out of the circle of credibility.

Students sometimes tell me that they are bogged down by the limitations of an individual or couple or small group therapy conversation in relation to the wish to play a part in addressing the systematic and structural injustice in the world. I am the last person to talk them out of such discomfort. Yes to rage. Yes, the therapy room is not in charge of writing policies for housing, banking, safety, policing, clean water, food security, education, child care, gun control, accessibility, bodily autonomy, and so on.[7] I tell my students that when I get mad enough, I sometimes entertain fantasies of becoming a lawyer or judge or policy maker until I quickly remember the attention to bureaucratic details like statutes and precedents that their work entails and concentrate my anger back to my own realm of influence.

Here's the thing: the knowledge of the limitations of a therapy conversation in relation to structural injustices does not excuse us to shrug off our call to imagine the therapy room as a site of political activism.[8] We as therapists *are* in charge of writing policy all the time at our agencies, policies that pertain to note taking, letter-writing, exclusion and inclusion criteria for clients, accessibility, sliding fee scales, session limits, outcome measures, supervision conversations – and most centrally, the policies and rules and incitements of how we show our regard to clients and what kinds of conversations we routinely invite them into. The therapy conversation is a unique space that can formidably lend itself to practice healing and accountability in inviting and encouraging responses to both individual incidents of injustice as well as helping to transform the oppressive conditions in our relationships to each other.[9] Yes to using our position as therapists to write letters to lawyers, parole officers, teachers, psychiatrists, family members and employers, as expressions of extending the reach of justice-doing beyond the porous walls of a little therapy room! Yes to using our position as therapists for community practice and witnessing ceremonies of all creative kinds to bring our clients' worlds into the little therapy room.

But before all that, yes to examining our practices of attention inside the four walls of the little therapy room: how does it come to be that stories of the lived experiences of folks who are currently defying our cultural norms of wholesomeness are too often met with silent discomfort?

I propose that the first step of political activism inside the therapy room begins with the consideration of the endowment of our narrative resources of time, attention, curiosity, respect, solidarity, enthusiasm, wordsmithing, good will, labor, and tenderness toward the life experiences of those whom our specific culture has deemed people whose voices we can do without. Even if that were all we have – it turns out to be a very great deal.[10]

And how can we as therapists start to question the widely accepted "Master's Narratives" that keep our imaginations about the lives that are possible in check?

I believe you are now hoping that I would take you through a reflection of all the Master Narratives that I have met in my work over the past 10 years, like a parade of diversity-dissection. "Where is the book on all the Master Narratives?" a student of mine once asked. I told her and all the others to begin writing it by thinking about their own answers to such sentences:

"A good woman is . . ."
"A good mother is . . ."
"A good partner is . . ."
"A good man is . . ."
"The right kind of sex is . . ."
"A good relationship is . . ."
"A good daughter is . . ."
"The proper way of working is . . ."

"The proper way of doing anything is . . ."
"A beautiful person is . . ."
"A proper house is . . ."

This did not help. So let me then do something even more annoying and turn the attention, not on some imaginary categories of marginalized persons and what may or may not be riding them forward or backward in their lives – but to ourselves – and by "ourselves" I mean this particular generation of therapists practicing in North America and some of our own cultural Master Narratives that we inflict on persons in real conversations every day in our little therapy rooms.

We* (*meaning the above-mentioned generation of therapists in a particular time and space) live and work in therapy rooms that have rightfully achieved some degree of an impassioned refusal of the paternalistic visions therapy in which experts (who were they?) heavy-handedly told hapless clients (and who were they?) what was wrong with them and how to live and breathe and behave. All our students have been fiercely vocal about wanting no part in this tradition. Moreover, our students come to us with an interest in the political chants outside of the therapy room like BLM, the chant of "no justice no peace" chants in the form of hashtags called MeToo and Time'sUp, chants about abortion rights, about Trans and Queer Rights, about disability rights – and chants led by people like Lorelei Williams raising their fists and marching with The MurderedAndMissingIndigen ousWomenAndGirls Marches.

My students tell me that they are listening. In 1851, Sojourner Truth asked "AIN'T I A WOMAN" and the white feminists and suffragettes said "no" then – but the shame about that response to augment the rights of white women at the backs of all others has arrived on our shores as well. The discomfort and longing students feel tell me of the shaking of the belief in the Just World Theory that no longer silences the murders of black and indigenous men and women and queers. If I read a headline to my students that states: "College basketball star heroically overcomes tragic rape he committed" or "RCMP officer drags 19-year-old indigenous student down a hallway and steps on her head after detaining the woman during a wellness check" or "While white women are breaking the glass ceiling who is left to pick up the shards of glass after them?" or "trans girl raises her hand in class and shouts 'self-care is a tool of shame!'" – they howl and raise their fists in resistance.

They are deeply intrigued by the question of "Whose story is it?"[11] and the exercise of how to make sure that the authority of responding belongs to a black woman, an indigenous woman, a trans person, a queer person, a poor woman, a single mom, a 13-year-old who is demi sexual. They welcome the idea of raising up protagonists of spare strange and unique stories with the claims of authorship, substance, moral agency, and credibility, same as the ones who have always had free access to such lofty ideals.

Up the rebellion, yes!

However, my students fresh off their university adventures to prepare them to be therapists also tell me that in this time of change their education is taking a conversative turn – even while the glorious dumpster fires are burning outside. Inside our therapy rooms, what we seem to have managed to muster is a vision of "unconditional validation" with the aim to "make people feel a bit more comfortable for a day" and the ubiquitous skill that leads us all there is none other than "niceness." This iron vision of nice white lady culture[12] comes to us directly from the textbooks of "Counselling Skills 101" proliferated in every educational institution and written as matter-of-factly as to ensure they are basic skills not open to questioning.

These textbooks have instructed all our students, and have perhaps instructed you too, to:

1. Validate everything clients say.
2. Translate these "validations" back to clients in endless paraphrases that are in keeping with the psychologizing understanding of clients' distress.
3. Consider that clients always know what is best for them in every single circumstance as they are the "experts."[13]
4. Promote a vision of comfort as the ultimate aim of therapy by insisting on working on the SMART goals or "little factories of life skill production."[14]

In this vision, what is taught to students is to temper their social justice passion into a dazed and nice passivity. And the most suspect creature is a therapist who dares to believe or act as if therapy were a location for storying change. Watching tapes of Michael White or David Epston at work makes students whisper "but they are so *active*." And what if I made it linguistically worse and spoke to you of possibilities of social justice, healing, transformation, revolution, dramatic turning points, hinge moments, and of our responsibility as therapists to imagine them and invite them forward.

Here's what a student once wrote to me:

> When I imagine my learning journey, I think of Client X. I imagine sitting in front of Client X after she bared her soul and thinking to myself, "where do I even go from here?" I imagine trying to remember all my training and still coming up with a blank. In the moment it will not be comfortable. But I learned that that is the point: my client and I should sit in the discomfort. At the end of the day, clients want someone to listen, be kind, and sit in the discomfort about the process to bearing their souls to us.

I do not blame her for her imagination, but it breaks my heart: to help clients better bear the discomfort of talking about their lives. It is an expression of the collective misogynist white-cis-hetero-patriarchal belittling of a field of women and queer therapists to not consider their work of much consequence, to not consider themselves in powerful roles as witnesses to change within the therapy room.

But we preside over a substantial realm of influence: the therapy room and the conversation that takes place there and the real effects of our doing and non-doing that takes place there.

If I can, as part of my therapy conversations, stop fists being put through dry-wall for one couple, trigger the conditions for a radical healing divorce with no stalking, help a trans-man experience the safety and trust of a new intimate relationship even though they started out just proclaiming that their last abusive partner who forced them into sex has left them entirely broken, accompany an indigenous woman to remember the spirit of her mom to help her invent her response to being taken advantage of, or help an immigrant woman to apply for a job of her dreams, or an abused woman leave her house and burn the mattress of her bed, or Alicia to talk to her grandmother in the manner of her choosing, then I am beginning to do my job in the realm of social justice. I could go on and on. The point is this: our people deserve a therapist who is emboldened to no longer dismiss their own practice as inconsequential, who believes in change and the original invention of healing initiatives for living.

I once had another student therapist in an interview for a practicum position say, in response to "What would be your shyest hope of what might be possible for a client to feel or think after speaking with you?" The student answered "Freedom." But never again and never before was a student so bold.

In narrative therapy, our aim is yes, freedom, but what does it mean?

It means very simply to have more "options and more say about how our lives go."[15]

Do you hear the revolution brewing? To have more options and more say about how our lives go. This is the definition of personal agency that leads to freedom.

Shall we join the above student and resist the binds of Master Narratives together for a moment?

Imagine for a moment, if you can. Imagine back to a moment in your life when you led the charge of your day with boldness and good will. Let me tell you, for many women and queer therapists who have long been punished out of such remembrances, their first rescuable moment often lies far back when they were young. Remember such a moment. What were you up to in leading the charge of your day? Perhaps you hid under the covers to deliciously ruin your eyes with reading all day, perhaps you clowned it up for all your friends at school, perhaps you tinkered with some creation all day long, maybe you ran like wild on the soccer field or in dance, maybe you talked to frogs on your way home, or perhaps you laid in the grass and told a story to an imaginary other. There are thousands of options, but the point is: think of a moment when you boldly led the charge of your day and felt the satisfying joy of "I'm just happy to be here." Imagine your-self in that moment. What were your talents, your interests, your passions, your manners in that moment? And what if it were possible to practice therapy in this way, from a place of defiantly, joyfully, boldly leading the charge with good will?

Now further imagine you leading the charge in your relationships with others when you were but young. Remember a moment when you perhaps decided to console a person, entertain a person, accompany a person, build a bridge to a person, stand in the way of a person's ruin, stay behind with a person when everyone else left, or reaffirm your affection to a person. What did you know to do, and was it humorous, or gentle, or silent, or witty, or fierce, or a thousand other options? What if these remembrances of your moments of enacting your virtues were not only allowable but yearned for and ethically needed for you to meet the clients you are about to meet?

Lorraine Hedtke and John Winslade wrote in their article about the story of Michael White's death that in the workshop on that day, Michael White asked: "If we were all born originals, how comes it to pass we die as copies?"[16] I want to entrust this question to you, dear student, to carry together with me. Master Narratives will attempt to make you a copy, a copy of someone who is of supposed more moral worth than you according to white-cis-hetero-patriarchal master ideologies that neither you nor I agreed to when we were born. Master Narratives will attempt to prescribe you into obligations and entitlements that can be whittled down to your everyday practice as acceptable performances of the social mores of a "good therapist." These Master Narratives that will attempt to make you a copy will see to it that neither you nor your client nor any ideas you two raise are open to questioning and that you will become a lifeless, checked-out, robotic interlocutor of the living stories that human beings are waiting to lay at your feet.

But dare I say it: we were all once born free, and are now "freer than we think."[17]

In order to support you, may I tell you a story of freedom and its attempted binding? Listen.

A Donkey (in the original, an Ass) happened to see a night-ingale, one day, and said to it, "Listen, my dear. They say you have a great mastery over song. But I trust my own standards only, and I have long wished very much to hear you sing, and to judge for myself as to whether your talent is really so great."

> *On this the Nightingale began*
> *And through its cadences it ran*
> *How tender and most soft*
> *Anon its voice it raised aloft*
> *It whistled in a thousand ways*
> *Chanted and cajoled the ways of our days*
> *It sobbed and cried soft sorrows into being*
> *And cooed tenderly at the time that is fleeting*
> *trilled and warbled a steady shower*
> *Of tiny notes over tree and flower*
> *And murmured to all the promise of the reeds*
> *To enchant us to the beauty of our deeds.*

At once, all listened to the favorite singer. The breezes died away, the feathered choir was hushed, the cattle lay down on the grass. Scarcely breathing, the shepherd reveled in it, and only now and then, as he listened to it, smiled on the shepherdess.

Then the Ass, bending its head towards the ground, observed, "It's tolerable. To speak the truth, one can listen to you without being bored. But it's a pity you don't know our rooster (in the original, our cock). You would sing a great deal better if you were to take a few lessons from him. He has a voice that really keeps folks quite awake."

Having heard such a judgment, the nightingale hung its head in sorrow and took to its wings and flew far away.[18]

In this story, the nightingale is you, dear student. Be greeted, here with me.

Now the "cooing, sobbing, chanting and warbling" denote the nightingale's moment of deciding to lead the charge with her talents, boldly and with good will, and in this case, to good effect. By contrast, the song of the rooster denotes the pronouncement of the Master Narrative that her virtues are compared to. This pronouncement is spoken into being by a powerful other, a representative of the unquestioned social mores in this fabled world in which keeping people awake is raised as a taken for granted measure of a person's moral value.

Students have come to me as the nightingale in this story to tell me of the Master Narratives' whips that bind them in their conversations with clients.

"I have ADHD, I have anxiety, I freeze, I am too much, I am not good enough etc." – in comparison to the "rooster's song" and the "donkey's pronouncement." I wonder: have your talents and virtues also been compared to a rooster's song, dear student? Or perhaps it is worse, perhaps they have never been acknowledged, born out to you, or dared to be named because women and queer therapists have been instructed by the Master Narratives of our time to never lay claim to any pride, or achievement, or original knowing. Claims to care, attention, affection, knowledge, credibility, leadership, decision-making, initiative, influence, charge, and power, and so on, belong to others as per the instructions of the obligations and entitlements of the little white-cis-hetero-patriarchal fabled world we begin our careers in.

This is my hope, dear student: if you are to meet a nightingale, in your own person, or in that of your client, that you would have bold ears to hear their song, despite the instructions of the Master Narratives to deafness. I further hope that you would have eyes to see how the nightingale's song was rendered painful to them and the powerful binds of this pain on the nightingale's spirit. I hope that in your conversations with the nightingale, you may dare to raise your gaze back at the contempt of Master Narratives to render persons unequally worthy and unequally capable to account and to respond to the world and its people. I hope that another world would indeed become possible in which the nightingale's song and their freedom of movement are restored back to them.

I hope to shape such a world with you, dear student. Should we be bound and hounded out of town in the near future by the low hum of the choirs of Master Narratives, may we always find our way back by listening closely to the sound of a clear song, carried our way by the wind.

Locate yourself within the bigger, puzzling, and sometimes hazardous world around you. You are invited to do this work.

You are already doing this work. What combination of facts and lies represent you? What spectrum of identities do you hold dear while the larger world tells you that these identities don't even exist? What personal and public rituals do you perform to be seen? What truths must you create to fill the gaps? And what will you (you and I both) do with the knowledge we have (or haven't) been given?

For me, these questions are the same as poetry. They save me.

When this paragraph ends, this story is all yours.[19]

Love,
Sanni

Notes

1 "Cultural imperialism insists on conformity to the norms of master narratives that constitute the dominant group's identity, setting the standards for who people must be." Lindemann Nelson, H. (2001). *Damaged Identities, Narrative Repair*. Cornell University Press, p. 112.

2 "Lying is the work of people who are told their truths have no value.
    The labour of survival is laden with myth and misunderstanding.
    Silence is the work of people who can't comprehend that change is possible.
    (I still moonlight at all of these jobs.)" Dawn, A. (2016). *How Poetry Saved My Life: A Hustler's Memoir*. Arsenal Pulp Press, p. 114.

3 "Laughter is reserved for philosophers and free spirits." Source unknown. I think I read or heard this somewhere but my investigations to find the source have proven fruitless. Nevertheless, I quote this to all my students to inspire them to attend to moments of laughter in their sessions about all matters that we shouldn't – properly speaking – find ourselves laughing with our clients about.

4 "A member of the audience asks why he is using the expression 'subordinate stories' rather than 'alternative stories.' Michael references Foucault and the notion of dominant knowledges, and subordinate knowledges and discloses his own experience of joy in reading Foucault. He comments that subordinate stories

are not subordinate by chance. They are the result of the operation of modern power. Therapy, he says, is always political but it is about politics with a small 'p'. He speaks about the rise of normalising judgement as a governing force in people's lives. He describes the ways in which people are judged (not so much on moral grounds) but on a series of continua that measure normality, for example, from personal adequacy to inadequacy, from independence to dependence. In these normalising judgements, people's lives are represented as singlestoried. But the subordinate stories, he says, are often the more remarkable stories of the rich texture of people's lives. These stories contain their hopes and dreams, the things they cherish and hold dear, and the expressions of what they value. What is remarkable, Michael teaches, is found in the particularities of these stories rather than in any universals." Winslade, J. and Hedtke, L. (2008). "Michael White: Fragments of an event." *The International Journal of Narrative Therapy and Community Work*, 2, p. 74.

5 "Rather than invoking master narratives as a means of moral justification, counterstories resist these narratives by attempting to uproot them and replace them with a better alternative. They operate on the supposition that the norms of the community are to be found not only in its foundational narratives, but also in stories that offer other vantage points from which to assess a community's social practices. The teller of a counterstory is bound to draw on the moral concepts found in the master narratives of her tradition, since these played a key role in her moral formation regardless of how problematic her place within that tradition has been, but she isn't restricted to just these concepts. To the extent that her experiences of life and considered judgments make them available, she can also help herself to alternative understandings of lying, heroism, fairness, or propriety, testing her conceptions of these things for adequacy against conceptions offered by people within both her found communities and her communities of choice. The narrative agent who tells counterstories thus commands a wider range of moral resources that are available on MacIntyre's account to persons who are unjustly subordinated or excluded by a community's foundational narratives." Lindemann Nelson, H. (2001). *Damaged Identities, Narrative Repair*. Cornell University Press, p. 67.

6 "The most subversive thing a woman can do is talk about her life as if it really matters," Eltahawy, M. (2020) *The Seven Necessary Sins for Women and Girls*. Beacon Press, p. 36.

7 "One of the biggest issues with mainstream feminist writing has been the way the idea of what constitutes a feminist issue is framed. We rarely talk about basic needs as a feminist issue. Food insecurity and access to quality education, safe neighborhoods, a living wage, and medical care are all feminist issues. Instead of a framework that focuses on helping women get basic needs met, all too often the focus is not on survival but on increasing privilege. For a movement that is meant to represent all women, it often centers on those who already have most of their needs met." Kendall, M. (2020). *Hood Feminism*. Penguin Random House, p. xiii.

8 "How do I justify therapy as the site of my activism?" Médiné, S. (2020). "Spoken Word Publication: A hope for intimate liberation – activism in the therapy room." *Journal of Contemporary Narrative Therapy, December Release*, p. 3.

9 "Violence does not happen in a vacuum and transformative justice works to connect incidences of violence to the conditions that create and perpetuate them. It acknowledges that we must work to end conditions such as capitalism, poverty, trauma, isolation, heterosexism, cis-sexism, white supremacy, misogyny, ableism, mass incarceration, displacement, war, gender oppression and xenophobia if we are truly going to end cycles of intimate and sexual violence. Transformative justice recognizes that we must transform the conditions which help to create acts of violence or make them possible. Often this includes transforming harmful oppressive dynamics, our relationships to each other, and our communities at large." Mingus, M. (2019). "Transformative justice: A brief description." https://transformharm.org/tj_resource/transformative-justice-a-brief-description/

10 "So, this is all that we have – our lived experience of the world. But this turns out to be a very great deal. We are rich in lived experience. To quote Geertz: 'We all have very much more of the stuff than we know what to do with, and if we fail to put it into some graspable form, the fault must lie in a lack of means, not substance.'" White, M. (1989/90). "Family therapy training and supervision in a world of experience and narrative." *Dulwich Centre Newsletter*, p. 373.

11 Solnit, R. (2019) *Whose Story Is This? Old Conflicts, New Chapters.* Haymarket Books.

12 "Nice white ladies, and our protection, are fundamental to American culture. And this fact is destroying all of us. You should know that I, myself, am a white woman. I have been helped in my career by many nice white ladies, some of them feminists. I come from a long line of white women ancestors, none of whom would have ever identified as a feminist. When I enter a room, people assume I am a nice white lady. No one can tell just by looking at me that I have spent half my life in a conscious effort to not be a nice white lady. I have learned to disavow the niceness that is a cover for ignoring the pain of others, to distance myself from the safe emptiness of whiteness, and to detox from the poison of ladyhood." Daniels, J. (2021). *Nice White Ladies: The Truth About White Supremacy, Our Role In It, and How We Can Help Dismantle It.* Hachette Book Group, p. 1.

13 "Ten years later, I am still trying to undo my validation and active listening training – because I want to listen not only actively but also mindfully, compassionately, critically, lovingly. And I want to be listened to in those ways by my friends and family . . . Sometimes, I want advice from my friends, because the truth is, I don't always know what is best for me. I want to know if my friends think I am doing something wrong. The part of my lived experience that I express in words very rarely reflects the entire picture of my life. When I say that I am angry and want to attack someone verbally, sometimes what I mean is that I am afraid and want to be safe. In the past, when I have said that I wanted to die, what I meant was that I wanted someone to offer me a way to have a different life. I believe that in our best, most fallible human moments, the urge to over-validated comes from our fear of crossing boundaries, of replicating the traumas that abusive families and social oppression have enacted on our developing selves. I have come to believe, though, that strong relationships, revolutionary relationships contain the capacity for complexity and tension. That in a loving place, I am able to hear a friend disagree with me and know that they still care for me. That I can receive their advice and know that I don't have to follow it. That there is enough trust between us that our differences will not shatter us. I don't want to be validated. I want to be loved." Cheng Thom, K. (2019a). *I Hope We Choose Love: A Trans Girl's Notes from the End of the World.* Arsenal Pulp Press, p. 34.

14 "Groups can easily become –
    Garbage bins of sadness –
    Or little factories of life-skill production." Zheng, Z. (2017) Personal communication.

15 "I experience inspiration from the steps that people take to dispossess perpetrators of their authority, the steps that people take in reclaiming the territories of their lives, in the refashioning of their lives, in having the 'last say' about who they are." White, M. (1995). *Reauthoring Lives: Interviews and Essays.* Dulwich Centre Publications, p. 86.

16 "In the workshop Michael is speaking about difference. He is moving through references to Foucault and Derrida and Barbara Myerhoff and mentions his more recent reading of Deleuze. Again he is pointing to where to look in opening subordinate stories. There is always difference, he states. There is always the territory of the known and familiar and there are always subordinate stories. His aim in therapeutic conversation is to open a process of deterritorialisation, of creating distance from the known and familiar, from the immediacy of a person's experience. He speaks of his interest in literary theory and recounts the development of the new criticism. After Lionel Trilling, he plays with the eighteenth-century aesthetician's Romantic question of: 'Born originals, how comes it to pass that we die as copies?' (Trilling, 1972, p. 93) and, with Clifford Geertz (1986, p. 380), turns it on its head so that it ends up as: 'It is the copying that originates'" Winslade, J. & Hedtke, L. (2008). "Michael White: Fragments of an event." *The International Journal of Narrative Therapy and Community Work*, 2, p. 76.

17 "I am not a writer, a philosopher, or a great figure of intellectual life. I am a teacher. . . . My role is to show people that they are much freer than they feel, that people accept as truth, as evidence, some themes that have been built up at a certain moment in history, and that this so-called evidence can be criticized and destroyed. To change something in the minds of people – that is the role of an intellectual." Martin, R. (1988). "Truth, Power, Self: An interview with Michel Foucault." In L. Martin et al. (eds) *Technologies of the Self*, pp. 10–11.

18 Krylov, I. (2017). *Krylov and His Fables* (W. Ralston, Trans.) Hanse Books (original work published 1869).

19 Dawn, A. (2016). *How Poetry Saved My Life: A Hustler's Memoir*. Arsenal Pulp Press, p. 119.

For Larissa
What Is This, Discipline?

Larissa
cram people and their worlds and words
into straight and lined boxes
of colonized European composures

be very afraid of the sounds they make Larissa

Larissa
linearize your own words
to our white-boarded superimpositions
and let us directive you out of suspicious intuitions

be very afraid of the sounds of your heart Larissa

Larissa
don't sit on the floors
of imaginations leaping to meet the world of another
we have our white straight world
and it has always served us well

what is this, discipline?
It sounds suspiciously a lot like
a fear invented by white people

— but she is not afraid.

When she was 6
she said
your sophistications embarrass me
because they are nothing but
racist
and racist is nothing but
afraid
of difference
and I am not afraid.

She is not afraid of difference
She is curious
about what the only 6-year-old Chinese girl's thoughts are saying
    about us
what the autistic boy's frustrations are saying about mopping
and what the psychotic man's eyes are saying about life
and what the young woman's words are saying about where
    she's been

she is not afraid of difference
between us fleshy humans:
she comes alive
to follow your rivers
to flex her reality
to leap to you with her heart
to make magic credit cards
to cry at your humiliation

cram her room with the chaos
of many voices
and the realities they speak of
her practice is to amp them up
to full capacity

and then watch her dance
to the beautiful
sounds

3

What's Problematic About Calling It a Problem?

A Letter About Dilemmas

Dear student,

I am really raring to "lead the charge" in my own particular manner towards an opening of the consideration of the kinds of stories that we find ourselves working in as therapists, broadly named Up Against Stories and Counterstories. This will be the heart of the labor of these letters. But before we step towards this work, we have to linger for a while, because if we don't, I fear that all the examples and explications I want to give you might fall onto the ground like toothless words.

So, we will dawdle here for the duration of two letters intended to explore why stories are sometimes hard for therapists to shape at all.

Don't worry, I will return to my passion for the craft of therapeutic stories with every vigor afterwards. This pause has been necessitated by my study of students' transcripts who have been willing to show me their work and explicate to me what it is that undoes their best knowing in their sessions with clients. For some time, I continued to operate undeterred under the impression

DOI: 10.4324/9781003478478-3

that what was urgently necessary to help students out of their undoing was more clarity about the qualities of stories and more transcript excerpts of mine to show story-prompting with my clients in motion. My students taught me I was wrong. What was urgently necessary was clarity about students' own obstacles to the principles of story-shaping.

Let me describe it to you: as you set out to shape stories with your first clients at your beginning, a dragon will roar. This roar can take on many different messages, depending on the particular worst fears whipped into their hearts by cultural training in Master Narratives for each student. The roar catches the part of students' souls that has stored up ideas about the Master Narrative-driven assumptions and practices of what a good life is and what an honorable person is, which is different and unique and (often) tacit for each student.

Regardless of the spare and strange differences in assumptions and practices by anyone's training in Master Narratives, one of the intended effects of the roars seems to appear more often, especially at the beginning: the dramatic shunning of problems. The roar seeks a Teflon-coated and active cancellation of problems of all kinds and seeks the act of a veritable "stepping over" of difficult life experiences laid at your feet by despairing clients. This silencing pertains to all manner of problems but especially firmly to the problems that are deeply human and relational, the ones that we might term "What Human Beings Routinely Do To Each Other," or WHBRDTEO, which is a very snazzy acronym to join the "Why Would You Tell This Story" one from the first letter. I hope that these acronyms might prove useful to you to title the storage spaces in your soul for such lost arts of understanding power relations. But not until I prove myself, I know.

See, students will readily let me beseech them that their first job, in fact, their only job, in first sessions is to "gain a measure of understanding of what a person has been up against in life."[1] But when it comes to actual questions to help such storytelling along, students often pause and then veer off in favor of other ventures like talking about the weather, clients' pets or loved ones, movies, books or random metaphors or a 15-minute interlude of thinly veiled praise for the fact that clients managed to come to therapy at all. And here is where students have generously come to my aid in shedding light on my veritable cognitive disability to understand why they do so: "it is impolite, Sanni"; "I have to do relationship building first, Sanni"; "I have to get to know my clients outside of their problems first, Sanni"; "I have to start with something lighter, Sanni"; "can't I deal with the struggles later, Sanni?"; "but isn't narrative therapy strength-based Sanni?" and so on.

In short, while students might agree with me about the necessity of understanding Up Against Stories, they also, at the same time, agree that it is inappropriate and harmful to shape such stories forward.

Students have not come up with these tacit agreements on their own. I am writing in a time and a place in which an interest in struggles pushes against the white and straight Master Narrative-driven entitlements to comfort, to ease, to niceness, to positivity, to doubtlessness, to shamelessness in our engagements with other humans – in short, a kind of "functional stupidity"[2] that relies on all the learned tropes and stock plots that reinforce the status quo of human life and therefore, feel satisfyingly "right" to deliver and expect. We all have been peddled a particular vision by the reigning Master Narratives of human life as the pinnacle of achievement: the image of a happy, comfortable, and obedient middle-class consumer from whom all questions have been removed.

In our realm as narrative therapists, the antidote to this dead-end vision is, and always was, "rich story development." "The most powerful therapeutic technique I know is rich story development," Michael White said.[3] Externalizing conversations were intended to help in this endeavor, but "externalizing" has since been rendered a cute technique in a toolbox to further simplify the aspirational richness of story development (I will address this turn of events in the next letter). For all the roaring of the dragons, we collectively continue to refuse to endow our narrative resources into the raising of interesting problems and by extension, interesting protagonists, who are currently rightfully losing sleep over these matters that most matter to them.

This problem of the kneejerk reaction of dramatic shunning of struggle is so widespread that perhaps we have to replace the word "problem" with the word "dilemma" to invite a new generation of narrative therapists to turn their significant talents and virtues to the venture of shaping interesting Up Against Stories, or, in other words, the venture of "posing problems well."[4] Our collective effort to learn to pose problems well is in direct relation to the extent of the radicalness of change for our clients. Our willingness to care for the problematic by our generous attention on the poetics and politics of its creation enriches our clients – people who have almost abandoned hope that another life outside of the meaningless re-arrangement of little lives, was possible after all. Put another way, the richness of the construction of an Up Against Story determines in no small part the richness of a Counterstory. If all we have for an Up Against Story is a narrative label, what we are left with is the work of replacing that label for a Counter Label. Our colleagues in Cognitive Behavior Therapy and in the Self-Help Industry are far better poised to do such work; they call this "reframing" of "cognitive distortions" or the replacing of "negative self-talk" with "positive affirmations." Our call as narrative therapists is to work in stories instead.

I'll try one more time, for your discerning mind, dear student – your only job in first sessions is – despite the dragon, because of the dragon, in defiant and joyful disobedience to the dragon – "to praise a mutilated world."[5] Any better?

Remember how I wrote that when I first heard the idea that "the person is not the problem, the problem is the problem" – my whole life changed? I was once asked to elaborate on my intellectual grasp of this felt excitement by a mentor, and exclaimed something like the following: "Well, what it of course means is that I, or the identities I was given by encounters with other people, were never the problem, which by itself is freeing, but that the problem is the problem, which means that the problems I have are free and interesting again!" I exclaimed. My mentor laughed, whether with me or at me, I still don't know.

I have searched my transcripts and all of my beginning session notes and I haven't found students' fears about the inappropriateness or harmfulness of the effort to construct rich Up Against

Stories from the minute I meet clients to be founded. Instead, here are the questions I found myself asking upon my review:

- What precisely is "polite" about pressing clients back into the cultural performances of the niceties of social conventions of white lady culture?
- Do there exist speech acts outside of small talk that better serve the aim of shaping a trusting relationship?
- If we perfectly promote the performances of empty niceties found at every coffee shop and kitchen table that ever left people yearning for real intimacy – why do we charge so much for therapy conversations?
- If we court and obey all Master Narratives in our own speech styles – can we ask clients to court disobedience?
- What ways of speaking give people the best chance to evaluate your trustworthiness to them?
- What is more "strength-based" that insisting on speaking about struggles strongly?
- Is "coming to therapy" really the pinnacle of the amazing achievements you believe your clients having been, and currently being capable of in life?
- What exactly are we saying when we are giving ourselves permission to sort people's life experiences into "light" and "dark" and then go on enforcing "lightness?" (Exactly.)
- Is there a way to speak about what is keeping your client up at night and for the client to walk away from that feeling like they are more interesting than they ever gave themselves credit for?
- What is an attitude you would want to nominate to be met with if you went to talk about something really important to a therapist?
- In a game of chicken, who is always more afraid – the therapist or the client?

(I stole these from fierce client reflections!)

And seriously – the racism of sorting clients' speech acts into "dark" and "light" aside for a minute – what about sorting clients' speech acts into "good" and "bad," "problem" and "alternative?" I ask you earnestly: how do you know if "anxiety" is "bad?" Do you know for certain "depression" is "bad?" Can you say for certain that a moment of the client driving their car and stopping in the middle of the road to scream and rage at the windshield can be sorted into a pile of "bad" by you, the decider of what constitutes "good" and "bad" living based on what authority? What if you just sorted a most interesting and life-giving "pre-therapy strategy of achieving personal power"[6] into a garbage dump?

Moreover, when we purposefully practice shunning of rich Up Against Stories, we stand to lose any chance for our clients to shine in the dangers of the world they live in, precisely positioned in the almost-impossible abyss of what they are up against: agency, initiative, character, purpose, poise, beauty, poetry, change – nay – revolution, freedom, wisdom.

I have a feeling that perhaps you are with me, dear student, but not convinced yet. A story about my practice is urgently necessary for you to verify mere words for yourself.

At the outset of this story, please imagine me at the time I dispatched Alicia to gherkins on trees and teacups on saucers and the anticipation of risky conversations among important women in her life. To my immense relief, I had new-found speech abilities by my side and notebooks on my lap in all my sessions (which likely helped with wrangling the damn raccoon of words!), and you could find me fiercely scribbling down words all my clients said. If you zoomed in further, you could see that my notebooks contained something else curious beside client words: all of them had a hand-drawn line in the middle of each page. I had been taught a practice of note-taking by other narrative therapists that involved me drawing a line in the middle of the page and sorting client talk into "problem moments" on one side and "unique outcome moments" on the other.

And what was most absurd about this is that if you zoomed in really close, you could see that I never wrote on the side of "problem moments" but that all client words were grouped into "unique outcome moments" – which meant that I practically wasted notebook after notebook in a stupefying exercise of only ever writing on half a page.

As my funds for buying new notebooks were slim – I did the obvious smart thing: I still drew a line in the middle of the page – I just started guiltily writing over it. I mean I'm a real quick-study when it comes to shaking off teachings that don't serve me.

I won't tell you how embarrassingly long I persisted with drawing that line, apparently hoping that one day I would become that narrative therapist who could finally sort client speech acts fast enough into "problems" and "alternatives" as they were speaking – until I finally lost the hope on purpose, and good riddance.

Let's look at a transcript excerpt of a conversation with a young woman from my "line drawing days" that stands as a good example of resisting "good/bad binaries" from the first words out.

That night, a young woman appeared at my office door, accompanied by her mom. I was glad to see the young woman turn up with a companion, as I had made it a habit to invite important others from my clients' lives to join in our sessions, at any time, no warnings needed. Very few clients ever took me up on this habitual invitation, but I thought this might be one of those rare times. Both the client and her mom, however, politely refused my query if they wished to speak together, and mom said: "I'm just a driver tonight. I'll wait here," motioning to a book she had brought to keep her company. The young woman, Em, and I proceeded to talk through introductions without mom but despite my efforts at welcoming her with warmth and humor and tea, she was shifting in her seat and breathing uncomfortably. One look at Em, whose hands were now shaking to the point of her struggling to hold the teacup I had offered, caused me pause. "I . . ." I began, and felt the distant yet familiar stutter start to form. Before it could overtake me, this is what happened:

Sanni:　　　This is fucking nerve-racking, hey? To come to talk to a stranger. Don't worry, we'll totally figure this out together. (*Softly*) So something in your life is freaking you the fuck out right now, do I have that right? What is it, can you tell me?

Em:	Oh my god. *(Bursting into tears)* I don't know if I heard all the things you just said, this happened with the psychiatrist too, I'm sorry . . .
Sanni:	Oh, I promise you missed absolutely nothing of any importance! I'll give you a copy of all the stuff on paper, so you can read it later, so absolutely zero worries. Em, I've totally got you, no worries. What is scaring the crap out of you right now?
Em:	I . . . I've been like this for the . . . well, I've been struggling with a lot of anxiety, especially over the past six months. It's been really bad so I just want to work through to see where it's coming from. Going forward, kind of how to take care of it, how to manage it. If that makes any sense.
Sanni:	Makes every sense, Em. So the name of the scary crap is anxiety and it started 6 months ago? See, I can listen, I've got it. *(Laughter)*
Sanni:	Okay, I have two questions. Let's see which one is less scary and more interesting to you? First question*(Softly)*: if I really wanted to understand it, from inside your life, what's the absolute worst of it, Em, the worst the anxiety has got. What's it been up to in your life Em, the absolute shittiest it's been to you?
	Other question, why the hell did it start six months ago, like what were you going through then in your life to set off this half-a-year of crap? Which one, do you think?
Em:	The thing six months ago. I mean, I guess . . . I don't know. My brother, he's been through a lot in life, and six months ago . . . he you know, committed suicide, well I mean he didn't die, I got him to the hospital in time.
Sanni:	Wait, so, you saved your brother's life, is that . . .?
Em:	No! That's what everyone says. That if it hadn't been for me checking in on him. But it's . . . that's not . . . *(Crying)*

Sanni:	Em, no worries. So it's not that simple, gotcha. I have all the time in the world for this, Em. Here's what I'm thinking but I might be daft like everyone else so you gotta correct me, okay? (*Laughter*)
Em:	Okay.
Sanni:	Okay, I'm thinking that something . . . like a question, or a dilemma, or an image, or something came to you that day six months ago with your brother when his life was on the line. I'm thinking that something came to you, something really important, something that hasn't let you go ever since then. And then, I'm now way out of line maybe, but it's so important, the question you've been living with that the anxiety has forced you here to come weigh it with me . . . What do you think of that?
Em:	An image . . .?
Sanni:	Yeah, an image or a question or dilemma that hasn't let you go . . .
Em:	It's . . . I don't know. It's . . . I never told anyone this. I . . . when I was with my brother waiting for the paramedics. (*Crying*)
Sanni:	Yeah?
Em:	Well he asked me, he asked me, why shouldn't I die, like what's the use . . . (*Crying*), and . . . I didn't . . . I didn't . . . I said something, but I didn't know, and I was just waiting for the paramedics to come faster. But I don't know, this isn't the first time he tried, you know . . . (*Covering her eyes crying*)
Sanni:	Wow, Em, I'm imagining you facing his question of "what's the use" then, that night, and maybe holding it still in some way . . . is that alright for me to think of it that way?
Em:	Yeah.
Sanni:	Okay, so then I am thinking the anxiety didn't just appear bam out of nowhere, for no important reason at all, but that something is really, really worth anxietying about! (*Laughter*)

Sanni:	You know . . . like, when you think back on your most anxious moments in the past 6 months, what's been there, what's it talking to you about, what's its message or idea or thing it likes to obsess about?
Em:	It's kinda the same you know, like I'm thinking what's the use. You know. What's the use of anything if it's gonna be this hard . . . It's been hard for me to do anything at all lately, like to focus, or get out of the house . . . *(Crying)*
Sanni (Softly):	So let me see if I've got you then, Em. Six months ago, you are with your brother when his life is on the line, you said . . . "you showed up," I do have some questions about that, how it came to be that you "just" showed up at the crucial time, but never mind that for now. You're there anyway, and that saves the man's life, but as he is sorta teetering on the edge, he asks you, all despairing "What's the use. What's it all for, Em?" Is that right? And then for the past six months you've been carrying this question with you, in your body, it's been shaking you up, making you really uncomfortable, hard to focus, making you wanna crawl out of your skin with discomfort, making you stay at home, is that right?
Em:	Yes.
Sanni:	But even as you've been at home, you've probably told yourself a million times to move on and get over it, but the question in your body and in your soul just shouts louder, which is really bloody uncomfortable but also kinda . . . brilliant, I don't know if you know what I'm trying to propose . . .
Em:	I do! They put me on, I forget what it's called . . . but anyway, that's been helping a little. Like it was okay to come here tonight, well, you know, mom drove, but still. I wanted to come.
Sanni:	Well good if it's helping. That's sorta the body side of it, hey? But the question, Em, what do you say now that you have come here, what do

you say if you and I don't abandon your question anyway. The question of "What's the use?" What someone like you might get up in the mornings for. What . . . kind of different world you've barely dared to imagine into being, maybe for you and your brother . . . What do you say, would that be a good question for our conversation?

What I want to tell you with this transcript excerpt, dear student, is that what I have learned is that every person we meet has a good reason for coming to therapy, and that the most respectful beginning is to meet our person right there.

See, the easiest way I can say it is that all the clients I have met over the past years have come to meet with me because of "upsetting things" going down in their lives. In general terms, that's the best summary. My therapy clients aren't in the midst of winning awards, enjoying a blooming love life, or riding some streak of deeply satisfying focus in all their endeavors, only to suddenly think "Oh, I should probably go talk to someone about this." These folks are busy living in their amazing season of luck with meaning and connection and belonging and daring initiatives and faithful companions with whom to laugh and cry. These folks come to therapy later, if it so happens that their "amazing season of luck" changes to "snowflake dancing toward purgatory-season." The therapy clients I have met come to therapy because confusing, chaotic, upsetting, hurting, strange, and interesting things happen to them repeatedly over months or sometimes years in response to which they think "oh, maybe I should probably go talk to someone about this." They have all stored up these sequences of moments for a long time, and the chaos of these stored up moments is always some version of "upsetting" to them. They have come to therapy, in particular, rather than, say, a massage, a pilgrimage, or beer with girlfriends because they have often exhausted all the regular listening ears about that upset, and have gathered that it is no longer exactly salon-acceptable to talk about this upset, and because the upset simply will not vanish.

I don't know about you, but I am finding that in this era, and on this continent, it isn't exactly a commonplace occurrence anymore that human beings look at each other across the table with a concentrated focus of attention that says "And you? What's with you, love?" with a palpable interest to spend the night enriching each other in this manner. Sometimes when I take in the contexts of the barrenness of intimacy in my clients' lives, I can't help but conclude that people also come to us as therapists to talk about what they can't talk about elsewhere. I have found myself to be incredibly privileged and lucky and rich in my job as a therapist: I get to talk to people whose souls have been discomforted by life and who wish to speak to me about it. Imagine that.

So, I ask you in all sincerity: what is a better, more polite, stronger, relationship building, or more trustworthy topic of conversation than "what keeps you up at night, what do you wrestle with, what's upsetting and disturbing to you, what is something you can't live with anymore? Wait, what? You're going through what? How come? What happened? I have all the time in the world for this, I'm here."

The spirit of this orientation matter of factly proposes to clients a counter idea to the medicalizing and pathologizing of clients' lived experiences of upset; the proposal is very simply that "I imagine something of significance happened to you for you to be upset." The spirit of this orientation is to reinstate and restore stories of lived experiences and the political context of their creation as the center weight of therapy.

Em taught me that night that caring for her means to expend the considerable narrative resources I might have at my disposal at any given time to richly story what she is currently up against in life, which is the reason she has come to therapy. All other questions can wait their turn. So far, so good?

I'll raise the stakes one more time before I conclude: would you consider what Em told me in the above transcript a "problem?" Would you quickly classify it as a "problem story" that Em needs to be re-authored, externalized, or even counterstoried out of? Have the dragon roars awakened for you? What exactly do we find "problematic" about a young woman who is wrestling with the meaning of life and love for her brother to the point of physical distress? Did you catch a glimpse of yourself hoping for more ease and middle-class comfort and drug- or therapy-induced removal of moral questions for Em? In this context, perhaps you can understand what I mean: I respect Em's "anxiety" a great deal and would not wish to cure it, but care for it. And perhaps you'll agree with me about terming the pulsing heart of the matter which is not anxiety but the question of "why live?" a dilemma, rather than a problem. A dilemma that is worthy of all our time on any given night.

Love,
Sanni

PS: The next letter is Part 2 in the realm of dragons. You will meet Juliet. And yes, she is akin to the Juliet you might have heard of, from *Romeo and Juliet*.

See in this letter, I wanted you to consider a question: if you ever had the chance to meet Juliet from literature in your office at a rather crucial turning point in her life, exactly how many minutes would you spare on interviewing Juliet about the weather or what her nurse respects about her, or who might be invited to her bridal shower?

In the next letter, the question is as follows: if you ever met a Juliet (and you will, if you have eyes to see), would you wish to capture Juliet's life's dilemmas regarding desire, defiance, and domination in a sweet little externalization for her to overcome?

Notes

1   "I'm really interested in people's accounts of their experience. I really want to understand what life has been like for them. So, I guess the first part of my work is to try to get some appreciation of what persons have been going through. I think it's important that I achieve a degree of understanding about this, and I think that it's important that persons are aware that I have achieved at least a degree of this understanding." White, M. (1995). *Reauthoring Lives: Interviews and Essays*. Dulwich Centre Publications, p. 21.

2   "Functional stupidity is [the] inability and/or unwillingness to use cognitive and reflective capacities in anything other than narrow and circumspect ways. It involves a lack of reflexivity, a disinclination to require or provide justification, and avoidance of substantive reasoning." Alvesson, M. & Spicer, A. (2012). "A stupidity-based theory of organizations." *Journal of Management Studies*, 49(7), pp. 1194–1220, 1201.

3   "The most powerful therapeutic technique I know is that of rich story development." White, M. (2004). *Workshop Notes*. Cambridge, MA.

4   "Insofar as the problematic insists on the edges of the present, insofar as it calls upon the attention and demands a response without ever saying what that response should be, it is not the solving but the posing of the problem, the very dynamic of invention devoted to the possibility of posing the problem well, that itself becomes the most vital element in any response." Savransky, M. (2020). "Problems all the way down." *Theory, Culture, and Society*, pp. 1–21, 8.

5   "You should praise the mutilated world.
          Remember the moments when we were together
          in a white room and the curtain fluttered.
          Return in thought to the concert where music flared.
          You gathered acorns in the park in autumn
          and leaves eddied over the earth's scars.
          Praise the mutilated world
          and the gray feather a thrush lost,
          and the gentle light that strays and vanishes
          and returns." Zagajewski, A. (2002). *Try to Praise the Mutilated World from Without End: New and Selected Poems*. Macmillan Publishers.

6   "While formal diagnostic systems are all simply socially constructed taxonomies for organizing and grouping people's manifest expressions of distress, diagnosis as practiced is commonly a strategy for reification and objectification of the client and pathologizes their pre-therapy strategies for increasing personal power by whatever means available." Brown, L.S. (2018). *Feminist Therapy* (2nd Ed.). APA, p. 67.

For Em
Some Other World

Into the dying of the light
Walked a pair of legs, a voice, and a rage:
No! I shouted
And I shook him
And me

Somewhere another world is possible
I added,
But my words fell to the ground
Into the dust of the ambulance lights

I will not suffer a prison
Not for him
And not for me

So I bought him a star
And said it was for both of us
And it will shine our stories
Like our future children
And the songs they get up for in the morning
And the strength of the legs
That walk into rooms
To rage and shake
At the dying of
Starlight

4

When Externalizing Internalizes

A Letter About Narrative Diagnoses

Dear student,

I promised you Juliet last we wrote. I even said that she is akin to the Juliet you might know from literature. Strictly speaking, I lied. Do not set out to "find" stories that you already know in your clients' lived experiences. Your clients are unlike each other and unlike anyone you have ever met, and this uniqueness invites us to reorient to an attitude of openness to experience, rather than an attitude of seeking the safety of what is already familiar. This may sound fearsome at the very beginning, but this attitude of seeking to meet a "radical other"[1] whose dilemmas and Counterstories do not resemble any other story plots is the antidote to the strange direction of our field toward the categorization of human beings and the manualization of our approaches.[2]

Students sometimes ask questions like the following:

- "How do I work narratively with people who have experienced sexual trauma?"
- "What are some ideas for working with people who are of Asian descent?"

DOI: 10.4324/9781003478478-4

- "How do I work with indigenous people?"
- "What are some of the metaphors in your toolbox that you often use in your work?"

These questions are a beginning portal to express the yearning to work in a culturally sensitive manner, but they take a hack-saw approach to our clients' unique social locations and lived experiences.

So you may understand that I counsel students to be cautious about siding with easy assumptions in listening to what uniquely hurts a person and how they have seen fit to respond to experiences of sexual trauma in their lives. It makes every difference to me to speak to a person and come to know what it was like for her to be born the first daughter of two in China and what happened to her as a result. And what is the difference between asking an indigenous woman "what it means to grow up Metis in Canada" in general, or listening to her descriptions of her mother's spirit in a house with nine children and the scent of Bannock tea that spelled a "quiet calming" in the midst of poverty?[3] And I would wish students to not pre-arm themselves with slick "metaphors in their toolboxes" but to come instead pre-armed with a focused concentration of attention and openness to the many metaphors clients litter their talk with in their own vernacular, and to pick up and weigh these original metaphors in their storytelling and meaning-making efforts.

What I have found in the many years of storying lived experiences is that seeking patterns of experience and classes of people will yield "patterned categorizations" and rote responses. Seeking unique lived experiences and particular dilemmas in life will yield interesting protagonists in the midst of unheard-of life. If you set out to meet such people, you will pay the price or earn the rich reward of an attitude of "always beginning" and "thinking anew" that regenerates itself in every encounter and honors your clients' wonderful strangenesses.

So, what I really meant by "akin to Juliet" is an invitation: do set out to shape stories and treat clients as if they were as interesting as protagonists in literature in the midst of life that is breathlessly and dangerously on the move. Whence have they come, what are their visions and what will they do next, pray tell?

The particular Juliet whom you will meet in the transcript excerpt below, the mafioso fathers have curiously absented themselves, but there are mothers on both sides of the lines, passionately arguing for the respect their grown children deserve with their partners, especially in regard to the age difference between Juliet and her older partner. You find Juliet having overcome the mothers' questions regarding her union together with her partner 12 years ago; she is now a wife and a stepmother of a 14-year-old boy, and regaining her strength after a serious pain condition as a result of an accident, and in the midst of contemplating going back to school to finish her degree that she put on hold during the years of pain. I hope to show this excerpt as a doorway to considering the roar of dragons that may wish to guide us to shut down rich explorations utilizing the more widely known "narrative intervention" of what sometimes is mistakenly called "externalization."[4] This transcript is slightly abbreviated in places to better shine a light on the point of this letter:

Sanni:	I'm really interested Juliet, in that thing that brought you here to talk to me tonight. I know you had to wait a while for this appointment, so you had some time to reconsider and say "ah fuck it." *(Laughter)* But here you are! So, here's my assumption: I'm thinking there is something in your life that really matters to you right now, that you're wrestling with, that is giving you grief or restless nights, or follows your feet through your days, like a question, or a dilemma, or an experience that just bloody matters to you. What do you think, am I out to lunch with thinking that way?
Juliet (Laughs):	No . . . I really need help managing my emotions, you know. It's been a long time coming and a lot of people have been telling me that I'm all over the place.
Sanni:	"Managing emotions" – that sounds intriguing. But who the hell are these people telling you, what was that . . . "all over the place" – I feel like

	you've been sent to "manager development" or some such *(Laughter)* . . . Do you agree with whoever those people are?
Juliet:	Yeah. I probably should have done it earlier. Maybe things wouldn't have gotten so bad. My mom and his mom, and everyone has been against this relationship. I don't know. Maybe they were right, but I didn't want to hear it. We did go to couple's counselling years ago, but that was a totally different situation, I was in so much pain . . .

Juliet proceeds to tell the story I summarized in the introduction earlier.

Sanni:	Wow, okay, so let me see if I got this right. Moms' opinions on both sides have been strong, but you and Rick defended your love over age divides, and won them over. You've been living together and parenting Ben together and Rick really cared for you after the accident when you were in pain. And even the couple's counsellor couldn't force you two apart, although he tried his best. *(Laughter)* And you're finally feeling your old strengths returning to you, and some new ones! Okay. But Juliet, what brings you here now, in this moment in your life, what is most restlessly on the move for you now, you said something earlier, about "managing emotions" and "things getting so bad" – I don't know what you meant by these expressions, or if these are the questions that brought you here.
Juliet:	I'm here, well, maybe I kinda need more like anger management, I think. Or general emotion management, or whatever.
Sanni:	Anger management? What do you mean? Have you been struggling with anger as of late?
Juliet:	Not just as of late. I don't know, there's different kinds of anger I guess, but there are times I get

angry and I pick fights, and I can hit below the belt, you know, be really mean.

Abbreviated . . .

	I mean Ben and I used to call it Stripey, it's based on a badger but that was a long time ago . . .
Sanni (Writing, laughing):	Wait, Stripey?
Juliet (Laughing):	Yeah, Stripey, it's like a badger that's sleepy and sort of in a bad mood, we used to talk about it like in the mornings, like being in a Stripey mood, like sleepy.
Sanni:	Okay so a Stripey mood, like a badger mood, alright, I'm following. It's a bad mood, you described it earlier as "mean," like how did you say, it "picks fights"?
Juliet:	Yeah. It's a . . . I snap, and then I pick a fight. Like I can go for a long time and be forgiving but then I snap, and I can't let it go. And then I just follow Rick around the house and refuse to let it go, and he sort of walks away from me, but I follow him, and am just badgering him.
Sanni:	And the idea is that I am supposed to help manage you out of this badger thing? Like badger management. *(Laughter)* That sounds like I don't want to do that. *(Laughter)* Okay, no seriously, Juliet, the reason I say I don't want to do that to you is this: let's say you have a good reason for the bad mood, for anger, for "badgering," or that this Stripey thing doesn't beset you out of nowhere, for no reason . . . What happens in your life that makes you angry, or . . . another way of saying that is . . . what's the badger's insistent message? *(Laughter)* Yeah, I mean it, Juliet: what are you trying to say in those moments – I really care about that.

Juliet: Well, the thing is, Rick and I . . . we've grown distant, I guess.

Abbreviated . . .

Sanni: Wait, is the badger Stripey thing there to bridge the distance between you and Rick? Is that what you are telling me? If the badger thing sounded like something, what does it sound like, what are its aims for you and Rick?

Juliet: What does it sound like? . . . Rick, well he doesn't talk to me anymore, he hasn't in a long time, like he rolls his eyes and pulls out his phone when I just want to talk, and he tells me it's because I'm in a bad mood, but I don't think I am, not at first. But when he says that it's like I say "we never talk anymore!" Like I yell "talk to me!" Like literally that's what I am yelling when I'm following him around. Like "be with me or stop walking away!"

Sanni (Softly):   Okay Juliet, do you see how the badger's message is so important? If you're saying "talk to me," "be with me," "don't walk away," "I miss you."

Juliet: Yeah. But it's not . . . it's too late now. It's not a thing anymore. I don't know that I miss him anymore. I was hoping he'd come with me to therapy, but there's no chance of that now, I don't think. Okay . . . He told me a couple of weeks ago that we're done, you know, that I should move out, you know, go to my sisters' place. But we have to at least figure out how to talk to Ben about it, but we aren't really talking at all now so. I don't know what to do.

Sanni: What do you want to do? If anything was possible . . .

Juliet: I don't know. It's gotten so bad, I don't know.

Sanni: That was maybe too general a question of me. You know, I asked what you want because I suppose I wonder if you are in agreement that you and Rick are done and that it's good to move out. I guess that's what he wants? (*Juliet: "Yeah."*) And maybe I don't know why that is what he wants now, but first I want to know what *you* want. Because it's a different conversation if you agree with him or strongly disagree with him now.

Juliet: Yeah, you know, I was so upset, if you had asked me a couple of weeks ago, I think I was in shock, so I don't know. But now I've had time to think over the past few weeks, and I guess, I'm coming to . . . I don't think I want to be with him anymore. I think I don't want to go back to that.

Sanni: Okay, this change of heart . . . is interesting to me. Are you telling me that it went down two weeks ago . . . that two people had a change of heart . . . in a house somewhere in the world?

Juliet: It was bad before then, but yeah, that's when it changed, or ended, I guess.

Abbreviated . . .

Sanni (Softly): Ended, okay. I hear that. I, of course, immediately want to ask you about that day when it happened, the change of heart, like the story of it, where you were and what happened between you two. But is that interesting to you? The other option if you'd rather not see it all over again, or you don't have questions about it anymore, then it might be more interesting to you to look at the story of what's about to happen, like for us to tell the story how you want to put this decision to not go back into motion now after the shock has abated. What do you think? Which story is better to work in?

Juliet: It's kind of upsetting, you know. I guess what happened is that I hit him, that's what happened. I'm not proud of it, you know. (*Crying*)

Sanni: You hit him?

Juliet:	Yeah, we were arguing, I picked a fight again, I was following him. I was yelling, you know, I was so angry, he kept walking away, so I followed him into the bathroom, and I had a glass of wine in my hand, so I threw it in his face, and then I hit him, I punched him in the face . . .
Sanni (Softly):	Okay Juliet, I know you said you're not proud of it, so I'm taking it that that was not okay with you . . . like how hits below the belt aren't okay, like you said before.
Juliet (Crying):	I know! I know.
Sanni:	Well then, beyond wine-throwing and punching him in the face . . . What was the effect of that on him? What happened next? Did he get mad in return?
Juliet:	No! That's the worst of it. He didn't. You know what he did. He just wiped off the wine and walked away. That was it. The next day he told me that it's over, you know, that we're done and I should move out. But even that, it wasn't a discussion, he wouldn't even really talk to me then, but I didn't want to get into it again, so I just sat there and said okay.
Sanni:	Okay, I want to get there, to you sitting there and agreeing and saying okay that you are done now. But before I do, can I ask you something more about what happened?
Juliet:	Yeah.
Sanni:	If it had a message, you know, and I know you're not proud of it, the hit and the wine, but if it carried a message to Rick, what were you saying to him in that moment when you hit him? What did you want to say?
Juliet:	I wanted to hurt him! Like how I've been hurting all this time when he's just walking away. I wanted *(crying)* this . . . it was like a last chance, you know? I know I shouldn't have hit him, but it was like a last chance, like a last-ditch effort, you know, if that makes any sense.

Sanni:	A last chance to say, stop, look at me, talk to me?
Juliet:	Yes! Yes. I just was fighting, you know, in a weird way I was fighting for us, you know, I can't explain it.
Sanni:	You're explaining it fine. You were fighting and yes, don't hit people, but you were fighting how you'd been fighting for a long time.
Juliet:	For the past, I don't know, five years. We haven't even kissed for . . . five years.
Sanni:	I think I get it, Juliet. No talking, no kissing for the past five years, and you following him on his footsteps, in the last years ever more badger-like, I get it, to intensify this ask of yours, "be with me, talk to me, where are you going, stop, be with me." Is that right? (*Juliet nods tearfully*) Juliet, if in that moment you were fighting for him to stop and be with you, and you fought dirty even, by hitting below the belt, what have you been saying for the past five years of what you don't want to live without any longer, and especially in that moment, even if it came out all wrong?
Juliet:	That I'm not done with romance. I want romance! I want love. I want to be kissed. I want to talk. I want someone to have my back, not just to support me and whatever, but to talk to me, like, to argue back, even.
Sanni:	Okay, you. I hear that: "I want romance. I want love. I want to be kissed. I want to argue. I want to talk." I hear that loud and clear, Juliet. And I gotta tell you, this is about the most beautiful thing I've heard in a long time. In a long time, Juliet. (*Juliet crying*)
	(*Softly, after pause*) What do you say, Juliet, we don't do anger management at all. That we don't think of this as an anger problem, but we care about this dilemma of yours properly . . . the despair a woman feels when love and attention vanish, and her spark of resistance to say "No. I don't want to live like this. I want love. I want

romance." And what that clarity means for her starting today. What do you think if we centered this dilemma as the heart of our work?

Now, dear student, I want to discuss the many invitations inside this transcript excerpt to shun stories of poetic and political dilemmas in clients' lives in the context of their creation only to seal them off into negative-sounding "problems" prematurely. But I have a hunch I have to take care of a worthy distraction first in order to not lose your attention with fretting wonderings. Unless you are very different from all the students I have met, your mind might currently be centered on the moment of hitting and its discussion inside the story, no?

I want to tell you that I owe a great debt of gratitude to a few particular women over and above the men I have met over the years in my therapy room who have come to talk to me about their pain of acting outside of their ethics in relationships, and about their use of violence to make a point with their family members, especially their kids. They have memorably instructed my soul to take positions against violence in human relationships of all kinds, including in this excerpt, violence against a man.

I owe these clients, the mothers and adult daughters in particular, who have spoken to me together and apart about their worst moments with each other, the clarity I have gained in discerning coercion of all kinds and the words to help me to embrace the moral pain coercion causes for the recipient, and differently but significantly, its user. Violence in relationships brings with it legal and ethical considerations in our therapy work, but I will help you with those matters should you ever find yourself in such a rare conversation with a client during your training time with us, so don't worry. The most important part for now is the moral discernments of violence. What I learned from the plain-spoken women who have accounted to me of their use of violence in their relationships with their kids, in particular, is that the work with human beings who act outside of their own best knowing to coerce and abuse other human beings is some of the most important work I have ever done. If we can care about the dilemmas and the moral pain in the wake of these actions, I believe we are doing some of the most significant feminist labor we might ever do in favor of making more equitable and healing relationships possible. The endeavor to care about violence and its creation and its users has taught me a strange lesson: the ones wielding it need our help in as urgent a manner, perhaps even more intensely than its recipients for its degrading effects on both their souls. The courage of the women whose ability to account we fought for together through thick and thin also instructed me to waste all my good will in efforts to heal all manner of "smaller-sounding" matters of "What Human Beings Routinely Do To Each Other."

Fair enough for now? I will return to conversations about people acting outside of their ethics in a later letter entitled No Angels.

Returning to our task at hand to study the above transcript together, how many invitations did you count to exchange storytelling efforts for the sealing off of a dilemma into a tidy "problem"? Why did I not begin my work on "emotions," on "anger," or at the latest, on a "badger named Stripey"? And why did I return to repeat the beginning question to Juliet of why she had come to talk to me that night after the story she told of some of the triumphant achievements of love of the past 12 years?

Remember the definition of a story we looked at? A story for our therapeutic purposes has a setting, characters, a protagonist, and a dilemma of everyday living and relating which answers the query of "why are you in therapy?"

We can do all manner of fancy footwork about "the badger named Stripey," get very interested in Juliet's relationship with Ben that gave rise to this memorable metaphor or grow confused with assumptions that Juliet came to talk to me because of the effects of the history of the two mothers' involvement in her relationship with Rick, or story the effects of pain on her body or the words of a couple's therapist. Session upon session can be on the raising of wondrous stories complete with a protagonist in a setting and a plotline – but our beginning stories should respond to the query of "why are you in therapy?" or "what is on your heart most restlessly right now in your life?" to avoid the weeds and wildflowers of random storytelling ventures called anecdotes of little consequence. I believe our clients deserve the labor of our fingerhold on this clarity, especially while they are a bit busy with the sweaty efforts of discerning trust while holding their lived experiences in a net of words for us during our meetings. For me, this faithfulness of ours to our clients' most urgent dilemmas makes the difference between therapy sessions that feel evasive and selfish in my favor, and those that feel urgently generous towards what most matters to clients. My trustworthiness in tracking stories out loud for clients assures them more convincingly of my listening and my respect for their living and tells them, ever so insistently, that the matters that matter to them, matter to me. The energetic transport of bothering to tell stories that are without a doubt worth the struggle to tell them is easily lost, but, also, found!

This point is so important that it earns an acronym: Excuse Me, But Do You Earn Your Keep, Dear Story? (EMBDYEYKDS) as a storied response to the query of why this person has come to therapy.

It sounds so simple, doesn't it? And yet it is the most difficult task of all. Why do we get distracted from telling such stories? I believe that as narrative therapists living in our particular time and space, we have been formidably instructed on what to attend to instead: to answer the query of why is this person in therapy with a narrative diagnosis. Hear the dragon roar: this particular distraction is a Master Narrative-driven dragon that instructs us to give way to our favorite game of binaries: the sorting of people and ideas and lives into "positive" and "negative" before anyone has a chance to say peep. Problems and solutions. Struggle and ease. Good and bad. Negative identity conclusions and positive identity conclusions. Problem stories and alternative stories. Anger and calm. Stripey Badger and Love Me Tender. And so it goes.

If I were to try to say it very simply, we have gone along with the investiture of the dominant variable of a positive-negative binary to judge human living. But for narrative therapists, the variable was always meant to be "richness." This is why I would rather speak of "stories" and "unstories" – they are not separated by the variable of a positive-negative binary but separated only by way of the variable of richness. Human living is better respected by the use of the variable of richness.

The usurping of narrative therapy practices by the positive-negative binary has caused a wide-spread depoliticizing of practice.[5] This depoliticization of narrative practice is happening in almost every transcript I review, and in every teaching and training context I attend, and I want to let you in on the secret: the favored means in the service of this development appears to be that thing called "externalizing." You see, dear student, narrative therapy found its beginnings in umbrage against the sweeping ethos of pathologizing and medicalizing of persons' lived experiences. But the success of the hidden Master Narrative-driven simplification pressures has caused narrative therapists to sign their practice over to the philosophical and political orientation of striving for no more than replacing clients' negative-sounding identity conclusions with positive-sounding identity conclusions. In other words, we have started using "externalizing" as a means to "internalize" both problems and unique outcomes.

Applied to the case of Juliet, the positive-negative binary would press me to follow marching orders in which I am to take up "anger" as the named "problem," especially by the time it is offered to me as a "Stripey Badger," although you can hopefully see that this distinction is merely the same as "depression" and a "muddy overflowing river" we met with Alicia. Together with Juliet and set to my heartbeat in the wake of a roar, I would be called to matter-of-factly investigate times when she has been able to "resist the Stripey Badger," times when she has "managed to be effective against the Stripey Badger" and "steps to reclaim her life from the Stripey Badger."

But wait, oh please, for the love of all the mutilated universe, please wait. What do we stand to lose when we go along with the practices of narrative diagnosis, for make no mistake, that is what we are doing when we treat Juliet's "Stripey Badger Anger" as a "problem" prior to its storying into a rich human dilemma? We lose every reason for her anger. We lose the context of its creation. We lose the politics and poetics of Juliet's lived experience of anger.

Whenever we give ourselves permission to conclude what our client's "problem" is, give it a name, even a punchy narrative externalizing name, and form a conversation to oppose this identified "problem" – we are powerfully enacting a narrative diagnosis of a problem. We are saying to your client who has come in weeping and raging in the chaos of her soul and with a slew of confusing words and in the midst of the richness of her lived experience that caused said emotions – "yes, THIS is your problem."

It is a powerful temptation to exclaim: "but those were Juliet's words! Juliet said 'I need anger management, I struggle with Stripey Badger!'" Yes. That's true. But when we go along with the clients' named label, we are still powerfully enacting a narrative diagnosis. We are putting our weight and considerable power in the service of opposing a particular problem. Shouldn't we care to know more details about clients' lived experiences before using our power in this manner? In addition, when we structure an entire conversation around how to stand against "Anger" we are also putting our easy agreement behind the fact that "Anger" is a negative thing, something Juliet ought to be effective against, something Juliet ought to reclaim her life away from. Do you see the Master Narratives hard at work here, to cause our collusion to discipline a woman's anger by way of tacit agreement with prejudiced assumptions about the stock plot of "angry women"?[6] The enactment of narrative diagnoses outside of richly storied human dilemmas will call therapists to rely on tropes of Master Narrative-driven stock plot prejudices about the constitution of good and proper living. Dear student: welcome to our time in which women's and queer people's anger and many of their most vivid life experiences are still not welcomed as human dramas worthy of our best imaginative and storytelling talents.

You see, I will not relent on this position: Juliet has come to my office to talk and cry and rage for very good reason. And yes, this is also a political position: I refuse to accept that her "anger" just "happens" to her.

Her "anger" is a response to something that is happening in her life.

Her "anger" is an expression of something that is happening in her life.

Her "anger" is an effect of something that is happening her life.

Her "anger" is a pre-therapy strategy for achieving personal power.[7]

My job is to care for the story of these happenings.

But the use of narrative diagnoses is often powerfully wielded to shut down any exploration of what her anger stands in relation to – and it doesn't matter if you take care to phrase these supposed "problems" with humorous or gentle narrative labels. If we do not take our narrative politics seriously to invest with every fiber of our being in finding out just why a person is crying in your office, or routinely experiencing "Anger" in her life, then we are left with internalizing labels in the form of narrative diagnoses.

See, dear student, here's the thing: Externalizing is not as much a technique or an intervention or a tool. It is a vernacular, a way of speaking. It is a means to rich storytelling of lived experiences of struggle, not an ends to vanquish or exorcise negative life experiences. "The purpose of this work is to encourage people to take up an observer position in relation to the events of their lives; to become a narrator of their lives," Michael White writes regarding externalizing conversations.[8]

Externalizing conversations are meant to help both Juliet and me speak of what Juliet is currently going through in life and help us word the context of the creation of Juliet's dilemma as richly as we possibly can.

Have a look at the following sentences from the transcript: "Let's say that this Stripey thing doesn't beset you out of nowhere, for no reason . . ."; "What's the badger's insistent message?"; "If the badger thing sounded like something, what does it sound like, what are its aims for you and Rick?" These are all questions that rely on the vernacular of externalizing conversations to help Juliet and me find our second-best words to give voice to the stories of anger in this moment in her life, in the setting of her own house in which love is fading, and the experience of the often-gendered withholding of attention by her partner, and the impoverishing effects of this on her person, her body, and her future. In the stories that Juliet tells in this session, the externalizing dialect or vernacular helps her become a narrator, an observer to the confusing happenings and actions in her house and in her own body and soul. Anger is complicated between Juliet and me to no longer be a simple problem as much as an expression of Juliet's embodied longing for her life and her relationships: "Okay Juliet, do you see how the badger's message is so important, if you're saying: *'talk to me, be with me, don't walk away, I miss you . . .'"*

Dear student, if this letter serves any purpose at all, let it be this: despite the dragons, because of the dragons, in willful defiance of the dragons, resist taking up unstories, especially the ones your clients propose to you about themselves, as your marching orders for the doing of therapy. Seek the rich stories of "what happened" instead. The stories that are neither positive nor negative, but rich as life.

Oddly, the telling of such stories, and the posing of worthy life dilemmas, will take you much further into healing work than you could imagine. How much is won by spending all our initial sessions in the effort of posing dilemmas well?

At the end of the session, Juliet sighed and packed up her bags. "Thank you," she said. "For what, my dear?" I asked. "I know what I'm gonna do now. I'm not going home, I'm going straight to my sister's now," she said, already scanning her phone for her sister's number. "Why?" I asked, not doubtfully, but curiously. "Because I'm not giving up on kissing at 35 years of age," she smiled at me, and dialed her sister's number while walking out the door.

We can come to story our clients' lives such that we can understand their anguish at this moment in time. The point is not to separate them from this anguish by way of labeling it away. The point is to richly story it. This is a narrative answer to the human problem of pain.

Love,
Sanni

PS: Shall we get to it then? The shaping of rich Up Against Stories with the clients you will meet? I can almost feel you challenging me to show you ways of doing so without pressing clients to take a survey of the insides of toilets: "and then this crap happened, and then this crap happened, and oh yeah, then this other crap happened too . . ."

Notes

1   "Simply put, his [Levinas] ethical response is to put the relation to the other before knowledge or theory of being. To be face to face with another person overwhelms all our concepts and theorizing, and evokes an infinite experience of responsibility: 'To be in relation with the other face to face is to be unable to kill' (Levinas, 2006, p. 9), which applies as much to thoughts and language that override the other as to murder" (p. 333). Larner, G. (2008). "Exploring Levinas: The ethical self in family therapy." *Journal of Family Therapy, 30(4)*, 331–352.

2   "Michael, I am now speaking to you almost eight years from the date in San Diego. We couldn't have predicted how the neo-liberal regimes of thought have infected all the therapies and the worlds we live in, and already artificial intelligence is working on robotizing therapy. My friend, David Codyre here in Auckland tells me that the Mental Health Services are investing in this and that potentially the Watson IBM computer could, believe it or not, be 'taught' to have narrative conversations with people! Let me tell you about a federally funded service for youth to which I consult in West Sydney. You will recall it as the most marginalized area in Sydney. Staff are provided with four manuals, and I am sure you could guess which they are: anxiety, depression, psychosis, and trauma. No manuals as you can see for poverty, racism, asylum seekers and refugees who have fled war zones, unemployment, or for the indigenous Aboriginal people. I am very sorry to tell you that 'maps' in some instances have been eclipsed by the trend to manualization of our avocations. And believe it or not, there was even a narrative therapy app in the United States for $5 or so. It seems to have failed to go viral I am glad to say. The watchword in every manualization of a practice is 'fidelity'. Do you get it? If the practice doesn't work, you have failed to be faithful to the manual. The genius and inventor of virtual reality, Jaron Lanier (2007), in the different context of musical composition, refers to this as being 'locked in,' which denies the imagination and forbids creativity. The 'freedoms' required for our imaginations to re-imagine narrative therapy have now become very precious. And we may have to secure them appealing to a greater fidelity to the 'spirits' I referred to in order to respond to

your concerns that narrative therapy had become uncreative and why everyone sounded just like you and not themselves." Epston, D. (2019). "Re-imagining narrative therapy: An ecology of magic and mystery for the maverick." *Journal of Narrative Family Therapy*, Release 3, pp. 1–18, 10.

3   "Maskihkiy defines cultural humility as approaching each individual/family/group/community in a humble, open way with a stance of "not knowing" in relation to culture and worldview. The Euro Western construct of competence implies that one has fully mastered knowledge. This is not entirely possible within Indigenous worldviews, as there is always more to learn & knowing is fluid & relational. Humility is a central cultural teaching for many Indigenous peoples." Fellner, K. (2023). "Maskihkiy wellness." www.maskihkiy.com/consulting

4   "Having just given you this formula, I have to give you a warning. If externalization is approached purely as a technique, it will probably not produce profound effects. If you don't believe, to the bottom of your soul, that people are not their problems and that their difficulties are social and personal constructions, then you won't be seeing these transformations. When Epston or White are in action, you can tell that they are absolutely convinced that people are not their problems. Their voices, their postures, their whole beings radiate possibility and hope." O'Hanlon, B. (1994). "The Third Wave: The promise of narrative." *Psychotherapy Networker*, November/December, pp. 19–29, 28.

5   "All of this psychologizing of personal experience, and all of these formal analyses, are deeply conservative. They are invariably pathologizing of the lives of those people who have been subject to abuse, and, in so doing, divert attention from the politics of the situation. As well, so many of the interpretations of this sort discriminate against women's ways of being in the world and champion dominant men's ways of being in the world." White, M. (1995). *Reauthoring Lives: Interviews and Essays*. Dulwich Centre Publications, p. 92.

6   "I honor that angry four-year-old girl. I honor her belief that she deserved to be free of molestation, free of interruption, free of a man who believed he deserved her time and attention. She was born with a pilot light of anger, tenacious and sure of its right to

flare whenever treated unjustly. I believe all girls are born with that pilot light of anger. What happens to it as they grow into women?" Eltahawy, M. (2019). *The Seven Necessary Sins for Women and Girls*. Beacon Press, p. 15.

7   "While formal diagnostic systems are all simply socially constructed taxonomies for organizing and grouping people's manifest expressions of distress, diagnosis as practiced is commonly a strategy for reification and objectification of the client and pathologizes their pre-therapy strategies for increasing personal power by whatever means available." Brown, L.S. (2018). *Feminist therapy* (2nd Ed.). APA, p. 67.

8   White, M. (1995) *Re-authoring Lives: Interviews and Essays*. Dulwich Centre Publications, p. 134.

For Juliet
For All the Wished for Kisses

For all the wished for kisses
all the longed-for words
All the wanted looks
That were lost
To us
In this house

I lay down my
Five-year
Fight
And
Lose
Us.

But the dream of
Love
And its kisses
Lives on
In every step
Of this parting
Of sweet sorrow.

5

Why Are You in Therapy?

A Letter About Up Against Stories

Dear student,

Imagine. Imagine yourself at your first session with a new client. You begin with a first question that seeks the interestingness of your client in a manner that does justice to your recent remembrances of those beautiful moments in your history when you have been "leading the charge with good will." You query the unstories your client will likely offer you by seeking a story of why this person has come to therapy. You're not a mindless machine so you do notice a shimmer of scale and the rumblings of a roar with a jolt of #UhOhHowWillWeEverFi gureTHISOut. But you do not shrink or shut down your client's stories of struggle to comfort yourself in response. You do not set out to separate the client from their lived experiences, but to care for them instead. You proceed, on purpose, to amplify the stories of their lived experiences to the size of the whole of your office. What is that? What happened? What is that like for you? What was it like to hear that? What were you saying in that moment? What do you think of this? Why? Why is this important to you? Why does it matter so much to you? Etc.

What kind of story will you now meet, some minutes in to your first conversation?

DOI: 10.4324/9781003478478-5

You will likely meet an Up Against Story. Some students really love it when I sweat concepts all the way down to something structural, and if you are not one of them, please ignore this: Up Against Stories can be summarized to take the form of "X Makes Me Feel Y"-stories. As a foreshadowing, a future letter entitled Counter Stories summarizes them to take the form of "Given X and Y, I Did Z On Purpose"-stories. These two different kinds of story forms are separated only by the direction of the flow of agency. But more on this later.

Now some clients of mine have told Counter Stories from the outset and have required very little help from me save for an edge-of-my-seat "what happened next?" or some witnessing responses to propel them along. But that is a rare experience in my therapy office, so it will serve you better to expect an Up Against Story. Our job at the outset is not to resist Up Against Stories, but to understand them. Michael White writes in response to a question about how he gets started:

> I'm really interested in people's accounts of their experience. I really want to understand what life has been like for them. So, I guess the first part of my work is to try to get some appreciation of what persons have been going through. I think it's important that I achieve a degree of understanding about this, and I think it's important that persons are aware that I have achieved at least a degree of this understanding.[1]

Some students have found it useful when I have translated the work of the beginning sessions to a clear message: don't set out to fix or solve anything here at the beginning. You will lose out on a rich Up Against Story, which, in turn, sets limits on the richness of the later Counter Story. Instead, set out to care for your client's Up Against Story enough to understand it as well as you can; know that by so doing, you are setting the conditions for expansive Counter Stories to become possible.

I have come to understand that students are anxious to discover mountains of misery that would render both them and their clients fragile and incapacitated were they to embrace such an endeavor. But the purposeful practice of setting out to shape interesting Up Against Stories forward has not born out this anxiety to me: to take an Up Against Story seriously in its interestingness has not yielded a barrage of intensifying complaining to me. Instead, I have come to discover what matters most to clients in their lives inside these unforgettable examples of Up Against Stories. Paradoxically, the focused concentration of attention on the complications of an Up Against Story and the political context of its creation and the moral bids to its teller cut through the shallow and opaque venting-validating cycles that many therapists and clients find so demoralizing in their conversations. The spark that I hope is never extinguished in the venture of therapeutic storytelling is the interestingness of human beings inside the most restless stories of their lived experiences.

So for your purposes, dear student, I want to introduce a few practices that will come to your aid in the effort to not solve struggle away at the outset, but to pose it well as a dilemma worth its struggle. Shall we do so from the vantage point of a session of mine in which a client and I are shaping an Up Against Story forward?

This letter is intended to take a closer look at practices that can help you shape these beginning conversations to respect your client's Up Against Story while also refusing to glamorize, valorize, or validate the supremacy of misery.

The practices for the purposes of this letter are:

1. Good faith practices
2. Storytelling questions
3. Witnessing responses
4. Imaginative leads
5. Moral reading prompts

Please know that the compartmentalization of these practices into distinct-sounding categories is not intended to invite you to memorize these into steps you are to follow in your future in the name of rehearsed spontaneity. I am only insisting on an artificial list of practices to help support your clarity in understanding the transcript excerpt I am about to show you. Without the attempt to clarify categories of practices, beginning students tell me that transcript excerpts have about as much edge and shape to their minds as a bowl of porridge.

The following transcript excerpt shows an attempt to shape an Up Against Story forward in a first session with a client whose name is Maddie. I will draw your attention to the above-named practices inside the transcript by noting them in brackets. I will elaborate on the first three of the practices after this read of practices-in-motion. The last two, the Imaginative leads and Moral reading prompts are noted for now but stand to be better elaborated on following letters. Fair enough?

Maddie, like Em, had shaky hands when we first began speaking. When I asked her what mattered to her above all else that night, she exhaled shakily and hesitated. Into the silence and her visible struggle, I said:

Sanni: Hey, it's okay, take all the time you want . . . I could also try a different question that's not so much in your face. [Good faith practice]

Maddie
(Smiling,
her eyes
filling with
tears):   No, it's not that. I'm just . . . trying to catch my breath . . .

Sanni: Right. Well, you know you find yourself in good company with lots of people sitting right here on that couch before you and just trying to breathe too, including me at times. *(Maddie smiles again through her tears)* So no worries. [Good faith practice] *(After pause)* Well, just thinking back on it now, I could tell you a bit about the very good reasons some of the people who have sat here have had for having a hard time breathing when they first arrived . . . Would that be good? [Good faith practice]

Maddie (Exhales, with determination):	Thank you. But . . . I can do this. This is just big for me. But I promised myself I would do this. I can do this.
Sanni (Gently):	Okay, Maddie. Whatever "this" is, I promise I'm here with you, I won't just let you crash and burn. *(Maddie smiles at me)* But help me out real quick, why did you promise yourself to do "this?" Why is it so important for you to do "this?" [Good faith practice]
Maddie:	Well, my girlfriend and I decided that I should probably talk about this.
Sanni:	Do you agree with your girlfriend about that? That you should talk? [Good faith practice]
Maddie:	Yes! I love her. So much. She's the reason I'm here. She's like the best person I've ever met, and she is so supportive of me. I feel safe for the first time in my life. We almost came together, you know, because you said on the phone that it's okay to bring someone, but then she thought she'd give me some space to talk about it however I wanted first. She might still come at another time though.
Sanni:	She's more than welcome. I just gotta remember to welcome her properly as, how did you say it? "The best person Maddie has ever met." *(Maddie laughs)* What's her name? [Witnessing response]
Maddie:	Her name is Mia.
Sanni:	Alright, welcome Mia. You know, I'm tempted to ask why Mia is the best person you've ever met. Should I ask you that? Or do you wish me to tend to your promise with you? [Good faith practice]
Maddie:	Yeah, the promise.
Sanni:	Okay, does Mia know you made this promise to do this tonight with me?
Maddie:	Yeah, she knows.

Sanni:	And Mia . . . she cares about whatever the "this" is? She is tending to it too? [Good faith practice]
Maddie:	Yeah. She is so supportive of me.
Sanni:	Okay. Maddie, I might not be at all on the right track here, please correct me if I'm wrong, but did you make this promise to do "this" because of your love for Mia? [Good faith practice]
Maddie (Tearfully):	Yes! That's such a nice way of saying it. *(Exhales)* Okay. I'm doing this for Mia.
Sanni (Choked up):	Okay. Here goes for love or nothing then. [Witnessing response]
Maddie (Laughs through tears):	Yeah. Okay here's what happened. Mia and I . . . the reason I called you was because Mia and I had a bad moment, we were in bed, and all of .a sudden I just freaked out, you know. I freaked out, and jumped out of bed, and I was crying and yelling, and I didn't even understand what was happening. And I couldn't calm down, and Mia . . . she was just asking me what was wrong, and don't know, I have never loved anyone like her, she makes me feel so safe, so why is this happening then . . .
Sanni:	Okay, so here you were, all safe in bed, and in love, with the best person, and then next thing you know, you're freaking out yelling, and everyone is confused like what is happening? [Witnessing response] Do you remember what you were yelling? [Storytelling question]
Maddie:	I yelled, "No, don't do that! I don't want that! I don't want that!"
Sanni:	Okay . . . was it in response to something that Mia was doing at the time, or did something else come to you that you were protesting? [Storytelling question]

Maddie:	I don't know . . . No. I do know. It's just hard to say. Mia wasn't doing anything wrong, . . . she was just touching my belly . . . *(Crying)*
Sanni:	Okay, okay. Well, before I ask anything else, how did the story go on, Maddie? You jumped out of bed, like "no! I don't want that" and Mia is all shocked and so are you, and then what happened? [Storytelling question]
Maddie:	I got dressed, as quickly as I could, and then I went to the living room, and just broke down on the couch, just crying.
Sanni (Softly):	And then what happened? [Storytelling question]
Maddie:	Mia came out and sat with me and held me. *(Crying)*
Sanni (Softly):	She did, hey? [Witnessing response] And then? How did the night end for you two, on the couch? [Storytelling question]
Maddie:	We just talked, and she held me. She was so good. And we . . . we figured out that I need some help. That it's probably related to the trauma of my first relationship. And that I've been broken for a long time now.
Sanni (Softly):	Is this the "this" that you promised to talk about tonight, the trauma of your first relationship? [Moral reading prompt]
Maddie:	Yeah.
Sanni:	And we'd be betraying your promise if I, say, asked a bunch of questions now about your and Mia's love instead? Like you'd be super disappointed in me for being a traitor to your promise, like "Why the fuck did she not help me to talk about it when I asked her to?" [Good faith practice]
Maddie:	*(Laughs a little)* Yeah.
Sanni:	Maddie, I'm here. I'm here all the way, for whatever it is. But we have many ways to talk about this. Do you want to look at some ways to talk together with me first, and then decide how you want to go about the talking – all of which

	still fulfill your promise? Or do you want to let what happened to you just tumble out, however it wants to, and we hold it together? [Good faith practice]
Maddie:	Maybe . . . think about it first . . . I like that. Thank you.
Sanni:	So say we did talk tonight, I wish I could ask you about both the worst and the best that might happen to you then . . . but maybe we'll start with the best . . . say your best hope would come true after talking, like you kept your promise, and the universe decided not to ignore you but to repay you, like that girl in that tale where the gold came raining down on her underneath the castle arch afterwards . . . what would happen to you, what's the best thing that could happen to you if you did keep your promise? What would come down on you like raining gold under the castle arch? [Imaginative lead]
Maddie (Smiling):	I'd be free. I'd confront it . . . and process it and be free . . . And I know the worst too.
Sanni:	You do? What's the worst?
Maddie:	That I'll be so embarrassed, and it won't help.
Sanni:	I see. Embarrassment is now my sworn enemy in this talk! [Witnessing response] *(Maddie laugh)* And we're doing this for love and freedom then. You're something else, you know. [Witnessing response] But anyway. What does it mean, though, to "confront it" and "process it"? [Moral reading prompt] Is this your "work" before freedom? [Moral reading prompt] You know lots of women have puzzled here about those very words.
Maddie:	You know I don't really know what it means. I was in this women's studies class, and that's what they kept saying, you have to process it. They spoke about it so easily, and I sat there thinking, oh my god, this is me, and I was so embarrassed, like how

is this the first time I am hearing the words consent and stuff, and they are talking about it so easily . . . like they were . . . self-assured . . . and I . . . You know, I knew what rape was. My aunt was raped when I was a teenager, and there was a court case and all that, and it happened . . . you know, it was a stranger in a parking lot, so I knew that that was rape. But how could I have been so stupid? It didn't even occur to me that it could happen in a relationship, you know, boyfriend girlfriend, or . . . I realized that it happened to me, and I felt so stupid.

Sanni: What if, Maddie, what if we could speak differently . . . somewhere beyond the self-assurance of those words like consent and rape and the advice for you to "process it" . . . what if we could attempt to not care about those words right now at all, and everything that they mean, and who is stupid if they don't know the right words . . . what if we spoke, just you and me, in your own words and who cares if they are the right ones, about what hurt you then and what hurts you now? Just that. And only in words that you know. [Good faith practice]

Maddie: Yeah!

Sanni (Softly): What hurt you then, Maddie, or what hurts you now thinking of it, or what hurts you in your relationship with Mia? What do you wish to be free of? [Moral reading prompt]

Maddie: That I took whatever was coming at me, in relationships. That I didn't know I could say no. I didn't even know to stop to ask whether I am ready. I just laid there! I didn't do anything! I just laid there. He was my first boyfriend and I still cringe thinking about it. He did whatever, and . . . he was mean. He laughed . . . (Choking up) . . . about my weight, he was awful.

Sanni: Okay, okay. As you laid there, Maddie, all silent, what was going on in your mind, your body, your soul? [Storytelling question]

Maddie:	I was . . . I didn't know, I was . . . revulsed. That's the word. But I didn't even know . . . to take that seriously.
Sanni:	You laid there, in a state of revulsion, noting him and his meanness and his jokes, and everything went "if this is what they call love, then eww." [Witnessing response]
Maddie:	Yes. *(Exhaling)* It's weird. It sounds better when you say it.
Sanni:	Why do you think, Maddie? What did I say differently? [Moral reading prompt]
Maddie:	No, it's like it doesn't feel as much like I did . . . nothing.
Sanni:	Then what else Maddie? What else did you think or feel? [Storytelling question]
Maddie:	No, that's it. I really did nothing, said nothing. I was so stupid.
Sanni:	This may be a weird question Maddie, and maybe not interesting to you, but how did you get out of that relationship with him? [Storytelling question]
Maddie:	I broke up with him, but I stayed for far too long. That's also the stupidity of it. I stayed for two years.
Sanni:	And after two years, what then? [Storytelling question]
Maddie:	It wasn't a dramatic break-up, you know, if that's what you're looking for. I just started to not have as much time for him, and we went our separate ways.
Sanni:	And after it ended, were you all broken up for a while, did you cry lots, or were you so relieved like "thank gawd that's over" or what happened to you? [Storytelling question]
Maddie:	I got into another relationship. *(Said shamefully)*
Sanni:	Oh sheesh, did you break all the rules of the rulebook that say we have to be single for x-many months to lick our wounds, and "process" shit or

	whatever before we are allowed to smile and hope and reach for a person? [Moral reading prompt]
Maddie:	(*Laughs a little*) I did.
Sanni:	What kind of a relationship did it turn out to be? [Storytelling question]
Maddie:	It was fine. It was . . . the only thing I knew then was that what I wanted was someone . . . sweet. That's the word.
Sanni (Quietly):	You wanted . . . sweet. [Witnessing response] What all does this word mean to you? Maddie, I want to know, I am flooded with the imagination of "sweet" as a counter to "mean" and "revolting" and someone who didn't stop to ask and wonder where you had gone and how you were in such a vulnerable intimate moment . . . Maddie, did you somehow know to seek the counter to "revolting," even when you were young and didn't know the supposed right words for it? [Moral reading prompt]
Maddie:	(*Crying*)
Sanni (Softly):	What are the tears trying to tell . . .? [Moral reading prompt]
Maddie:	That I . . . maybe I . . . maybe I wasn't so stupid.
Sanni:	Holy crap! "Maybe?" [Witnessing response] (*Maddie smiling through tears*) Well, say you were me and you heard this story about a young 15-to-17-year-old with her first boyfriend, and the awful meanness of their most intimate moments, and her frozen revulsion, and how he did not attend to her in those moments, and she told you that she didn't know how to make any sense of any of that at the time, and all the words for it had been hidden to her, but the one thing, the one thing she knew for sure, and she could say, the one thing was "NO to this mean awful frozen revolting thing" and "YES to something sweet." And this she knew without any doubt, and at all of 17, with all the pressures to be some good girl at that age,

in her whole body she knew this, and sought to shape this knowing forward in stepping towards a sweet person.[2] [Witnessing response] What say you? In what kind of universe would you, as me, think, "well she's stupid?" [Moral reading prompt]

Maddie
(Laughing):   I wouldn't!

Sanni
(Laughing):   Yeah, even if you were just some dumbass therapist like me . . . this you'd know. Okay, Maddie, do you want to keep talking about you at age 17 and what you knew to do in response to awfulness that wasn't stupid at all but actually really interesting? See, the thing is, I am still holding Mia close to us, so the other option is that we talk about your lack of stupidity now and look at you jumping out of bed and crying and yelling "I don't want that" with her and assume that it's probably for a very interesting reason you did that? Which option would be more interesting to you? [Good faith practice]

Maddie: Huh. I think, I'd really like to talk about what happened with Mia and me . . .

Dear student, I hope you can see in this transcript excerpt that the categories of questions and responses have much overlap with each other. The important point is that all of these practices are in service of shaping a story forward, a story with an interesting protagonist, memorable characters, a unique setting, a political context, and a plotline in which substantial dilemmas of everyday living are at stake and which answers the query of "why are you in therapy?"

As an aside, I wonder what you were thinking and feeling when the transcript cut off just as Maddie and I were about to explore her jumping out of bed and crying and yelling? Did you have an anxious or frustrated sense that we should rather find a more "positive" avenue for our conversation? Did you wonder if

Maddie would leave our conversation more fragile or burdened if we set out to look at her moment of freaking out with her partner Mia?

This might be a good lead into studying some of these practices of storytelling together.

First, I know you are probably wondering what I meant by "Good faith practices." I could have also named them collaborative practices or practices of consent, both of which would very much be fitting. Practices of collaboration and consent require a person to have the capacity to collaborate and consent, which is a belief that is often only paid shallow homage in our field in the form of consent paperwork at the outset or "how was this conversation for you?" questions at the end of sessions.

Not only do I believe that the persons who come to see me in therapy can collaborate with me and give their consent to a line of inquiry, I invite them to discern, think, and feel into the ongoing negotiations of collaboration alongside me at every turn. I seek, not just consent, but to cede influence over our conversations to clients by making sincere proposals, asking their opinion about which way we might turn, inquiring why we should speak this way or that, and offering my discernments and queries about the ways they have chosen as honestly as I can. I do not mean to blindly follow clients, nor do I mean to set out to mutely lead; I mean to negotiate with clients, in good faith. I mean to invite them to negotiate with me, in good faith.

This means that I have sometimes said "I don't think this is the best way of speaking about this and here's why I think that. What do you think?"

Good faith is a practice of belief, not in innate virtues of mine or those of clients, but in the relational practice of human beings routinely showing up in good faith to negotiate and figure out dilemmas together.

Here are some examples of such Good faith practices from the conversation with Maddie:

- "Hey, it's okay, take all the time you want . . . I could also try a different question that's not so much in your face . . ."

- "Well you know you find yourself in good company with lots of people sitting right here on that couch before you and just trying to breathe too, including me at times. *(Maddie smiles again through her tears)* So no worries. Well, just thinking back on it now, I could tell you a bit about the very good reasons some of the people who have sat here have had for having a hard time breathing when they first arrived . . . Would that be good?"
- "Maddie, I'm here. I'm here all the way, for whatever it is. But we have many ways to talk about this. Do you want to look at some ways to talk together with me first, and then decide how you want to go about the talking, all of which still fulfill your promise? Or do you want to let what happened to you just tumble out, however it wants to, and we hold it together?"
- "Okay, Maddie, do you want to keep talking about you at age 17 and what you knew to do in response to awfulness that wasn't stupid at all but actually really interesting? See, the thing is I am still holding Mia close to us, so the other option is that we talk about your lack of stupidity now and look at you jumping out of bed and crying and yelling 'I don't want that' with her and assume that it's probably for a very interesting reason you did that? Which option would be more interesting to you?"

In reading the above questions, what do you notice? Who is Maddie to me, as exemplified in the spirit of these questions? What are these questions telling Maddie about my beliefs in her capacities to discern good ways of speaking?

I believe there is no greater practice of respect that I can embrace than treating clients as substantial shapers of our conversations and asking them to make difficult moral discernments out loud and in good faith that have the power to change the course of my mind and our entire conversation. Again, I stress, this does not mean that I agree to all that clients ask of me during the course of therapy. I mean to negotiate, sincerely, and in good faith, and with an eye on the real effects of our speaking for both my client and me.

Good faith practices also express my sincere practice of belief that:

- all clients are agents in their own lives,
- currently struggling with oppressive binds (practical, systemic, cultural, and ideological) to their freedom of movement in the world,
- wondrously capable of originating preferred experiences into being if granted the narrative resources to do so,
- endowed with unique manners of expression to convincingly pass on what matters so much to them in their lives,
- and with living moral codes that can be discerned and known.

These beliefs are a practice of antivenom to the belittling of persons that is a result of our collective engagement with Master Narratives about clients' lives. Belittling Master Narratives that dominate our speaking and our beliefs can hold our words and imaginations hostage in shaping interesting conversation about interesting lives going forward. This happens particularly often in our work with marginalized clients, with women, and with queer folks.

Okay look, perhaps a joke from my colleagues will enlighten this point for you: "Sanni's clients always think and ponder and have deeply felt questions all the time . . .", they say, laughing. It's true, they do. And it's also true, I invite this of them. See the power of our beliefs in action? I know of no greater treachery as a therapist than to stand in a long line-up of people who tacitly believe that women and queer people exist for little lives of little consequence save for little acts of sharing and caring.

The Storytelling questions and Witnessing responses in the above transcript serve the same purpose to shape a rich story forward. I am interested in all the ways we might make a story linger in its complications and all the ways we might resist a slide back into our usual eyes in the instant of the words leaving a speaker's mouth to make their way to a listener. My hope is to help shape a story that is not in any way representative of anyone but her, in this case Maddie.

Let's look at some examples together of these two interlinked practices:

- "Okay, so here you were, all safe in bed, and in love, with the best person, and then next thing you know, you're freaking out yelling, and everyone is confused like 'what is happening'? [Witnessing response] Do you remember what you were yelling?" [Storytelling question]
- "Okay, okay. Well, before I ask anything else, how did the story go on, Maddie? You jumped out of bed, like "no! I don't want that!" and Mia is all shocked and so are you, and then what happened?" [Storytelling question]
- "(softly) She did, hey. [Witnessing response] And then? How did the night end for you two, on the couch?" [Storytelling question]
- "Okay, okay. As you laid there, Maddie, all silent, what was going on in your mind, your body, your soul?" [Storytelling question]
- "You laid there, in a state of revulsion, noting him and his meanness and his jokes, and everything went 'if this is what they call love, then eww'." [Witnessing response]
- "And after it ended, were you all broken up for a while, did you cry lots, or were you so relieved like 'thank gawd that's over' or what happened to you?" [Storytelling question]
- "(quietly) You wanted . . . sweet." [Witnessing response]

Both Storytelling questions and Witnessing responses aim to be entirely unique and particular to this client, phrased in their vernacular, following an important story that has earned its keep as a response to the "why are you in therapy"-question well enough to understand it, down to the couch, and the actual words exchanged, and the lived experience of the protagonist. Uttering unique-to-each-client (rather than disposable) Storytelling questions backs clients in their effort to find their own second-best words[3] out of the confusion and pain of experiencing the world in a nameless way. In Maddie's, as well as many other clients' cases, it can represent a relief to be this frankly approached on a subject she deemed unapproachable when first we met. And there is more: as clients are invited to tell of their own important lived experiences and are met with an effort to bestow all possible narrative resources to them, clients become narrators who can pick and choose what themes the protagonist will focus their attention on. This, I believe, is one of the important ethics of externalizing: to become both Narrators and protagonists with the ability to think and discern while being swept up in the emotional, sensory, physical and moral whirl of life.

And there is one more thing: although it is for now but a secret foreshadowing of the purposes of Imaginative leads and Moral reading prompts: when the stories that matter are shaped forward, embedded in the words and grammar of that collaborative effort is action, because the metaphors and words we use to denote our lives invariably speak to us about what to do next. In other words, a narrator will chaperone a protagonist into the next chapter, and the hair-raising question: "What will she do next?" All without you having to control your client into the good life advice of the professional academy. But shh, that's a secret no one will believe.

Guess what Maddie's invention was? Maddie and Mia held each other after our conversation on that couch of theirs as Maddie recounted our conversation to Mia. Somehow, somewhere, in interesting-queer-women-land, the night ended with melted chocolate streaked on both their bellies. That's all I'll say.

Until next time,

Love,

Sanni

Notes

1 "I'm really interested in people's accounts of their experience. I really want to understand what life has been like for them. So, I guess the first part of my work is to try to get some appreciation of what persons have been going through. I think it's important that I achieve a degree of understanding about this, and I think that it's important that persons are aware that I have achieved at least a degree of this

understanding." White, M. (1995). *Reauthoring Lives: Interviews and Essays*. Dulwich Centre Publications, p. 21.

2   "Negative resistance can be seen as the embodied refusal which precedes not only a particular discourse but discourse as such. It is negative in the sense that it entails a refusal of a current identity but since it lacks its own narrative or discursive contents it cannot yet affirm what the person should become. It is a bodily resistance, not yet a discursive one." Guilfoyle, M. (2014). *The Person in Narrative Therapy: A Poststructural Foucauldian Account*. Palgrave, p. 116.

3   "Words fail us when we need them to say exactly what we need them to say. Poetry reminds us that we are at our best when we risk everything with our second-best words on behalf of the life of another." Surkan, N. (2018). *Can Poetry Save a Life? Presentation to Alberta Health Services*, Calgary.

For Maddie
Love Is Sweeter

Let me just catch my breath
To say this one thing:
When I laid there
I said nothing.
When I got up and walked away
I knew one thing
Clear as day:

Love is
tender
and sweeter
than all my training allowed me
To hope.
And love
always
lets me
take a breath
and asks me
what my inhales and exhales
might be saying

6

Whose Story Is It?

A Letter About the Agentive Turn

Dear student,

We have arrived at both a conceptual and grammatical hinge moment in these letters. I will attempt to explicate the idea of an agentive turn conceptually and practically by way of looking at a transcript.

But I begin with a story that has been useful to students in the past in their understanding of the agentive turn. On this particular afternoon, a friend and colleague called me requesting a "quick poetry consult." She told me that her client was arriving in a few moments, and that she had been wrestling with her notes and the poem she had written for the client in preparation. She made quick work to catch me up that the previous session had been a tearful conversation about an experience of rape that the client had suffered and that the client had said sentences like "I lost my virginity" and "he took everything from me." She told me that she had tried to write a therapeutic poem from these expressions, but that she was stumbling over printing "I lost my virginity" and the word "everything" in black and white letters. "Help me," she said, "I want to reflect what happened, but I can't write it like that. What is wrong with this?" We sat with the

DOI: 10.4324/9781003478478-6

client's expressions together for a minute and then came to: the agent is all wrong. The sequence of who did what is reversed: in these sentences the client appeared to be the agent in deeds not her own and to disappear as the agent of her very own responses to these deeds.

I'll spell it out: *she* did not "lose" her virginity, like one loses an umbrella or keys. She was neither the decider nor the agent in her own rape, however much patriarchy would instruct us to phrase it that way. Instead, *he* was the agent in that decision and deed.

Furthermore, he most certainly did not "take everything" from her as patriarchal notions would have us believe about the totalizing power of abusers. She responded to the rape in particular and purposeful ways all her own. So we pitched to begin the poem thus:

> *"He took what was mine to give*
> *And wanted to be my everything*
> *But I . . ."*

My friend filled in the rest of the poem with the client's unique ways of responding.

This letter is an elaboration of the discovery in the above conversation. (By the way, I did remove all expletives that were exchanged between my colleague and I about what happened to this youngen, but please do consider them absent but implicit. You're welcome.)

What may sound to you like a subtle syntactical shift is both an expression of a political stance against victim-blaming as well as a therapeutic stance of attention to clients' responses to all happenings. I might say it another way: no rapist shall have the final say of how a person's life goes. This authority belongs to your client and can be reinstated to her with the aid of an agentive turn in your conversations. Just as the externalizing grammar suggests to your client that "you are in the presence of an ideology and you should know this," the agentive grammar suggests to your client that "the other side can do magic too."

The agentive turn denotes a shift of your attention from "life happenings" or "the actions of others" toward your clients' responses to life and others. It's a shift from a previous sense of "everything merely happened to me" towards the attention on "but I also happened to everything."

This flip of attention is denoted in the structure of stories; whereas beginning therapy stories, or Up Against Stories take the form of "X makes me feel Y," the construction of Counterstories takes the form of "Given X and Y, I did Z on purpose." These two kinds of story structures differ in the direction of the flow of agency. See?

If you hate algorithms, I can say it another way. The agentive turn relies on an act of raising an active protagonist above the fray of the passive happenings in her life. By way of both a flip of attention as well as a flip of grammar, it resists the ghosting of your client to the role of a passive side character in her book, an "object to whom things only happen to." An agentive turn is an undeterred commitment of interest to the question "and what does she do next?" It promotes your client to the position of the Narrator and the protagonist of the book of her own life, and transfers to her both shaping rights and responsive powers.

As just an example, please contrast this to the stock plots of demotion of persons assigned female throughout much of the history of Western art: they appear as inconsequential decoration in paintings with one discernible attribute of some constancy which is the curious inability to get dressed in the morning;[1] or, consider how certain persons in literature seem to appear all too often as tropes of companions, love interests, or side characters with no discernible inner life and whose central purpose is to provide the real protagonist with important life lessons.

So, clearly the acronym of this letter is: Whose Story Is It? or WSII. I might as well say, are we there yet?

But if you had a living response to sentences that might read "basketball start heroically overcomes tragic rape he committed" or "I lost my virginity" – we are starting to get closer to the justice of the question: "Whose story is it?"[2]

In our little therapy rooms, we can resist the injustice of inequitable narration in which only certain persons' actions and lives are afforded our interest. If we wish to oppose this oppressive legacy of Master Narratives in which only some persons are considered credible and interesting shapers of worlds – we have to turn our attention to how these oppressions have also taken hold of our own imaginations, and the metaphors and questions we tell and ask of our clients. We can bring about a radical change about by looking at our questions, our stories, our poems and other documents, and our metaphors with our clients and amending them by way of a principle of agency.

This agentive turn was invited with Maddie in the previous letter in the form of the question:

"Okay, okay. As you laid there, Maddie, all silent, what was going on in your mind, your body, your soul?" and invited with beginning of "but I . . ." in the story of the poetry consult. Michael puts it this way:

> This is an account of personal agency. An account that emphasizes what could be called a person's agentive self. It includes details about what the person has been up against in the performance of this personal agency and against this background, emphasizes the significance of any more recent steps the person has been taking toward having more to say about how their life goes.[3]

What does a commitment to a protagonist sound like?

Here are the political and grammatical hints for your consideration:

1. Sort out the responsibility for actions in an unfolding story justly. Do not let your clients slip into a protagonist position in a sequence of decisions and actions undertaken by others ("I lost my virginity"), and conversely, do not let others usurp your clients' responsivenesses ("he took everything I had"). Stubborn guilt and shame about the actions of others in your client's life often denote a confusion about the agentive rights and responsibilities of others. Stubborn resentfulness and vindictiveness

often denote a confusion about the agentive rights and responsibilities of your client.

2. Speak and write forthrightly about the actions of others ("he yelled," "they expelled me," "she rolled her eyes in contempt"). But do not let entire sessions or entire therapeutic documents be dominated by an unjust fascination with others' actions. Attend to, ask, and write about your clients' responses in turn. I say this to you emphatically because there is nothing as heartbreaking as entire sessions and documents that have elevated a person who did great harm to your client to the protagonist position in her life.

3. Favor verbs over adjectives. The agentive grammar denotes an interest in a sequence of actions, not a sequence of identity conclusions. It is different to try to convince your client of a positive identity conclusion in the form of adjectives like "you were so courageous" or "you are such a patient mom" than it is to speak and write of the actual actions of your client that denote what you would label as "courageous" or "patient." I ask students to replace their temptations toward descriptive identity-adjectives with reflecting on the sequence of verbs of the clients' story instead. Hannah Arendt writes that "story-telling reveals meaning without the error of defining it."[4] We might extend this to say "sequences of actions reveal character without the error of labeling her." This preference for verbs applies both to your client and descriptions of other people in her life. Do not spend entire sessions or therapeutic documents pondering people's "toxic moms," "narcissistic exes," "manipulative friends," or "domineering bosses." Ask for stories that contain verbs of the actions behind these dead-end adjectives. Sequences of action in the form of verbs make it far more possible (compared to adjectives) to help you and your client toward the clarity of discernment. Note that Maddie and I came together to characterize her ex's actions (his jokes) as "mean" and further called out the action of "he didn't stop to ask and wonder where you

had gone and how you were in such a vulnerable intimate moment . . ."

4. Amplify mere descriptions of emotion to "declarations of want." Behind every strong emotion is a longing of what your client was looking for at the time. Think of words like want, willfulness, desire, purpose, longing, hope, intention, conviction, motive. Agentive actions that drive stories forward are always stories of what a person wanted, given the happenings they were placed in. And what was it that you wanted in that moment? Or: what did you decide then? Or: and what did you resolve next? are all questions that seek to understand the desire that drove a particular sequence of actions. In narrative terms – motives or desire drive the plot of a story forward. The naming of your clients' purposes or desires in any given moment are the translation of raw emotion toward an "agentive turn in the plotline that is driven by felt conviction." In this way, emotions do not become the protagonist, but are instead fueling the agentive turn of your protagonist in a significant impassioned manner. Maddie phrased it this way: "the only thing I knew then was that what I wanted was someone . . . sweet."

5. Beware of a passive orientation to the use of verbs that brings the action of an interesting story to a screeching halt. This is perhaps most starkly visible in the favored verb of our current therapist training: "navigating . . ." In therapy training, clients are always "navigating" something or other as if they were all nautical engineers and sailors; alternatively, clients might also be "coping," "dealing with," and "managing." How do these verbs drive storytelling forward? Just one example: the word "navigating" in its progressive "ing-form" is so ubiquitous that I have sometimes jokingly said: unless your client is Carola Rackete, who is an actual ship captain (and who by the way, was arrested for docking a migrant rescue ship without permission in a port in Italy), please find any other verb to reflect your clients' actions than "navigating." And even if Rackete is your client, choose

among the agentive verbs of: "she picked up 53 migrants in the Mediterranean sea," "she took a measured look at their exhausted condition," "she steered toward the closest safe port," "she asked for permission to dock," and "when she was refused authorization, she first grew very quiet, then turned on the engines and . . ."

6.   And what if your client maintains she did nothing, in the moment when it matters? That, my dear, is impossible. Remember Maddie who maintained that she did nothing and "just laid there?" The question that insisted that "doing nothing" is impossible relied on my conviction that clients are always already responding, and that silent responses count: "as you laid there, Maddie, all silent, what was going on in your mind, your body, your soul?" Maddie was "revulsed," "she observed his mean-ness" and she thought "if that is what they call love, then eww."

Have a look at other silent examples: "She sighed"; "She pondered"; "She felt the soda bubbles of joy rise in her blood"; "She resolved"; "She opened her mind"; "She curled her hand to a fist"; "She held it all back"; "She took a breath"; "She narrowed her eyes"; "She took a measured look"; "She imagined"; "She dreamt"; "She decided"; "She closed her eyes"; "She chose"; and so forth. These are all descriptions of "invisible action" in significant moments of clients' lives to hone in on the point of the commitment to discern all manner of responsivenesses – the quiet ones that don't even make a sound, all the way to the jolting sounds of a clanging hammer that surely caught your attention as your client tells the story of dismantling the marital bed in a triumph of freedom. The point is: here she was, here and here; your protagonist in the story. Can you see her yet? And if you can, are you interested to know – what does she do next?

Now please understand me, dear student. The concept of an agentive turn and the resulting commitment of your attention to the actions of your protagonist are not meant to foreclose the telling of rich Up Against Stories. Remember, it was precisely important for my colleague and me to find a way to speak more justly of rape, not shun its telling!

I know I've said it before, but I'll bore you again: both the X and the Y in the construction of Up Against Stories are important to witness and understand in meeting struggling clients. What happened? And what has it been like for you? are our first obligations to come to understand what our clients have been up against. The concept of the agentive turn grows from these queries to remind us not to cede our commitment to an interesting protagonist in the midst of her life happenings. Our clients are not to be confused with witless and hapless mannequins who are but moved around or bound by the power of other persons, or of inequitable systems that back up other persons, or by supposedly inevitable relations inspired by Master Narratives; power is a relation that touches all things, but it has an incomplete hold of every person. "Where there is power, there is resistance," Foucault writes as a vivid reminder to me of the possibilities that exist at the very moment clients have chosen to come to therapy to speak of that which happened.[5]

In addition, regardless of what the neoliberal master controllers would have us believe, much of the worst of what happens to human beings is not in your nor in your clients' control at all: poverty, cancer, death, pain, abuse, control, neglect, hate, contempt, inequality, dismissal, disregard, and so on.

In other words, the protagonist at the center of your story is not sovereign; she is not in control of life, her birth into positions of privilege or lack thereof, or the actions of other persons. It would be cruel, arrogant, and, further, oppressive to treat her such. She has, however, the ability to account (she is account-able) and the ability to respond (she is response-able) to the important happenings of her life – and this turns out to be a very great deal. It means she is a bearer of agency, not only constituted by life and other persons, but constitutive of life and other persons in her own right. I hope we might never confuse experiences of oppression in persons' lives with their real effects traceable through generations, with our clients' power to object and act in response to these conditions and experiences.

I believe in inviting clients' abilities to account and their abilities to respond to the important experiences of their lives as conferring unto them the moral obligation to respect the dignity of persons. The agentive turn is a space of attitude that finds itself entirely uninterested in either blame or pity as suspect poles of more common attitudes of disrespect.

I mean to be interested in the client's Up Against Story, precisely because it illuminates to me the wondrous capacity for agency in all human life. Another way of saying this is, the richer and more dangerous the abyss, the more fascinating and humbling any gestures of the person standing in front of it.

Okay, ready for a practical example?

I have this transcript excerpt from a few years back that was a real turning point for me in my thinking and that helped me understand the agentive turn that changes a story from an "X makes me feel Y-story" (which is the structure of an Up Against Story) to a "Given X and Y, I did Z on Purpose-Story" (which is the structure of a Counterstory). This particular conversation between Fabiola and me was a dramatic demonstration to me of the empowerment of shaping an agentive turn in relation to the particular Up Against Story that brought this person to therapy. I am really hopeful that it will be of use to you, as it was to me!

The conversation is not a first session but happens after a break in our therapy work as Fabiola had been experiencing some relief from what had first brought her to therapy. In the previous sessions prior to this break, Fabiola and I had been grappling with the real effects of three years of Fabiola's ex-boyfriend stalking her in response to his disagreement with her separation from him. We had been speaking about the Master Narratives that specified his entitlements and her obligations to continue the relationship with him and his three-year flying rages at her refusal. We had spoken of her agentive responses to his attempts to coerce her attention and care and Fabiola had been coming to life in an astonishing manner, taking up traveling, writing, working, and dating with new-found joy, and we had spread out the frequency of our conversations accordingly.

Into this hopeful space, I received a rather urgent email from Fabiola to make an appointment with me; she walked into our conversation in a low and quiet manner. Here's where we begin:

Fabiola
(Speaking in
a low manner): Ah, you know, I don't even know where to begin. It's been a difficult past few weeks.

Sunni: Yeah, you said in your email, you said "hectic and dark," and I've been wondering what you meant.

Fabiola: Is that what I said? It has been dark. A lot of things have happened. I don't know where to begin. *(Low, slumped together)*

Sanni: We'll figure it out. It doesn't matter where you begin, you'll find a way to tell me what's important – we'll do it together, and we can always go back and forth if we forget something important . . .

Fabiola: It feels like I have been living in a nightmare . . .

Sanni
(Quietly): Oh no, my dear.

Fabiola: Very dark stuff has come to light over the past few weeks. Let me see, I guess it started on November 1, is that the Day of the Dead? Or All Saints Day? Well around that time it's the Day of the Dead.

Sanni: Right.

Fabiola: That's very apt anyway. I went to a friend's house that day to play cards, and when I came back at midnight, I could see when I pulled up that there was something on my doorstep. So I went inside and tried to look at it through the window, it was a weird object, I was looking through the window. Like is that a doll?

Sanni: A doll?

Fabiola: Yeah, like a voodoo doll, you know what those are? . . .

Sanni: Yeah, I know those little ones, but this sounds . . . how big was it?

Fabiola
(Gesturing
about shoulder
length apart): It was laid there, there was some kind of drizzle on it, the police later said it might have been olive

	oil with some glow stuff in it, there were chicken bones laid out and beads attached to it.
Sanni:	Beads?
Fabiola:	Yes, like those alphabet beads. The beads around its neck, I had to bend down really closely to read them, they spelled my name, and I thought, oh my god it's my name, the beads were like choking her. Before I saw that, I thought maybe it's something left over from Halloween or something, that maybe someone discarded some decoration or something, but then I knew when I saw the beads . . .
Sanni:	They spelled your name?
Fabiola:	Yeah. And also other things, like Die Alone, Fuck Whore, Sick Baby, Cancer, Virus – I don't remember them all . . .
Sanni:	What the fuck. I'm sorry Fabiola, my dear, I'm so sorry you had to find that . . .
Fabiola:	Yeah, I have been really shaken up. *(Tearful)* And I don't believe in voodoo dolls, you know.
Sanni:	Right. What in the world did you do when you realized what the fuck that thing was?
Fabiola:	I called the police . . .

Abbreviated . . .

Okay, so in this excerpt that Up Against Story is clear and dramatic: a doll left by her ex on her doorstep with aims for Fabiola's life spelled out in alphabet beads. Sometimes, the aims of the actions of important persons are not as helpfully spelled out for us, but we have to puzzle over them. Here the aims are to lay a curse, to wish her ill, to cause a visceral response, to punish her, to cause her to live in fear, to impose on her time and attention again, and so on.

In the above excerpt, notice that I am not doing much except for trying to understand the Up Against Story. Fabiola and I share a familiar and trusting relationship at this point, so my questions above are brief as they don't require as many protections from misunderstanding or more help to ease her speaking in the context of our knowing each other.

In the "X makes me feel Y"-story structure I am trying to understand the X, and in the next segment below, trying to understand the Y of it all, the real effects of the doll on her life.

Sanni:	How did it go with the police when they came?
Fabiola:	I think I was in shock. You know, I keep thinking about people and their hatred. I keep thinking: do people find me an easy target? Why do they hate me?
Sanni:	I see, this question found you again, well it literally arrived at your doorstep now and translated into this question . . .
Fabiola:	Yeah well, on some days I think maybe I am doing something really right. But mostly I can't believe he is hell-bent on bringing me down.
Sanni:	That's a great way to say that, he sure is bloody trying his best to bend you to hell! What do you mean, you are doing something really right?
Fabiola:	I don't know. I don't know. I have been going up and down with depression. I feel so down, it has been this slump. I keep asking myself, how do I manage this darkness and ugliness?
Sanni:	Here you are, doing something really right, living your life, and then this hate arrives, literally to your doorstep, and slumps you, drags you into ugliness, hey? Because someone had nothing better to do with their life than craft this piece of hate . . . What else has been the effect of hate on your life?
Fabiola:	It took me 10 days just to start sleeping better.
Sanni:	What did you do instead of sleeping, my dear? Did you wake restlessly, did you pace, did you pose more questions into the universe, or distract your mind until sleep found you?
Fabiola:	I don't know. I feel like the world is getting smaller. And people just disgust me, everywhere I look I think, are you all unkind people? I have

	such a distaste for unkind people. It feels like a rotten city. The only thing that gets me out of bed is the thought of leaving.
Sanni:	Leaving this city? What do you imagine when you lay there?
Fabiola:	I imagine other cities as a destination. I just want to get away from all the people. I am wondering about everyone, if they know and what they know, and I don't want to talk to people, no one is removed enough.
Sanni:	You know, I am imagining that. You thinking about leaving, leaving it all the fuck behind, the whole city, and longing to be out of this small circle of hate, and wondering what all has been touched by it. It makes all the sense in the world to me that you are longing for another world as a response to this . . .
Fabiola:	The hardest thing has been that the world is becoming smaller and smaller. The world I can be happy and immune in, from this toxin, this poison. The world I can function in unlike these people, the world where the purity of love and honesty wouldn't be tainted . . .
Sanni:	Oh Fabiola, and it was so! You were on your way! You were just on your way, remember, we both saw it last we spoke. Remember? You were off into the big world of your design, the world of honesty and love. And remember that time in the summer when you were so distraught because the lawyer told you that they had given him your address because of the restraining order. Remember how you cried all the way home? Remember how upset you were? This bloody thing wouldn't have happened if he didn't have your address. Bloody hell! So now, thanks to the intricacies of restraining orders, he knows just exactly where to find you and pull you into this shrinkage world of hate.
Fabiola:	I want my freedom back.

Sanni:	The freedom to be . . .
Fabiola:	Not bothered by this. I want to move on with my life. But without fail, every few months, I am being pulled into it. I have specifically told everyone, don't even bother telling me any news of him, or news of any more of his smear campaigns. But now they come to my doorstep. *(Tearful)* These are problems that are not mine to absorb. The productive years of my life are being wasted by this . . . *(Tearful)*

Abbreviated . . .

In the above excerpt, you are probably on to me now as I am following up the effects of the doll on Fabiola's person, inside the details of her house, her mind, her body, her nights. This is the Y part of the "X Makes Me Feel Y"-story. What do you notice about the effects of hate? Sleeplessness, withdrawing, a longing to leave, a shrinking, a recoiling from unkindness, questions about her person. I surely could have asked many more elegant questions, and witnessed the effects on her more articulately, so please don't get too hung up on the details of the questions. The point is this: does your imagination fly with me to Fabiola's side? Are you open to wonder and think and feel and viscerally accompany her in what the arrival of a doll like this to her front step might have been like? Can you experience some measure of the doll's effects in the context of a three-year stalking situation that Fabiola was only just shyly beginning to explore her steps out of? Does her despair resonate in your soul or in your mind or in your body in some way? Remember to focus your concentration of attention on Fabiola's ways of relating to the doll, even as your person might be strongly relating to it in some particular way of your own.

In the next excerpt, a particular question aids an agentive turn in this conversation that was leading Fabiola and me deep into despair. Please take note of this agentive turn in the conversation below, what words aided it, and the effect of this turn on Fabiola's spirits:

Sanni:	Okay. (*Long pause*) You know, Fabiola, you know how we as human beings have . . . hopes for each other. I know we have been speaking about how your ex is not participating in this part of high hopes or conscientious break-ups, but is continuing to decide to take poisonous stances that are beneath him and beneath you. But I'm not talking about him now. Do you know how when we as human beings meet each other in the world, when we grow close to each other . . . how we hope to have a hand in how the other person might feel or think about themselves as a result of our encounter; how we hope to have a hand in the story others tell about themselves from here on after; how we long to accompany others well past our actual contact with each other, how we want to say "I wish you well" and speak that into being in some way, like, will each other on in some way, into the future . . .
Fabiola:	(*Looking up at me*)
Sanni:	Well, so what do you think, what would you say, what were the hopes of this doll in your life? If she arrived on your doorstep with hopes and aims and intentions for you . . . I mean some of them were spelled out right there in the beads, but what was to be the intended effect of the doll? What are her intentions, her hopes for you and your future? And say, you really, very seriously took up the doll's hopes in your life, and committed to pondering them for some months, maybe many months, maybe for years, or maybe even a lifetime, and kept them close to you, and kept conversing with her, kept her counsel, and always checked in with her before making any significant life decisions, and carefully arranged your life according to her aims for you . . . where would she hope to take your life then . . . ?

Fabiola:	*(Long pause. Not looking at me. Shaking her head as if in a lively conversation with herself)*
Sanni (Smiling, after waiting a while for a worded response which did not come):	I can see that you have an opinion on the doll's hopes for your life!
Fabiola:	Yeah. I have more courage than that! In fact, I have more courage now than I ever had!
Sanni (Smiling, in great surprise at the above words):	The doll had hopes, and you're like: "No . . . Nope. HELL no. I have more courage than that thing [the doll] knew. Like thank you heaps, I can see where you want to take me for the rest of my life and no. Just no." Is that what it sounds like?
Fabiola:	Yeah. Something . . . let go of me. Something here has been liberating for me. There was some last hold he had over me, and it . . . let go.
Sanni (Motioning a letting go with her hands):	Something . . . what? This doll arrived and instead of . . . *(Motioning a tightening grip with her hands)* something . . . let go instead?
Fabiola:	Yeah. You know I called the police . . . Actually, I have been calling a few people. This is maybe unrelated, but I called the place where I volunteer . . . I have been wanting to leave for a while,

but now I called them, and I just resigned from these tasks that I don't want to do. They are not a good use of my time, and I have known that for a while now. It felt so good to be free of that.

Sanni
(Laughing):   Ha, you've been making some phone calls. The doll was meant to just freak you the fuck out, and make your circle smaller, but instead, you've been busy freeing yourself, not just with the police, but the liberation was like: "What else? What the fuck else? Ah, here's this volunteering thing she doesn't want to do, let's get rid of that in this spirit too . . ."

Fabiola
(Laughing):   Yeah. A knot came undone for me . . . something un-knotted. And my speech is looser too.

Sanni
(Laughing):   Wait, wait, knots and speech were let loose . . . there's a poem . . . or a quote . . . "if a piece of knotted string can unleash the wind . . ."

Fabiola
(Interrupting):   I have lived for so long in fearful anticipation of what's next. And now it's time for more big picture thinking.

Sanni: But what does it mean: your speech is looser too? Have you been speaking?

Fabiola: Yes, I went to (name of organization) – I gave a speech on domestic violence. It's now become a strong personal essay, my friend said to me, it's an essay now . . . There is less fear now, I want to contact the media, and talk to them about being stalked, I don't care anymore, I have free license to speak, to tell my story.

Sanni
(Smiling):   The doll thinks she can draw you into a private hell-bent conversation with her for all your life, and here you are speaking, publicly and privately . . . on the matter of freedom?

Fabiola:	Yes, my heart is flitting free now.
Sanni:	Flitting free. Man, did the doll ever miscalculate!
Fabiola:	Yes, she was mis-aimed.
Sanni:	She was aiming alright . . . and then she was like: "Huh? What happened? How did I miss?" Maybe the doll had no idea that you don't live in her world at all. That your world is about the design of freedom, of words, of beauty, of love, of well-wishes . . . What happens when a doll of hate arrives at such a person's doorstep?
Fabiola:	I am kicking it to the curb. She is not mine.
Sanni:	To the curb . . . coincidentally, Fabiola, what DID happen to the doll? Where did she go, I mean in real life?
Fabiola:	She was thrown into a plastic bag, I think.
Sanni:	Did you do that?
Fabiola:	No, I watched. The police did it. They picked her up and put her into a plastic bag . . . I guess that's where things go sometimes.
Sanni (Choked up):	They do! Although I didn't quite know that until you just taught me that. Quite unceremoniously I imagine, she went . . .
Fabiola:	Actually, I remember now, the plastic bag had writing on it . . . it said "Property of Police Department."
Sanni (Laughing out loud):	It did not! You're shitting me. Well, if that isn't poetic justice. The doll IS the property of the police department, isn't it, because of the absurd nature of a restraining order and how they then give away the address of the person they are restraining you from, just so he can now find you, so this really is between your ex and the police, and the doll really is their property . . .
Fabiola (Laughing	

out loud):	Yes, like a discarded item . . . they just put her in the bag. And the negative aura disappeared.
Sanni:	In the bag. Fabiola, your laughter is so awesome to hear. I imagine that laughter is one of the things that the doll was aiming for. And so your laughter here is . . . another insurrection.
Fabiola:	Yes, there hasn't been a lot of laughter lately. But I am turning the story of the doll into something inappropriate, and I am reclaiming my innocence. They are turning me into a powerhouse, and they don't even realize it.
Sanni:	Wait. The doll arrived at your house and in response you're growing a powerhouse?
Fabiola:	Yes, something is pumping through my veins now, something is pulsing, it's hard to keep it down now, it's anger. And I have actually been thinking, in response to the "die alone"; I am gonna go back to the dating app now.

Abbreviated . . .

I wanted to show this transcript excerpt to you because it is one example of a dramatic agentive turn in a conversation. I'll show you a few other conversations that also center an agentive turn in the next few letters so that you can consider this concept from a few different angles and inside different stories.

But back to a brief reflection on the conversation with Fabiola. You probably noticed that Fabiola's conclusions regarding the effects of the doll on her life changed quite dramatically in the above excerpt, especially in comparison to her previous descriptions. Her manner of speaking also changed from a tearful low to laughter and empowered, emphatic claims. What, then, did Fabiola make herself available to consider, up against and in willful defiance of "die whore" in this moment with me? What change did she embrace?

Now notice I had done nothing to even gesture in the dir- ection that Fabiola and I might somehow attempt to speak of constructing more literal and concrete safety for her in relation to her ex's continued trespasses on her person. This decision was born out of respect for Fabiola's efforts, as she had already exhausted the sensible and important legal possibilities with restraining orders and police involvement. The construction of a fruitless dialogue about an illusion of control over her ex's past, present, or future actions in her life would be further blaming and exhausting of Fabiola and not a source of empowerment in this context. In fact, this Master Narrative-driven victim-blaming ideological shorthand had been a such a significant source of pain in Fabiola's life already: well-meaning witnesses and allies in her life routinely responded to her accounts with incitements to "have firmer boundaries" or to "make sure that she chose dating partners more carefully in the future" or to "think about what she could do to keep herself safer" in these moments of trespasses. Fabiola's ex's decisions are his, not hers; she cannot ultimately choose the material or events that he continues to place on her literal doorstep. And yet she can narrate them and respond to them; and this turns out to be a very great deal.

The change in Fabiola's account and response to the doll happened after this line of questioning:

> What are (the doll's) intentions, her hopes for you and your future? And say, you really, very seriously took up the doll's hopes in your life, and committed to pondering them for some months, maybe many months, maybe for years, or maybe even a lifetime, and kept them close to you, and kept conversing with her, kept her counsel, and always checked in with her before making any significant life decisions, and carefully arranged your life according to her aims for you . . . where would she hope to take your life then . . .?

After this query, you might have noticed that I only served as a rather gentle and receptive witness to the achievement of the position she embraced over the rest of the conversation, helping perhaps to find words to befit this new development in an improvised, slightly incredulous manner: Something . . . what? This doll arrived and instead of . . . (Motioning a tightening grip with her hands) Something . . . let go instead? (Laughing) Or, wait,

wait, knots and speech were let loose . . . there's a poem . . . or a quote . . . "if a piece of knotted string can unleash the wind . . ." Or: Fabiola, your laughter is so awesome to hear . . . and so on.

So, what happened to Fabiola then? Here are some tentative considerations for you to ponder and take into your sessions in the future to verify with your own clients:

What is the felt effect of being treated as a passive recipient to life happenings?

What is the felt effect of an invitation to an agentive response to life happenings?

In other words: what is the difference between ceding our ground to the thing that stands before us, and responding to it from the ground that we have chosen to stand on instead?

In this conversation, Fabiola taught me an important lesson: there is a spark of life in a person's agentive moral decision to respond to life happenings strongly and purposefully. This spark denotes the energetic change from "X Makes Me Feel Y Stories" or Up Against Stories to "Given X and Y, I Did Z on Purpose Stories" or Counterstories. That is the heart of the agentive turn. Who are you in relation to this life happening? Perhaps, I'm "freer than I feel."[6]

Love,
Sanni

Notes

1   "The history of Western art is just the history of men painting women like they're flesh vases for their dick flowers," Gadsby, H. (2018). *Nanette*. Netflix.

2   Solnit, R. (2019). *Whose Story Is This? Old Conflicts, New Chapters*. Haymarket Books.

3   "This is an account of personal agency. An account that emphasizes what could be called a person's agentive self. It includes details about what the person has been up against in the performance of this personal agency and against this background, emphasizes the significance of any more recent steps the person has been taking toward having more to say about how their life goes." White,

M. (1995). *Re-authoring Lives: Interviews and Essays*. Dulwich Centre Publications, p. 143.

4   "It is true that storytelling reveals meaning without committing the error of defining it, that it brings about consent and reconciliation with things as they really are, and that we may even trust it to contain eventually by implication that last word which we expect from the Day of Judgement." Arendt, H. (1970). *Men in Dark Times*. Mariner Books, p. 105.

5   "Where there is power, there is resistance." Foucault, M. (1978). *History of Sexuality, Volume 1: An Introduction*. Vintage, p. 95.

6   "My role – and that is too emphatic a word – is to show people that they are much freer than they feel, that people accept as truth as evidence, some themes which have been built up at a certain moment during history, and that this so-called evidence can be criticized and destroyed. To change something in the minds of people – that's the role of an intellectual." Martin, R. (1988) "Truth, Power, Self: An interview with Michel Foucault." In L. Martin et al. (eds) *Technologies of the Self*. University of Massachusetts Press, pp. 10–11.

For Fabiola
On the Day of the Dead

on the Day of the Dead
or All Saints Day
whichever
a doll arrived with hell-bent wishes
straight out of a nightmare

it hoped to slump
the purity of honesty and love
in its erratic project of hate
and to shrink
my world
to the small quarters of
disgust at the rot:
are the years of my life to be thus wasted?

I looked at her closely
and saw my life unfolding in conversation with her;
and I sat up many nights with our imaginary dialogue.
But then came the dawn
When I told her gently:

sorry hon,
wrong address
this is the one the police gave you
but I don't live here

did you know
my house is a powerhouse
with more courage than I ever had
in it
my heart is flitting free
and my speech is loose
and my knots are undone
and my partner in conversation
is not a voodoo doll

but the whole wide world
of innocent laughter
well-wishes
and
that of
love
love
love

7

What Do You Want to Do, Love?

A Letter About Counterstories

Dear student,

I promised you a letter about leads into Counterstories, so here we are then, in the midst of surprise joy: the re-charge of story-work. I cannot tell you all the revolutions of my mind when I first heard David Epston speak of Counterstories, which is a term borrowed from Hilda Lindemann Nelson,[1] but suffice it to say, I felt that joy. David Epston asked, "What is a Counterstory and what does it counter?" and I sat back and thought: "Well, this changes everything." In one fell swoop, I could see how I had long engaged in a bit of a snow job to cover up the rickety hinges between unrelated anecdotes of alternative stories or snippets of unique outcome moments in my work. I could not wait to default on the vicissitudes of wishful thinking à la "I know a good story when I hear one" and get to work all over again, but this time, on purpose.

Here's how I might say it today: a Counterstory takes an Up Against Story, or "X Makes Me Feel Y" Story, and changes it to a "Given X and Y, I did Z on Purpose" Story. A Counterstory is a kind of Alternative Story if you like, but the important part is that it is a particular Alternative Story that answers back to (or "counters") both the hurt and the Master Narrative-driven

DOI: 10.4324/9781003478478-7

identity conclusions of the Up Against Story. A Counterstory is precisely not a random positive anecdote of living of which there are thousands and thousands to choose from in your client's life: a Counterstory earns its keep as a story of therapeutic merit if it is a convincing pièce de resistance to the hurt and to the binds on a person's freedom spelled out by the details of their Up Against Story.

An example: remember that strange fable about the nightingale and the rooster somewhere toward the beginning of these letters? Don't worry if you don't, the point was that a nightingale's song was compared to a rooster's song and was judged wanting for its lack of "effectiveness in keeping people awake." At the end of the fable, upon having heard such judgment of its efforts, the nightingale "hung its head and flew far, far away." Now here's how it relates to Counterstories: say the nightingale were your client and had been judged and unbelonged for the uselessness of her song and was now reeling with hurt in the aftermath of this experience – what kind of a story would we need to counter her hurt? Sometimes, we as therapists might be tempted to try and convince her of a story of her pretty flying skills or perhaps nominate the iridescence of her feathers as sites of alternative inquiry, cleverly poised to make her feel better. Spoiler alert: it might, make her feel better, that is. But the idea of a Counterstory challenges us to consider telling a Counterstory about the nightingale's song if we wish to really care for her hurt. Do you see? The aim of narrative therapy isn't to make a person momentarily feel better by way of convincing her of a positive alternative, or by way of smartly disguised flattery of her other random talents. The aim of narrative therapy is for your client to have "more options and more say about how her life goes," especially given what she has just been through and what she now believes about herself as a result of it.

Another example, but this time in the form of my favorite two-line therapeutic poem I ever wrote in my whole time as a therapist:

I come from a long line of store-bought everything
But I have been baking as of late.

In this poem, the Up Against Story – the "store-bought everything" and its Counterstory, the "I have been baking as of late" are not compartmentalized islands – but in direct relation to each other, the Up Against Story constitutes the necessary backdrop against which the agentive initiative of a protagonist in a Counterstory pops.

Students at the beginning of their work often become somewhat anxious diggers of a kind. They ask me "How do I find the Counterstory?" as if it would be found, unearthed, dug up, discovered, revealed, like truth, or a magic trick. The answer is far less truthy and magical, unless you know the revolution of someone attending to you during the long night as if they had no better place to be as a kind of magic of the ordinary: Counterstories arise from the details of Up Against Stories by way of your focused concentration of attention, down to the words, preferably pen in hand.

So perhaps you might now forgive me for all the belaboring of your attention on rich Up Against Stories in the previous chapters, and all the warnings against foreclosure of this task. Both the happening (the "X") and its effects on the person (the "Y") are so important to care for because they tell you what is uniquely hurting and binding your client. How will you otherwise be able to notice or attend to the unique Counter-initiative that your client undertook in her life, on purpose, despite what hurt her, in relation to what hurt her, in willful defiance of what hurt her: "Given X and Y, I did Z on Purpose." Clients will seldom proudly announce a Counter-initiative to you, in the form of "Sanni! Guess what I did! Wait until you hear this, it's a real doozy!" – although sometimes they do. I will never forget a woman who came to a session and silently showed me a brand-new key in her curled fist. We had been speaking of her life in the windowless basement downstairs from her parents who alternatively drank and fought and made sure to whittle down my clients' dreams and initiatives for living (like moving out or publishing an essay) in tirades of anxious doom-spiral bickering: "What if you can't handle it" and "Your mental health is not stable enough" and "We're very concerned about you being alone," and so on. Up against her fears, in spite of the fretful doubts that beset her, and in willful defiance of her reputation as a fragile failure in life, she resolved to tour a little apartment with windows, and to sign a rental agreement and take possession of a key. When she showed me the key in her slightly shaky hand, I jumped to my feet in an inarticulate moment of "Whaaaaat?" and we hugged and laughed.

But Counter-initiatives are not always so obvious. Mostly clients will strew little sketches of their boldest Counter-initiatives into the flow of the session as matter of factly as "la-dee-da, nothing to see here." It is your focused concentration of attention that will help guide you in those moments to ask: "Wait, you did what?"

Here's an example of a young woman doing just that, describing a Counter-initiative and strewing it in to a therapy session without much fanfare. This woman had come to therapy in the aftermath of her breakup with her long-time boyfriend Ryan, whom she continued to live with for stability for her son Dylan. In particular, this young woman's dilemma centered on a long-time struggle to express her thoughts and wants to Ryan in their relationship. Listen:

> And then like the end of Saturday night, I don't know, it was like five o'clock in the morning, and I was really drunk and high on blow. And I started . . . I was freaking out and I started crying and just getting really upset about life and how like, I'm okay with me and Ryan separating, but I would like it if we could be "companions" somehow without being romantic, but being really good friends still, because Dylan has a really good family unit. And it's not fair that Dylan has to give up his stepdad. And like, I can't even remember all of it. It was just like, it was a mess. And I was bawling my eyes out. And then all of a sudden, in the middle of me crying and finally telling him all this, Ryan grabs my hands and asks me if I want to fuck. And I was mad, I was so mad, I was like "WHAT part of ALL of what I just said makes you think I want THAT?"

This is a Counter-initiative moment that would rightfully raise every fist in the air among my colleagues! Can you hear her, the protagonist who just changed her life dilemma with the help of one pilot light of anger and a breathtaking question? Can you see how your diligent attention to this young woman's Up Against Story and the dilemma of her living on the ground of her life, would guide your attention to hear this Counter-initiative pop? If you cannot, then this young woman's Counter-initiative will stay just a snippet of an initiative that "also happened." Lived experiences of Counter-initiatives can become Counterstories in therapy, but only if you hear them. Margaret Atwood puts it this way:

> when you are in the middle of a story, it isn't a story at all, but only a confusion; a dark roaring, a blindness, a wreckage of shattered glass and splintered wood; like a house in a whirlwind, or else a boat crushed by the icebergs or swept over the rapids, and all aboard powerless to stop it. It's only afterwards that it becomes anything like a story at all. When you are telling it, to yourself or someone else.[2]

This is what guided my attention in the transcript in the Up Against Chapter with Maddie. Do you remember, she remarked "it's weird; it sounds better when you say it." But what had I said that prompted this acknowledgment? I had made a witnessing response, constructed out of exactly what Maddie told me: "You laid there, in a state of revulsion, noting him and his meanness and his jokes, and everything went 'if this is what they call love, then eww'." Did I make that up, draw that out of thin air, or some therapist toolbox of favored life knowings about rape? No. I know this down to the words, Maddie, because you told me so. The proposal of "If this is what they call love, then eww" is an attempt to translate Maddie's own word of "revulsion" into living inner dialogue, nothing more. It is the attention on what had happened to Maddie (the X) and the effects on her (her revulsion) as well as the Master Narrative-driven conclusion that "I did nothing, I'm so stupid" (the Y) that guided my attention to speak with her about what she did next, on purpose. I'll remind you of what she said: "the only thing I knew then was that what I wanted was someone . . . sweet. That's the word."

And what is "sweet" in relation to "revulsion"? What is the visible and plausible evidence of "this is what I did" in relation to "I did nothing, I'm so stupid"? And, humorously, what is chocolate then in response to "revulsion" and to mean jokes about Maddie's body? See?

I even blabbed it all out loud to Maddie herself:

> You wanted . . . sweet. What all does this word mean to you? Maddie, I want to know, I am flooded with the imagination of "sweet" as a counter to "mean" and "revolting" and someone who didn't stop to ask and wonder where you had gone and how you were in such a vulnerable intimate moment . . . Maddie, did you somehow know to seek the counter to "revolting," even when you were young and didn't know the supposed right words for it?"

Maddie responds to me by crying and when I ask about the message of the tears, she proposes "That I . . . maybe I . . . maybe I wasn't so stupid."

"I know this because you told me so." And did we get it all wrong already, my dear, shall we utilize the power of the delete button on these words now in favor of others more in line of what you are meaning, in this thoroughly rotten and magical venture of depicting lived experiences by way of imperfect words exchanged among imperfect persons?

So far, so good? On to the practice part of this letter then. I am reprinting a lengthier piece of transcript from its beginning, as students often request to see the conversation play out rather than just observing an isolated piece from the middle. In this session, Tom and I are conversing with a client by the name of Syrie in our second session with her. It also shows the poetic summary of the first conversation as it was presented to Syrie herself. The transcript is perhaps self-explanatory, but please attend to the ways in which a Counterstory takes shape based on painful memories that Syrie has been carrying for a while, and is imagined into being inside the details of Syrie's life. Remember, Counterstories take an "X makes me feel Y Story" and change it to a "Given X and Y, I did Z on Purpose Story." That is the work of this excerpt and the Counterstory work begins at a 13-year-old on a windowsill. The excerpt does not represent the entirety of the session as we continue speaking with Syrie for a bit longer. I have also attached a letter we wrote for Syrie to witness the reach of this session in an imaginative manner.

Sanni:	Okay, so we've prepared some things for you.
Syrie:	Okay.
Sanni:	One of them is a bit of a thought piece from your words from our conversation last time . . . I'd like to read it to you, if that's okay.
Syrie:	Okay.
Sanni:	And then maybe Tom, I know we ended last time with thinking about how to pose Syrie's dilemma well. Would it be okay if I read this to both of you and then we can chat the three of us together and see if we can make this dilemma interesting?
Syrie:	Yeah, okay. Yeah, that sounds great.
Sanni:	Okay, good. Okay, I'll just, I'll just read first this because it might set the stage for some . . . for some of the thoughts. It's entitled, "In a woman's life."
Syrie:	Okay.
Sanni:	

In a woman's life

> *There is a time to sparkle*
> *A time to be the funny one*
> *A time to joke and tease*
> *A time to be left alone with decisions*
>
> *There is a time for self-consciousness*
> *A time to try*
> *A time to hesitate*
> *A time to linger*
>
> *There is a time to leave*
> *A time for fabulous*

A time to really feel it
A time for confidence

There is a time to come into my own
And not worry about counter culture at all
A time to dive right in
And pop back up

There is a time for longing
A time to pass on
A time for meaning

And always
There is a time to say
"I want to claim my time
On this earth"

And all the mothers come before us
Shall be felt
To smile in wonder at our hopes

Syrie:	Nice. I like that.
Sanni:	The reason I wrote it like that was because you spoke spontaneously about the different times in your life, the time when you were funny, and sparkly, the times growing up when you felt alone, and the time of hesitation for leaving that relationship after the time for lingering. You said that the time of leaving was "fabulous," and you said: "No, this is a wonderful time to come into my own." So that's how the phrase in the poem was born! And at the end I wanted to honor the memory of your mom, Syrie, and the tenderness for her that you spoke with last time.
Syrie:	Thank you.
Sanni:	Okay . . . So for me the extension of this note is maybe the question, "Well, what is it time for now, then? What shall it be time for now, my dear?"

Syrie: Yeah, yeah, yeah, I had a bunch of thoughts over the week. So I don't know if you want me to share those now . . .

Sanni: Yes! We want to know everything about your thoughts!

Syrie: Oh ok, ok. Um, well, a couple things that I wanted to mention is I remembered I was talking to my brother and, and then I remembered like, part of the reason why I wanted to start therapy again, I guess I forgot it, you know, it happens, right. And I, but I do think it all ties together, of course. My brother reminded me when we were talking about when we were younger, when we were teenagers, and my immediate reaction was "Ew. No. I don't want to think about that time." And then I remembered that over the past year or so, when I look back at myself when I was younger, like a teenager, I have really . . . I don't want to think about it. Like I have terrible thoughts about that time. I feel really ashamed. And it's really, I just moved back to Alberta and the memories are starting to come up, it's this really strange thing that's kind of been burdening me, and I can't, I try to be like . . . compassionate towards myself. But then, I'm having a hard time being compassionate because I'm more, I didn't like who I was back then, like in hindsight, and yeah so that's really, that was part of the huge reason why I wanted to figure all that out, but I obviously think it. Everything ties together with the decision to have a child.

Sanni: Thank you for the thoughts, this is a really wonderful complication to our conversation last time about your decision to have a child. Something that has been burdening you, memories of not liking yourself, that have been coming up . . .

Syrie: Yeah. Yeah, exactly.

Sanni: I wonder, is it too hard a question . . . would you like to try your hand at answering this . . . like what is your hunch about the bridge between what you struggled with in remembering your teenage years and the decisions about having a child that you are considering? Do you have any hunches about the bridge between those things? It's okay if you don't, we can also listen for the bridges together as we speak . . .

Syrie: Yeah, I do, I do. Like, I mean, I have been thinking about my nephew. I feel really sad when I think about him because I worry so much about him and what he's going to go through as he, you know, gets older and gets into his teenage years because of what I went through. So I think there's a link between like, you know, bringing a child into the world or taking care of a child knowing that, how hard it is for them. And like, how, how would I be able to help or how can I make sure they don't have to go through that, you know?

Tom (To Sanni): Yeah, well I'm thinking about what Syrie talked about last time –her mom, whom she described as loving and compassionate, but also not providing "any kind of guidance," I think she said . . . *(To Syrie)* maybe you were talking about your mom in those teenage years in particular?

Sanni: Does that resonate with you?

Syrie: Yeah, it does! In the past two years or so, I realized, in talking to others, that I really had no guidance at all. Although it feels kind of bad to say because I don't want to blame my parents or my mom, but I really just felt like I was ignored in a lot of ways because there was so much going on, you know . . .

Sanni: Okay, can we make a pact, Syrie, in favor of your freedom of consideration? What if we promised that we have no interest of bringing dishonor to your parents and your mom in particular – no mother-blaming – none of that. But instead to look for understanding and some sturdy words

	for, "well my mom wasn't great at this guidance thing," you know?
Syrie:	Yeah. Thank you.
Sanni:	Would it be in line with your hopes if we all band together to still love and honor her and understand what she was talented at and good at, and understand her legacy in your life and hold her really dear – all the while also saying, "AND you messed this part up, mom" . . . would that free you up to speak?
Syrie:	Yeah. Okay. I feel really sad when I think about my nephew, and a part of that is how alone I was growing up . . . I just didn't like who I was, and it makes the whole time kinda painful.
Sanni:	What . . . were you left alone with, Syrie? What were you longing guidance from your mom about? . . . Tom, she said twice . . . "I didn't like who I was" and I wonder about the charge against her person in that.
Tom:	Yeah. That's a strong charge. And I wonder if that has a lot to do with "I didn't like how I was left alone" . . . How aloneness complicates whether I like myself or not.
Syrie:	Wow . . .
Sanni:	I want to be very gentle about this part of it because your mom died and sometimes it's hard for us to take up things with people who die, right? You're meant to be pious and honor her. But perhaps we can ask a question about "why didn't Syrie like herself in her teenage years and why was she left alone with that?" in a way that honors both Syrie and her mom and what they were up against at the time.
Syrie:	Yeah . . . That's interesting because I didn't look at it, I haven't looked at it that way . . . yeah, I think I would kind of like to look at, you know, exploring the teenage years and that I didn't like who

	I was, because it's, it is. I guess I just kind of want to look at it a different way and clear it up so I can stop thinking about it, really.
Tom:	Yeah, especially in light of what I was thinking when I heard what Sanni called the "charge" against you.
Syrie:	Right. Yeah.
Tom:	And in light of the statement of "I really like myself now."
Syrie:	Right. And I didn't like myself then, but I like myself now so that's why I'm happier now.
Tom:	Well how in the world did you get to such a place, given that you were left on your own, without guidance. She must have done a whole bunch of guidance of her own to get her there.
Syrie:	Yeah, no, it's true. It's true. Yeah.
Sanni:	Did you kind of mother yourself into liking yourself, Syrie, at some point?
Syrie:	Yeah, at some point yeah, right, I did, I think. I think that was what happened. Yeah.
Tom:	And that's also worth not forgetting and holding on to together as we have the conversation about the hard times.
Syrie:	Yeah, yeah, for sure. Well, I guess that's one of the key things I want to work on actually is having compassion for myself in those years, because I want to learn to just accept myself for everything, so that I can relieve myself of the, like, the criticisms I have about myself in those years, you know.
Sanni:	Yeah, we could go about this in two different ways: One of them is to listen to the criticism that you have for yourself, what it sounds like, what you are being accused of, what it sounds like in your conversations with yourself. The other way is to think of particular images, or memories, stories of particular days that keep coming up for

	you that are hard to hold. Which way do you like better?
Syrie:	Yeah. Stories is better.
Tom:	Okay, what kinds of stories are the memories wanting to remind you of?
Syrie:	Um, well, like, I know that I did a lot of the things, the stories which I'll say in a minute because I just wanted attention, you know, and I wanted love. I started drinking with my friends at, you know, age 13, and sneaking out of the house and meeting up with boys and, you know, I guess a lot of the, a lot of the stories are about meeting up with boys. And some of those times I remember with a lot of shame. Like I regret some of the boys . . . well, a couple, that weren't really that nice. One time I got really, really drunk and blacked out and then we got caught and they brought me home to my parents. I don't remember anything. Um, yeah. And so just like being, I was bad, you know? And I, you know, tried different drugs and I was just, it just felt so lost, and like I didn't know what I wanted so I just kind of did . . . I look back and think, did I even want that or did I do all these things looking for attention or looking for love . . . somebody to pay attention to me . . .
Sanni (Softly):	Syrie, did the boys hurt you. Is that one of the most burdensome . . .?
Syrie:	No, no, that's not one of the most burdensome, I guess it's just, I think it's like meeting up with boys and having sex, but feeling lonely and confused . . .
Sanni:	Okay, the reason I am asking is if the boys hurt you, Syrie, is because then we're in a particular realm of accusations against young women and girls and I don't want to confuse the two. But if it was more confusing and lonely . . . then the charges sound a bit different . . . like "sneaking out" and "being bad."

Tom (To Sanni):	And "drinking" and "meeting up with boys."
Sanni (To Tom):	Right, like to, to make sure, it sounds suspicious, right?
Tom (To Sanni):	Like a case to be made . . . Charges against her to make a case that she was bad in some way. There's like two stories here, probably more, but one are the accusations that "here's all the flashbacks of her doing suspicious things."
Sanni (To Tom):	That the memories are telling her she's morally questionable in some way as opposed to "phew, dodged a bullet, did some 13-year-old stuff and got away with it."
Tom:	You know the other part that I hear, I heard, is "well, I was attention-seeking." *(To Syrie)* Does that also come as an accusation?
Syrie:	This is wild, listening to you. Yeah, it does. It does.
Tom:	Well, Syrie, you tell me – what the hell is wrong with someone who's 13 years old and left, left to go out on her own, longing for connection and love . . . What is wrong with seeking . . . love?
Syrie:	Oh my god! Exactly. It's so negative, you know all the thinking is negative, like I don't think about wow I did dodge a bullet because nothing bad really ever happened to me. I never was, I was never like attacked. I was never like, you know, I was never hurt. Um, I actually did pretty well like considering the circumstances.
Tom:	Like, if we can replace the moral of the story in some way . . .
Sanni (To Tom):	Eek! It's too early.
Tom (To Sanni):	Is it too early?
Sanni (To Tom, making motion of throwing a ball far):	Yeah.
Tom (To Sanni, laughing):	Okay.

Sanni:	Syrie, he's gonna work on replacing the moral of the story but I'd like to ask you something before he does . . . Can you shelve his question for a minute, I promise you, we can return, but this "sneaking out" – I'm literally imagining somebody climbing out the window like . . .
Syrie:	Which I did, I did.
Sanni:	Really? You did? Oh fucking brilliant. *(Laughter)* Okay. So, if we were to imagine you, all of age 13, a bit lonely, a bit lost, a bit spirited . . . climbing out the window . . . can you see it, can you see . . . her?
Syrie:	Yes.
Sanni:	Okay I'm imagining her on the window ledge of your room right before climbing out . . . maybe she's taking a last look at her room, to think of all the things that good girls should be doing . . . and then turning to the window again . . . .what was her hope, her longing that she climbed out of the window for?
Tom:	Yeah, something that was driving . . .if we could imagine her in that moment right before she climbs out the window . . . and press play . . . What is driving her? Maybe a wish to feel something or risk for something . . .
Syrie:	Oh my gosh, I mean, almost . . . think all of that but like the risk for sure. The risk. Like almost just like something to do . . . like it was exciting.
Tom:	So: excitement.
Sanni:	As opposed to being nicely your bedroom and doing homework or something.
Syrie:	Yeah, like, or maybe it was like, because it was like the dichotomy between being a good girl, and then doing something bad, maybe something, something like wanting to be liked, wanting to have fun . . . I think she wanted to be liked, or to risk for that.
Tom:	A desire to be liked, is that similar to a desire to belong?

Syrie: Yeah, yeah, yeah, I think so, to belong in the group of friends. Yeah. Um, but also just like, I mean, I look back and I just think there was a sense of "why not?"

Tom
(In humor): That sounds really terrible.

Syrie
(Laughing): I know, I know! It's just in my mind it's so bad, like, why did I do all these things, but it's not even bad so I don't know why I'm feeling that way it's just, it's so like . . .

Sanni: What is the standard you're being compared against, like . . . What is a good girl?

Syrie: Good question . . .

Sanni: I know what it sounds like for me, like in my life as a 13-year-old, I guess a good girl isn't sexual, a good girl wouldn't want to get drunk with life . . . a good girl doesn't have thoughts like that.

Syrie: Yeah. Like, I think that's, that's a lot of it, a good girl isn't like, doesn't get drunk or isn't sexual. Yeah, I think so. And I grew up with the Catholic Church so it's like a sin it's, you know, it's a sin, essentially, before you're married. I think that did seep into my . . . thoughts . . .

Tom: Is that particular Catholic idea of "good girl-ness" seeping into the charges against you, you think?

Syrie: Yeah. I never thought of it. I know I learned that through my father and through the church.

Tom: And everywhere else in the world?

Syrie: Yeah, yeah!

Tom: I couldn't help but think that I don't know if I've ever talked to a man who has done exactly the same things as you, and looks back and feel badly like there's something morally wrong for having explored and adventured and risked because those are the words we use for boys.

Syrie: Yeah, yeah, I think that it mainly is, well the drinking and the sexual stuff, because I wasn't,

I wasn't bad. It wasn't bad. I felt alive, and I was seeking . . .

Sanni: Wow. You felt alive. Well the story goes, whichever church you can apply that to . . . a good girl is sexy, never sexual . . . she's sexy to look at and sweet, but she doesn't have desire or interest, or a body . . .

Syrie: Yeah, yeah.

Tom: Well, I'm thinking back to the question, your question about the kind of like, the feelings of climbing out the window, what they were in anticipation of, in search of? If they weren't evidence of badness, but in search of something? You know your words "excitement" and "desire to be liked" and doing something fun.

Syrie: Another thing that came to mind is just freedom.

Tom: Ah, okay. Yeah.

Syrie: Yeah.

Sanni: Wow. Freedom. I am keeping all these words here in a list, Syrie. You know, we started out with an imagination of a girl on the window sill, about to climb out of the window. And we asked you to imagine her in that moment, and why she would climb out there, what she was running toward in that moment, what she was listening to. And you have now said, she was listening to the sound of excitement, of fun, of a desire to experience being liked, of aliveness to the sound of freedom calling to her. And she chose these things instead of obedience to the rules of being a "good girl." I want to ask you a question, Syrie. Are you with me so far?

Syrie: Yeah, go ahead.

Sanni: Okay, Tom, maybe I'll need your help in asking this. You know, Tom, what I want to ask her, and I want you, Syrie, to think alongside the thinking of this question so you can understand why I am asking you this. What's on my mind is that girls' lives, especially starting at the age of 13, are often so focused on obedience, on being "good," on following the rules, on not taking up space, on not listening to the sounds of excitement or desire or freedom. And I think people might say that because they worry about young girls getting in trouble, getting hurt. And you said, Syrie, that not all the experiences you had with boys and with drinking when you did climb out that window were wonderful, some of them weren't nice. So the worry is real. But what I want to ask Syrie about is – in all the worry about girls getting hurt – do we ever worry about young girls being too obedient, and burying all their desires and their longing for excitement and freedom six feet under to follow the rules? Do we worry about that? What I am trying to ask about is . . . the cost of that. And that perhaps we should worry a bit more about girls who follow all the rules, and what happens to them if they deaden their own aliveness in favor of safer lives . . . Tom, can you help me . . .

Tom: Yeah, I think you asked it. Does the set-up of the question make sense to you, Syrie? Sanni is asking, what would the cost have been to your life, Syrie, if your 13-year-old self had buried her dreams for freedom and desire and excitement and decided to become a follower of the good girl rules?

Syrie: Oh my gosh, now that you are asking me that, all I feel is that, no I don't want that! I mean I think of, you know, I just had a movie flash before me, you know where the girl was young and, you know, 13 and she, yeah I mean, what else are you going to do? It's . . . It seems perfectly normal and . . . kind of interesting.

Tom: Okay. Would you dare, I ask, I don't know, would you even kind of *like* such a girl?

Syrie: Yeah, I would. Yeah, I did. I had this like, almost, like admired her. Yeah, for sure. This really resonates. It's weird. I could see a flash of her so differently now.

Tom:	How much, if you look at that girl, Syrie, that 13-year-old girl who escaped the bars and opened the window and in search of those things. How much did that 13-year-old girl have to . . . What part do you think she might have played in your eventual escape from this relationship that you decided you no longer wanted and you coming to like yourself?
Syrie:	Yeah, that's true. I mean I guess all of the things that I named are similar to how I would name the escape of the relationship, which is like excitement, risk, freedom, and desire, all of those. So, yeah, yeah, I think there's obviously there's some parallels there. I do think, that not being such a good girl growing up helped me leave that relationship. It feels similar, you know . . . now that I think of it. Leaving that dead relationship to go travel instead. It . . . felt similar. I left a dead relationship to go search for freedom and desire. Wow.
Tom:	I wonder, Syrie, is this aliveness in you the very thing that maybe is behind your current efforts in life too, the reason you came to talk to us? You know, with that question of, "well, what is it time for now in life?" Which are, as you've mentioned, are also about excitement and desire and freedom. Right.
Syrie:	Yeah. Yeah.
Tom:	That those very things have somehow been cast as a problem and what if they're not problems? But . . .
Syrie:	But . . . they're part of like, they're kind of part of who I've become, more of who I am, you know.
Tom:	And who you have been?
Syrie:	Yeah, yeah, who I have been all along because I'm looking at those things and I'm like, that's me. You know?
Sanni:	Like climbing out the window like this? *(Points middle finger)* Oh, yeah?

Syrie: Yeah. And just like all the words I said, as those are all things that, like, those are all my . . . part of my values. So I get to express them in different ways. I'm thinking about how I have taken them up in so many ways.

Sanni: Oh, what do you mean? What do you see? Are there other realms in which you've taken up values of excitement and risk and desire and freedom?

Syrie: Yeah, yeah, for sure. Like, I mean, it's kind of just . . . They're how I live my life, I feel minus the like . . . well I mean of course I need to be liked but not as strong as obviously when I was little. But the freedom and the desire and the excitement and everything that's kind of . . . those are the top, top ones for me, you know, in most areas I feel. Yeah, yeah.

Sanni: Have they guided your, your work life in some way or more . . .?

Syrie: They have, like right now . . . Yeah, no, they have really guided my work life. I have taken risks with work, and that's probably the reason I'm happy with, with my work right now anyway. But they guide me in other ways too . . . like travel and just meeting people and . . .

Tom: Leaving relationships . . . thinking about fostering?

Syrie: Yeah, exactly.

Sanni: This chokes me up, Syrie. You see, when women come to see me at age 40 or age 50 and say I'm, I'm happy with my job or I'm happy with the way I travel. I'm happy with the way I relate. I'm happy with my life, my life is fabulous. I'm like *gasps*, because I can see, I can see it as many decision points in which that woman gave herself permission, to ask themselves what they wanted. It means to me that their "want" wasn't dead. They did not allow the want to be deadened even though pressure to deaden the want is formidable.

Tom: The want is not dead.

Sanni:	The hunger for living is not . . .
Tom:	Freedom and . . .
Syrie:	Oh, I love that. I have to write this all down.

You may or may not find yourself interested now to follow a discussion of the above transcript, its reaches and invitations for thinking may already be clear to you. In that case, please skip the following lines.

There are many ideas, both small and large, that you and I might find ourselves deliberating on inside the conversation with Syrie. There might be ideas about the pitfalls and possibilities in working with a co-therapist, thinking-out-loud practices in front of clients, or better ways to wordsmith questions to elaborate on and so forth – all of which are rich and interesting matters. But this letter is about Counterstories. If you find yourself interested in a particular way to capture the above work that might serve your understanding of Counterstories leaving other thinking threads dangling for now, I might summarize it for you in this manner:

Syrie's Unstory

I was some sort of "bad girl" growing up. This is either because I am an attention-seeker or because I didn't receive proper guidance from my parents.

Syrie's Up Against Story (or X Makes Me Feel Y Story)

The remembrance of my teenage years is upsetting to me. In particular, the remembrance of sneaking out and drinking and being with boys when I was 13, is causing break and enters of shame in my ordinary waking life right now. The remembrance is causing me to feel suspicious of myself and others. The remembrance of my longing for attention makes me feel worthy of dislike in this world. This remembrance is visiting me now and is creating obstacles to dreaming my near future life into being: in particular, this remembrance is meddling with the ways I look upon myself, my nephew, my parents, and in particular, my considerations of the Counter-initiative of fostering a child.

Master Narratives in Syrie's Story:

A girl doesn't have interesting reasons for the initiatives she sets in motion.
A girl does things for indecipherable reasons or because others didn't give her better reasons.

A girl's aims for living aren't worthy of moral consideration.

A girl is either a good girl or a bad girl.

A good girl is not sexual, does not drink, does not sneak out at night.

A good girl does not have interesting longings and ethics that can be queried and discovered.

A good girl is not interested in adventure, risk, excitement, desire, liking, freedom.

A good girl does not desire, insist on, or seek attention from others.

In our conversation with Syrie, Tom and I spent some time and words to query the easy Master Narrative-driven conclusion that there is necessarily something wrong with "sneaking out" and "drinking" and "being with boys." Tom also took on the Master Narrative that "attention-seeking" necessarily be a dirty ambition for a good girl. We called these Master Narratives "charges" and "accusations" against Syrie in an effort to invite a position of a bit more freedom for Syrie's own deliberation of why she did what she did. It is sometimes difficult for any client to find words for their desires that drove their actions, but it is nigh impossible when Master Narratives of proper behavior are limiting what is even possible to imagine as interesting desires for original actions. If Tom and I had not expended some effort to downgrade the Master Narrative-driven conclusions about proper personhood for a good girl, Counterstory questions about what Syrie was seeking would have been difficult for her to contemplate or respond to. "I did it because I sought attention" she might have well continued to answer, ruefully. "And what is wrong with seeking attention?" is a necessary query to put the Master Narratives in their place such that Syrie can be a bit freer to reconsider her motives without this noise. So: Counterstories resist Master Narrative-driven identity conclusions.

After this interlude, Tom and I invited a Counterstory in the following manner:

> *"So if we were to imagine you, all of age 13, a bit lonely, a bit lost, a bit spirited . . . climbing out the window . . . can you see it, can you see . . . her?"*
>
> *" . . . I'm imagining her on the window ledge of your room right before climbing out . . . maybe she's taking a last look at*

her room, to think of all the things that good girls should be doing . . . and then turning to the window again . . . what was her hope, her longing that she climbed out of the window for?"

" . . . Yeah, something that was driving . . . if we could imagine her in that moment right before she climbs out the window . . . and press play . . . What is driving her? Maybe a wish to feel something or risk for something . . ."

Based on these ideas, here is a first pitch of Counterstory that we wrote for Syrie based on the above conversation and read to her at the beginning of the next session:

Syrie's Counterstory Letter

Dear 13-year-old,

I see you. I see you there on the windowsill, about to climb out at night and in secret. Maybe I am like a kind of godmother, or good aunt, and I want to talk to you for a second – I know you don't have a lot of grown women talking to you in their own voices about what it's like for them, so I'll try to do that.

See I haven't forgotten what it's like. And don't worry, I won't try to talk you out of what you are about to do. The choir of dusty Catholics might be suspicious of you and talk to you about "good girls" and "bad girls." Other dusty people might take one look at you there, all of 13, and hovering on the windowsill and they might try labels onto you like "attention-seeking" or "thrill-seeking" or even "foolish" – but they are just applying the tired idea that youngens are supposedly so very different from adults. They don't know you. I know you.

You're not foolish, so you'll understand this idea: the seeking of attention or thrill, to feel ourselves the center of a focused attention and come alive with thrill – why would that be bad thing? It's a real question.

Only one thing, dear 13-year-old beloved beautiful brave one – keep your eyes open, in both attention and thrill – keep your wits about you and know that you are free to say no and yes. Do not let your soul be trespassed by either attention or thrill, discern it with clear eyes, my dear, and then dare say yes and no loudly and with conviction and purpose and compassion – that is the greatest thrill of all, I tell you.

I know that maybe you won't believe me for this moment, as I am just some 40-some-year-old woman to you. So I smile at you, because I know you and I will meet at age 42 and even then, I'll just smile and abstain from saying "I told you so" when you finally tell me "you were right that time when I was 13."

Anyway. Here's what I really wanted to say. I see you. In search of excitement, risk, freedom, and desire, even young as you are. That's why you are going to climb out in a moment and run towards the search of all things alive. I know that now. And I want to say thank you, little one. If you hadn't done what you did, I would have a hell of a lot of work now to do to unearth the spirits of excitement, risk, freedom and desire from 6 feet under as a 40-some-year-old.

So go ahead, young one. Go and live and discover your desire, your freedom, the thrill of all the world. The world awaits. I know you can do it while keeping two eyes open, one compassion gold thread to your mother in respect for her, and 1heart full and wild.

I'm waiting for you on the other side.

Until we meet again,

Love

Your godmother Syrie

Perhaps you can see the reach of this letter, dear student. Counterstories that take on the form of "Given X and Y, I Did Z on Purpose" offer a thrice-over resistance:

1. Counterstories are stories. (I shall refrain from reprinting the definition of a story again. Remember the thing that has a protagonist, a setting, and plotline blahblahblah. I hope you can fill in the blanks by now!) By nature of being stories, Counterstories resist unstories.
2. Counterstories are particular kinds of stories. They are agentive stories, and their plotline is driven by the desire of your protagonist. By nature of being agentive stories, Counterstories resist the format of Up Against Stories ("X makes me feel Y stories") and take on a format of "Given X and Y, I did Z on Purpose Stories."
3. Counterstories resist the Master Narrative-driven identity conclusions that are binding of the credibility and

freedom of movement of persons.[3] When the impressiveness of Master Narratives is downgraded, a person's conscience and their desires are upgraded to be more freely considered and discerned as to their intention and effect, on herself, others, and the future of her life and relationships.

That's it for tonight, dear student. I'm off, out the window now. I shall return with a look at how the protagonists walking through Counterstories are not angels, but humans.

Love,
Sanni

Notes

1 "A counterstory is a story that resists an oppressive identity and attempts to replace it with one that commands respect." Lindemann Nelson, H. (2001). *Damaged Identities, Narrative Repair.* Cornell University Press, p. 6.
2 Atwood, M. (1996) *Alias Grace.* Knoph Doubleday Publishing, p. 289.
3 "The counterstory positions itself against master narratives. It is the master narratives that the counterstory resists." Lindemann Nelson, H. (2001). *Damaged Identities, Narrative Repair.* Cornell University Press, p. 6.

8

What if We Aren't Angels?

A Letter About Interesting Protagonists

Dear student,

Michael White wrote: "I would argue that, because our lives are multi-storied, then we are all multi-motived, and that some of our motives have positive real effects in terms of our lives and in terms of our relationship ecologies, and some very clearly have very negative real effects."[1]

This letter is an encouragement to you to consider the richness of both the motivations of your protagonists, as well as the richness of the real effects of their actions, anew. In short, clients are no angels.

In my work with students over the years, I have noticed a curious stifling of students' attitude of curiosity in all matters of "what human beings routinely do to each other" – in favour of an a-priori assumption of perfection in motive, and thereby a suspension of the discernment of the real effects of clients' actions. On the flip side, the same press on curiosity seems to be operating on students to invest in the antagonists in clients' lives all manner of easy badness and veritable villainy, until neither protagonists nor antagonists can be raised to resemble real persons entangled in rich relational and life contexts anymore.

DOI: 10.4324/9781003478478-8

Clients are unquestioned angels and challenging people in their lives are unquestioned assholes. It makes sense: the construction of angels necessarily begs the creation of monsters to oppose them, right? But I ask you: what is a better "memorable character" in a clients' story than an obstinate and challenging antagonist? But shh, that's a secret. We as therapists are pressed to take our place in a new era of caricatures.

And I understand – students have made an important ethical commitment to refuse the history of pathologizing of persons, and the kind of authoritarian streak in therapy rooms in which expert therapists tell people what is wrong with them and what to do about it. Yes, good riddance to such excesses.

But the therapy training of students leaves them with a very flimsy alternative: the well-worn practices of summarizing, paraphrasing, normalizing, and validating taught and touted matter-of-factly as "neutral" basic counselling skills. Kai Cheng Thom writes that these practices have long escaped the therapy training labs out into the world to be understood as "love."[2]

Within the task of story-shaping central to narrative therapy, basic counseling skills like validation have an effect of stifling a dynamic conversation by forbidding the attitude of open curiosity, thereby foreclosing the telling of rich stories into foregone conclusions.

Michel Foucault writes this about curiosity:

Curiosity is a vice that has been stigmatized in turn by Christianity, by philosophy, and even by a certain conception of science. Curiosity is seen as futility. However, I like the word; it suggests something quite different to me. It evokes "care"; it evokes the care one takes of what exists and what might exist; a sharpened sense of reality, but one that is never immobilized before it; a readiness to find what surrounds us strange and odd; a certain determination to throw off familiar ways of thought and to look at the same things in a different way; a passion for seizing what is happening now and what is disappearing; a lack of respect for the traditional hierarchies of what is important and fundamental. I dream of a new age of curiosity.[3]

I can ask you this: how can we shape the conditions for rich story development if we cannot query the multitude of motivations of actions as well as the multitude of real effects of actions, both for our protagonists and antagonists in stories with all the boldness of curiosity? I can also ask it this way: why must clients be perfect little Victorian angels who neither knew what they were doing, did only what others made them do, or did nothing whatsoever of any consequence at all? Or: How long must clients suffer our lack of curiosity in the interestingness of their many motives, the shape and shocking effect of their colourful deeds, and the complexity of their character and history?

Bear with me a moment longer if you can: just as I am sick and tired of "clients navigating" their lives, I confess I am sick and tired of hearing the same proposals for Counter Story themes over and over again: clients are "caring" and clients are "loving" and clients value "fun" and "belonging in community." If you have an ear for Master Narratives yet, you can easily see how these values coincide with the culturally sanctioned wants of women and queers. Family, community, and care are supposed to be our highest ideals – these are the currently sanctioned wants authorized by the good housekeeping seal by the proper-person police themselves. In fact, women and queer folks who show little interest in these values have to defend themselves and court painful mis-readings of their character. If you are interested in hitherto un-sanctioned wants or motives: what about sex, power, attention, ambition, achievement, respect, legacy?[4] What about the desire to have "more say and more options as to how our lives go?" Hell, what about demanding a space at the table where decisions about my life are being made? What about shaping the proceedings of the world according to an unbidden design of my own? What about being a decider of moments, days, months, and years? What about the longing to give and receive a good experience, to engage in a frank and honest debate about the clash of wants with interesting antagonists? What about leaving the world better than you found it?

The comfort-seeking cis-hetero-white patriarchy has done an admirable job of ensuring that women and queer people hedge themselves into oblivion, that all words for "and what is it that you wanted?" are lost to us. But it is my tireless hope to amplify women and queer people's wants to the size of my little therapy room and beyond, to ask each of them: if you want your wants to be considered, then we must first endeavor to say them.

I want to show you three different transcript excerpts that will hopefully clarify what I mean for your own deliberation. But before I do, I want to entrust you with a consideration: what I have learned over the years is that both the construction of clients as unquestioned angels and the construction of other persons as unquestioned assholes can severely constrain the visibility of the moral spark of agency that reveals character. To have "more options and more say" about how our life goes is what we might summarize as "agency." "More options" relates to the concepts of the "ability to respond" (response-ability) and "more say" relates to the ability to account (account-ability) for your clients.

I have come to be taught that life initiatives, of word and deed, are part of a moral striving towards agency; part of a particularly human yearning to not just survive and then die, but to be able to say, at the end of the day, or the end of a life, that we contributed something, something that mattered to us. Or that we died trying. I will never forget a client of mine firmly reiterating to me in conditions of grave anxiety and near-guaranteed failure of her initiatives that "I want to be able to say that I fucking tried." As therapists and clients, we struggle for words to denote lived experiences and often end up overstating or understating both trespass and agency in evasive and ill-considered ways. The effects are a shaping of thin stories in which your client will appear not further interesting than a leaf in the wind, blown this way and that by random life happenings.

I believe that in the service of an agentive turn, "languages of suffering"[5] that aid the ability to account for wrongdoing, trespass, harm – our own and that of others – is an endeavour to create language that better respects our clients. Hearing descriptions of what human beings routinely do to each other freely queried, handled, and negotiated can present an unhoped-for unusual relief for clients. Questioning the meanings of moments that have appeared to us as unquestionable, both our clients and us may be presented with a possibility to imagine the experiences of others and ourselves with renewed, reenergized, and more specific language that in turn carries a spark for nuanced responses that had not occurred to us before in that way.

In the same way, adroitness of language that refuses to praise clients for achievements that required no effort and no risk on their parts is a formidable act of avoiding our clients' humiliation. Let me ask you this way: have you ever received congratulatory credit for something that cost you nothing to do? And how did that feel to you? Did you much wish to keep speaking to your applauder? And on the flip side, have you at times felt yourself to be entirely invisibilized and stepped over in any meaningful acknowledgment of the efforts of initiative that you did, in fact, bend and sweat over until your back ached and you smelled strange?

And let me complicate the matter one step further in considering that most of my clients have been women, poor women, queer women, trans women, women with disabilities, women of colour, marginalized women. What exactly are we conveying about our beliefs about the capacities of these clients, in particular, by the veritable performances of empty social mores of "validation training" in response to their experiences and enactments of trespass? And what are we saying about our regard for them if we cannot bring them questions, proposals, disagreements in earnest and a suspenseful expectation that their minds and lives and choices can be accounted for and responded to?

Imagine if we were to refuse the caricature of our clients as pure angelic victims full of sage life advice in every word they speak? And if we were to meet another, who clearly has enacted trespasses in her life by her own telling, could we also set out to refuse the remaining caricatures of an "angry difficult woman" in need of teaching to control her emotions and that of a "helplessly unskilled woman" in need of teaching how to live? Outside of caricatures, could we expect to negotiate agency with actual living humans? And if we did explore all manner of actions and motives that have destructive real effects – would we really have to haze persons in contempt or would we catch a glimpse of what is in them being in us as well and reach for non-contemptuous language to stretch over the loneliest moments of shame, together?

The following practice examples are purposefully set in the life territories of difficulty: in particular, in the territories of anger, and expressions of threats by clients as such realms have been problematized by the therapeutic persuasion as areas of suspicion in women's lives. These conversational realms have also often been at the center of disciplinary actions (heavy-handed suicide interventions, threat assessments, the diagnosing and pathologizing of life experiences, censure, involuntary hospitalization, etc.) against women, queer clients, and persons of colour. It is my belief that these disciplinary actions have been derived in agreement with taken for granted, prejudiced views of persons. In order to propose a counter-practice to such prejudice, I am reaching for a reorientation in the view of a person, which then necessarily makes some practices impossible and begs others to be invented. All three transcripts are brief, but self-explanatory excerpts of particular "hinge-moments" in conversations in which the clients were asked, in various ways, to deliberate, discern, and deliver decisions on "proposals" put to them by significant others and by significant experiences of impassioned moments in their lives.

Transcript 1: "Anne and the Leprechaun"

Anne:	The borderline is back. Full force.
Sanni (Gently):	Tell me of the miserable come-back then.
Anne:	I was freaking out about the presentation. All day. And then I called my friend. As we were talking, I grew more and more despairing, just so anxious and angry, I had the sense that I just can't do this, that it's too much for me, the presentation. I should just cancel. My friend tried to help but it wasn't helpful. The borderline just kept growing, like a rush . . .
Sanni:	Wait, "the borderline" kept growing then, during the phone call? How do you discern that, what happens to you?

Anne:	It takes over, it's like a rush that I can't resist. I wanted to cut so badly, and I started making jokes about dying.
Sanni:	Wait, I want to understand all of this! An almost irresistible rush that takes over you, and it comes with great counsel too – "hurt yourself, Anne," or "die, Anne?" *(In a grim voice)*
Anne:	Yeah, it's like a rise . . . like a river that keeps rising. I made a bunch of mean jokes to my friend about wanting to die. After a while, my friend said "that's not funny," and hung up on me.
Sanni:	Mean jokes . . . whose word is that, "mean," is that yours or your friend's?
Anne:	Mine.
Sanni:	Why do you call it "mean"? Is this an evaluation you stand by, now in hindsight? Is this a fair evaluation?
Anne:	Yeah, I don't think that's cool to do to someone. And I love my friend, I don't want to talk like that. I was just so . . . fuck! Why is it so hard?
Sanni:	Forgive me for going back so much today, and screech at me if you think I am not asking the right questions, will you. My question is this: why is it mean and uncool to make jokes about dying to someone you love? Is this something you have decided in life? What messages might jokes about dying send to your friend that you're not okay with?
Anne:	Well . . . I know what it feels like, you know. I know how serious it is, and when you love someone, it makes you feel so helpless if they are messing with the seriousness of it.
Sanni:	Okay, there's two questions now, can you help me and pick the one that interests you more. Or say, neither, Sanni and give me a third option. *(Anne laughing)* I could ask about the time you made this decision, to not mess with those you love and about all that went into such a decision. OR: We can go back to the phone call and your

	discernments and discoveries there. Does either interest you?

Anne: Yeah, the phone call!

Sanni: Okay, so let's see, am I getting it right, that there was a rush, and then a mean voice, mean to you, and mean to your friend, and it was just talking, spewing things . . . (*Anne nodding*) I know this is hard – but my question is why? Why should you be taken over by meanness, all of a sudden in that moment? Why was it talking to you about hurting you? Why was the meanness messing with you like that? Why should you be taken out of life or get hurt, or lost to your friend, what does that meanness not want you to think about or experience?

Anne: What?

Sanni: Right, I'm sorry, that was a mess. I was thinking . . . and maybe this is not true, you tell me: I had this thought that if something mean steps in and all of a sudden has a great suggestion like "die, Anne" and uses that much power to "take you over" – maybe you were just on the verge of discovering something *worth* taking over. Do you know what I mean? What were you just figuring out or perhaps on the verge of figuring out that afternoon before it felt the need to shut you the fuck up in a mean way?

Anne: What was I figuring out? I was thinking about the presentation . . .

Sanni: Do you think . . . is it in the arena of presentations that it would like to seize upon with its grabby hands? What were you thinking about before it stepped in like that?

Anne: I was thinking . . . I was thinking about the applause after. Or being there, at all, everyone looking at me. Or the criticism. I don't know.

Sanni: I know it's hard, Anne. Is it possible you were having some rather interesting, new ideas about who you are before it felt the need to supply its own old and predictable ideas to stop you from thinking further? Is this possible? If that's what happened, I want to know exactly what you were on to thinking about and make sure we don't lose it again to this tiring fucker. That's why I am annoying you with this.

Anne (Laughing):	It's okay. It's interesting. So, what's the question?
Sanni:	Well, I can do better than what I asked . . . Remember how you talked about the men in that group being "story stealers"? I liked that term so much . . . what if the mean fucker that stepped in was the ultimate story stealer? What story that you were on to just right before did it try to steal?
Anne:	I guess it's about presenting. Standing there and having something to say in front of all those people. The idea that they might like it, or not like it, it doesn't even matter which.
Sanni:	Standing there and having something to say, huh. Anne, would you be surprised if I said I'd be rich if I took a penny each time a client here tells me that standing and speaking in front of an audience is a real rose garden?
Anne:	No, that wouldn't surprise me.
Sanni:	And if you and all the other women and queer people here were to do it anyway, claim the equal right to stand there and have something to say, despite all the ancient forces that would have us shut up, where would that take us? What were you about to claim, perhaps, for yourself, when you were contemplating the presentation?
Anne (Growing tearful):	The possibility that I could make a contribution. That the thoughts are original and interesting. That I could . . . belong, for a moment.
Sanni (Tearful in turn):	Imagine that, Anne. Imagine. Imagine what a daring idea that is to think that you could make

an original contribution to a community you care about, is it possible that this imagination is so beautiful that it might be worth stealing from you, distracting you from with a bunch of mean noise?

Anne (Crying)

Sanni

(After a

silence):   Can I tell the next person here who is struggling with a presentation what you just said, "the possibility that I could make an original and interesting contribution"? Can I tell her that you dared to think that that's possible? (*Anne nodding*)

Sanni (After

a silence):   Can I ask you one more thing?

Anne (Smiling)

Sanni: The rush you felt, the "river rising" that you talked about, *before* the meanness, was that a good thing, or was it already part of the meanness?

Anne: No, I think that was just feeling. Like something rising with feeling, like a river.

Sanni: Are you okay with the feeling of the river rising, like looking at it, or is there something dreadful about it as you look at the rise, are you afraid?

Anne: No! It's alive, the river is alive, it's okay. It can flow.

Sanni: If the river rising is just aliveness, and possibility, what is this thing that you call borderline then, what is that in this image with the river? As you stand there and look at the flow and delight in its aliveness, how do I imagine the borderline thing then?

Anne: The borderline is like a . . . maniacal leprechaun, flying around and screeching.
        (*Both of us bursting out laughing at her image*)

Sanni: A maniacal leprechaun, alright, flying over the rising river and screeching something maniacal at you? I didn't expect that . . . You know, I am imagining you there alone now, because your

	friend hung up on you. Did the leprechaun keep screeching all night at you, afterwards?
Anne:	Yeah, for a while. I contemplated all the things, drinking, cutting, all of them.
Sanni:	And did you take its counsel?
Anne:	No.
Sanni:	You didn't? Why? What did you do instead?
Anne:	What did I do? I don't know. I sat there. After a while I guess I just calmed down.
Sanni:	I like how you're saying "just," when both of us probably know there is no "just" about it. What in the world was calming to you, that night, as you stood there at the edge of the river?
Anne:	I don't know. I talked to it. Yeah. I talked to borderline.
Sanni:	See, they don't write about that, the story stealers, the . . . border agents. I guess they don't know much about rivers and leprechauns, after all . . . You talked to borderline. What on earth did you say?
Anne:	I said, "we're not going to do this."
Sanni:	Was this like a firm command, or was it said gently, or did you have to raise your voice or what?
Anne:	No, it was quiet. Like soft.
Sanni (Softly):	We are not going to do this. What was the "this," what was this soft consideration re-considering?
Anne:	Hurting my friend. Raging. We don't need to do that.
Sanni:	And do leprechauns listen?
Anne:	Yeah. After a while. I talked to it gently. It changed to . . . peaceful.

. . .

Transcript 2: "Tess and the Stormy Story"

Tess:	And then I wanted to die again. Jump off the balcony. I went out there and looked down and thought how much I want to do it.
Sanni:	Wait. How did it go from arguing with mom over

the advent calendar, to "DIE, Tess!" *(Said in menacing voice)*

Tess: I don't know. That's how it goes. I just want to die.

Sanni: Wait, can you help me understand it though. What happened? Here you were, arguing with Mom, and mom being slow as usual, and not hearing you and not getting it, and how in the world does it go from you being in that moment, to something telling you "DIE TESS, YOU SHOULD DIE!" Why would something tell you that, all of a sudden? Why should you die over this? Why was it telling you that?

Tess *(More intrigued now)*: I don't know. Maybe . . . well I got so angry. I don't know if I should get so angry with mom, she just doesn't get it.

Sanni: Okay, so Mom doesn't get it, and then the dilemma is: is she being mean to me, or does she really not get what I'm saying? If I asked your anger, what would it tell me what you were thinking?

Tess: What?

Sanni: Sorry, that didn't make any sense. I mean: you got angry in the argument with mom. Does this thing step in to tell you you deserve to die because you got angry at your mom? Because we as women aren't allowed to feel that way? Or that you ought to die because Mom deserves compassion, and you can't always give it to her? Or that you deserve to die because Mom got to you again, even though you had sworn to stay calm? I want to know why you deserve to die in that moment? Why are you given a death sentence in that moment?

Tess: I think, I think because I lost my cool! I don't want her to be able to push my buttons, how many times have I been here, and yet I'm unable to stay cool!

Sanni: Okay, help me out. Is this because you believe in a code of coolness towards your mom? Or maybe,

	because you claim the authority over your buttons, and don't want to cede them to your mom?
Tess:	Yeah! I do want that authority!
Sanni (Smiling):	It's tough to do, hey? Damn, we are so helplessly dependent on each other hey, damn other people getting under our skin and fishing around and finding the damn buttons anyway. Is it possible, you think, with your mom? Has this happened before, when you totally stayed cool like yoghurt or something . . .

Tess (*Laughing. Then tells me story of a moment when it was possible. Afterwards we return to this*)

Sanni:	But I still want to understand how it went from "shit I lost my cool" to "DIE TESS." Is this a button you'd like to look at too, because I really do, I'm really interested in that, in your authority over THAT one.
Tess:	Yeah. I don't know . . .
Sanni:	Is losing your cool a crime for which you would sentence another woman to die? Even if she told you that she'd vowed she wouldn't lose her cool, and NOW she failed again.
Tess:	No!
Sanni (Smiling):	What would you tell another woman?
Tess:	I'm not even going to answer that. YOU know.
Sanni:	I DO know! You'd give a whole speech about compassion and say this is just her first try at life too, and how important the task is to remain cool with Mom, of all people, but Mom is the hardest of all challenges, and that the failure means just how hard it is. And that if she succeeds, WITH MOM, then it's like she's got the power of the universe to stay cool WITH ANYONE EVER. I know you'd tell her that.
Tess (Smiles):	Or something like that.
Sanni:	Are we allowed in life to make mistakes? To fail? EVEN at the most precious of our relational

	commitments? To fail at them, like ALL the time, or every other day at least?
Tess (Very quietly):	Yeah. I am thinking about it. It's much harder to allow that for me, than for someone else.
Sanni (Very quietly):	I know.
Tess:	It's just . . . it's all such a shitstorm when I get angry . . .
Sanni (Quietly):	I understand. *(After a pause)* Hey, before when you were talking . . . the "shitstorm" is just reminding me, about something you said last time . . . you were saying something about a "first stormy story" . . . Is the "DIE TESS" like a first stormy story?
Tess (Smiling):	Yes!
Sanni:	So you get angry, and then you are disappointed at yourself, which triggers a whole shitstorm! And in the storm, it roars "DIE TESS" for your FAILURE.
Tess (Laughing):	Yes.
Sanni:	But if it's just the "FIRST" STORMY STORY, if it didn't roar so loudly to drown out all other stories, what else would you hear? Is there another story here, maybe one about Mom, or about you having permission to be angry, or compassion for failure, or I don't know what, but is there another story here that is lost in the storm? Something that maybe we should make sure we don't lose.
Tess (Suddenly tearful):	Yes.
Sanni:	Will you tell me? Can you tell me?
Tess:	Yes. *(Tearfully)* Underneath the storm, all I feel is "I don't want to be alone."
Sanni (Choked up, gently):	"I don't want to be alone"? That's what the storm wants to hide from you? That longing, that desire

	to BE with someone, maybe Mom, but maybe it isn't possible with mom, someone who SEES you anyway, do I have that right?
Tess:	Yes. I never thought of it this way.
Sanni:	Why should you not know that you long to be with someone, really be with someone? Why should you not get to say that in life? Why should you die, rather than say that?
Tess:	Because . . . then I'd be admitting how much I need other people.

. . . .

Transcript 3: "Valerie and the Counsel of Anger"

Valerie (In a raised voice):	I am so angry! I could just . . . aagh, I am furious! I think he slept with my roommate! That's the one thing, the one thing – I am so angry right now! I have been shaking all afternoon, just wanting to . . . I want to hurt him back! I want to hurt him like he hurt me.
Sanni (Quietly):	I'm sorry, Valerie . . . did you just find out about this today . . .?
Valerie (Yelling):	Yes! Fuck! He knows that that's the one thing to do to me to hurt me – he did this on purpose! And she, how could she . . . but I don't even want to think about her, I just want to hurt him now, I want to show him how much this hurts.
Sanni:	Okay, I, . . .
Valerie (Interrupting):	You know what I have been thinking about. I can get back at him, I have been thinking about the ways to do it, I'll make it so he walks in on me kissing Ryan, I can do it today when he comes home from work!
Sanni:	And then what? Would that do –

Valerie
(Interrupting): Ryan would be only too happy to do it too. He's been wanting to sleep with me for a long time, so now is the perfect time, he'd be over in a flash.

Sanni: And if your boyfriend walked in on you and Ryan would that –

Valerie
(Interrupting): But maybe that's not enough! Maybe I don't know, I just want to hit him or push him like really get in his face.

Sanni: Get under his skin too, have him feel you and the hurt of this, and if you did manage that would that do justice?

Valerie
(Yelling): I don't fucking care! I just want him to feel my anger, to really feel it, I just want to –

Sanni: Right, can you help me understand what the anger –

Valerie
(Yelling): No! I don't care! I need a way to hurt him, to unleash this anger on him, I don't even know if that thing with Ryan, I don't know if I can stand it, so an easier way, I can just wait for him and push him down the stairs when he comes home, I'll kick his legs out from underneath him and all I'll say to him is –

Sanni
(Interrupting,
very softly): I don't believe you.

Valerie
(Yelling): WHAT???

Sanni
(Very softly): I don't believe you. I don't believe, not for a second, that that's why you came here today. You don't need my help to figure out exactly how to hurt him. I don't believe that's why you asked for this meeting. *(Valerie looking at me. Continuing softly)*

I *know* you're angry, I know, but I refuse to believe this about you, that the anger's only counsel is to hurt him, or that you'd follow such counsel without a few questions of your own. Will you let me care about your anger and ask you what your anger is fighting for, what matters so much to you in this?

Valerie
(Suddenly
starting to cry): Fuck. I don't want to hurt him. Fuck.

Sanni (Softly): Can I ask the anger directly, if you could tell me the words, what are you fighting for right now in Valerie's life?

Valerie (After
a long pause): I'm fighting for us. For what we have. I trusted him *(Crying).*

Sanni
(Choking up): Tell me everything. Teach me what you know.

Valerie: I am just, I have been remembering all the talks we had lately, you know. Sitting together on the couch and he was holding me, we were so close. And I trusted him with all that shit about my ex *(Crying).* He knew! He knew! That's why this is so hard . . . Actually, I am not 100% sure he slept with her. I just saw one of the texts on his phone earlier today, and I freaked out *(Crying).*

Sanni (Softly): Can you tell me why, if something happened between them, why would this be the worst?

Valerie: Because then it would be gone. All gone. I couldn't. I couldn't . . . even if we were to stay together, I couldn't trust him like I did . . .

Sanni: Is the anger roaring about . . . I don't know . . . the rarity, the beauty of intimacy and trust . . .? Is it worth roaring about?

Valerie
(Nodding): Yes.

. . ..

I hope that these transcript excerpts, although far from perfect, impress upon you the possibility of sturdy negotiations of agency in clients' lives who need not meet any prerequisites of angel-hood to deserve our most caring attention. I further hope that you might have caught a glimpse of what can become possible if we as therapists dare to make invitations to accountability and responsibility (in short, agency) in those moments when the rubber meets the road in our clients' lives.

If there is a tentative conclusion to this chapter, perhaps it may be this: in what kind of a story does your protagonist have more character?

Love,
Sanni

Notes

1   White, M. (1995). *Re-authoring Lives: Interviews and Essays*. Dulwich Centre Publications, p. 32.

2   "I was formally trained (some might say indoctrinated) in validation culture as a volunteer in a radical leftist/queer crisis support service when I first entered community as a teenager . . . The volunteer training taught me, over a forty-hour series of very intense workshops, that validation in the form of active listening (validating someone's experience unquestioningly, assuming that they always know what is best for them, repeating what they are saying word for word, never being directive or giving an opinion, and saying 'mhmmmm' a lot) was the foremost aspect of being a supportive person . . . I have seen countless such situations where queers, struggling to be supportive of friends and community members, have adopted validation as the easiest and most politically correct approach. We tell people that they are always right, that their perception of minor conflict as life-threatening is accurate. We tell them that their substance use is fine and their choice, even when it is affecting us or others very negatively. We tell them that they know themselves best, and if self-harm is the only option that they see for survival, then they should go on doing it, no questions asked. In short, we enable. I don't think we do this because we are

bad friends or lazy people (though bad friends and lazy people exist in community. I have personally been both bad and lazy too many times). I think we do it because this is how queer counterculture has trained us to love." Cheng Thom, K. (2019a). *I Hope We Choose Love: A Trans Girl's Notes from the End of the World.* Arsenal Pulp Press, pp. 32–33.

3   Foucault, M. (1997). "The Masked Philosopher." In *Essential works of Foucault, Volume 1: Ethics, Subjectivity, and Truth*, p. 325.

4   "The Gospel of Mona presents instead the seven necessary sins women and girls need to employ to defy, disobey, and disrupt the patriarchy: anger, attention, profanity, ambition, power, violence, and lust. I call them 'sins' but of course they are not. That are what women and girls are not supposed to be or do or want. They are condemned as 'sins' by a patriarchy that demands we acquiesce to, not destroy, its dictates." Eltahawy, M. (2019). *The Seven Necessary Sins For Women and Girls.* Beacon Press Boston, p. 10.

5   "Writing in 2017, the Danish scholar Sven Brinkmann reminds us that 'as human beings, we do not simply suffer in a simple, physical way but are also capable of understanding our pains and miseries in and through the languages and vocabularies we have. The most powerful tool to mediate our understanding of suffering has arguably become the psychiatric diagnoses, serving as a widespread language of suffering'. . . . I have no doubt that is so. And with this in mind, I want to remind you of what I am calling 'characterizing' and remind you of its languages and vocabularies and histories that are far older and more inspiring than contemporary psychiatry. As well, I might refer to these as languages of unsuffering." Epston, D. (2022). "Coming to know young people as promising characters." *Journal of Contemporary Narrative Therapy*, April, pp. 38–55, 40–41.

For Anne
What Is a Good Life?

And don't give me normal
Because normal is a setting on the dryer
But help me:
I want to know the mystery, the conflict, the anguish
The risk and the rest of this
Human deliberation
Of how to live well

9

What is Beautiful and What is Ugly?

A Letter About Moral Reading Prompts

Dear student,

I know we have taken some conceptual leaps and bounds together over the course of the last letters. This is the concluding letter to the conceptual reaches for your beginning. The final letter following this one will be less conceptual and intended more as a siren call to write to your clients after they leave your sessions.

Cheryl Mattingly writes: "A poetic narrative imitates action and experience through clarification and condensation, revealing causal connections between motive, deed, and consequence, which allows for a moral reading of events. The purpose of a narrative is not simply to tell what happened, but to provide a moral perspective on past events."[1]

So there it is. The details of a story are like the trees, and the meaning of the story is like the forest, so let's not lose the forest for the trees. Otherwise, we may wander all day among all details of a beautiful story, all the way down to the color of her dress that day, the make of the get-away car, and the weather on

DOI: 10.4324/9781003478478-9

the day she decided to live after all to become mere chronological oddities that serve little clear purpose. But imagine you never lost sight of that fact that it was the day she first wore a dress in her transition to a woman and that the redness of it was precisely to call attention to her right to dress her body and challenge every passer-by with a half-defiant-half-scared-to-death look to boot. Imagine he escaped his first relationship with a man that had become controlling down to the details of every part of his day, and he had a five-minute window to pile what was most important to him into a get-away car, in a breathless moment of "my mind is my own and cannot be owned." Imagine you knew the weather on the day she decided to raise her head above the fray and stop and turn the wheels of her own destruction around to vow to look for the destroyers upstream instead. So, details and meaning, trees and forest – that makes sense right? The details together with the meaning raise a storytelling venture above a webcam-approach to the chronology or facts of encounters and happenings in this ordinary extraordinary world.

I love the phrase of "allowing for a moral reading" in Mattingly's quote above. See, dear student, the secret about narrative therapy is that our efforts are all poised to "reposition the person" (to allow for a moral reading) not "reframe the problem." Do you see it? The insistence on storytelling, the externalizing vernacular, the attitude of curiosity, the questioning of Master Narratives and labels inside stories, the agentive turn – all poised to reposition the person. The central question is this: how can we trigger the conditions in our conversations to invite a person to step out of the midst of a burning dumpster fire of lived experiences, to look with us at the fire from a position of "outsight," from a position of authorial rights and agency to weigh and discern the fire. Is it a toxic fire or one for warmth, who lit it and for what purpose, what were the motives of the fire-lighter, and, if I were to let this fire burn, what would it do to my life, my relationships, my future?

These questions make up the heart of "moral reading." The "moral reading" of events, complete with "motive, deed, and consequence" (or in other words, "purposefulness, action, and real effect"), is what we wish to invite persons to engage with in our conversations, but not from the usual position inside their lived experiences with the familiar judgments upon them to render the "moral reading" stale and traditional. The breaking up of the familiar positions is what allows for "more options and more say" to appear.

Michael White puts it this way: "when people break their lives from the very negative stories of their identity and then they have the opportunity to stand in a different territory of their life . . . this reinterpretation facilitates a different expression of their experience."[2]

A look at the practices that we have traced so far over the course of these letters will show you that each of them is meant to be but an aid in responsive story-shaping:

1. Good faith practices
2. Storytelling questions
3. Witnessing responses
4. Imaginative leads
5. Moral reading prompts

Each of these is intended to help you and your client not just to shape a story forward together but to also reach for the meaning of stories that can help your client to step into an agentive role in relation to their lived experiences. So, the act of storytelling can reveal meaning, but only if we tend to its meaning. It is a powerful act to tend to the meaning of stories, and we as therapists need to be accountable and responsible for this practice. It is rather difficult to embrace our accountability or responsibility if we do not first acknowledge that "tending" to meaning of stories means that our minds are engaged in "reading" clients' stories. My urgent plea is to make this "reading" both present and explicit and visible to clients in the form of "moral reading prompts" put forth to your clients in your sessions. Moral reading prompts are aimed to invite clients to leap into an authorial position to claim the meaning of their story. An agentive authorial position of a story in which one is also the protagonist is a position of great moral authority. Your client is invited to call what is beautiful and what is ugly, from a position of a different "territory outside of negative conclusions about their identity," as Michael White puts it.

Are you ready for a practice example to show the above ideas in motion?

Okay, we'll study all of this together once more with the help of a story and a transcript excerpt with a young woman named Kara. I'll summarize the work of coming to understand Kara's Up Against Story and show you the transcript excerpt of the moment it changed in the midst of the attempt to understand it and explicate to you the above-mentioned practices that helped set the stage for Kara's radical agentive change. I'll particularly draw your attention to the weaving in and out between storytelling and asking for the meaning of Kara's story. In other words, you'll see all the story-shaping practices together with the invitations to moral reading prompts in the below transcript.

I met Kara a few years ago. Kara answered questions thoughtfully and in a quiet manner, barely above a whisper. To my surprise and great movement of my own heart, she often breathed a shy "thank you" in response to my witnessing responses to her stories. By and by, I came to understand her initial whisper as an effect of what Kara was up against in her life and her appreciation of something as simple as my sincere attempts to take her experiences seriously.

"When I was scared, you wanted to know what was scaring me, and when I was despairing, you asked what despaired me. This made me think of my own experiences as more reasonable, and made me look at what was scaring me differently. I started to think that I wasn't crazy or unreasonable," Kara reflected to me as one of her summaries of the effects our conversations.

In our first session, Kara told me that she was referred to me by her family doctor for her struggle with what had been termed "stress migraines." As a sidenote, I have come to have many thoughts about the practice of family doctors referring clients to me for the effects of a primarily medical matter, most commonly the experience of pain. I have no expertise in alleviating physical pain from migraines, endometriosis, fibromyalgia, broken backs, and so on. Sometimes, I have learned that the doctors behind these referrals believe in active and holistic care for their patients which I very much appreciate. At other times, the referral is a thinly veiled code for one of two beliefs: a. "the pain is imaginary" or b. "if this person had a more positive attitude, the pain wouldn't be so bad." These are the kind of garbage prejudices that women and queer people still very commonly face in their search for proper medical care that have significant binding effects on their persons, lives, and futures. These beliefs are very close cousins, or siblings, to the belief that "if she had worn different clothing, this other thing wouldn't have happened to her." Do you see the relation to victim-blaming rape culture that we sometimes get invested in in order to construct a semblance of safety from physical pain or rape? It goes like this: "Oh that's scary. But what the rest of us have to do is practice positivity, or self-defense, or proper dress, or proper vigilance when in public spaces so that the 'painful happenings' can be contained in a kind of explanation that gives us a semblance of control." Beware of this idea, this insidious Master Narrative hiding in plain sight in all manner of places, and forwarded to you often in the expressions of despair of those who are contending with death, cancer, rape, broken backs, break-and-enters, and so on.

But notwithstanding my sometimes-founded wariness about some "pain management" referrals, here she was, this woman named Kara, with "stress migraines" as her first stated reason for therapy. Although I do not believe I can do much about migraines, I asked about her experience of pain from the migraines, their frequency and intensity, and a bit later, whether she agreed that they were in some substantial rela-tionship with stress. When she agreed, I of course wanted to know what "stress" meant to Kara: "there's stress, and stress, and stress. What kind of stress is this, what stresses you out to the point of migraines in life?" I asked. In Kara's descriptions of "stress migraines," I noted that she mentioned "locking herself into the bathroom" a few times when she had migraines and asked her rather matter-of-factly if and why the bathroom was a good place, and wondered whether she maybe had baths to help with pain. To my surprise, Kara grew quiet and tearful instead of responding. Gently, I proposed out loud if maybe she lived with other people and whether there were limits on her privacy and whether that was why locking herself in a quiet place like a bathroom was a relief to her.

In response to these proposals, Kara introduced me to her Up Against Story over the next few sessions in which her partner Matt stood outside the bathroom door – the bathroom being the only room in the house with a lock, of course – pounding on the door and demanding that she come out. Kara never used the term "abuse," but that is neither here nor there as the term "abuse" is very much, shall we say, "abused" in our current time and space such that it has all but lost its meaningfulness in making discernments about what human beings routinely do to each other. So instead of bandying terms, Kara told me stories of the names she was routinely called, "lazy" and "slut," especially when Matt came home to a chore or other that was not finished. Kara's lists of chores were lengthy despite her working a full-time job of the same hours as Matt. Kara told me stories of tripping on stairs, and on the sidewalk and not being certain whether she was pushed by Matt or a "klutz." Stories of spending hours upon hours in enclosed spaces, most often the car or her own bedroom from which she could not escape, and Matt speeding the car to dangerous speeds or ripping apart bedsheets until she quickly ceded a disagreement in his favor. She told me of being the recipient of screaming rants by Matt that went on for hours during which she said nothing, looked at the floor and could not move or tried to appease him with doing as he demanded for the fear of further escalation.

In our first conversations, upon telling these stories, Kara often whispered "is that reasonable?" as her question to me. When I queried what she meant, she said, "Well, maybe I should have cleaned the closet," or the "grout in the bathroom," or been "more careful about not making mistakes on the spreadsheet" – maybe "I angered him," she said, despairingly, "maybe I could have helped him better in some way."

In response to these stories, Kara and I arrived at the posing of a dilemma: how do I live with the stress of threats and pressures that another person constantly causes for me? We worked through the pragmatic options: separating from Matt, talking to Matt, inviting

Matt to join us in our sessions to help address this dilemma that was not in Kara's hands with him. Matt, however, declined all invitations. I found myself distressed about the depth of injustice that Kara was suffering at home that seemed unethical to address with Kara by herself in therapy. I received good counsel from Loree Stout who listened to my ethical queries and said, "Well, if Matt won't come to therapy for now, then work in the meanwhile to strengthen Kara's voice." This I resolved to do: "from whisper to a strong voice then," I thought. I wrote poems for Kara about her experiences. One of these reads as follows and hopefully gives you a sense of the Up Against Work we were doing:

Can I have a witness

> *If he pretends that it didn't happen*
> *If he doesn't admit to it*
> *If he avoids apologies*
> *If he gets mad enough to shut me up*
> *If he blames me for bringing it up*
> *If he says that I imagined it*
> *If he pretends he is the victim of this stress*
> *If he switches from shouting to "what-are-we-making-for-supper"*
> *even faster*
> *If he bullies me into silence*

> *What happens then*
> *To that thing that happened?*
> *Who remembers it?*
> *Who learns from it?*
> *Where is the memory stored?*
> *And who can ever speak to it?*

> *Do the things that happened*
> *Find a way to live some place?*
> *My body*
> *My mind*
> *My imagination*
> *My idea of love?*

I live between safety and threat
Between out-of-control screaming
And what's-for-dinner-honey?
These words are my only witness

In response to Kara's experiences of threats and coercion and intimidation at home, and her reach to make discernments about these moments in the form of her question: "Is this reasonable?" I started telling the moments back to her, enacting them as she had just told them to me, and proceeded to ask her: "Is this reasonable then?" at the end of my antics. To my baffled surprise, most of the time, when I did these play-by-plays of the dialogue and the actions that she had just imparted to me, she responded by laughing out loud! Sometimes she started laughing even while I was replaying the scene for her: "here you are, in the middle of the baking isle, and he says this . . . and then you say this . . ." I remember these retellings and Kara's responses so vividly as her laughter stood in grave contrast to the sound of her early whisper, and because these were the first times I had ever heard her laugh. Kara taught me an invaluable lesson I have never since forgotten: laughter is reserved for philosophers and free spirits. Laughter is a great crack of a beginning.

Dear student, you might already see some sketches of the Storytelling questions, and the Witnessing responses practices that I promised to write to you about in the above summary. But I know you will not be satisfied until you see such practices in transcript form in order to evaluate these practices in motion inside a session. So, below is the transcript that started out much like my previous Up Against Conversations with Kara. I will draw your attention to various Storytelling questions, Witnessing responses, Good faith practices, Imaginative leads and Moral reading prompts in brackets inside the transcript excerpt and elaborate on them after this read.

Kara:	But there is something that I wanted to talk to you about . . . (*Pauses*)
Sanni:	Alright.
Kara:	Something happened. And I don't know how to even process it . . . (*Pauses*)

Sanni:	Alright, no worries. We'll figure it out.
Kara:	It's about Matt. He . . . it came out of nowhere. It had been getting better, and I . . . was hopeful. He was a lot calmer the last while. And then . . . *(Pauses; looking at the floor, losing her words)*
Sanni:	No worries, Kara. I remember your hopefulness last time, remember, we were both hopeful together, because of the set-up of the new apartment and how having two exits backed you up, and I remember how he was listening to you differently. [Witnessing response] But I take it, something changed, and some crap went down? [Storytelling question]
Kara (Looking up at me):	Yeah . . .
Sanni:	Some very anti-calm crap, what is that, like *loud* crap, or *tense* crap, wow I'm really good with my words today aren't I? [Storytelling question]
Kara (Smiling):	Yeah . . . He . . . we were supposed to clean up the office for my friend to move in, the roommate.
Sanni:	Oh, she's moving back?
Kara:	Yeah, she'll be back for six weeks to do this practicum.
Sani:	When is she getting here?
Kara:	In a week.
Sanni:	Alright, sorry, I'm asking because I know her presence in the house can change the crap too. But anyway, you were supposed to clean up the office . . . [Storytelling question]
Kara:	Yeah, we were cleaning out the office, well, *we* weren't, *I* was, because Matt wasn't helping. The office is a mess, and he's been stressed out about it for weeks, going on and on about how we have to clean it up before Allison gets here, and on Sunday I started. It's a mess, I mean the closet is full of Matt's university stuff, papers, and books, and I don't even know what, there were like Ziploc bags with pens that don't work anymore, anyway,

	and there's clothes everywhere and the desk in there is the same, so someone has to go through it all and sort it.
Sanni:	Ugh, I can imagine the job! How was it decided that it was yours to do? [Moral reading prompt]
Kara:	It wasn't. He went on and on about it, so I already made up my mind to do it on Sunday, just get it done.
Sanni:	So here you are, on some Sunday afternoon? *(She nods yes)* Sorting through Ziploc bags with defunct pens? *(Kara laughs)* Good lord, alright. [Witnessing response] So then what happened? [Storytelling question]
Kara:	Well, he came in all of a sudden and started screaming.
Sanni:	What do you mean, screaming? Like screaming words? Like at you? Or at the mess? . . . [Storytelling question]
Kara:	No, at me. I don't remember a lot of it, like that I am lazy, and that I'm too slow, and that I should have done it sooner, and that it'll never get done, and . . . mostly that I'm lazy, and stupid . . . *(Voice trailing off, almost inaudible)*
Sanni (Softly):	Oh, Kara, I'm sorry. You don't have to repeat it all, I am guessing the usual insults and swearwords at your person. [Witnessing response]
Kara (Not looking at me, distressed):	Yeah.
Sanni:	Yeah, I've heard them before, never mind. [Witnessing response] So here you are, sorting HIS damn stuff after he had been going on and on about it and stressing you the fuck out about it for weeks, and then INSTEAD of helping, he thinks he'll give you a real helpful little pep talk? *(Said with sarcasm)* [Witnessing response]
Kara (Laughs. Then answers	

in a strong voice):	Yes! He stood there in the doorway and screamed at me. He literally locked me in a room for an hour and half to rage at me.
Sanni:	He locked you in? [Storytelling question]
Kara:	Yes! He stood in the doorway; he wouldn't let me pass.
Sanni:	You tried to pass? [Storytelling question]
Kara:	Yes. I tried to leave a number of times, and go past him, and he wouldn't let me. He was raging, "No, I'm talking to you."
Sanni:	Gosh. Okay. You tried to get past him, I know, this is what we had just been celebrating, that you could get out, but now you're in this room and he's standing in the doorway, blocking you. [Witnessing response] What did you do then? [Storytelling question]
Kara:	At first, I tried to reason with him. I tried to soothe him, like we'll get it done, there's a week left. But it didn't matter, he kept screaming, it almost made it worse. At one point, I just sat down on the bed and cried . . . *(Trailing off)*
Sanni:	You tried to get out, you tried to talk to him . . . I understand, Kara! And then when nothing worked, you sat down, and the tears came. [Witnessing response] What were the tears saying . . . gosh, I hate how that question sounds, but more like, the trying to get out, the things you said, and your tears, what were they all trying to tell him? [Moral reading prompt]
Kara:	To stop!
Sanni:	Stop, Matt, just stop, you're scaring me . . .? [Moral reading prompt] Is that what . . .?
Kara:	Yeah, you're scaring me, you know how he gets, and I didn't know if he was going to do something, I mean at one point, he took the nearest box of stuff and just threw it and emptied it.

Sanni:	Threw it in the room, at you, like in anger, like with force? [Storytelling question]
Kara:	Yeah, not at me, but he was picking up boxes and emptying them, like how he threw the tomatoes . . .
Sanni:	I remember the tomatoes . . . *(Imitating a throwing smashing motion)* Was that when you tried to leave? [Storytelling question]
Kara:	No, I was just sitting there frozen, and I said, "Stop, Matt, stop."
Sanni:	And then you were crying, on the bed. [Witnessing response] Did he the fuck stop? [Storytelling question]
Kara:	Not for an hour and a half. At one point he screamed for me to stop crying because the neighbors might hear.
Sanni:	Fuck. So that's what we're worrying about now. Not that you're scaring your wife, and hurting her, and she's going: "Oh my god, oh my god, where's this going and I can't get out," but we're worrying that the fucking NEIGHBORS MIGHT HEAR? [Moral reading prompt]
Kara:	Yeah. At that point, I got angry.
Sanni:	Did you? [Storytelling question]
Kara:	Yes, I got angry and said I would call the police if he didn't stop.
Sanni:	Did you? Wow, Kara. [Witnessing response] Did you get up or did you say this sitting down? [Storytelling question]
Kara:	I got up, and when he didn't let me pass, I said that to him. And then he turned around and left.
Sanni:	He left the house? [Storytelling question]
Kara:	Yeah, he went to the gym, I could hear him grab his stuff and slam the door.
Sanni:	And you? When you heard the door slam and were left to your own devices, what came to you, what happened? [Storytelling question]
Kara:	At first, when I was sure he had left, I was sad. I cried again. But then I got angry.

Sanni:	Help me, Kara, help me with the anger, I want to understand how it came to you, to stand by your side as you were now alone in the house, how anger formed and gave you counsel. Can you tell me anything about the counsel of anger in that moment? What did it say to you about what had just happened? [Moral reading prompt]
Kara:	It said that . . . he crossed a line.
Sanni:	Matt crossed a line. An important line, a line beyond which there's anger to come to your aid, is it . . . does the anger tell me that it's an inviolable line then? [Moral reading prompt]
Kara:	Yes! It was an important line . . . I have never been so angry before.
Sanni:	This is really important to me, Kara . . . I want to understand this moment in your life . . . if the anger had words, what would it have said to you in that moment? [Moral reading prompt]
Kara:	It said . . . how dare he.

Sanni (Barely audibly, moved, repeating "how dare he")

Kara:	And you know, when he came back from the gym, he was all calm and happy, and you know what he said to me, he said "I'm sorry that happened." And I was so angry when he said that, and he noticed it, and said I needed to relax.
Sanni (Snorting derisively):	I'm sorry that happened, just relax, sweetheart? [Witnessing response]
Kara:	Yes! That's laughable! How about "I'm sorry I did this you." But even that is flat. How about: I am sorry I did this to you . . . on purpose! I'm sorry I chose to continue even when you were crying. I understand stress and I understand anxiety. Neither of those things locks me in a room and screams at me for an hour and a half straight. That is not something that happens.

Sanni:	Wow, Kara, I'm scrambling to write that down. Neither of those things . . . locks me in . . . that is not something that happens . . . [Witnessing response] Kara, the reason I want to know everything about this moment in your life is . . . Do you remember how you at first always questioned yourself in these moments, and then brought the question to me, afterwards, and the question was "Is this reasonable?" Remember how when we first met, you would tell me something that had happened between you two, and then your question was "Is this reasonable?" (*Kara nods*) How come it's different now? [Moral reading prompt] How come you are not asking me or yourself that question? [Moral reading prompt] You went from "is this reasonable?" to "how dare he," and "I'll call the police," and this act of yours, to call it out like that, tells me about some shift, some insurrection that has quietly taken place here. [Moral reading prompt] Am I wrong to think of it this way? Have you undergone an insurrection, all quietly, and now it's suddenly become visible, on an ordinary Sunday with a Ziploc bag of defunct pens in your hand? [Moral reading prompt]
Kara:	Yeah. Even he can sense that something has changed.
Sanni:	What do you mean? [Moral reading prompt]
Kara:	Well, he has been following me on my heels for days, he's insisting and pushing and asking, like, what are you thinking, what are you thinking? I am not answering him, because honestly, I don't know yet. I am just furious, and I need to think it through before answering him.
Sanni:	Here it is again, "fury," and you called it "anger" before. You know, you said the fury came to you before he left the house already. Is there something important to understand about the moment the

fury stepped to your side, because that had never happened before? You said, am I wrong, you said, that you first got angry when he yelled at you to stop crying because the neighbors might hear. [Witnessing response] Is that important? [Moral reading prompt] Why did fury come to you then? Is there a line here that he crossed that is important to understand? [Moral reading prompt]

Kara: Yeah . . . I think, in that moment I saw that he won't protect me, that he doesn't care about me. That he doesn't care, even if I cry. And it made me think of all the things I have done to protect him. And something just broke inside of me . . .

Sanni: Something broke . . . maybe something broke free, can I say that, your binds of protection of him, and you cried in that breaking. And I know you rarely cry, Kara. [Witnessing response]

Kara: It was the first time I did.

Sanni: Is it okay if I try to understand that a bit more fully, Kara? *(Kara nods)* You called his actions already in the moment. You were clear already then. The fury is part of that, and your tears. You were no longer wondering, is this unreasonable, or did I do something wrong that might explain this, you were clear in calling it. [Witnessing response] Is it right for me to put it that way? That on this Sunday, moral clarity was there, strong and alive? [Moral reading prompt]

Kara: Yeah. *(Tearful)*

Sanni
(Pausing a while): If you had to call his actions on Sunday something, if you had to rate . . . them . . . judge them on a scale of sorts, and on the lowest rung of that scale is the word "unreasonable" . . . and on the highest rung – what would be on the highest rung on a scale of moral crimes that a person might commit? Would that be a crime . . . a crime, what

	would it be a crime against . . . [Moral reading prompt]
Kara:	A crime against human dignity.
Sanni:	Phew! Thank you. Alright! [Witnessing response] The scale starts out with "unreasonable" on the bottom and "a crime against human dignity" on the highest rung. What are some middle rungs then? To judge an action by . . . Like "unfair" might show up somewhere . . . [Moral reading prompt]
Kara:	Unloving.
Sanni:	Unloving. I like that. What else? [Moral reading prompt]
Kara:	Unjust.
Sanni:	Unjust. *(Pausing to think)* Then: if you had to judge Matt's actions on Sunday, of locking you in for an hour and half, and proceeding to scare you by throwing things, and screaming at you, screaming accusations at you, according to your moral deliberation, what kind of crime was this, Kara? It was unreasonable, yes, but beyond that, what would you call it? [Moral reading prompt]
Kara:	It was a crime against . . . my dignity.
Sanni (Moved, softly):	And you knew that then, didn't you? [Witnessing response]
Kara:	Yeah.
Sanni:	Now, Kara, what would a person have to have done to deserve an assault on their dignity? What are the circumstances in which you would say a person's dignity is now forfeit? [Moral reading prompt]
Kara:	I don't understand.
Sanni:	Eek, right. Maybe I'm getting ahead. Let's see . . . maybe this doesn't interest you. But you know, on the way here, when I was driving, I was listening to CBC and they were discussing a handover of prisoners from a Canadian ship to a U.S. ship and

they were debating whether or not the Canadians had the moral right to hand them over because they couldn't be assured how they would be treated on a U.S. ship . . . [Imaginative lead]

Kara: I heard the same thing.

Sanni: Did you? Alright! Well, the thing I am trying to so poorly word is that the prisoners were all drug dealers, like heroine dealers, so they had indeed committed crimes. Heroine dealing is not kid's play. But still they were discussing the fair treatment of prisoners according to the Geneva convention, like how can the Canadians be assured of the ethical treatment of these prisoners going forward, that they won't be tortured or sleep deprived and will be given adequate sleeping quarters and medicine and food and all that. [Imaginative lead] Now what crime would a person have to commit to not fall under the Geneva convention? [Moral reading prompt] I mean, is there a crime Kara that you can imagine that would warrant erratic screaming and threatening with an uncertain outcome? [Moral reading prompt] What crime would a person have to commit in order for you to sentence them to your conditions on Sunday? [Moral reading prompt]

Kara (Long
silence): There is no such crime.

I hope you can see the interweaving of practices to help shape the story of a Sunday afternoon in Kara's life, and the subsequent queries to invite her to a moral reading of the events mentioned. Kara is invited to give meaning to her tears, to her anger, her fury, the "breaking of something inside her" – all of which are embodied effects of Matt's treatment of her (in contrast to these effects being thought of as "problems"). Kara is invited to give meaning to these effects:

- ◆ What were the tears saying . . . gosh, I hate how that question sounds, but more like, the trying to get out, the things you said, and your tears, what were they all trying to tell him?
- ◆ Can you tell me anything about the counsel of anger in that moment? What did it say to you about what had just happened?
- ◆ Does the anger tell me that it's an inviolable line then?
- ◆ I want to understand this moment in your life . . . if the anger had words, what would it have said to you in that moment?
- ◆ Remember how when we first met, you would tell me something that had happened between you two, and then your question was "Is this reasonable?" How come it's different now? How come you are not asking me or yourself that question?
- ◆ Then: if you had to judge Matt's actions on Sunday, of locking you in for an hour and half, and proceeding to scare you by throwing things, and screaming at you, screaming accusations at you, according to your moral deliberation, what kind of crime was this, Kara? It was unreasonable, yes, but beyond that, what would you call it?

Perhaps you are wondering why I initiated the conversation about the radio program and introduced the metaphor about the treatment of prisoners on ships in this session. The aim of this introduction is to help Kara with the task of the moral reading of the events of her Sunday. It is but an imaginative lead that is meant to help lift both of us above her Sunday afternoon for but a moment and then allow us to return to her Sunday with renewed clarity. Why is this important? What is the work of this question and why is it so weighty as to even endow it with a metaphor of a ship and the rules of engagement for prisoner exchange?

Remember that over the course of all the preceding many months, Kara's recurring question to me in our conversation had been: "is this reasonable?" Embedded in the cadence of this question lies the realm of another both metaphorical and real realm of patriarchal assumptions about "women's part" in the abuse that their partners are choosing to enact against them. These assumptions are ripe with the popular Master Narratives of our time in which within heterosexual relationships women partners are assumed to be obligated to remain responsible for the emotional labor and tenor of relationships, and men partners are entitled to this care without corresponding responsibilities to consider the effects of their actions. In her question, Kara is searching for the "reasonable reason" for Matt's treatment of her, which invariably can be traced back to her: "Perhaps I didn't clean up fast enough?" and so on.

In this conversation, the imaginative lead of the metaphor of the prisoner exchange came to our assistance to help Kara to account for and respond to Matt's actions on an ordinary Sunday afternoon in an extraordinary moment of allowing Kara's moral reading of events. "There is no such crime . . . (a person could commit that I would sentence them to my conditions of erratic screaming and threatening with an uncertain outcome)" is a speech act in stark opposition to "is this reasonable?" and of grave moral consequence for Kara's life, as it turns out.

Why? What happened? I did not know this at the time, but Kara went home that night and informed Matt that she could take no more of his, in her own words, "scaring" of her, and that she had informed her parents and his parents of what was going on, and that she proposed the possibility of his attending therapy to address these difficulties and make a lasting change in their relationship. To my great surprise, Matt agreed to therapy, contacted Tom, and negotiated consent for therapy in which Kara was the supervisor to Matt's and Tom's efforts to account for the effects of his actions and make a radical change.

I said once before that I don't believe in participating in noxious "they lived happily ever after" fables as accounts of therapy, but below is part of poem that was written for both Kara and Matt after a joint couple's session in their future after a summer that might have been a nightmare, but was not.

This poem is a reflection of both Matt's and Kara's joint descriptions of their summer which involved travel to various places for work. This summer of travel, including packing and preparations and planning that it involved would have previously been experienced by both of them as "nightmarish" for the anxieties that it would have invited Matt into and his practices of anxiety that he gave himself permission to unleash upon Kara. The summer of travel then became a fitting realm of Matt's Counter initiatives that were storied in this session we had one fine September day after the summer. In the poem, you will see "what the summer could have been" – you will see reflections of Matt's struggle with anxiety, phrased as "angst" in front of a "taskmaster" which had been whipping him forward to try to control the vicissitudes of change and chaos. It turns out that Matt's "to-do-list" in life, let alone a moment of travel was literally, not metaphorically, 4000 items long, and that he spent every moment of his life anxious and guilty about his failures to meet these tasks as a sign of his failure of proper personhood. You will also see that Matt had been invited into a "moral reading" of the relational effects of tormenting Kara in his fits of anxiety by way of control and screaming and forceful coercion to have her "help" him stay the chaos of life in obedience with the taskmaster's demands. In the end, Matt reflected that previous therapies that were aimed to help him with his "compulsions" had not brought him relief beyond a few techniques, but that it was his understanding of the effects of his actions on Kara whom he loved, that helped him discern and take a position regarding his relationship to "anxiety." This, Matt phrased in his own words as "a sacred step back." Kara affirmed the observations of Matt's change, both in writing and in joint meetings. Tenderly, she added at the end of one of these sessions, "when we go home tonight, we'll delete the list of 4000 items off your computer." When he had a shadow of panic cross his face in response, she said "THIS I'll help you with, don't worry." He remarked, "If it's for me, then I don't know. If it's for you, it's a different story." So it is.

Perhaps the poem will stand as a good enough ending here, about the hope of the effects of triggering the conditions for moral readings to occur. In Kara's and Matt's words, here is what happened next:

For Kara and Matt
The Summer of Love

This is the summer of
gripping angst on a wheel
a summer of endless packing and preparations
a summer of panic and attacks on Kara
a summer of bellowed rule
by the Lord Commander of us all
the almighty taskmaster
in front of whom we are but mice
on a wheel.
-

Or
-

This is the summer of travel
the summer when the world opens
to our dominion:
the tacos in Tijuana
the pool in San Diego
the theatre in Montreal
and the little restaurant in New York.
This is the summer of my remembrance of my
hopes for Kara
when we first met up:
her crossleggedness on the beach
the contentedness of her soul
the curiosity of her eyes in a good story.

Kara my love
this is the summer of my sacred step back:
I shall build you a shelter
May it span both our bodies
May it keep us
from the disaster of the taco truck that moved before our dinner
Or from the rain in New York
That wrecks all well-made plans.
Under this shelter

and in the New York rain
I would whisper a toast to you Kara
it's okay my love we'll go with the rain if it pleases you
because this shelter by my hands
will not just stop the wheel
but break it
all for the light in your eyes shimmering in the New York rain.

So, dear student, we tell stories in narrative therapy, not to label or validate persons or to reframe problems. Story-shaping in therapy is a most powerful therapeutic technique, because good stories, unlike advice, strategies, praise, evaluations, judgments, and rules, tell people what to do. In more surprising ways than you or I could ever imagine. Hannah Arendt writes: "the manifestation of the wind of thought is not knowledge; it is the ability to tell right from wrong, beautiful from ugly. And this at rare moments when the stakes are on the table, may indeed prevent catastrophies."[3]

Love,
Sanni

Notes

1   Mattingly, C. (1998). *Healing Dramas Clinical Plots: The Narrative Structure of Experience*. Cambridge University Press, p. 29.
2   White, M. (1995). *Re-authoring Lives: Interviews and Essays*. Dulwich Centre Publication, p. 84.
3   Bernstein, R. (1978). "Thinking on Thought." *The New York Times*, www.nytimes.com/1978/05/28/archives/thinking-on-thought-thinking.html

10

What Now?

A Letter About Writing
as an Antidote to Despair

Dear student,

You may have wondered about the poems attached to the preceding letters throughout this book or you may have ignored
them as a bit of decorative noise on the margins. Either way, this
letter is meant as an encouragement to you. I'm here to urge you:
write. Write to your clients.

I could tell you to write to clients, because it's a sure tell-tale
sign of a narrative therapist, invoking the tradition argument
of "because that's how it's always been done." It's true that
narrative therapists in particular are known for therapeutic
document writing as a unique practice. David Epston, a master
of this practice, once said in a presentation on therapeutic letter
writing, that one of the reasons to write is "to make amends"
for the missed understandings in a particular session.[1] Isn't that
something.

I could tell you to write to clients if you're interested in
being a renegade to the "ways it's always been done". Writing
letters to clients subverts the usual power dynamics in therapy
by opposing the dominant practice of "reporting on persons"

DOI: 10.4324/9781003478478-10

via secret session notes and turning reports into relational documents for your clients' deliberation instead. It's true that narrative therapists in particular are known for their interest in the creative use of power in the therapy room. In a presentation, Navid Zamani laughed that most narrative therapists he meets seem to "operate somewhere on the continuum of civil disobedience."[2] Now that's a claim.

I could also tell you to write to clients because clients like the documents. Insert all manner of expressions for "like" such as "feel so heard by," "look forward to," "appreciate," "benefit from," and so on. A client once put it this way: "you just broke through 364 days of loneliness." An often-quoted survey of the perceptions of the value of narrative letters by clients, the average worth of a letter was equal to "more than three individual sessions."[3] So suffice it to say, clients LIKE letters.

These are all true reasons, but there is another. Write to your clients for you. The practice of writing therapeutic documents will supervise you into becoming a narrative therapist faster than anything I or anyone else can do for you.

When I started working as a narrative therapist, I asked my boss for permission to write letters to clients as my session notes and I remain grateful for her indulgence of me. At the time, all I knew is I would benefit from thinking through my sessions on paper and organizing my thoughts into a written document for the person whom they concerned. It seemed like a tremendous waste of time to write other session notes that wouldn't serve the next conversation directly.

I had no idea what would happen to my practice as a result. See, I sometimes sit with therapists who tell me "my clients ramble on and on in a confusing fashion" or "sometimes I'm unsure why this person is coming to therapy" or "they tell me of the latest bit of life happening and I don't know what to do" or "my clients pressure me to fix them or give them advice and strategies". Therapists ask for my advice for how I deal with such dilemmas and sometimes I'm frank and tell them what I'm telling you now: write to your clients.

In *Re-authoring Lives*, Michael White wrote:

> If we acknowledge that it is the stories that have been negotiated about our lives that make up or shape or constitute our lives, and if in therapy we collaborate with persons in the further negotiation or renegotiation of the stories of persons' lives, then we really are in a position of having to face and to accept, more than ever, a responsibility for the real effects of our interactions on the lives of others.[4]

For me, writing to clients is a visible initiative in the endeavor to step into our collective responsibility with a willingness to try. I do not mean to offend you, but I hope you take great offense anyway: there is a seductive spirit of settling into easy privileges afoot that might beckon you "not try harder than a mediocre white man." "Therapy is just about sitting with people," or "I shouldn't work harder than my client," I hear far too often. But if we're going to practice narrative therapy in a story-impoverished, quick-fix saturated world, we need convincing and effective means to invite clients outside of the domineering cultural scripts of "what to expect when you go to therapy." Reading a letter or poem to the client at the beginning of each session that cites them and centers their lived experience is among the very best means to this counter-invitation.

It shouldn't be surprising that clients meander and struggle in the effort to describe their lives, that they are more familiar and comfortable with the use of short-hand of labels and identity conclusions, that they seek for the absolution of validation, and ask for expert solutions and strategies at the end of sessions. All of this is in keeping with the current dominant cultural Master Narratives that set clients' expectations for the task of therapy as well as both of your roles in it. So: surprise them. Make a confident narrative counter-offer. I have never met a client yet who wasn't immediately interested in a surprise description of themselves as a protagonist in a vivid moment of a complex and nuanced moral dilemma, set in motion in their own vernacular. A letter or poem is the easiest way to help you issue this counter-offer.

One more true tale for your consideration: in the absence of a narrative document, what is the most common therapy session opener across practices of all persuasions? "And what would you like to talk about today?" Welcome to the opening lines of most clinical practices.

Therapists sometimes try to sell me on this question as an example of "collaborative practice" and perhaps it is so. But more often than not, this session opener will lead you and your client straight into the above-mentioned booby traps of the usual confusions and little frustrations of therapy practice.

See, I would rather see a therapist announcing, brightly or shyly, "I wrote something for you from your words and my thoughts about our last conversation, would you like to hear it?" at the outset of sessions. Imagine what that would be like. But it doesn't stop there. I would further see a therapist read a note that pulls the story threads from the previous sessions and commits to the client as an interesting protagonist wrestling with wicked dilemmas and on the cusp of making an extraordinary move on the ground of their ordinary life, all presented in the client's own imagery and vernacular. Imagine what that would be like.

But it doesn't stop there. As a last step, I would like to imagine a therapist making a proposal to the client about a question or topic of conversation for today's session that is in direct relation to and a satisfyingly logical extension of the questions and topics they've held together thus far.

Can you imagine that? Most of my session beginnings end in a question like: "does my proposal interest you, is it fair, based on last time? Or perhaps there is something important that has come to you since we spoke that you want to bring in to enrich our understanding so far?"

In my work, I have come to understand session openings as a crucial, urgent, and important site of negotiating the topic and tone of the whole next hour in a collaborative fashion. I've learned that the work of resisting all manner of booby traps happens at *session beginnings*, in particular. In short, I believe that it is the thoughtfulness of the beginning of sessions that is one of the things in therapy work that makes the crucial *difference that makes a difference.*[5] A letter or poem will help you do this work.

And if I haven't convinced you yet, then write, for the reason because it will make you so happy. "There is a pleasure in the pathless woods." A sociologist at the University of Moscow, Grigori Yudin, writes: "I always remember something Putin said in mid-2021. He said, completely unprovoked, that there's no happiness in life . . . It's as if he says 'the world is a bad, unjust, difficult place, where the only way to exist is to struggle constantly, to fight, and, at the outer limit, kill.'"[6] The practice of studying your clients' descriptions of daily life because "apparently I have to write a note based on this now" will surprise you: it will tell you much and more about what human beings routinely do to each other: you will start to hear threats, lies, ridicule, wheedling, demands, cajoling, taunting, stonewalling, and all manner of rich human conflict and its technicolor effects on our lives. But, if you look and listen (read your notes!) you will also start to hear decency, mercy, kindness, pride, quietness, softness, compassion, humor, tenderness, sweetness, love, and liking.

So my dear: write to your clients, write after every session.

Take your clients' words from session notes (and write better session notes if you're not already!) and turn them into letters, post-it notes, postcards, poems, certificates, interviews, role-plays, stories, spoken word summaries, cartoon sketches.

If you don't have the time, write in the time you do have. Session notes are a legal requirement, so use that time.

Even if you are a terrible writer, write.

If you don't know where to begin, take out blank paper and start lifting the words of your client from your session note onto it. You will be amazed at what you have, I swear.

And if it still seems risky and dangerous – it is no more so than the next session will already be. Only this time, you'll be one letter less foolish than you were yesterday. You're welcome.

I once asked the therapists at the Collective why they wrote to clients. Below is a summary representation of their thoughts.

Notes

1 "I considered that the exercise of writing a letter of a session afforded me the opportunity to reflexively appreciate it, or in some cases, when I found it difficult to write a letter, I realized I needed to make amends for a meeting that didn't go anywhere and the letter allowed me to do so. In some ways, writing letters was for me the best sort of supervision, not only as I mentioned above but as well being able with the client's feedback to evaluate my own practice." Epston, D. (2018). "Friendly introduction to this special release on letterwriting." *Journal of Narrative Family Therapy, Special Release*, pp 1–3, 1.

2 Zamani, N. (2022). *Couples Therapy and Narrative Ethics.* Narrative Live Workshop Series, San Diego, CA.

3 Nylund, D. & Thomas, J. (1994). "The economics of narrative." *Psychotherapy Networker, November/December*, pp. 38–39.

4 White, M. (1995) *Re-authoring Lives: Interviews and Essays*. Dulwich Centre Publication, p. 15.

5 "Bateson maintains that information, or rather, the elementary unit of information is the difference that makes a difference." Bateson, G. (1972) *Steps to an Ecology of Mind.* University of Chicago Press, p. 459.

6 Luitova, M. (2023, February 24). "'Russia ends nowhere', they say. Sociologist Grigory Yuduin discussed a year of war and what comes next." Abridged translation by E. Laskin. *Support*, Meduza. https://meduza.io/en/feature/2023/02/25/russia-ends-nowhere-they-say

7 Paljakka, S., Green, S., Luhtanen, T., Morton, C., Saxton, T., Szlavik, L., & Vincent, C. (2021). "On the pedagogy of poetics." *Journal of Contemporary Narrative Therapy, August Release*, pp. 32–54, 34–37.

A Response to the First Time Sanni Assigned an Easy Task: Why Did You Start Writing Poems for Clients?

As I sat there, listening
I had been shriveling up like a rose bush in dry soil
In the barren landscape of counsellor training:
The parrots were squawking:
Microskills, lists of good questions, SMART goals
Was THIS what I had put my mind to doing?

The spirit of the poems grabbed me:
This was real talk defending clients' honor
And there were swear words, metaphors and rich details
The poems were alive
Subversive, moving, funny, and human
And I was enlivened:
Alert, vibrating, writing notes vigorously
It opened a door to another world.

It was the poems that set me ablaze
They were my way into narrative therapy
And I don't understand how therapists who don't do poems do it
Because how do you do narrative therapy
Without this: People's. Words. Matter.

You have to understand:
I too, have the whole: "I-Am-Not-a-Writer
And-I-Certainly-Have-Never-Tried-To-Write-Anything-Since-A-
Grade-7-Haiku" -thing
Going on,
I still don't think of myself as a poet,
I would never win a poetry writing contest.
But I listen for the poetry in a person's story without trying
And how do you explain THAT?

I am not a preschooler,
And I believe that my mind has the ability to create

Something unique
I do not need a "fill in the blanks" form.

So I hit the ground running and never looked back.
And here's what I learned:

Lesson number 1:
Writing poems catches people by surprise!
I saw them weep, laugh, and beam with pride.
They print them, post them, share them, bind them, re-write them
in calligraphy
They wait with bated breath for the next one
And I can't wait to read them to him or her or them
"Yes, read it now!" People say,
"Aahhh, I have a poem written just for me."
"You just broke through 364 days of loneliness."

And if I had a nickel for every time a person says
"I feel so heard"
I wouldn't have to work anymore
But I would choose to do THIS.
And that's not all.

Lesson Number 2:
It takes a lot of attention
To put together a piece of writing
In a way that has the potential to be healing.
If I don't pay attention
People can feel like shit.
People's responses to the poems
Are the ultimate supervisor of my work:
They require me to puzzle on clients' dilemmas
Sweat with power dynamics
Choose poignant moments

The poems are a gauge
And they keep me on my toes:
They show me

What I did in session,
And if I struggle writing a poem,
It tells me it was a shit session:
Creativity arises most readily from creativity.

The form of the poem invites it to be dynamic
and changeable
unlike a letter that is so paragraphy:
Prose cannot capture a love that is at risk of a rebel coup,
or the discovery of childlike gentleness as an "anti-anxiety"
strategy
Or what a pill bottle by the bed spells:
"and when I see it
by the bedside
it's a pill container full of trust"
Would we have the same thing to work with
If I had read a dreary letter saying
and then you did this, and then you did this . . .?
A poem does not bind you to the conventions of regular speak.
You can convey big ideas and big feelings in so few words.
You can be weird and interesting.
Perhaps the thing I love most about poetry is the tonality:
In a way, you have the ability to infect someone with a vibe,
like a give-no-shits vibe,
or a this shit is a s-trrrr-r-u-g-g-l-e vibe,
or a frantic vibe,
or a calm vibe,
or an 'I got this!' vibe,
or a deep with feeling vibe . . .

And if people can be moved by their own stories
Laugh at their oppressors
Feel empowered
And catch a glimpse like this:
"if that person is the person I am in this world
then goddamn why do I hate myself so much sometimes?
I loved the person in the poem, she was jovial yet deep.

She is handling and coping with the situation beautifully with grace
and dignity."

Then
I say,
Go ahead
Don't trust me
Write a poem instead,
Listen carefully to their response.
And
Then wait for what it is that will come next

Shh:
If it's neither
Unicorns
Nor
Vicarious Trauma
*Would you be surprised?*⁷

Love,
Sanni

Epilogue

Dear student,

I now have the unenviable task to summarize this whole she-bang to you and to say goodbye.

You can't hear it but I'm muttering "rats" under my breath. My talents at soundbites and goodbyes are the worst.

But okay, it did once happen at a workshop that I took a marker and sketched out the important conceptual pieces of this book on a whiteboard. It looked something like this:

From
("what's keeping you up at night, dear?")

Unstory

To
("what happened, dear?")

Up Against Story (X Makes Me Feel Y Story)

Over

Attention to

Master Narratives
("wait, why?")

And

The Agentive Turn
("what do you want, dear?")

To

Counter Story (Given X and Y, I did, am doing,
will do Z on Purpose)
("hell yes!")

That's it. When I saw this drawing on the whiteboard, I muttered "ta-daa" under my breath, and when people in the first row caught me at a rare moment of un-Finn-like personal celebration of sweaty endeavors and laughed at me, I added "that only took me a few years." Most of the lovely people at the workshop then celebrated with me, save for one dude who wanted to punch me in the face over the break because I hadn't satisfied his wish for tips and tricks and also, because I may have said the word "penis" jokingly during the course of the workshop, and let's face it, that's just not okay.

But that penis joke reminds me to tell you that a man (it's very disrespectful to call him a penis, I know, I was convincingly yelled at about this!) was consulted a lot during the writing of these letters. His name is Tom, and he is rightfully my co-author even though I turned around and wrote the letters in my voice and my hand. Tom read every draft of every word I wrote and laughed at all my jokes. That's the most important thing. The second-most important thing is he listened to a lifetime of practice stories from me, and sorted wheat from chaff in my talk and helped me choose my transcript examples and case stories to clarify concepts. Third, he asked me questions like "who is the person in narrative therapy, Sanni?" and "well, who is the person *not*?" or "where are we aiming in narrative therapy, Sanni?" and we talked far into the night. He taught me all manner of things about the narrative turn and why he was excited about the agentive turn and then he interrupted me to read from all sorts of narrative texts to me while I was writing. So forget the joke, what is really disrespectful is to say "that only took *me* a few years." But he's alright, he's named as my co-author, so don't worry about him.

Okay, back to the regularly scheduled programming after that little infomercial.

I know that the above template does not soothe all your anxieties, dear student. It does not provide you with tips and tricks, this is true – it does something that I believe is a better beginning: it insists on a re-orientation of your attention from a self-conscious focus of "what should I do?" to "what is the client saying?"

I did that on purpose. I believe that this relational re-orientation of your attention is a healing effort whenever an anxious self-focus is threatening to drown you in tears or despair or the wanton wish that some authority figure would swoop in and karaoke you out of the session. I believe that in such a moment, a relational re-orientation toward your client is a labor of love. A Russian philosopher and literary critic by the name of Mikhail Bakhtin once wrote this definition of love that is still the best one I have heard: "Love is the concentrated focus of attention that enriches the beloved over time."[1]

Please don't get me wrong. The wordsmithing competition of "what would be the most charming or intelligent way to ask a question about this?" is a worthy ambition and I engage in it with all my students and colleagues until the cows come home. However, the first step is always the question of "where in the story are we?" In short, this way we can avoid all individualist and neoliberal incitements toward measuring your work by wearing a "girl boss T-shirt" to your sessions and calling it in otherwise.

When my students hand in their weekly transcripts for supervision, I ask them to write me an introduction of their case conceptualization on the basis of the following questions, which are basically deepening inquiries into the main question of "where in the story are we?"

Why is this person in therapy?

What is their restless question of living?

What is happening to this person right now on the ground of their life?

What is the political, cultural, and relational context of this happening?

What are some of the Master Narratives that are whipping them forward or keeping them stuck?

What does this person long for?

What is this person most afraid of in this longing?

What are this person's surprise initiatives of living that directly respond back to their dilemma?

I wrote these out as a scare-crow for you to consider before you rush in to apply for a practicum at the Collective!

Joking aside, these questions aren't easy, and require thinking and attention and pouring over words. And still, the effort of this attention, on your clients' words, will reward you differently than the effort otherwise poured into your own sense of self-worth. Don't feed the dragon, dear, they will bite your arm off and set your mattress on fire when you're asleep.

Okay, now I have accompanied you as far as I can to prepare you for your beginning. And now, we must part as the world is waiting for you.

I tried to find some fitting words for such an occasion and here they are. They come to you from the Finnish translation of Dumas' *The Count of Monte Cristo* that my kiddos and I listened to in audiobook format on car-rides. You must forgive me the odd choice – Finnish is a weird language spoken by a tiny country of introverted people and that's why there's only a minuscule selection of sketchy books in Finnish to choose from in audiobook format. Here's what I heard in Finnish – I stopped the car to take notes even:

> I can teach you what I know in 2 years, but I can't teach you to use it. That is a matter of philosophy. Knowledge is not the same as wisdom. The world is full of learned people. But wisdom is another matter.[2]

I scribbled down the quote a little wrong as it turns out, but I thought it beautiful to not console you out of the adventure that is about to begin: the next years of putting mere knowledge to practice to forge it into wisdom.

"But wait, Sanni. You can't just say to turn my attention to clients' stories. I don't want to do any navel-gazing, but surely I must look at my words as well! What if I harm clients on the way?" you ask me wide-eyed. Well, you'll do this at times. I don't believe that you'll shout at people or call them names or judge them out loud as most therapists are too polite for any of that. But you'll do your harm in more insidious ways by setting all kinds of bad ideas in motion – bad ideas that carry your judgments and your frustration with clients into the room and to your client under your breath. Do you want to know what is truly awful about that? Only a very small minority of clients will protest out loud when you forge ahead to make bad ideas set in like concrete over their lives and persons. So, we must invoke some additional measure of harm to follow our bids of conscience toward a better idea. If a session leaves your conscience churning about your own words, I encourage you to consider the following questions to help you in your moments of felt failure:

◆ What am I doing?
◆ Why am I doing it?
◆ What could I do that's better?
◆ Why would that be better?[3]

Can you bear with me for one more secret to ponder that might make failure an interesting site of inquiry for you? One of the things I have learned from accompanying my students in their work is that you will not suck by accident. Your kind of sucking will bear some relation to your virtues and talents – which makes a "study of suck" interesting, at least to me. You will suck in your very own unique and spare and strange way. For example, if you are talented at the virtue of boldness, you will be tempted to foreclose healing stories with practices of advice-giving; if you are talented at the virtue of gentleness, you will be tempted to forego prompting healing stories in favor of passive listening in your sessions; if you are talented at the virtue of hopefulness, you will be tempted to shun interesting stories with practices of convincing your clients out of their dilemmas; if you are talented at the virtue of compassion, you will be tempted to forget about agentive stories in favor of reflecting anxious feelings with your clients until they have a panic attack at minute 40 and so on.

My most recent student and I had this conversation just last week: she talked to me about getting into a compulsive habit of making lengthy witnessing summaries to her clients inside her sessions ("what am I doing?"). She had felt a tug of her conscience in observing the real effects of this practice on her clients, which were boredom and confusion and a stalling of the energy of the exchanges. When I asked her why she was doing it, she thought a while and then proposed that it was an expression of her commitment to making clients feel "heard" in her sessions ("why am I doing it?") Makes sense, right?

We experimented a very short while with the practical question of other possible ways to satisfy the value of clients' feeling "heard" ("what could I do that's better?"). We talked through moments of when she felt "heard" in her life, and whether lengthy summaries of repeating every word she had said would best achieve this for her. She laughed at the imagination of the effect of having every detail reflected back to her in a conversation with a friend or therapist – "ew . . . that would make me crazy!" ("why would that be better?"). That's as far as we got. In her following transcripts, I saw with great interest that she has very resolutely resolved to forego her lengthy summaries in favor of apparently practicing asking a question in the clients' vernacular about the "heart of the matter."[4]

See, failures are not dangerous to your health. They are interesting! Let your moral restlessness advise you into a lifetime of the most ambitious and raucous competition there is: to fail a little better next time. I'll meet you there.[5]

> *And, yes, poetry must be thanked too.*
> *The written word can be a faithful witness*
> *if you're willing to show yourself.*
> *Moreover, poetry reunited me with the girl*
> *who didn't mind the endless backwoods tree line*
> *and was thrilled by the sound of coyotes screaming at night.*
> *Someday I'll write about her.*[6]

Love,
Sanni

Notes

1   Mikhail Bakhtin: "what marks a true love experience is nothing possessive or erotic . . . Rather, love is an urgent curiosity, almost more cognitive than emotional, it is an intensification and concentration of attention that enriches the beloved over time with

a high quantity of individuated responses." Emerson, C. (1996) "Keeping the self intact during the culture wars; a centennial essay for Mikhail Bakhtin." *New Literary History*, *27*(1) pp. 107–126, 113.

2   Edmond Dantes: "'Now, it will scarcely require two years for me to communicate to you the stock of learning I possess.'

'Two years!' exclaimed Dantes; 'do you really believe I can acquire all these things in so short a time?'

'Not their application, certainly, but their principles you may; to learn is not to know; there are the learners and the learned. Memory makes the one, philosophy the other.'

'But cannot one learn philosophy?'

'Philosophy cannot be taught; it is the application of the sciences to truth.'" Dumas, A. (1844). *The Count of Monte Cristo*. Penguin Books, p. 185.

3   "This question – how can we live a more ethical life? – has plagued people for thousands of years, but it's never been tougher to answer than it is now, thanks to challenges great and small that flood our day-to-day lives and threaten to overwhelm us with impossible decisions and complicated results that have unintended consequences. Plus, being anything close to an 'ethical person' requires daily thought and introspection and hard work; we have to think about how we can be good not, you know, once a month, but literally all the time. To make it a little less overwhelming, this book hopes to boil down the whole confusing morass into four simple questions that we can ask ourselves whenever we encounter any ethical dilemma, great or small:

What are we doing?

Why are we doing it?

Is there something we could do that's better?

Why is it better?

That's moral philosophy and ethics in a nutshell – the search for answers to those four questions." Schur, M. (2022). *How To Be Perfect: The Correct Answer to Every Moral Question*. Simon & Schuster, p. 4.

4   "How can we assist people to name their experience? How do we ask questions in such a way that words come alive for people? How can we ask questions in such a way that people make such vocabularies of experience their own? How do we allow people to

decide what words resonate for them? What words are capturing of experience and, in particular, that experience that has not had words before? Experience that has not been rendered in to an event before? What do we do in therapy talk that generates the new rather than merely reiterating the old? Shouldn't we take an interest in words that are alive with association? Shouldn't we think about the poetics of language and concern ourselves with how words feel to people?" David Epston as quoted in Paljakka, S. (2021). "Christina and the Robin: A decidedly narrative response to rape." *Journal of Contemporary Narrative Therapy, September release*, pp. 6–31, 10.

5   "In this work, I do come up against my own personal limitations, which I then want to explore. These are limitations with regard to language, limitations in my awareness of relational politics, limitations in my capacity to negotiate some of the personal dilemmas that we are confronted with at every turn in this work, limitations of experience, limitations in my perception of options for the expression of certain values that open space for new possibilities, and so on. I want to explore these limitations, by talking about them with those persons who seek my help, and by talking about them with other therapists, and through personal reflection, through reading, and so on. In exploring these limitations in this way, I can extend what for me were the previously known limits of my work." White, M. (1995). *Re-authoring lives: Interviews and essays*. Dulwich Centre Publications. p. 38.

6   Dawn, A. (2013). *How Poetry Saved My Life: A Hustler's Memoir*, Arsenal Pulp Press, p. 56.

References

Ahmed, S. (2014). "Sweaty concepts," *Feminist Killjoys*, https://feministkilljoys.com/2014/02/22/sweaty-concepts/

Alvesson, M. & Spicer, A. (2012). "A stupidity-based theory of organizations," *Journal of Management Studies*, *49*(7), 1194–1220.

Annas, J. (2013). *Intelligent Virtue*. Oxford University Press.

Arendt, H. (1970). *Men in Dark Times*. Mariner Books.

Atwood, M. (1996) *Alias Grace*. Knoph Doubleday Publishing.

Barnard, F. (2010). "It's Marvellous to Be Free." *The Independent*. www.independent.co.uk/news/world/africa/it-s-marvellous-to-be-free-says-abducted-aid-worker-frans-barnard-2111508.html

Bateson, G. (1972) *Steps to an Ecology of Mind*. University of Chicago Press.

Bauman, Z. (1993). *Postmodern Ethics*. Blackwell.

Beckett, S. (1989). *Nohow On; Company, ill seen ill said, worstward ho!*, Grove Press.

Bernstein, R. (1978). "Thinking on Thought." *The New York Times*, www.nytimes.com/1978/05/28/archives/thinking-on-thought-thinking.html

Brown, L. S. (2018). *Feminist Therapy (2nd Ed.)* APA.

Canadian Psychological Association & The Psychology Foundation of Canada. (2018). *Psychology's Response to the Truth and Reconciliation Commission of Canada's Report*.

Carlson, T. (2020). "What is a good story? Counterstorying in narrative therapy." *Journal of Narrative Family Therapy*. April Release.

Cheng Thom, K. (2019a). *I Hope We Choose Love: A Trans Girl's Notes from the End of the World*. Arsenal Pulp Press.

Cheng Thom, K. (2019b). "Storytelling and poetry workshops." https://moosejawpride.ca/event/kai-cheng-thom-storytelling-poetry-workshops/

Daniels, J. (2021). *Nice White Ladies: The Truth About White Supremacy, Our Role in It, and How We Can Help Dismantle It*. Hachette Book Group.

Dawn, A. (2016). *How Poetry Saved My Life: A Hustler's Memoir*. Arsenal Pulp Press.

Dumas, A. (1844). *The Count of Monte Cristo*. Penguin Books.

Eltahawy, M. (2019). *The Seven Necessary Sins for Women and Girls.* Beacon Press.

Emerson, C. (1996) "Keeping the self intact during the culture wars; a centennial essay for Mikhail Bakhtin." *New Literary History*, *27*(1), 107–126.

Epston, D. (2018). Personal communication.

Epston, D. (2019). "Re-imagining narrative therapy: An ecology of magic and mystery for the maverick." *Journal of Narrative Family Therapy*, Release 3, 1–18.

Epston, D. (2022). "Coming to know young people as promising characters." *Journal of Contemporary Narrative Therapy*, April, 38–55.

Fellner, K. (2023). "Maskihkiy wellness." www.maskihkiy.com/consulting

Foucault, M. (1978). *History of sexuality, Volume 1: An introduction.* Vintage.

Foucault, M. (1997). "The Masked Philosopher." In *Essential Works of Foucault, Volume 1: Ethics, Subjectivity, and Truth.* New Press.

Frankfurt, H. G. (2005). *On Bullshit.* Princeton University Press.

Gadsby, H. (2018) *Nanette.* Netflix.

Guilfoyle, M. (2014). *The Person in Narrative Therapy: A Post-structural Foucauldian Account.* Palgrave Macmillan.

Kendall, M. (2020). *Hood Feminism.* Penguin Random House.

Krylov, I. (2017). *Krylov and His Fables* (W. Ralston, Trans.) Hanse Books (original work published 1869).

Le Guin, U. (2019). *Words Are My Matter: Writing of Life and Books.* Mariner Books.

Lindemann Nelson, H. (2001). *Damaged Identities, Narrative Repair.* Cornell University Press.

Lorde, A (1984). *Sister Outsider: Essays and Speeches.* Crossing Press.

Luitova, M. (2023, February 24). "'Russia ends nowhere', they say. Sociologist Grigory Yuduin discussed a year of war and what comes next." Abridged translation by E. Laskin. *Support*, Meduza. https://meduza.io/en/feature/2023/02/25/russia-ends-nowhere-they-say

Martin, R. (1988). "Truth, Power, Self: An interview with Michel Foucault." In L. Martin et al. (eds) *Technologies of the Self.* University of Massachusetts.

Mattingly, C. (1998). *Healing Dramas Clinical Plots: The Narrative Structure of Experience.* Cambridge University Press.

May, T. (2017). "The stories we tell ourselves." *New York Times.*

McCann, C (2017). *Letters to a Young Writer: Some Practical and Philosophical Advice*. Random House.

Médiné, S. (2020). "Spoken Word Publication: A hope for intimate liberation – activism in the therapy room." *Journal of Contemporary Narrative Therapy, December Release*.

Mingus, M. (2019). "Transformative Justice: A brief description." https://transformharm.org/tj_resource/transformative-justice-a-brief-description/

O'Hanlon, B. (1994). "The Third Wave: The promise of narrative." *Psychotherapy Networker,* November/December, 19–29.

Paljakka, S. (2021). "Christina and the Robin: A decidedly narrative response to rape." *Journal of Contemporary Narrative Therapy*, September release, 6–31.

Paljakka, S., Green, S., Luhtanen, T., Morton, C., Saxton, T., Szlavik, L., & Vincent, C. (2021). "On the pedagogy of poetics." *Journal of Contemporary Narrative Therapy, August Release*, 32–54.

Patrick, D. (2015). "Poetry of the Week: 'Sorry' by Ntozake Shange." https://alchemist.brenau.edu/2015/11/03/poetry-of-the-week-sorry-by-ntozake-shange/

Reynold, V. (2016). "Hate kills: A social justice response to suicide." In J. White, I. Marsh, M. J. Kral & J. Morris (Eds.), *Critical Suicidology: Transforming Suicide Research and Prevention for the 21st Century*. University of Chicago Press, pp. 169–187.

Rilke, R.M. (1992). *Letters to a Young Poet*. W.W. Norton.

Saunders, G. (2021). *A Swim in a Pond in the Rain*. Random House.

Savransky, M. (2020). "Problems all the way down." *Theory, Culture, and Society*, 1–21.

Schur, M. (2022). *How to Be Perfect: The Correct Answer to Every Moral Question*. Simon & Schuster.

Solnit, R. (2014). *The Nearby Faraway*. New York: Penguin Books.

Solnit, R. (2019) *Whose Story Is This? Old Conflicts, New Chapters*. Haymarket Books.

Surkan, N. (2018). "Can poetry save a life?" Presentation to Alberta Health Services, Calgary.

White, M. (1989/90). "Family therapy training and supervision in a world of experience and narrative." *Dulwich Centre Newsletter*.

White, M. (1995). *Re-authoring Lives: Interviews and Essays*. Dulwich Centre Publications.

White, M. (2004). *Workshop Notes*. Cambridge, MA.

White, M. (2007). *Maps of Narrative Practice*. W.W. Norton.

White, M. & Epston, D. (1990). *Narrative Means to Therapeutic Ends*. W.W. Norton.

Winslade, J. & Hedtke, L. (2008). "Michael White: Fragments of an event." *The International Journal of Narrative Therapy and Community Work*, No. 2.

Zagajewski, A. (2002). *Try to Praise the Mutilated World from Without End: New and Selected Poems*.

Zheng, Z. (2017) Personal communication.

Index